Geocryology

A survey of periglacial processes and
environments

Dedicated to the memory of

Professor Jan Dylik (1905–1973)
    Founder of the *Biuletyn Peryglacjalny*
Professor Richard Foster Flint (1902–1976)
    Colleague and Mentor

Geocryology

A survey of periglacial processes and environments

A. L. Washburn

Quaternary Research Center, University of Washington

Edward Arnold

This is the second edition of
Periglacial Processes and Environments,
first published 1973 by
Edward Arnold (Publishers) Ltd
41 Bedford Square, London WC1B 3DQ

Washburn, Albert Lincoln
  Geocryology. – 2nd ed.
  1. Glaciers
  I. Title   II. Periglacial processes and
  environments
  551.3′8   QE576

  ISBN 0–7131–6119–1

Printed in Great Britain by
Fletcher & Sons Ltd, Norwich

Contents

Acknowledgements

The author and publishers wish to thank the following for permission to use or modify copyright material. Figure numbers refer to this book, detailed citations may be found by consulting captions and references. Figures not otherwise credited are by the author.

Academic Press (NY) for Figs. 4.33, 4.34, 4.35, 5.39, 6.18, 6.22, 7.5, 12.5 and 12.6; Akademie Nauk USSR for Fig. 14.16; Alfred-Bentz-Hans for Fig. 5.1; American Association for the Advancement of Science for Fig. 4.14 © 1965; American Geophysical Union for Figs. 3.30 and 3.34; D. M. Anderson for Table 4.1; G. Andreas for Fig. 5.12; Arctic Institute of North America for Figs. 3.29 and 6.21; Association of American Geographers for Figs. 3.3 and 3.18; I Ya Baranov for Fig. 3.9; J. B. Benedict for Figs. 6.18 and 6.22; R. F. Black for Figs. 4.33, 4.34 and 4.35; J. M. Blackwell for Fig. 10.3; J. Brown for Fig. 3.16; R. J. E. Brown for Figs. 3.2, 3.4, 3.14 and 3.27; Butterworths, London for Fig. 4.1 and Table 4.2; K. L. Carey for Figs. 3.19 and 3.20; Ceskoslovenske Academia for Figs. 7.3, 7.4 and 7.5; A. E. Corte for Fig. 4.6 and Table 14.2; T. Czudek for Figs. 12.5 and 12.6; J. Demek for Figs. 12.5 and 12.6; Department of Physical Geography, University of Uppsala for Table 6.3; Dowden, Hutchinson & Ross, Inc. for Table 2.1; Ferd. Duemmlers Verlag for Figs. 5.1, 11.1 and 14.3–14.8; Franz Steiner Verlag, Wiesbaden for Figs. 5.8, 5.20, 5.32, 6.23 and 7.2; B. Frenzel for Figs. 5.8, 5.20, 5.32, 5.54, 6.23, 7.2 and 14.12; Friedr. Vieweg & Sohn, Wiesbaden for Figs. 5.54 and 14.12; Y. Fujii for Tables 3.3 and 3.4; Geo Abstracts Ltd., University of East Anglia for Fig. 6.28; Geological Society of America for Figs. 4.6 and 4.15; Geological Survey of Greenland and Meddelelser om Grønland for Fig. 3.8; Geologiska Föreningen, Stockholm for Fig. 4.13; R. P. Goldthwait for Fig. 5.39, M. A. Gonzalez for Table 14.2; K. Herz for Fig. 5.12; K. Higuchi for Tables 3.3 and

3.4; The Industrial Press (NY) for Figs. 3.1 and 3.32; Japan Association for Quaternary Research for Fig. 5.3; K. Kaiser for Figs. 14.11; Longman Group Ltd for Fig. 14.1 and Table 9.1; J. H. Leeman Ltd for Fig. 14.2; J. R. Mackay for Figs. 3.31, 3.32, 3.36, 4.5, 5.40 and 14.21; Methuen & Co. Ltd for Table 2.1; F. Müller for Fig. 5.50; National Academy of Sciences, Washington DC for Figs. 4.29, 5.21 and 5.22 and 5.23 reproduced from *Permafrost International Conference* (1966), pages 103, 78 and 79 respectively, and Figs. 3.5, 3.7, 3.17, 3.29, 5.52, 5.53 and 14.20 and Table 14.1 from *Permafrost Second International Conference* (1973), pages 38, 74, 225, 10, 82, 81, 90 and 92 respectively; National Research Council, Canada for Figs. 3.2, 4.2, 4.3, 4.23, 5.46 and 12.1 and Tables 3.3, 3.4 and 3.6; T. L. Péwé for Figs. 3.4, 5.52, 7.5 and 14.22; H. Poser for Figs. 14.3–14.8; Regents of the University of Colorado for Fig. 5.40; R. D. Reger for Fig. 7.5; S. Reiger for Table 2.1; J. C. Rogers for Fig. 3.13; Rutgers University Press, © 1977 for Table 2.2; P. V. Sellman for Fig. 3.16; R. P. Sharp for Fig. 5.33; Springer-Verlag, Wian for Fig. 11.1; L. Strömquist for Table 6.3; Tokyo Metropolitan University for Fig. 5.2; University of Chicago Press for Figs. 11.1 and 12.3; University of Toronto Press for Figs. 3.2 and 3.27; US Department of Agriculture, Forest Service for Table 2.3; A. A. Velichko for Figs. 14.13–14.16; Verlag Vandenhoeck & Ruprecht for Figs. 5.13 and 5.25; M. Vigdorchik for Fig. 3.12; E. Watson for Fig. 5.55; John Wiley & Sons (NY) for Fig. 14.1; R. G. B. Williams for Fig. 14.2 and *Zeitschrift für Geomorphologie* for Fig. 14.19.

Preface

This book is intended as a rather comprehensive overview of periglacial processes and their effects, present and past. Geocryology is included in the title to emphasize the pervasive influence of ice and its phase changes in these processes. The book is neither a formal text nor a reference manual but something of both and a guide to the enormous literature. Stress has been laid on the most recent publications, and on documentation by specific page citations for ease of reference. Because of the writer's personal interests and background, the coverage is uneven but he hopes the present edition is an improvement over the first in this respect.

Except as otherwise noted, temperatures are in degrees Celsius and the metric system is used throughout. Equivalent measurements in other units are given in parentheses if the original observations were reported in these units. A conversion table for SI units is given on the inside front cover.

Radiocarbon dates are given in years BP (Before Present, taken to mean before 1950).

The term soil, alone – as distinct from soil horizon, buried soil, or other usage where the context is clear – is used in a general sense as in engineering (cf. Legget, 1973) and does not necessarily imply profile development as in pedology.

In general capitalization, or lack of it, of terms such as Late v. late (Wisconsin, etc.), or Last Glaciation v. last glaciation, follows the author cited. Usages vary, and it should not be assumed that capitalization necessarily implies a formal time frame with well-established boundaries. Pending a universally accepted convention, and unless otherwise indicated, the present writer has adopted the informal lower-case style for his own work.

Much of this book was inspired by Professor Richard Foster Flint of Yale University. As a teacher just prior to World War II, as a fellow officer during that war, as a colleague at Yale many years later and throughout as a friend, Dick Flint encouraged the writer in many ways. An inter-disciplinary Pleistocene seminar at Yale in the early 1960s was also a stimulant in providing a fruitful discussion of periglacial processes and environments.

Professor Emeritus Hugh M. Raup of Harvard University and later of The Johns Hopkins University is another close friend who is responsible for the orientation of this book. As colleagues in field-work in Northeast Greenland in the period 1956–66, we had the opportunity to study and discuss many periglacial problems in detail. Numerous facets of this book benefited from this collaboration and some are based on the resulting publications in the *Meddelelser om Grønland*.

The writer is also indebted to Dr Duwayne Anderson, then of the US Army Cold Regions Research and Engineering Laboratory and presently of the State University of New York at Buffalo, and to Professor Larry W. Price of Portland State University for their suggestions relative to the first edition, and to Dr R. J. E. Brown of the National Research Council of Canada, Professor J. Ross Mackay of the University of British Columbia, and to many colleagues the world over whose suggestions benefitted the second. The writer's students at the University of Washington and Dr Chester Burrous of its Periglacial Laboratory at the Quaternary Research Center contributed more than they realized to both editions.

The publishers simplified the writer's task in many ways, and he deeply appreciates the cooperative attitude they exhibited throughout. Mrs Brenda Hall contributed greatly by preparing the index of both editions. Mrs Ruth Hertz, then on the staff of the Quaternary Research Center, typed the draft of the first edition and facilitated it in many other ways. Similarly, the second edition owes its appearance to the excellent work of Ms Patricia Leaming as editorial assistant whose sharp eye was invaluable, and of Ms Anne Swithinbank as library researcher and cartographer, both of the Quaternary Research Center.

Finally neither edition could have been written without Tahoe Washburn's wifely understanding and patient acceptance of too many uncommunicative evenings and weekends.

Conversion Table

Quantity	SI units	Conversion factors
Area	m^2	$1\ cm^2 = 10^{-4}\ m^2$ $1\ hectare = 10^4\ m^2$ $1\ acre = 4.047 \times 10^3\ m^2$ $1\ ft^2 = 9.290 \times 10^{-2}\ m^2$
Conductive capacity	$J\ m^{-2}\ s^{-\frac{1}{2}}$	$1\ cal\ cm^{-2}\ s^{-\frac{1}{2}} = 4.187 \times 10^4\ J\ m^{-2}\ s^{-\frac{1}{2}}$
Density	$kg\ m^{-3}$	$1\ g\ cm^{-3} = 10^3\ kg\ m^{-3}$ $1\ lb\ in^{-3} = 2.768 \times 10^4\ kg\ m^{-3}$
Energy, work, heat	J	$1\ cal = 4.187\ J$ $1\ BTU = 1.055 \times 10^3\ J$ $1\ ft\ lb = 1.356\ J$
Latent heat	$J\ kg^{-1}$	$1\ cal\ g^{-1} = 4.187 \times 10^3\ J\ kg^{-1}$ $1\ BTU\ lb^{-1} = 2.326 \times 10^3\ J\ kg^{-1}$
Length	m	$1\ \text{Å} = 10^{-10}\ m$ $1\ ft = 0.305\ m$
Mass	kg	$1\ lb = 0.454\ kg$ $1\ ton = 9.072 \times 10^2\ kg$
Pressure	$N\ m^{-2}$	$1\ dyne\ cm^{-2} = 10^{-1}\ N\ m^{-2}$ $1\ lb\ in^{-2} = 6.895 \times 10^3\ N\ m^{-2}$ $1\ mm\ Hg\ (0°C) = 1.333 \times 10^2\ N\ m^{-2}$ $1\ atm = 1.013 \times 10^5\ N\ M^{-2}$
Shearing stress	$N\ m^{-2}$	$1\ bar = 10^5\ N\ m^{-2}\ (= 14.5\ lb\ in^{-2})$
Specific heat	$J\ kg^{-1}\ K^{-1}$	$1\ cal\ g^{-1}\ °C^{-1} = 4.187 \times 10^3\ J\ kg^{-1}\ K^{-1}$ $1\ BTU\ lb^{-1}\ °F^{-1} = 4.187 \times 10^3\ J\ kg^{-1}\ K^{-1}$
Surface tension	$N\ m^{-1}$	$1\ lb\ ft^{-1} = 1.459 \times 10\ N\ m^{-1}$
Temperature	K	$°C = K - 273.16$ $°F = 1.8\ (K - 273.16) + 32$
Tensile strength	$kg\ m^{-2}$	$1\ kg\ cm^{-2} = 10^4\ kg\ m^{-2}$
Thermal conductivity	$W\ m^{-1}\ K^{-1}$	$1\ cal\ cm^{-1}\ s^{-1}\ °C^{-1} = 4.187 \times 10^2\ W\ m^{-1}\ K^{-1}$
Thermal diffusivity	$m^2\ s^{-1}$	$1\ m^2\ h^{-1} = 2.778 \times 10^{-4}\ m^2\ s^{-1}$

1 Introduction

I Definition of geocryology and periglacial: *1 Geocryology; 2 Periglacial.* **II Objectives: III processes:** *1 General; 2 Frost action; 3 Other processes.* **IV Environments:** *1 General; 2 Polar lowlands; 3 Subpolar lowlands; 4 Middle-latitude lowlands; 5 Highlands.* **V References:** *1 General; 2 Frost action*

Geocryology or periglacial research as defined below began with early reports of permafrost in the USSR but the 'father' of periglacial studies was Łoziński in Poland, who in 1909 first discussed the paleoclimatic implications and coined the term periglacial. Intense impetus to development of the subject grew out of the 11th International Geological Congress excursions to Spitsbergen in 1910, whose participants included some of the most prominent geologists of the time (De Geer, 1912*b*). Their reports of the then little-known periglacial features stimulated much interest that has led to an enormous interdisciplinary literature comprising geographic, geologic, engineering and, most recently, environmental interests.

Descriptive, regional surveys and process-oriented studies have emphasized frost wedging, frost cracking, frost heaving, solifluction (gelifluction), and such features as cryoplanation terraces, pingos, and the many forms of patterned ground. Laboratory studies of frost heaving began with Taber in 1929, and since then laboratory research has become increasingly important because of the engineering implications of frost heaving and permafrost. Government research centres established for such studies now include the Centre de Géomorphologie (CNRS) at Caen in France, the Permafrost Institute at Yakutsk in the USSR, and the US Army Cold Regions Research & Engineering Laboratory (CRREL) at Hanover, NH in the United States.

Advances in regional and process studies have also advanced the paleo-environmental aspects of periglacial research, and it is now clear that the most important features pertinent to climatic and other environmental reconstructions are those indicative of permafrost. The temperature significance of these features varies somewhat with their nature, and the approximations now available remain to be refined.

Despite the accomplishments of the past hundred years, numerous aspects of periglacial studies still need to be investigated, many of which bear significantly on practical problems as well as on purely scientific questions.

I Definition of geocryology and periglacial

1 Geocryology

Most definitions of geocryology, although emphasizing frozen ground, do not exclude glaciers, and it is clear from the *English-Russian geocryological dictionary* (Fyodorov and Ivanov, 1974) that the Russian term geokriologiya is highly generalized. As defined by R. J. E. Brown and Kupsch (1974, 13), geocryology is 'The study of earth materials having a temperature below 0°C.' According to the American Geological Institute's *Glossary of geology*, geocryology is 'The study of ice and snow on the Earth, esp. the study of permafrost' (Gary, McAfee, and Wolf, 1972, 290). Permafrost is also emphasized by Poppe and Brown (1976) in equating the term geokriologiya with 'geocryology, permafrost studies', and in some instances (cf. Markuse, 1976, 118) geocryology (Geokriologie) is strictly equated with study of permafrost (Dauerfrostbodenkunde). Cryo-prefixes are increasingly common, especially in Russian, in a complex and sometimes confusing array of terms in the translated literature.

Because the term geocryology is widespread, it is used in the title of this book but in the restricted sense as applying to frozen ground (seasonally frozen ground as well as permafrost) but not to glaciers. It has the advantage of being less of a tongue twister than periglaciology; on the other hand the adjective periglacial, discussed below, is useful because it is widely employed, specifically excludes glaciers, and is less cumbersome than the adjective geocryological. Although geocryologic could have been adopted, periglacial is used in the following to accord with its wide acceptance and with the previous edition of this book.

2   Periglacial

The term periglacial was introduced by Łoziński (1909, 10–18) to designate the climate and the climatically controlled features adjacent to the Pleistocene ice sheets.

Many investigators have extended the term to designate nonglacial processes and features of cold climates regardless of age and of any proximity to glaciers. As a result there have been varying usages (cf. Butzer, 1964, 105; Dylik, 1964a; 1964b). The restricted definition is followed in the USSR where it is applied to features bordering former ice sheets in the European section, and adjectives such as geocryologic or cryogenic are employed for contemporary features. As discussed by Jahn (1975, 3–4), this usage presumably reflects the fact that frozen ground was studied in Siberia (where present or former glacial influences are largely absent) long before the term periglacial was proposed. Nevertheless, and despite criticism because of its lack of precision (Linton, 1969), the term is being widely adopted in the extended sense, as here, because of its comprehensiveness and climatic implications (cf., for example, Embleton and King, 1975; French, 1976a; Hamelin and Cook, 1967; Jahn, 1970; 1975; Tricart, 1967; and numerous articles in the *Biuletyn Peryglacjalny*).

The term has no generally recognized quantitative parameters, although some rough estimates of precipitation and temperature limits have been given. According to Peltier's (1950, 215, Table 1) estimate, his periglacial morphogenetic region is characterized by an average annual temperature ranging from $-15°$ (5°F) to $-1°$ (30°F) and an average annual rainfall (excluding snow) ranging from 127 mm (5 in) to 1397 mm (55 in). Peltier's (1950, 222, Figure 7) diagram of morphogenetic regions, which

has been widely reproduced, is reasonably consistent with these figures for the periglacial region except that the lower limit of rainfall is given as 0 mm (0 in). Peltier also proposed a periglacial erosion cycle dominated by frost action and solifluction as a counterpart to Davis' fluvial ('normal') cycle (W. M. Davis, 1909). However, the concept is highly idealized and fails to take adequate account of running water (cf. French, 1976a, 164). According to Lee Wilson's (1968a, 723, Figure 9; 1969, 308, Figure 3) morphogenetic classification the periglacial domain has a precipitation range from some 50 to 1250 mm and a temperature range from some $-12°$ (10°F) to $2°$ (35°F) in variable combinations. French (1976a, 5) defined the periglacial domain as including '... all areas where the mean annual air temperature is less than $+3°C$'. However useful for particular purposes, any such definition is arbitrary because the periglacial concept itself is sufficiently broad and imprecise to defy quantification. The diagnostic and necessary criterion is a climate characterized by intense frost action and snow-free ground for part of the year. As stated by Tricart (1967, 9), '*les pays froids sont ceux où l'action géomorphologique de l'eau est commandée par son existence à l'état solide, permanente ou périodique.*'

Tricart (1967, 29, Figure 4; 30) stressed permafrost (whether or not it is in balance with the present climate) as the primary characteristic of the periglacial domain. However, he recognized a distribution of minor periglacial features, such as earth

1 Regions of accumulation with underlying syngenetic permafrost containing ground ice   2 Regions of accumulation with underlying seasonal frost only   3 Regions of equilibrium between processes of accumulation and denudation on syngenetically frozen rocky substrate containing ice wedges   4 Flat regions of equilibrium between processes of accumulation and denudation (on syngenetically frozen rocky substrate)   5 Regions of cryoplanation, stable with respect to accumulation and denudation (on epigenetically frozen rocky substrate)   6 Regions stable with respect to accumulation and denudation (on seasonally frozen rocky substrate)   7 Regions without frozen-ground processes modelling the landscape   8 Regions of dominant denudation with underlying permafrost   9 Regions of dominant denudation with underlying seasonal frost only   10 Glaciers   11 Polygons with ice veins   12 Polygons with ice veins, associated with [agréments] thermokarst forms   13 Peat bogs with flat hummocks [buttes gazonées]   14 Baydjarakhs   15 Alases   16 Polygons with soil veins   17 Reduced (deformed) polygons with soil veins – forms infilled with cover loam   18 Forms similar to hummocks [buttes gazonées] with gaps [enfoncements] resulting from degradation   19 Pseudo kames   20 Hummocks [buttes gazonées] – mounds   21 Cryoplanation terraces   22 Nonsorted circles [formes tachetées – médaillons]   23 Sorted polygons [polygons de pierres], circles, and other forms sorted by freezing   24 Solifluction stripes on slopes   25 Stratified icings ['Nalédi']   26 Seasonal hummocks [buttes gazonées] resulting from soil heaving   27 Perennial hummocks [buttes gazonées] resulting from soil heaving   28 Hummocks [buttes gazonées] due to water migration toward the frozen surface   29 Solifluction forms related to soil flow   30 Present limit of permafrost

1.1 Present and upper Holocene periglacial features in USSR (*after Markov, 1961; Popov, 1961; key translated. For commentary and somewhat different interpretations, cf. Jahn, 1976a, 131, Figure 14, 132–3*)

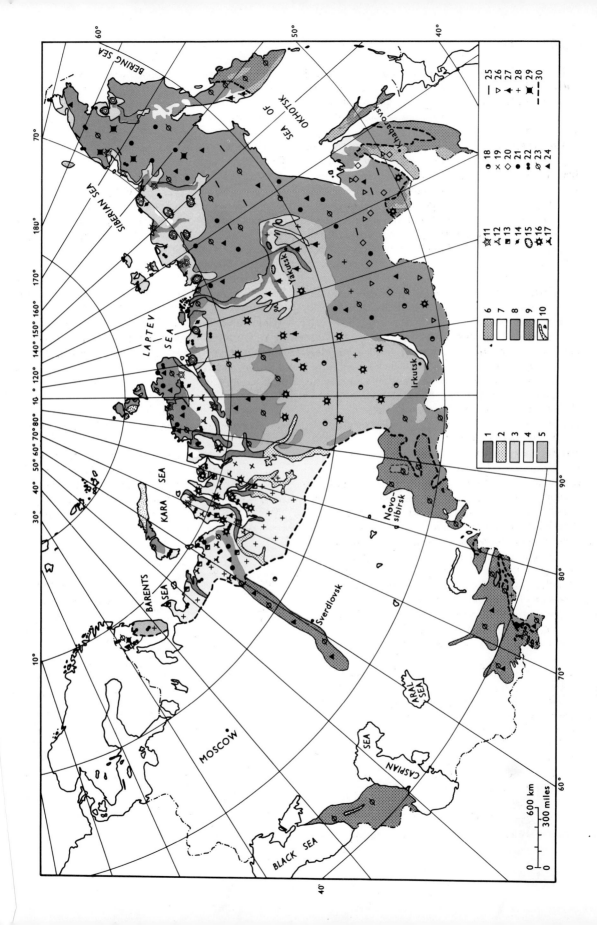

hummocks, as lying outside the periglacial domain as defined by permafrost; also Tricart (1967, 56–67) defined as periglacial some climates lacking permafrost, and it is clear that he did not consider permafrost to be a necessary condition in a definition of periglacial. Péwé (1969, 2, 4) came closer to regarding permafrost as a necessary criterion, and French (1976a, 2) considered it a diagnostic but not a necessary criterion. Jahn (1975, 11) specifically excluded permafrost as a criterion, although he accepted, in general, a mean annual temperature of $-1°$ as the equatorward boundary of his periglacial zone, a boundary that in many places is in fact applicable to permafrost as noted in the next chapter and recognized by Jahn (1975, 28). Clearly, the periglacial and permafrost environments have much in common but to make them synonymous is overly restrictive, since many features such as gelifluction, frost creep, and several forms of patterned ground that are related to frost action, are commonly regarded as periglacial but are not necessarily associated with permafrost. Furthermore, it is common to speak of former periglacial environments; yet in the present state of our knowledge there are very few criteria by which a former permafrost condition can be proved. Although a cold-climate process, glaciation is not periglacial by definition. Glaciation and 'periglaciation', where both are present, are complementary aspects of cold environments. As noted by Jahn (1975, 21–2), this does not necessarily imply synchroneity in effects, since the response time of glaciers to climate change (Meier, 1965) may be very different than that of periglacial phenomena, especially to cooling. It has been argued that it may take 15 000–30 000 years for an ice sheet to build up (Weertman, 1964), yet the region fronting the eventual border could have been responding to the same climate change by developing permafrost throughout the glacier's advance.

As used in the following, the term periglacial designates primarily terrestrial,[1] non-glacial processes and features of cold climates characterized by intense frost action, regardless of age or proximity to glaciers (Figures 1.1–1.2).

II   Objectives

The objectives of periglacial research are to (1) determine the exact mechanism of periglacial processes, (2) determine the environmental significance of the processes, (3) apply the information to reconstruct Quaternary environments, and (4) use these historical and process approaches as an aid in predicting environmental changes.

III   Processes

1   General

Many different processes are responsible for periglacial effects, but for the most part these processes are not peculiar to periglacial environments. Rather they are common to many environments that have a climate sufficiently cold to leave physical evidence of its influence. It is the combination and intensity of these processes that characterize periglacial environments.

Given low enough temperatures both the poleward and upper altitudinal limits of most periglacial features, where limits exist, are determined by precipitation blanketing the land with perennial snow and ice; on mountains this limit is the snowline. Permafrost can be an exception, depending on the

1 Regions of maximum glaciation   2 Regions of sediment accumulation corresponding to maximum glaciation (with underlying syngenetic permafrost) 3 Regions of equilibrium relative to processes of sediment accumulation and transportation corresponding to maximum glaciation (with underlying epigenetic permafrost)   4 Regions of dominant transportation processes corresponding to maximum glaciation (with underlying permafrost of destructive origin ['d'origine génétique destructive'])   5 Extension of ocean during period of maximum glaciation   6 Regions of Valdai Glaciation   7 Regions of accumulation corresponding to Valdai Glaciation (with underlying syngenetic permafrost) 8 Regions of equilibrium relative to accumulation and transportation corresponding to Valdai Glaciation (with underlying epigenetic permafrost)   9 Regions of dominant transportation corresponding to Valdai Glaciation (with underlying permafrost of destructive origin ['d'origine génétique destructive'])   10 Regions without permafrost phenomena   11 Maximum limit of glaciation   12 Southern limit of permafrost during maximum glaciation   13 Shoreline during maximum glaciation   14 Fossil ice-wedge polygons ['Polygones a fissures remplies de glace fossile'] developed during maximum glaciation   15 Soil-wedge polygons ['Polygones a fissures remplies de sol'] corresponding to maximum glaciation 16 Hummocks [buttes gazonées] and concave forms corresponding to maximum glaciation   17 Cryoplanation terraces corresponding to maximum glaciation 18 Nonsorted circles ['formes (taches) en médaillon'] corresponding to maximum glaciation   19 Sorted polygons ['Polygones de pierres'], circles, etc., corresponding to maximum glaciation   20 Solifluction stripes ['Bandes de solifluxion sur versant'] corresponding to maximum glaciation   21 Perennial hummocks [buttes gazonées], due to frost heaving, corresponding to maximum glaciation   22 Solifluction forms corresponding to maximum glaciation 23 Limits of Valdai Glaciation   24 Southern limit of permafrost during Valdai Glaciation   25 Fossil ice-wedge polygons ['Polygones à fissures remplies de glace fossile'] developed during Valdai Glaciation   26 Soil-wedged polygons ['Polygones a fissures remplies de sol'] corresponding to Valdai Glaciation 27 Degrading polygons with soil veins (pebbles in cover loam) corresponding to Valdai Glaciation   28 Nonsorted circles ['formes (taches) en médaillon'] corresponding to Valdai Glaciation   29 Solifluction forms corresponding to Valdai Glaciation   30 Discovery site of Upper Paleolithic fauna   31 Periglacial fossil macroflora   32 Periglacial fossil pollen and spores   33 Mammoth remains in permafrost ['tjäle']   34 Rhinoceros remains in permafrost ['tjäle']   35 Horse (*Equus caballus*) in permafrost ['tjäle']   36 Regions of loess ['d'apparition de loess et de formations loessiques']   37 Regions with ancient continental dunes

1.2 Pleistocene periglacial features in USSR (*after Markov, 1961; Popov, 1961; key translated*)

---

[1] Marine ice-shove ridges and offshore (submarine or subsea) permafrost are included as periglacial features by the present writer.

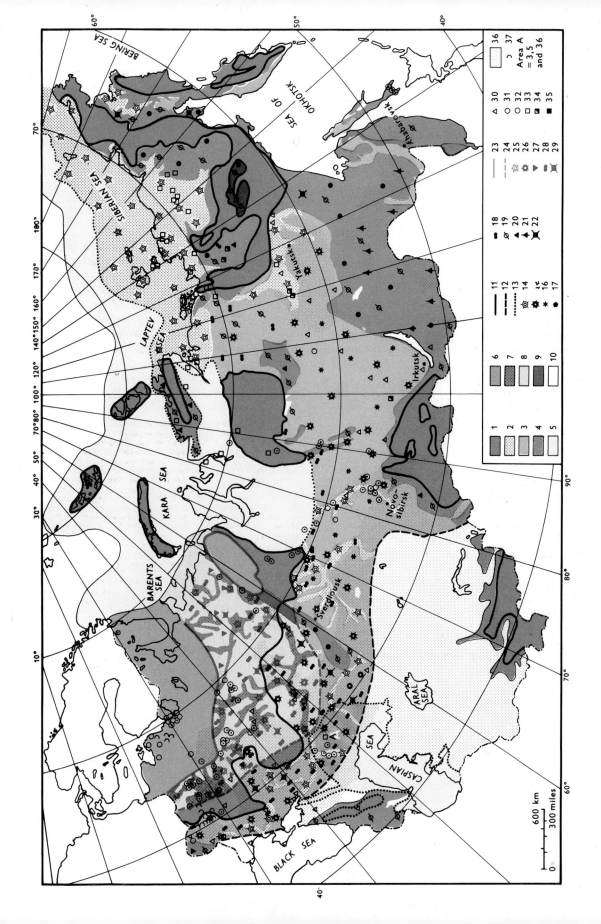

thickness of the snow or ice cover. On the other hand the equatorial and lower altitudinal limits are limited by temperature rather than precipitation.

2   Frost action

By far the most widespread and important periglacial process is frost action. Actually, frost action is a 'catch-all' term for a complex of processes involving freezing and thawing including, especially, frost cracking, frost wedging, frost heaving, and frost sorting.

3   Other processes

In addition to frost action, certain aspects of mass-wasting, nivation, fluvial action, lacustrine action, marine action, and wind may produce characteristic periglacial features.

IV   **Environments**

1   General

Many factors make up environment but for periglacial environments the overriding controls are regional climate and topography. Local factors may modify the regional climate and in this and other ways strongly influence processes.

Various periglacial zones based on different criteria reflecting cold climates have been suggested. Büdel (1948) recognized two zones – a frost-debris zone (Frostschuttzone) characterized by barren stony surfaces, and a tundra zone (Tundrenzone) characterized by treeless vegetated surfaces. Corbel (1961, 19, Figure 11) stressed the role of precipitation by outlining 3 arctic periglacial zones based on differences in precipitation. In Jahn's (1975, 11) opinion

The periglacial zone is unquestionably a climatic zone. Its boundaries are determined neither by winter (frozen ground) nor summer (vegetation) temperatures but rather by the mean annual temperature curve. The annual isotherm of $-1°C$ is perhaps nearest to that limit [outer boundary], although periglacial phenomena often overstep this line, especially where residual Pleistocene permafrost is present in the ground in areas under forest, like in Siberia.

In the present writer's view, except for a few phenomena that are demonstrably dependent on permafrost, the distribution and climatic implications of features that most investigators would readily accept as being periglacial are still too poorly known to permit grouping the features into generally applicable zones. Detailed local zonation studies should eventually lead the way to broader applications. The following types of climate and their influence on processes illustrate some of the complexities involved.

Tricart (1967, 44–67; 1969, 19–27) recognized the following periglacial climates:

A   *Cold dry climate with severe winters*
Encompasses elements of the D and E climates of Köppen, discussed later. Characteristics include (*a*) very low winter temperatures, (*b*) short summers, (*c*) permafrost, (*d*) low precipitation, (*e*) violent winds. Consequently there is (*a*) intense freezing, (*b*) reduced or even negligible activity of running water, and (*c*) important wind action.

B   *Cold humid climates with severe winters*
(i) *Arctic type:* Corresponds to the most humid parts of Köppen's ET climates, excluding those without marked seasons. Characteristics include (*a*) similar mean temperatures to A but with a tendency for smaller annual range, (*b*) permafrost, (*c*) great climatic irregularities that tend to be masked by mean figures, (*d*) greater humidity with annual precipitation totals almost always exceeding 300 mm, resulting in appreciable snow cover and some rain. Consequently, as compared with A, (*a*) freezing is less intense and less long, (*b*) wind action is reduced by the snow cover, (*c*) running water is more important.

(ii) *Mountain type:* Corresponds to prairie-alpine zone of temperate zone. Characteristics include (*a*) monthly temperature trends similar to B(i) but with higher means, (*b*) precipitation is much greater than in B(i) and tends to inhibit wind action and deep penetration of frost, (*c*) freeze-thaw cycles are less common in the summit areas than in the valleys. Consequently (*a*) frost action is important but permafrost is commonly lacking, (*b*) running water is an important geologic agent, (*c*) wind action is slight.

C   *Cold climates with small annual temperature range*
(i) *High-latitude island type:* Characteristics include (*a*) mean annual temperature near 0° with small annual range (generally on the order of 10°), numerous freeze-thaw cycles, (*b*) instability of weather, (*c*) considerable precipitation, generally exceeding 400 mm, which tends to inhibit wind effects. Consequently (*a*) frost action is characterized by many freeze-thaw cycles of short duration and slight

penetration into the ground, (*b*) wind action is slight.

(ii) *Low-latitude mountain type:* Characteristics include (*a*) lack of seasonal temperature variations, (*b*) considerable variations in diurnal temperature, far exceeding the seasonal variations, (*c*) high precipitation except in arid mountains such as Puna de Atacama. Consequently (*a*) there is considerable frost action because of the frequent freeze-thaw cycles, (*b*) only slight frost penetration into the ground, (*c*) absence of permafrost, (*d*) lack of wind action in high-precipitation areas.

As stated by J. L. Davies (1969, 13–14), this classification brings out some marked contrasts. Permafrost is characteristic of A, absent in C. Annual temperature cycles are of large amplitude and extend to considerable depth in A; the opposite is true in C. Freeze-thaw cycles are much fewer in A than in C. Running water is much less important in A than in C. In each case conditions in B tend to be intermediate.

In some respects Tricart's classification may be superior to Köppen's but the latter has the advantage of more precise boundaries between categories.

The Köppen classification of climates (Köppen, 1936; Köppen-Geiger, 1954; cf. Strahler, 1969, 224–30) is widely used and is the one on which the following scheme is based for polar, subpolar, and middle-latitude lowland climates and geographic zones. That these climates and zones can be similarly designated arises from the fact that the zones are climatically defined. The addition of highlands is to emphasize the role of altitude in determining environmental factors. Only the most general kind of classification of climates and zones is used here in view of the limited knowledge concerning the distribution and frequency of periglacial processes and features. Gerdel (1969) has cited some of the practical limitations to climatic classifications and has presented a useful general description of the characteristics of cold regions, especially in the Northern Hemisphere.

2   Polar lowlands

In the polar zone, the average temperature of the coldest month is $< -3°$ and of the warmest month $< 10°$. The zone, which is controlled by polar and arctic air masses in the Northern Hemisphere, lies roughly north of lat. 55°N and south of lat. 50°S. It includes an ice cap (Köppen's EF) climate, dominated by arctic and antarctic air masses, in which the average temperature of the warmest month is $< 0°$, and, of more concern to periglacial research, a tundra (Köppen's ET) climate in which the average temperature of this month is $> 0°$.

The zone is characterized by ice caps, bare rock, and/or vegetation of tundra types – mainly grasses, sedges, small flowering plants, and in places herbaceous shrubs.

3   Subpolar lowlands

In the subpolar zone the average temperature of the coldest month is $< -3°$, the temperature of the warmest month is $> 10°$ but there are less than four months above this temperature (Köppen's Dfc, Dfd, Dwc, Dwd climates). The zone is controlled by continental air masses and extends roughly from lat. 50° to 70°N (northern taiga zone) and from lat. 45° to 60°S. Because the climatic gradients vary with longitude there is some latitudinal overlap.

The 10° isotherm for the warmest summer month, which is accepted here as the boundary between polar and subpolar lowlands, tends to coincide with treeline in the Northern Hemisphere. Characteristically, coniferous forest predominates. However, treeline (i.e. the northern limit of scattered trees) can be some $1\frac{1}{2}$ degrees of latitude north of the coniferous forest (cf. Washburn, 1951, 270–1).

4   Middle-latitude lowlands

In the middle-latitude zone the average temperature of the coldest month is $< -3°$ but there are more than four months with average temperature $> 10°$ (Köppen's Dfa, Dfb, Dwa, Dwb climates). This zone is controlled by both polar and tropical air masses and extends roughly from lat. 35° to 60°N.

5   Highlands

A highland climate may differ in important ways from climates primarily controlled by latitude. For instance, highlands are commonly colder than lowlands in the same zone, some highlands have stronger diurnal than seasonal temperature changes, and the orientation of mountain slopes tends to exert a strong climatic influence. Particularly in periglacial environments where topography is a critical factor, it is desirable to make a marked distinction between lowland zones controlled primarily by latitude and highland zones controlled by altitude as well as latitude.

In a general way counterparts to cold lowland zones can be found in highlands, and parallels are frequently cited between arctic and alpine zones and between subarctic and subalpine zones. However, there is considerable difference of opinion regarding zone classifications, many of which are based on biological considerations reflecting climate rather than being based directly on climatic parameters (Löve, 1970). No consistent attempt is made in the following to adopt any particular highland zonation except to recognize that the altitudinal treeline, like the latitudinal treeline, is a critical boundary for certain periglacial processes. Rather, highlands are cited in relation to polar, subpolar, middle-latitude, and low-latitude zones to indicate that certain periglacial processes and features are more common in highlands than in lowlands of the same latitudinal zone, either because of a more rigorous climate in the highlands or because of topography. As an order of magnitude, 1000 m will be considered the minimum altitude difference between a highland and a lowland in the same general region.

V   References

1   General

Although periglacial research began to emerge as a recognized field following Łoziński's (1909) work, only recently has the subject been comprehensively treated. The Polish and Russian literature is particularly rich but the present writer has had to forego much of it that has not yet been translated. General references include: Büdel (1977, 37–91), Cailleux and Taylor (1954), J. L. Davies (1969), Embleton and King (1975), French (1976a), Jahn (1970; 1975); Kaplina (1965), Kudryavtsev (1978) Lliboutry (1965, 927–1007), Peltier (1950), Poser (1977), L. W. Price (1972), Romanovskiy (1977), Tricart (1963; 1967; 1969), Troll (1944; 1958), Vtyurina (1974; 1976), and the *Biuletyn Peryglacjalny* (1954–). The most detailed updated monograph in English at the research level is the one by Jahn (1975), which is

especially valuable because of its discussion of the extensive Russian literature, much of it unavailable in translation. A booklet of permafrost definitions has been compiled by R. J. E. Brown and Kupsch (1974) and an extensively illustrated English and French glossary of periglacial features by Hamelin and Cook (1967). There is also a very useful *English-Russian geocryological dictionary* (Fyodorov and Ivanov, 1974) and *Russian-English glossary of permafrost terms* (Poppe and Brown, 1976).

Regional studies of contemporary periglacial phenomena include Bird (1967, 161–270), Bout (1953), Boyé (1950), Büdel (1960), Kelletat (1969), Jan Lundqvist (1962), Malaurie (1968), Markov and Bodina (1961; 1966), Popov (1961), Schunke (1975a), Sekyra (1960), and many others (cf. Poser, 1974a; 1977). An album of contemporary permafrost features for use in university courses was compiled by Popov (1973).

A continuing survey and bibliography of field investigations concerning the nature and rate of periglacial processes has been started by the Co-ordinating Committee for Periglacial Research, a committee of the International Geographical Union's (IGU) Commission on Present-Day Geomorphological Processes (French, 1976c).

2   Frost action

Frost action, including permafrost, is at the centre of much periglacial research. The literature here is voluminous and only some of the most comprehensive and helpful bibliographies and references are cited below. (Cf. also references cited under Significance in section on Permafrost in the chapter on Frozen ground.)

Akademiya Nauk SSSR (1973), American Meteorological Society (1953), Andersland and Anderson (1978), Arctic Institute of North America (1953–1975), Beskow (1935; 1947), Canada National Research Council (1978)[2], Carleton University and École Nationale des Ponts et Chaussées (1978), Dostovalov and Kudryavtsev (1967), Highway Research Board (1948; 1952a; 1952b; 1957; 1959; 1962; 1963; 1969; 1970; 1972), Jessberger (1978), T. C.

---

[2] *Third International Conference on Permafrost – Proceedings.* Volume 1 consists of 139 submitted papers, Volume 2, of 8 review papers, 4 Chinese contributions, and other material. Field guides and 2 volumes of English translations of foreign-language papers complete the publication series. Volume 1 of the *Proceedings* was issued after the present work had been submitted to the publisher but, through the kindness of the authors, provision had been made to cite selected papers, and references were added in proof. The other publications had not yet appeared by the time this note was added.

Johnson *et al.* (1975), Jumikis (1977), Melnikov and Tolstikhin (1974), S. W. Muller (1947), National Academy of Sciences (1973; 1978³), National Academy Sciences-National Research Council (1966), Norges Teknisk-Naturvitenskapelige Forskningsråd og Statens Vegvesens Utvalg for Frost i Jord (1970–), Organisation for Economic Co-operation and Development (1973), Popov (1967; 1969–), Protas'yeva (1967; 1975), Terzaghi (1952), US Army Corps of Engineers Cold Regions Research and Engineering Laboratory (1951–; 1973), J. R. Williams (1965), P. J. Williams (1967).

In parallel with the survey of field investigations, noted in the previous section, a continuing survey and bibliography of laboratory investigations of frost action has also been activated by the IGU Co-ordinating Committee for Periglacial Research (Pissart *et al.*, 1976).

---

³ *USSR Contribution Permafrost Second International Conference.* This volume, consisting of papers translated from the Russian, appeared after the present work had been submitted to the publisher. Although it has been possible to insert references to several of the translations, a number of pertinent ones had to be omitted.

2   Environmental factors

I   Introduction

The most nearly independent environmental factors influencing the various periglacial processes and the development and distribution of frozen ground are climate, topography, and rock material. Their degree of independence depends on the scale. For instance, zonal climate – a climate determined by the largest-scale factors of latitude, atmospheric circulation systems, and widespread highlands – is an independent factor. However, smaller-scale climatic effects are dependent on superimposed influences, some of which may result in an azonal climate – a climate that is atypical and not truly representative on a zonal basis.

The mutual interaction of climate, topography, and rock material is illustrated by countless examples. Thus topography, through altitude and exposure, modifies zonal climate but climate can modify topography by determining the processes acting on a region – for example, glaciation produces cirques that create variations of exposure and thereby influence the local climate. The nature of a rock influences the effect climate may have on it but climate may determine the kind of rock developed – for example, an evaporite formed in an arid basin.

In contrast to the foregoing factors, snow cover, liquid moisture, and vegetation are always dependent. However, they can be critical controls of periglacial processes, especially frost action, since they can determine whether or not a process is climatically zonal or azonal in its effect.

The following generalized review is to set the stage for later discussion of the environmental implications of periglacial processes. Environmental factors with special reference to permafrost have been reviewed

in some detail by R. J. E. Brown (1978) and R. J. E. Brown and Péwé (1973, 72–80).

II   Basic factors

1   Climate

a Scale   In discussing the influence of climate, it is useful to recognize three scales of climate: zonal climate, local climate, and microclimate. As noted, zonal climate reflects only large-scale effects such as latitude and widespread highlands; it is the critical element to establish when determining past climates and reconstructing climatic changes. Local climate represents the combined influence of zonal climate and local topography. Microclimate is at a still smaller scale in that it incorporates the additional influence of ground-surface characteristics such as vegetation, moisture, and air-earth or air-water interface effects (Geiger, 1965). As indicated below, local and microclimatic influences can be highly significant and must be evaluated in reconstructing past zonal climates from biologic and geologic evidence.

b Climatic parameters   The most important climatic parameters controlling periglacial processes are temperature, precipitation, wind, and their seasonal distribution. The past influence of wind may be readily discernible through erosional and depositional effects but temperature and precipitation may interact so that their relative importance may be difficult to determine. For instance, both sufficiently low temperature and sufficiently high moisture are required for glaciation and for certain frost-action effects, but within limits one parameter may substitute for the other. Thus increased snow accumulation and nivation may result from lower summer temperatures and less melting as well as from increased winter snowfall. On the other hand, increased permafrost may be due as much to lower winter snowfall, and hence less insulation of the ground, as to lower air temperatures.

c Zonal climate Present-day climate can be studied and described by quantitative observations that allow for local and microclimatic effects in describing zonal climate. This is much more difficult to do in reconstructing past zonal climates, since the evidence at any one place may be very strongly influenced and in some cases dominated by purely local and microclimatic effects.

d Local climate The extent to which local topography can influence climate is seen in the differing precipitation regime of windward and lee slopes and in the differing temperature regime of north and south slopes of mountains, especially in high latitudes. For instance, on the Jungfrau in Switzerland there were 196 days yr^{-1} with air fluctuations through the freezing point on the south slope whereas on the north slope, 127 m lower, the number was only 22 (Mathys, 1974, 57). Other factors being constant, depth of thawing in the Northern Hemisphere may be 50 to 60 per cent greater on south-facing than on north-facing slopes (Zhestkova *et al.*, 1961, 50; 1969, 8), although the opposite effect has also been observed, presumably because the lower near-surface soil temperatures on the north slope studied resulted in slower evaporation after rainfall (Hannell, 1973, 181). Such differences can strongly affect the nature and distribution of periglacial processes, even in the same valley as described, for example, by P. G. Johnson (1975). A study of Chitistone Pass, Alaska, showed that in this alpine permafrost environment, 'Topographic factors – slope and aspect – explain more of the spatial-temporal variations in active layer thermal regimes than the thermal and radiative properties of specific sites' (Brazel, 1972, 57).

e Microclimate Meteorological observations at the common shelter height (1.7–2.0 m in Europe, 4 ft in the United States) do not adequately indicate the climate at the ground surface, yet this is where some critical processes occur. For instance, the number of freeze-thaw cycles at the surface can be very different from those a little higher in a shelter. Wind velocities are also significantly different. Microclimate can vary within very short distances laterally as well as vertically, and is strongly influenced by differences in distribution of snow and vegetation. In this sense it is more of a dependent than basic factor (cf. below) but the interaction can

be complicated. Microclimatic effects in periglacial environments can be very large as illustrated by observations in Antarctica (MacNamara, 1973, 201–4; 220–31). In the Mackenzie Delta, northern Canada, where mean annual air temperatures are −9° to −10°, microclimatic factors can cause the mean annual ground-surface temperature to be 5° to 10° warmer over a distance of about 160 m (M. W. Smith, 1975, 1423).

2 Topography

The way topography can influence climate and thereby, indirectly, process is outlined above. Topography can also exert a direct effect on process. For instance, the configuration and gradient of a hillside can determine whether the dominant mass-wasting process is landsliding as opposed to frost creep or solifluction. A very low-angle slope favours retention of moisture and development of certain forms of patterned ground.

3 Rock material

a General The term rock material as used here covers structure, mineral composition, texture, and colour, and includes both bedrock and unconsolidated material.

Bedrock is essentially an independent factor in relation to periglacial processes, since it is usually a pre-existing, very much older feature. Its influence on the distribution of periglacial features can be critical as illustrated by numerous studies, including Brosche's (1977, 180–8; 1978, 56–61) discussion relative to the Iberian Peninsula, and Graf's (1971, 48–51; 110–13) studies in the Swiss Alps and the Bolivian and Peruvian Andes.

From an engineering viewpoint unconsolidated material derived from bedrock is simply soil. Formerly, for soil to qualify as such in pedologic usage, it had to contain humus but many pedologists now accept mineral soil (i.e. lacking humus) as a true soil (Tedrow, 1977; Ugolini, 1979). Examples of such soil are common in some cold environments. Several classifications of cold-climate soils are in use (Tables 2.1–2.2). Because of its logical (although complex) construction, the soil classification system of the US Department of Agriculture (Soil Survey Staff, 1975) is tending to replace the older Zonal Classification in the United States but there is as yet no universally accepted classification. Tedrow (1977, 267–81) has recently proposed a genetic system for polar soils, based on first order

Table 2.1. Classification of cold-climate soils (*after Rieger, 1974; F. C. Ugolini personal communication, 1977; Ugolini, 1979*)

Cryic Great Subgroup of US Department of Agriculture Soil Classification System[1]

Name	Description
Pergelic Cryaquepts	Poorly drained grey or mottled soils with thin (usually less than 20 cm) surface organic mats
Histic Pergelic Cryaquepts	Poorly drained grey or mottled soils with thick surface organic mats
Ruptic-Histic Pergelic Cryaquepts	Poorly drained grey or mottled soils with a thick organic mat in part of each pedon (as in a trough between polygons) and a thin or no mat in other parts; pedons commonly include frost circles
Pergelic Cryaquolls	Poorly drained grey or mottled soils with dark base-rich upper horizons
Pergelic Sphagno-fibrists	Fibrous *sphagnum* moss peat, at least 60 cm thick
Pergelic Cryofibrists	Fibrous sedge peat, at least 40 cm thick
Pergelic Cryohemists	Partially decomposed peat, at least 40 cm thick
Pergelic Cryorthents	Well-drained loamy, clayey or gravelly soils with no developed horizons
Pergelic Cryopsamments	Well-drained sandy soils with no developed horizons
Pergelic Cryochrepts	Well-drained brown soils with very thin or no dark upper horizons
Pergelic Cryumbrepts	Well-drained soils with dark, acid upper horizons and, usually, brown subsoil horizons
Pergelic Cryoborolls	Well-drained soils with dark, base-rich upper horizons
Pergelic Cryorthods	Well-drained soils with a thin bleached upper horizon and a brown or reddish brown subsoil horizon in which aluminum, iron and organic carbon have accumulated
Ruptic-Entic Pergelic Cryochrepts, Ruptic-Entic Pergelic Cryumbrepts, etc.	Well-drained soils in which part of each pedon is a frost scar with no recognizable horizons
Lithic Cryorthents, Lithic Cryochrepts, etc.	Well-drained soils with bedrock less than 50 cm deep

Great Soil Groups of Zonal Classification

Name	Description
Lithosols	Well-drained soils with shallow bedrock (Lithic Cryorthents)
Regosols	Deep, well-drained soils with no genetic horizons (Pergelic Cryorthents and Cryopsamments)
Polar Desert soils	Well-drained soils of the Polar Desert Zone with no genetic horizons (Pergelic Cryorthents)
Arctic Brown soils — Normal phase	Well-drained brown soils with or without dark upper horizons (Pergelic Cryochrepts, Cryumbrepts, and Cryoborolls)
Shallow phase	Soils as above with shallow bedrock (Lithic Cryochrepts, Cryumbrepts, and Cryoborolls)
Moderately-well drained phase	Soil gradational between normal phase and Tundra soils, with gleying in lower horizons (Pergelic Cryaquepts)
Rendzinas	Dark soils over shallow calcareous rock (Lithic Cryoborolls)
Podzols or Podzol-like soils	Soils with bleached upper horizons and brown lower horizons (Pergelic Cryorthods)

[1] Five orders of the US Department of Agriculture soil classification system (Soil Survey Staff, 1975) occur in the Arctic: Entisols – soils with little or no evidence of pedogenesis (formative element –*ent*, applied in various combinations); Histosols – soils derived mainly from organic matter (formative element –*ist*); Inceptisols – soils affected by pedogenesis and loss of minerals without appreciable accumulation horizons of clay or mixtures of Al, Fe, and organic C (formative element –*ept*); Mollisols – soils with dark base-rich upper mineral horizon (formative element –*oll*); and Spodosols – soils with an accumulation horizon of amorphous mixtures of Al, Fe, and organic C (formative element –*od*).

Table 2.1. Classification of cold-climate soils—*cont*

Tundra soils Upland Tundra	Relatively drier poorly drained soils with brown and yellow colouration (mottles) in the upper horizon and with thin organic mats (Pergelic Cryaquepts and Cryaquolls)
Meadow Tundra	Wetter soils with dark grey colours and somewhat thicker organic mats (Pergelic Cryaquepts)
Half-Bog soils	Saturated soils with organic mats approximately 15 cm to 30 cm thick (Histic Pergelic Cryaquepts)
Bog soils	Saturated soils with thick organic accumulations (all Histosols)

Gleysolic Static Cryosols	Poorly drained soils with mottles and low chromas at mineral surface. Peaty layers common at surface.
Organic soils Fibric Organic Cryosols	Soils with organic layers greater than 1 m thick and composed of fibric material in control section
Mesic Organic Cryosols	Soils similar to fibric but organic material is dominated by mesic material in control section
Humic Organic Cryosols	Soils similar to Mesic Organic Cryosols except that organic content is dominated by humic material in control section

Cryosolic Order of Canadian Classification

Name	Description
Turbic	Mineral soils strongly cryoturbated and generally associated with patterned ground
Orthic Turbic Cryosols	Soils strongly cryoturbated with tongues of intermixed mineral and organic material, imperfectly to moderately well drained (equivalent to Upland tundra and Forest tundra)
Brunisolic Turbic Cryosols	Soils less affected by cryoturbation and having unbroken Bm horizon similar to the Orthics
Regosolic Turbic Cryosols	Soils lacking Bmy or Bm horizons, surface organic horizon may be present but organic intrusions are lacking in subsoil (generally in high Arctic or Alpine areas)
Gleysolic Turbic Cryosols	Poorly drained soils with mottles and low chromas, and Cg or Bg horizons at mineral surface
Static	Mineral soils without strong cryoturbations
Orthic Static Crysols	Soils having a gleyed Bm horizon above the permafrost table
Brunisolic Static Cryosols	Similar to Orthic subgroup but with thicker Bm horizons. (Equivalent to Arctic Brown)
Regosolic Static Cryosols	Except for Ah horizon, soils without pedogenic horizons

Arctic and Tundra Zones Genetic Soil Groups and Subgroups of USSR Soil Classification

Name	Description
Arctic soils	Mean July air temperature for A horizon: 2°–6°
Desert- Arctic	Well-drained primitive turfy soils with slight accumulation of humus. Some weak gley in active layer. Associated with polygonal terrain (Regosols? Pergelic Cryorthents)
Sod-Arctic	Well-drained soils developed under sodded tracts (polar steppe) with Al and B horizons. Carbonates may be present in profile (Arctic Brown Soils, Pergelic Cryochrepts, Cryumbrepts, and Cryoborolls)
Tundra soils	Mean July air temperature for A horizon: 6°–10°
Sod Tundra	Well-drained soils of uplands (Pergelic Cryorthents, Cryochrepts, Cryumbrepts, Cryoborolls)
Alluvial Tundra-Sod	Well-drained soils of flood plains (Pergelic Cryorthents)
Gley Tundra	Poorly drained soils (Pergelic Cryaquepts)
Water- logged (moss-gley) Tundra	Poorly drained soils of low positions (Histic Pergelic Cryaquepts)
Bog Tundra	Soils with standing free water (Histosol and Histic Pergelic Cryaquepts)

Table 2.2. Polar soil classification (*after Tedrow, 1977, 267–81, Tables 15–2a—15–2d*)[1]

First order	Tundra soil zone	Polar desert soil zone	Subpolar desert soil zone	Cold desert soil zone (Antarctica)

Second order	Well-drained soils
	  Arctic Brown soil
	  Podzol-like soil
	Mineral Gley soils
	  Upland Tundra soil
	  Meadow Tundra soil
	Organic soils
	  Bog soils
	Other soils
	  Ranker soil
	  Rendzina soil
	  Shungite soil
	  Grumusols
	  Lithosols
	  Regosols
	  Soils of the
	    solifluction slopes
		Well-drained soils
		  Polar Desert soil
		  Arctic Brown soil
		Mineral Gley soils
		  Upland Tundra
		  Meadow Tundra
		Soils of the
		  hummocky ground
		Soils of the polar
		  desert-tundra
		  interjacence
		Organic soils
		  Bog soils
		Other soils
		  Regosols
		  Lithosols
		  Soils of the
		    solifluction
		    slopes
		    (may be a form of
		    gley soil but usually
		    well drained)
			Well-drained soils
			  Polar Desert soil
			  Arctic Brown soil
			Mineral Gley soils
			  Upland Tundra
			  Meadow Tundra
			Soils of the
			  hummocky ground
			Soils of the polar
			  desert-tundra
			  interjacence
			Organic soils
			  Bog soils
			Other soils
			  Regosols
			  Lithosols
			  Soils of the
			    solifluction slopes
				Ahumic (frigic) soils
				  Ultraxerous
				  Xerous
				  Subxerous
				  Ahumisol
				Evaporite soils
				Ornithogenic (Avian)
				  soils
				Other soils
				  Protoranker
				  Algae peats
				  Hydrothermal soils
				  Regosols (recent
				    soils)
				  Lithosols

[1] The present table differs from the original in capitalizing such terms as Arctic Brown, etc., in conformity with Table 2.1.

soil zones and second order Great Soil Groups, followed by three further orders. His classification (Table 2.2) is thus very similar to the Great Soil Groups of the Zonal Classification but provides for genetic types of the Great Soil Groups to appear in more than one soil zone. Recent surveys of polar soils include Tedrow's (1977) monographic overview based on a lifetime's research, and Walton's (1972) worldwide review of High Arctic soils. Many fine-grained, cold-climate soils are characterized by a vesicular texture as a result of freezing, thawing, and drying (Bunting, 1977).

Soil can be a dependent or independent factor. Since it results from processes acting on bedrock, it is dependent on climate to the extent that the processes are climatically controlled. However, in many places the soil, like bedrock, predates the situation being considered and in this sense is independent. Furthermore there is such a variety of ways in which the unconsolidated material can vary that even though it is a function of one periglacial process it could be considered an independent factor in relation to a different periglacial process.

Rock material is less dependent on topography than vice versa, and in this respect is also essentially an independent factor. Therefore despite intimate interactions between climate, topography, and rock material, it is not circular reasoning to regard rock material as an independent as well as a dependent variable in considering periglacial processes.

As noted above under Microclimate and below under Colour, ground-surface temperatures can be very different from air temperatures. Another illustration of this is provided by observations in an alpine permafrost environment where the rock-surface measured 42°, four times the air temperature 10 cm above the rock (Mathys, 1974, 49, 61, Figure 5).

A convenient quantity by which to compare the ability of materials to adjust to cyclical changes of temperature is known as conductive capacity or contact coefficient, and is defined as $\sqrt{\rho c K}$ (cal cm^{-2} s$^{-\frac{1}{2}}$), where ρ = density, c = mass specific heat, K = thermal conductivity. Rock material can vary considerably in conductive capacity. Thus a cyclical temperature change, other conditions remaining constant, will reach some three times deeper in sandy clay (conductive capacity = 0.037 cal cm^{-2} s$^{-\frac{1}{2}}$) than in dry sand (conductive capacity = 0.011 cal cm^{-2} s$^{-\frac{1}{2}}$) (cf. Gold and Lachenbruch, 1973, 5–6).

b Structure   The structure of rock material comprises its gross features: consolidated bedrock or unconsolidated sediments, nature of jointing in bedrock or of fissuring in unconsolidated sediments, or attitude of stratification planes in either case. In many places structure determines the relative importance of processes. For instance, joints and fissures favour ingress of moisture and they localize weathering and subsequent erosion, and the attitude of joints and fissures determines the direction and inclination of the resulting features.

c Mineral composition The mineral composition of rock material strongly influences its reaction to weathering and abrasion and thus to erosion. Bedrock or sediment in a warm humid climate may weather very differently from material of the identical mineral composition in a cold dry climate, and chemically different materials in the same climate may have diverse reactions. In either case the quantitative effect on a given geomorphic process may be very significant. Wind action may produce numerous ventifacts in a limestone region but comparatively few where there are only harder rocks.

d Texture The texture of rock material (i.e. the grain size and arrangement of its mineral constituents or particles) strongly influences its characteristics in any climate, whether the material is bedrock or unconsolidated. In general, fine-grained bedrock is more resistant than coarse grained to weathering in a periglacial environment, other conditions being comparable. On the other hand, fine-grained unconsolidated material may be more subject than coarse to frost action. For instance, as discussed under Freezing process in the chapter on General frost-action processes, it is common engineering practice to regard soils as susceptible to frost heaving if they contain several per cent of particles finer than 0.07 mm (passing the 200 mesh screen, *c*. the upper limit of silt) (Terzaghi, 1952, 14), or (Casagrande criterion) containing more than 3 to 10 per cent particles finer than 0.02 mm (Casagrande, 1932, 169).

Some of the many different ways in which texture can interact with periglacial processes have been discussed by Schunke (1975a, 214–26) with specific reference to Iceland. Frost action, in turn, can strongly influence texture by rearranging mineral particles as, for instance, in frost sorting. Another example is the change that can be effected in hydraulic conductivity of soil by freeze-thaw cycles (Benoit, 1975).

e Colour Although colour is a function of mineral composition and texture, it is worth listing separately to emphasize its effect on temperature: namely dark-coloured materials absorb radiant heat and warm their surroundings whereas the opposite is true for light-coloured materials. For instance, thawing of snow has been observed adjacent to dark-coloured objects at air temperatures as low as −16.5° (G. Taylor, 1922, 47) and even −20° (Souchez, 1967, 295).

4 Time

Time is an independent factor that is usually overlooked in periglacial studies. However, in the opinion of P. J. Williams (1961, 346) 'observed variations in density of occurrence of specific fossil frozen ground phenomena are not so much indicative of particular climatic conditions but of length of time during which the features could be formed'. This view is based on the belief that the processes responsible for periglacial phenomena such as solifluction and patterned ground are very slow; for instance, that it would take, say, 12 000 years for solifluction to move materials 10–20 m downslope.

On the other hand, observations in Northeast Greenland indicate that present rates of solifluction on a slope of 10°–14° could move materials 9–37 m downslope in 1000 years, the amount depending on the moisture conditions, which varied along the contour. These and other observations on solifluction by various investigators (cf. Washburn, 1967, 93–8, 118) and the rapidity with which some forms of patterned ground can develop (as indicated by occurrences on recently emerged shores and by direct observation), show that more information is required before quantitative inferences can be made regarding the extent to which time affects the processes. In any event, as stressed by Williams, the time factor certainly merits consideration in paleoclimatic reconstructions.

5 Human activities

Human activities are so widespread as to constitute another independent factor. Landscape changes caused by such activities are legion and can so change the natural environment that periglacial processes and features can be destroyed, modified, or enhanced over wide areas. Deforestation is an example that can lead to considerable uncertainty as to whether or not the altitude of some periglacial forms with respect to tree line is climatically determined (cf. Hagedorn, 1977, 232; Höllermann and Poser, 1977, 340). Domestication of grazing animals that tend to disrupt the vegetation is another example (cf. Höllermann, 1977, 250; Kelletat, 1977b, 218). The effects of human activities can be sufficiently ancient that their impress on an area is no longer immediately obvious.

III   Dependent factors

1   Snow cover and ice cover

The amount and distribution of snow cover are functions of climate and topography. The effect of the latter is particularly important in the distribution of snowdrifts.

The insulating effect of snow cover on the temperature at the snow ground interface can be estimated in cgs units by applying Lachenbruch's (1959, 28; Plate 1) equation

$$\Delta \tilde{\Theta} = \frac{1}{\pi} A^*[1 - A(X)/A^*]$$

where $\Delta \tilde{\Theta}$ = change in mean annual temperature at the ground surface as the result of the snow, A^* = amplitude of mean annual temperature at bare-ground surface in summer and upper snow surface in winter, $A(X)$ = steady amplitude resulting from wave with amplitude A^* passing through snow of thickness X.

By reducing the amplitude and depth of daily and seasonal temperature changes in the ground, snow cover can reduce frost action, even if the thermal properties of the snow are no different than those of the soil, although in many places snow is a good insulator as a result of high porosity. However, the effect of snow cover on heat exchange between the air and ground is complicated. Among other considerations snow, because of decreasing thermal conductivity, tends to be a better insulator the lower the temperature (Pavlov, 1978, 71–2), especially in forested areas of strongly continental climate where loose snow and depth hoar containing large air spaces are common. In windswept tundra and alpine environments having hard-packed snow, depth hoar tends to be lacking, porosity is reduced, and the temperature dependency is much less apparent or absent (S. I. Outcalt, personal communication, 1978; cf. Goodwin and Outcalt, 1975). In areas where permafrost is discontinuous, variations in snow distribution and thickness can be the critical factor controlling its presence or absence (cf. Nicholson and Granberg, 1973). In places variability of snow cover can reverse the effect of exposure in different years (Table 2.3.).

By moderating the ground temperature, snow can protect vegetation from frost action; it can also provide the critical moisture for growth. Thus snow favours vegetation in several ways. However, if lasting too long, snow can also completely inhibit vegetation. In addition, thawing of snow, especially snowdrifts, can control the amount of moisture and thereby the nature of many periglacial processes in an area (cf. Shunke, 1975a, 201). Each of the above effects can be critical for a given periglacial process.

Ice is a less effective insulator than an equal thickness of snow but it will prevent a sufficiently deep river or lake from freezing to the bottom and thus inhibit the development of underlying permafrost in areas where it is otherwise present (cf. Gold and Lachenbruch, 1973, 17–18). A glacier with a bottom temperature of 0° will also inhibit development of permafrost but frozen ground would form beneath glaciers whose base was below the PT melting point as in places beneath the Antarctic Ice Sheet (Gow, Ueda, and Garfield, 1968) and Greenland Ice Sheet (B. L. Hansen and Langway, 1966).

2   Liquid moisture

The amount and nature of liquid moisture and its seasonal distribution are functions of climate, topography, and rock material. Climate exerts the large-scale control, topography modifies the large-scale climatic influence, and rock material can determine the amount of moisture entering and remaining in the ground. The texture of the rock material may be diagnostic in this respect and either favour or

Table 2.3. Snow and frost depths on north and south slopes, Coulee Experimental Forest, southwestern Wisconsin (*after Sartz, 1973, 2, Table 1*)

Date	Snow depth (cm)		Frost depth (cm)		Remarks
	North slope	South slope	North slope	South slope	Snow depth to nearest 5 cm
25 Feb 1970	25	0	8	11	
25 Feb 1971	55	25	0	0	Snowpack formed before ground froze
10 Mar 1972	45	25	16	23	
1 Mar 1973	5	0	21	12	Little snow throughout winter

inhibit frost action and vegetation, each of which also influences the other in complex ways. Péwé (1974, 42, Figure 3.11) has suggested that a mean annual precipitation between <5 cm and 25 cm (*c.* 15 cm) is a critical minimum for many periglacial processes and features. The nature of the liquid moisture is important because solutes can lower the freezing point appreciably. For instance, the salinity of soils in the Antarctic dry valleys depresses the freezing point sufficiently to prevent freezing and permit movement of pore fluids at depths where the temperature remains about −18° throughout the year (McGinnis, Nakao, and Clark, 1973, 140).

3 Vegetation

Vegetation, like liquid moisture, is a function of climate, topography, and rock material. Some plant species are much more sensitive to climate than others. Zonal climate sets the broad pattern but local climate can dominate a given situation, and there is an intimate interaction and mutual influence between vegetation and microclimate. For instance, the mean annual air temperature above '... unvegetated dry soils can be 1° or 2° warmer than a nearby site underlain by vegetated wet soils' (Ferrians, Kachadoorian, and Greene, 1969, 8). Also, plant remains in the form of peat pose special conditions because of the low thermal conductivity of peat compared with mineral soil. Thus frozen ground can be present in peatland but absent elsewhere (R. J. E. Brown and Williams, 1972). Destruc-

tion of a peat or vegetation cover can have far-reaching effects, especially where there is permafrost (Jerry Brown, Rickard, and Vietor, 1969; Rickard and Brown, 1974; Viereck, 1973). Near Fairbanks, Alaska, permafrost in test plots from which the vegetation had been stripped to different degrees in 1946 continued to thaw until at least 1973 without reaching equilibrium (Linell, 1973). Following destruction of vegetation by fire at Inuvik in northern Canada in 1968, thawing continued until at least 1976 (Mackay, 1977*e*).

Aside from its influence on local climate, topography is critical in that the angle of slope affects the continuing contest between the ability of plants to take root and opposing processes such as mass-wasting. Soil characteristics also exert a strong influence on the ability of plants to survive opposing processes.

Thus the especially important ways in which vegetation influences periglacial processes are through its insulating effect and its binding effect on soils. In Kryuchkov's (1976*a*, 38–41; 1976*b*, 33–6) view, the nature and extent of vegetation as primarily determined by climate are the dominant factors controlling some varieties of patterned ground. Their influence on the altitudinal zonation of periglacial features was emphasized by Karrasch (1977, 174). Details of the interaction between vegetation and frost action in soils are complicated but can be of critical importance. The problems are discussed by Balobayev (1964; 1973), Benninghoff (1952; 1966), R. J. E. Brown (1966*a*), Kryuchkov (1976*a*; 1976*b*), Raup (1965; 1969; 1971*a*), and Tyrtikov (1973; 1978) among others.

3 Frozen ground

I Introduction

Frozen ground is central to a consideration of periglacial processes and environments. In particular,
knowledge of its nature sets the stage for discussing
the various processes collectively known as frost
action, considered in the next chapter.

In many respects it is important to distinguish
between seasonally frozen ground and permafrost.
The latter is especially significant in periglacial
studies and is therefore discussed in some detail.

II Seasonally frozen ground

1 General

Seasonally frozen ground '. . . is ground frozen by low
seasonal temperatures and remaining frozen only
through the winter' (S. W. Muller, 1947, 221).
Some investigators (cf. S. W. Muller, 1947, 6, 213)
include the zone of annual freezing and thawing
above permafrost (i.e. active layer), others (cf. Black,
1954, 839) restrict or appear to restrict the term to
a non-permafrost environment. Because of the ambiguity and difficulty of trying to differentiate between
two zones of annual freezing and thawing by a
descriptive term equally applicable to both, the
present writer uses the term seasonally frozen ground
in the broad sense and specifies, where necessary,
whether the reference is to a permafrost or non-
permafrost environment. In most instances, use of

the term active layer (discussed later) for the seasonally frozen ground above permafrost helps to avoid
ambiguity. Ground freezing has been estimated to
occur over almost half (48 per cent) of the land
mass of the Northern Hemisphere, the southern limit
being near lat. 40°N and characterized by frost
penetration to a depth of about 30.5 cm (1 ft) once
in 10 years (Bates and Bilello, 1966, 6). On this
basis, the area of seasonally frozen ground as distinct
from permafrost constitutes about 26 per cent of the
land area, assuming that permafrost, exclusive of
mountains, covers 22.35×10^6 km² (Table 3.1), or
22 per cent of the total land area in the Northern
Hemisphere (100.30×10^6 km²). Many soil properties and processes are radically changed as the result
of ground freezing, with many practical consequences. For instance, permeability is commonly
decreased and therefore run-off increased by frozen
ground, with vast implications for flooding, water
supply, and other hydrologic problems. However,
it would be a mistake to assume, as often done, that
frozen ground is necessarily impermeable; rather a
wide range of conditions is possible (Dingman,
1975, 28–47).

2 Depth of seasonal freezing and thawing

Depth of freezing and thawing is controlled by many
of the factors reviewed previously, and is therefore
subject to considerable local variation. On a larger
scale, there is a clear increase in depth of seasonal

freezing with increasing latitude in a non-permafrost environment, the range being from a few millimetres to over 1.8 m (72 in) in the United States (Figure 3.1) and up to a depth of 3 m in Canada (Crawford and Johnston, 1971, 237). In a permafrost environment the trend is towards a decrease in depth with increasing latitude, and the permafrost tends to approach the surface of the ground.

The climatic parameters responsible for deep seasonally frozen ground and for permafrost may be very similar in the transition zone, but the resulting products can be quite different in a non-permafrost or a permafrost environment. Construction problems, for instance, can be many times more difficult in a permafrost environment than where there is seasonally frozen ground without permafrost.

It may not always be practicable to obtain direct measurements of the depths to which seasonal freezing and thawing extend, and several indirect approaches have been suggested. Some of these involve a number of variables, and others are relatively simple yet useful.

For example, the depth of winter freezing can be estimated by the Stefan equation, which in English units has usually the form (Yong and Warkentin, 1975, 391–7)

$$x = \sqrt{\frac{48k_f F}{L}} \qquad (1)$$

where x = depth of freezing (ft), k_f = thermal conductivity of frozen soil (0.9 (BTU h^{-1} ft^{-1} $°F^{-1}$), F = freezing index (number of F degree days below freezing), and L = latent heat of water in the soil = $L_w w \gamma_d$ (where L_w = latent heat of water [143.4 BTU lb^{-1} water], w = amount of water in soil [in decimal fraction of dry wt. of soil], and γ_d = unit weight of dry soil [in lb ft^{-3}]).

The same equation can be used in metric units by appropriate substitutions, with x = cm, k_f = cal h^{-1} cm^{-1} $°C^{-1}$, F = C degree days below freezing, and $L_w w \gamma_d$ is of the form where L_w = 80 Cal g^{-1}, ω is decimal fraction of dry wt. of soil, and γ_d is in g cm^{-3}.

Or, following Jumikis (1977, 206–7)

$$\xi = \sqrt{\frac{48K \cdot F}{Q_L}} \qquad (2)$$

where ξ = depth of freezing (m), K = coefficient of thermal conductivity of frozen soil (Cal m^{-1} h^{-1} $°C^{-1}$), F = surface freezing index (C degree days), Q_L volumetric latent heat of water (Cal m^{-3}).

If the index F is in C degree hours, (2) assumes the form

$$\xi = \sqrt{\frac{2KF}{Q_L}} \qquad (3)$$

By inserting appropriate figures for the thawing period (thermal conductivity of thawed soil, thawing index), the equations can also be used to estimate depth of summer thawing. They apply to both permafrost and non-permafrost environments. A number of refinements and other expressions are available whose complexity increases with the accuracy required (cf. Andersland and Anderson, 1978, 131–49).

The equations cited have several potential sources of error. A serious problem is that the soil properties may be difficult to estimate without drilling and laboratory testing. Also in the absence of ground-surface temperatures, error is introduced by substituting air temperatures without allowing for the effect of snow or vegetation cover and the influence of colour on radiant heating. Lachenbruch's (1959, 28, Plate 1) equation, cited under discussion of Dependent factors in the chapter on Environmental factors, can be applied to evaluate the effect of snow cover. In addition to the mean annual air temperature, this equation also considers the amplitude of the temperature change from summer to winter, which can be a critical variable in evaluating the effect of a disturbance of the ground surface on depth of thawing. 'The larger the amplitude, the worse the thawing problem' (Lachenbruch, 1970b, J5). However, the thawing problem is minimal where summer temperatures are low and are above freezing for only a brief period as in the Antarctic dry valleys. Sophisticated computer programs and modeling are being increasingly widely used to evaluate the response of the ground to temperature perturbations (cf. Kliewer, 1973).

III Permafrost

1 General

The term permafrost, also known as pergelisol (Bryan, 1946, 635, 640) and perennially frozen ground, was first defined by S. W. Muller (1947, 3, cf. 219)

Permanently frozen ground or permafrost is defined as a thickness of soil or other superficial deposit, or even of bedrock, at a variable depth beneath the surface of the earth in which a temperature below freezing has existed

MAXIMUM DEPTH OF FROST PENETRATION (CM; CONVERTED FROM INCHES)

This map is reasonably accurate for most parts
of the United States but is necessarily highly
generalized, and consequently not too accurate
in mountainous regions, particularly in the Rockies

3.1 Maximum depth of frost penetration in United States (*after Strock and Koral, 1959, 1–102*)

continually for a long time (from two [years] to tens of thousands of years). Permanently frozen ground is defined exclusively on the basis of temperature, irrespective of texture, degree of induration, water content, or lithologic character.

The Institut Merzlotovedeniya in Yakutsk, USSR, one of the world's leading permafrost research agencies, specifies three years or more in accordance with the recommendations of the Commission on Terminology of the Institut Merzlotovedeniya im. V. A. Obrucheva (1956, 10; 1960, 6; Corte, 1969a, 130). On the other hand, some Canadian workers adopt a minimum period of one year only – i.e. ground remaining frozen throughout at least one summer (R. J. E. Brown, 1967a; French, 1976a, 47). The present writer follows Muller's original definition, specifying at least two years.

The term permafrost originated from what Muller described as a slip of the tongue that he was encouraged to formalize (S. W. Muller, personal communication, 1959). The term has been widely adopted but defined as perennially frozen rather than permanently frozen ground, since changes in climate and surface conditions can cause rapid thawing of permafrost. Muller's definition in which texture, induration, water content, and lithology are eliminated as factors (cf. also S. W. Muller, 1947, 30) supports the view that 0° is the basic criterion rather than the exact freezing point, which can vary with such factors including salts and pressure. Most workers cite a temperature below 0° in defining permafrost (cf. R. J. E. Brown, 1967a; R. J. E. Brown and Kupsch, 1974, 25; Ferrians, 1965), but in practice it is doubtful if ground that rises to 0° for part of the year would be excluded if the temperature rose no higher and its mean annual temperature remained below 0°. In fact a temperature as high as 0° is specifically accepted by the Bol'shaya Sovetskaya Entsiklopediya (Sharbatyan, 1974a, 370; 1975, 1) and by French (1976a, 49). Amended definitions of permafrost have been suggested. According to Stearns (1966, 1–2)

The term 'permafrost' is defined as a condition existing below ground surface, irrespective of texture, water content, or geological character, in which:
a. The temperature in the material has remained below 0°C continuously for more than 2 years, and
b. If pore water is present in the material a sufficiently high percentage is frozen to cement the mineral and organic particles.

A temperature definition alone is not considered sufficient, for often a geothermal situation exists in which a frozen, or cemented, state is not obtained even though the temperature of the material is well below 0°C....

The definition of permafrost given above includes both dry-frozen and wet-frozen ground. In the dry-frozen, or dry frost, condition there is very little or no water contained in the pores so that temperature becomes the only criterion. In the wet-frozen condition some cementing ice must be present.

It remains to be seen to what extent Stearns' redefinition is accepted. It is followed by Carey (1973, 8), and from the viewpoint of engineering requirements it has the merit of specifying cementation where there is wet-frozen ground. However, the last two paragraphs pinpoint a problem in that temperature alone is not considered sufficient for wet-frozen ground but is the only criterion for dry-frozen ground. Furthermore cementation is a subjective criterion unless physical parameters defining cementation are established. Muller's definition is more precise and for mapping purposes, at least, the easier to apply, and is followed here.

In an attempt to eliminate ambiguities, van Everdingen (1976) proposed that 'unfrozen' be replaced by nonfrozen, that 'frozen' be used only if ice is present, and that 'cryotic' be adopted to indicate a temperature <0° without implication as to phases in a soil/water system as indicated in Table 3.2. As with Stearns' redefinition of permafrost, van Everdingen's proposal must stand the test of time. The terminological problems addressed are real, and regardless of the terms used it behoves authors to be very specific if there is any possibility of misunderstanding.

Glaciers whose temperature does not reach 0° are permafrost by either definition but they are usually omitted from discussions of permafrost and treated separately (Ferrians and Hobson, 1973, 486). They will be mentioned here only incidentally.

Permafrost in the Soviet Union was mentioned as early as 1642 in Siberian military reports by Glebov and Golovin (Tsytovich, 1966), and was recognized in northern Canada by at least 100 years later (Middleton, 1743, 159; Rich and Johnson, 1949, 67, 71; cf. Legget, 1966, 3).

2 Significance

Civil engineering problems connected with building and highway construction, sewage and waste disposal, water supply, and hot-oil and cold-gas pipelines are magnified in permafrost regions, and much attention has been devoted to permafrost as a result (cf. Alter, 1969a; 1969b; Anisimova *et al.*, 1973a; 1973b; 1978; Jerry Brown, Rickard, and Vietor,

1969; R. J. E. Brown, 1970; Burdick and Johnson, 1977; Carlson, 1977; Cheng, 1975; Cohen, 1973; Corte, 1969a; Crawford and Johnston, 1971; Dingman, 1975; Environment Canada, 1974; Ferrians, Kachadoorian, and Greene, 1969; Finn, Yong, and Lee (1978); Gold *et al.*, 1972; Kachadoorian and Ferrians, 1973; Kudryavtsev, 1974; 1978; Kudryavtsev *et al.*, 1977; Lachenbruch, 1970a; Linell and Johnston, 1973; Melnikov and Tolstikhin, 1979; Sanger, 1969; Tsytovich, 1975; Vyalov *et al.*, 1962; 1965; 1973a; 1973b; 1978; Vyalov, Dokuchayev, and Sheynkman, 1976; J. R. Williams, 1970; and J. R. Williams and van Everdingen, 1973. A number of the references noted in the section on References in the first chapter are also highly pertinent.)

The problems are both civilian and military. The military interest that began with World War II led to the establishment of the US Army Snow, Ice, and Permafrost Research Establishment (SIPRE), later renamed the US Army Cold Regions Research and Engineering Laboratory (CRREL), which has been responsible for much of the basic research on permafrost in the United States. The US Geological Survey has also been very active in permafrost research. Yet our knowledge is still inadequate as illustrated by the controversy aroused by the trans-Alaska pipeline system before it was approved – and by the permafrost-related research that has been

advocated as a result of the north-south transportation route provided by the system (National Academy of Sciences, 1975). Increasingly, permafrost is becoming recognized as an important factor in Arctic land-use planning and its environmental impact (Andrews, 1978; Jerry Brown, 1973; Jerry Brown and Grave, 1979a; 1979b; Grave and Sukhodrovskiy, 1978; McVee, 1973; Melnikov, 1977). Not only are contemporary frost action and permafrost of engineering concern but fossil features resulting from past periglacial activity can also be important (Higginbottom and Fookes, 1971).

Aside from the engineering aspects, there are many intriguing scientific questions related to permafrost – its origin, climatic implications, its effect on life (Péwé, 1966c), and what it can tell us about past, present, and future environments. Much of the record is there to be read, and part of the task of periglacial research is to interpret that record.

3   Distribution

Because permafrost becomes thinner and breaks up into patches as it merges with seasonally frozen ground where permafrost is lacking, permafrost is commonly classified into continuous permafrost and

Table 3.1. Distribution of permafrost (*after Stearns, 1966, 9–10*)[1]

	Continuous		Discontinuous		Total	
	km² × 10⁶	mi² × 10⁶	km² × 10⁶	mi² × 10⁶	km² × 10⁶	mi² × 10⁶
Northern Hemisphere	7.64	2.95	14.71	5.68	22.35	8.63
Antarctica	13.21	5.10	—	—	13.21	5.10
Mountains	—	—	2.59[2]	1.00	2.59	1.00
Totals	20.85	8.05	17.30	6.68	38.15	14.73

	Land surface (per cent)	Permafrost	
		km² × 10⁶	mi² × 10⁶
Alaska	80	1.30	0.5
Canada	40–50	3.89–4.92	1.5–1.9
Greenland	99	1.68	0.65
USSR[3]	50	11.14	4.3
		19.04 (max)	7.35 (max)

[1] Few (if any) of the estimates include offshore permafrost.

[2] Regarded as too low a figure by Barsch (1977d, 124).

[3] USSR estimates cited in 1976 are similar (Melnikov and Balobayev, 1977). However, the USSR permafrost area having the bulk of visible ice (c. 19000 km³ at depths <50 m) has been estimated by Vtyurin (1973, 17, Table 1; 1978, 164, Table 1) to be only about 7000000 km², mainly in Siberia and the Soviet Northeast as shown in three informative sketch maps by Vtyurin.

Table 3.2. Terminology describing ground temperature and water phases in a soil/water system (*after van Everdingen, 1976, 864, Figure 2*)

H₂O content→ / Temperature ↓	No H₂O (except chemically bound and adsorbed)	Some H₂O (less than porosity)	Pore spaces filled with H₂O	Containing 'excess' H₂O	Zone descriptions — Phase	Zone descriptions — Temp.
T > 0°C	Dry, noncryotic	Moist or unsaturated, noncryotic	Wet or saturated, noncryotic	---	Nonfrozen	Noncryotic
— 0°C — Cryo point —						
T < 0°C / Initial freezing point of soil system	Dry, cryotic	Moist or unsaturated, cryotic	Wet or saturated, cryotic	--- , --		Cryotic
		Ice-poor, partially frozen	Partially frozen	Ice-rich, partially frozen	Partially frozen	
T < 0°C		Ice-poor, frozen	Frozen	Ice-rich, frozen	Frozen	

Vertical spanning labels (cryotic region): Dry permafrost (1); Moist permafrost (1); Ice-poor permafrost (1); Wet permafrost (1); Permafrost (1); Ice-rich permafrost (1).

(1) If temperature is perennially below 0°

discontinuous permafrost (Figure 3.2). Some investigators have followed Black (1950, 248–9, Figure 1) and S. W. Muller (1947, 6, after Sumgin) in also recognizing a sporadic permafrost zone, which appears to be convenient for alpine environments (Barsch, 1977*d*; 1978; J. D. Ives, 1974, 166) but perhaps less so elsewhere (R. J. E. Brown, 1967*a*).

The distribution of present-day permafrost in the Northern Hemisphere, disregarding high-altitude occurrences, is illustrated in Figures 3.3–3.11. This distribution correlates, at least in part, with the paths of anticyclonic polar air masses and hence cold, dry winters. A close correlation between the southern extent of the continuous permafrost zone and the southern (winter) position of the Arctic frontal zone is suggested by a map of Bryson's (1966, 266, Figure 33). In general the permafrost realm is characterized by days with temperatures < 0° for three fourths of the year, < − 10° for half the year, and rarely > 20°; precipitation is characteristically < 100 mm in winter and < 300 mm in summer (Velichko, 1975, 100). However, significant departures can occur. For instance, at Mesters Vig, Northeast Greenland, the mean precipitation for the 9

months (Sept.–May) having mean temperatures < 0° is about 300 mm rather than < 100 mm (Washburn, 1965, 52, Table 2). Stearns (1966, 9–10), following Black (1954, 839–42), concluded that permafrost underlies some 26 per cent of the land surface (including glaciers) of the world, based on figures compiled from various sources (Table 3.1).

Offshore (also known as submarine or subsea) permafrost, reported in Spitsbergen over 50 years ago (Werenskiold, 1922; cf. 1953), is common off some Arctic coasts (Figures 3.3, 3.12–3.13) (F. E. Are, 1976; 1978*a*; 1978*b*; Barnes and Hopkins, 1978, 117–22; Grigor'yev, 1966, 92–119; 1973; 1976, 95–126; 1978; Hobson *et al.*, 1977; Hunter *et al.*, 1976; Judge, 1974; 1975; 1977; Lachenbruch and Marshall, 1977, 1–6; Lewellen, 1974; Mackay, 1972*a*, 19–20; 1972*d*; Molochushkin, 1973; 1978; National Academy of Sciences, 1976; Osterkamp and Harrison, 1976; 1977; Rogers, 1976; 1977; Rogers and Morack, 1977*a*; 1977*b*; 1978; Sellmann *et al.*, 1976; Zhigarev and Plakht, 1974; 1976). It very probably also occurs in the Antarctic. Water temperatures in McMurdo Sound are about − 1.8°, and on the basis of observed

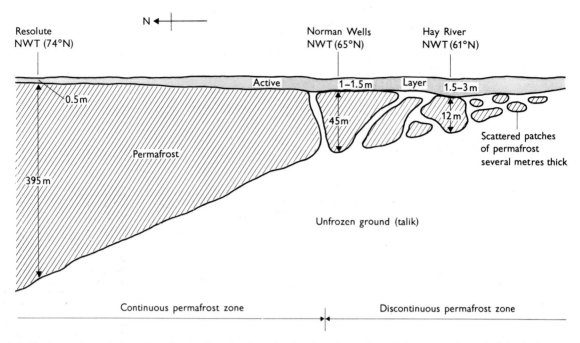

3.2 North-south vertical profile of permafrost in Canada showing decreasing thickness southward and relation to continuous and discontinuous permafrost zones (*after R. J. E. Brown, 1970, 8, Figure 4*)

low pore-water salinity in sediments off New Harbor and compressional velocities of sediments in Terra Nova Bay, the likelihood of ice-cemented permafrost to depths of at least 150 m below sea level in Terra Nova Bay is considered high by L. D. McGinnis (personal communication, 1978). There is now intensive research on Arctic offshore permafrost because of the difficulties it poses for offshore oil development (cf. Hunter *et al.*, 1976). Much of the offshore permafrost probably originated on land and became submerged by postglacial rise of sealevel and coastal erosion. Fossil permafrost beneath water depths of 70–80 m in the Laptev and East Siberian Seas has been reported but without details (Markuse, 1976, 120). Some ice-cemented, offshore permafrost is probably in equilibrium with sea-bottom temperatures ($c. -1.3°$ near shore in the Beaufort Sea – Lewellen, 1974, 420, 426; $c. -1.5°$ to $-1.8°$ near Little Cornwallis Island

in the Queen Elizabeth Islands – Judge, 1974, 432). Ice-bearing offshore permafrost may be widely distributed on the Beaufort Sea continental shelf (Barnes and Hopkins, 1978, 117–22). A comprehensive bibliography of offshore permafrost is being compiled by Vigdorchik (1977a; 1977b).

The difference in Table 3.1 in the figure for the Northern Hemisphere (22.35×10^6 km^2) and the total for Alaska, Canada, Greenland, and the USSR (19.04×10^6 km^2) is due to the different methods of estimating the figures. If cementation is accepted as a criterion of permafrost, and glaciers are excluded, a major reduction in the estimate for Antarctica and Greenland will apply if it turns out that the ice sheets here are extensively underlain by unfrozen material.[1] Excluding glaciers, Shumskii, Krenke, and Zotikov (1964), cited by Grave (1968a, 48; 1968b, 2), estimated that 14.1 per cent of the land

---

[1] Grave (1968a, 51–2; 1968b, 6) and Zotikov (1963) argued that the central part of the Antarctic is free of frozen ground, and drilling confirms the presence of water at the base of the Ice Sheet near Byrd Station (Gow, Ueda, and Garfield, 1968). However, the estimated PT melting point of $-1.6°$ at the drilling site would still be indicative of permafrost according to Muller's definition. Assuming that the Antarctic Ice Sheet is in a steady state, Budd, Jenssen, and Radok (1970, 301–5; 1971, 117–26) calculated, contrary to Grave and Zotikov, that the central area would be frozen and only relatively small marginal areas would be thawed beneath the ice; their assumption is subject to question but Budd, Jenssen, and Radok concluded that the central area would be characterized by permafrost even if ice gain exceeded ice loss by a factor of two. According to similar calculations, most of the Greenland Ice Sheet has a basal temperature below $-5°$ (W. Budd, personal communication, 1971).

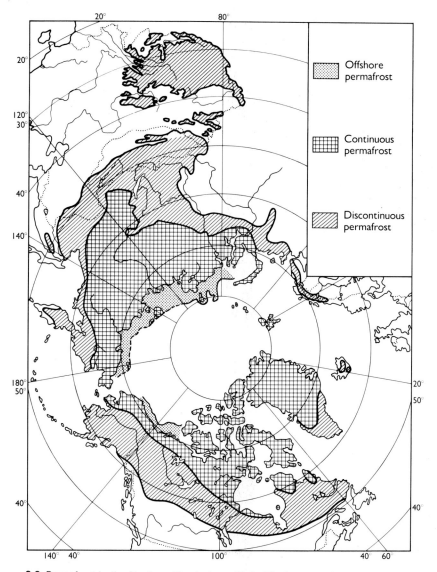

Sources: Alaska, land (T. L. Péwé, personal communication, 1978), offshore (Barnes and Hopkins, 1978); Canada, land (R. J. E. Brown, 1978b), offshore (Hunter et al., 1976); Greenland (Weidick, 1968; O. Oleson, personal communication, 1976); Iceland (Thorleifur Einarsson, personal communication, 1966; Priesnitz and Schunke, 1978); Norway (B. J. Andersen, personal communication, 1966; H. Svensson, personal communication, 1966); Svalbard (Liestøl, 1977); Sweden (Rapp and Annersten, 1969); Mongolia (Gravis et al., 1973, 1978); USSR, land (Karpov and Puzanov, 1970); offshore (M. Vigdorchik, personal communication, 1978); China (Academia Sinica, 1975, 20–1). The offshore permafrost extent shown for North America is very probably a minimum, judging from near-shore water temperatures among Canada's arctic islands (J. R. Mackay, personal communication, 1978)

3.3 Permafrost in the Northern Hemisphere (*T. L. Péwé, personal communication, 1978 with sources as indicated*)

surface is underlain by permafrost. R. J. E. Brown (1970, 7; cf. 27–8) and Péwé (1971; personal communication, 1972) estimated 20 per cent (excluding glaciers), and Péwé in agreement with Stearns (Table 3.1) indicated that permafrost characterizes 80 per cent of Alaska and 50 per cent of Canada. Ferrians and Hobson (1973, 479) agree except for raising the estimate for Alaska to 85 per cent. Jerry Brown (1967, 18, Table 4) estimated that the coastal plain of Arctic Alaska to a depth of 300 m contains 1675 km³ of ground ice, including pore ice (cf. Péwé, 1975, 49). The volume of ground ice in the USSR (probably excluding pore ice) is calculated to be 1900 km³ (Vtyurin, 1975; cited by Grave, 1977, 4). These estimates are based on numerous assumptions. According to Grave's earlier report, the world's permafrost comprises 0.83 per cent of the total freshwater ice. Most of the remainder is in the Antarctic Ice Sheet (>90 per cent), the Greenland Ice Sheet, and other glaciers. Together this ice constitutes some 75 per cent of the freshwater resources of the earth. The amount represented by permafrost is very small but its significance far transcends its quantity.

In northern latitudes there is commonly a difference of 1°–6° between the mean annual temperature

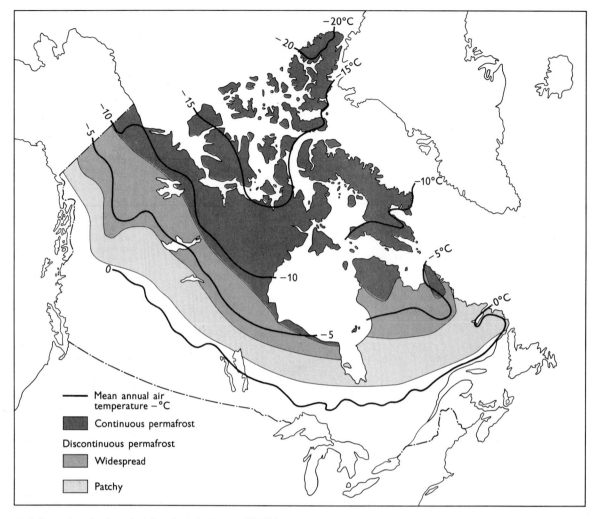

3.4 Permafrost in Canada (*after R. J. E. Brown, 1978b*)

of the air and that of the ground surface (Gold and Lachenbruch, 1973, 5). The latter is commonly warmer, and the difference is generally greatest where there is a thick winter snow cover (G. P. Williams and Gold, 1976, 180–3, Figure 3). The mean northward displacement of the ground-surface isotherm relative to the equivalent air temperature in northern Canada is reported to be 3.3° (Judge, 1973a, 38) (Figure 3.4). Comparison with climatic maps shows that the southern limit of continuous permafrost is not necessarily parallel to air isotherms. It lies north of the −6° or −8° mean annual air isotherm in Alaska (R. J. E. Brown and Péwé, 1973, 74, Figure 2; Péwé, 1966a, 78–9; 1966b, 68) (Figure 3.7). In Canada it is commonly north of −8.5° (R. J. E. Brown and Péwé, 1973, 75), al-

though in places discontinuous permafrost is still present at −9.4° (15°F) (R. J. E. Brown, 1969, 52). According to R. J. E. Brown (1967a) and Crawford and Johnston (1971, 237), the discontinuous and continuous zones merge where the mean annual air temperature is about −8.3°. This accords with arbitrarily mapping the boundary where the mean annual air temperature corresponds to a mean annual ground temperature of −5° at the depth of zero annual amplitude as discussed below. Probably about −7° approximates the southern boundary of continuous permafrost in the USSR (Baranov, 1959, Figure 24 opposite 201; 1964, 83, Figure 24; Akademiya Nauk SSSR, 1964, 234), but permafrost is absent in the Turukhansk region of USSR at −7° (Shvetsov, 1959, 79; 1964, 5–6; who

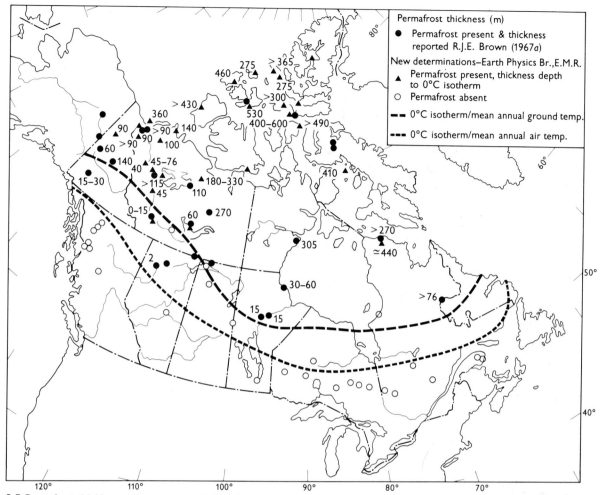

3.5 Permafrost thickness and mean annual air and ground–surface isotherms in Canada (*after Judge, 1973a, 38, Figure 3*)

cited Yachevskii (1889)[2] or −7.4° (Schostakowitsch, 1927, 396). R. J. E. Brown (1967*b*) has pointed out in an excellent review that there are significant differences as well as many similarities in environmental factors and the distribution of permafrost in North America and the USSR.

The distribution of permafrost can also be evaluated with respect to the temperature at the depth of zero annual amplitude – i.e. the depth to which seasonal changes of temperature extend into permafrost – which closely approximates the mean annual ground temperature at the surface (National Academy of Sciences, 1974, 40). Black (1954, 843) noted that permafrost at this depth is generally below −5° in the continuous zone, and according

to R. J. E. Brown (1967*a*; 1970, 8) the criterion of −5° at the depth of zero annual amplitude has been adopted in both North America and the USSR as an arbitrary boundary between continuous and discontinuous permafrost. The difficulties of accurately mapping permafrost are discussed by J. D. Ives (1974, 166–71) and Vostokova (1973; 1978).

Permafrost is in a particularly delicate ecological balance in the Subarctic whose southern limit can also be considered as the southern limit of discontinuous permafrost (R. J. E. Brown, 1966*b*). Possibly much of the discontinuous zone is in disequilibrium with the present climate but detailed observations are needed. According to a number of investigators the southern limit of discontinuous

[2] Yachevskii (1889, 346) gave −8° (*c.*) for Turukhansk, which lies in an area of insular permafrost (Akademiya Nauk SSSR, 1964, Plate 234). Kendrew (1941, 212), referring to lack of permafrost at Turukhansk, gave a temperature of 17°F (i.e. also about −8°C).

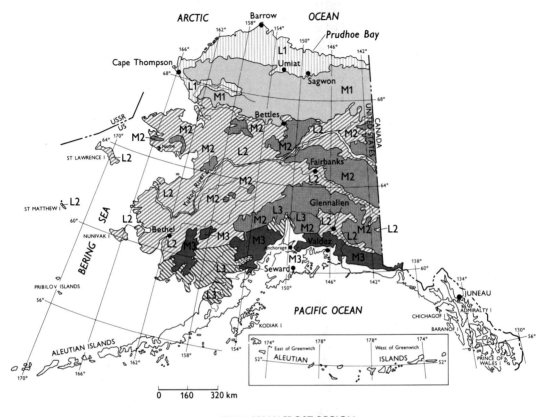

AREAS WITHIN PERMAFROST REGION

Mountainous areas, generally underlain by bedrock at or near the surface

 M1 Underlain by continuous permafrost

M2 Underlain by discontinuous permafrost

Underlain by isolated masses of permafrost

Lowland areas, generally underlain by thick unconsolidated deposits

L1 Underlain by thick permafrost in areas of either fine-grained or coarse-grained deposits

L2 Underlain by moderately thick to thin permafrost in areas of fine-grained deposits, and by discontinuous or isolated masses of permafrost in areas of coarse-grained deposits

Underlain by isolated masses of permafrost in areas of fine-grained deposits, and generally free of permafrost in areas of coarse-grained deposits

AREAS OUTSIDE PERMAFROST REGION

Generally free of permafrost, but a few small isolated masses of permafrost occur at high altitudes, and in lowland areas where ground insulation is high and ground insolation is low, especially near the border of the permafrost region

3.6 Permafrost in Alaska (*after Ferrians, Kachadoorian, and Greene, 1969, 3, Figure 2, cf. Ferrians, 1965*)

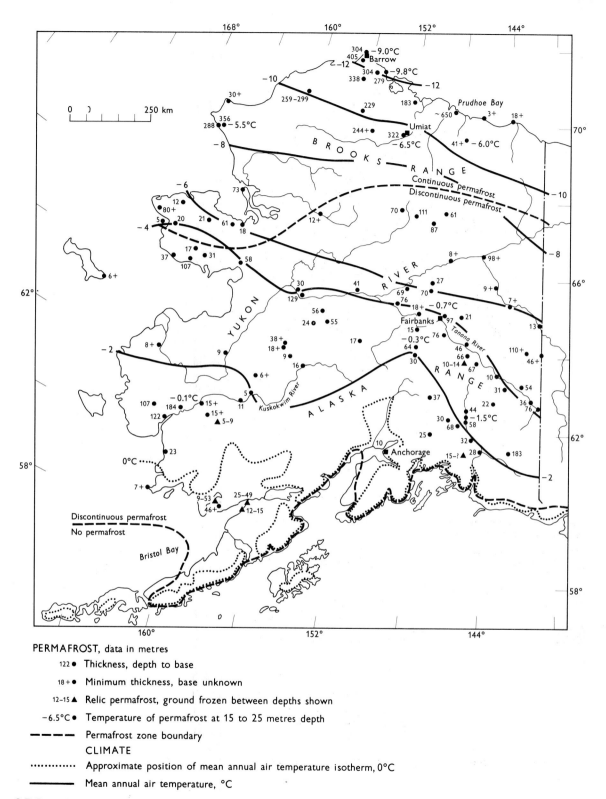

PERMAFROST, data in metres

122 ● Thickness, depth to base

18+● Minimum thickness, base unknown

12–15 ▲ Relic permafrost, ground frozen between depths shown

–6.5°C ● Temperature of permafrost at 15 to 25 metres depth

— — — Permafrost zone boundary

CLIMATE

.... Approximate position of mean annual air temperature isotherm, 0°C

——— Mean annual air temperature, °C

3.7 Permafrost thickness and temperature in relation to mean annual air-temperature isotherms in Alaska (*after R. J. E. Brown and Péwé, 1973, 74, Figure 2*)

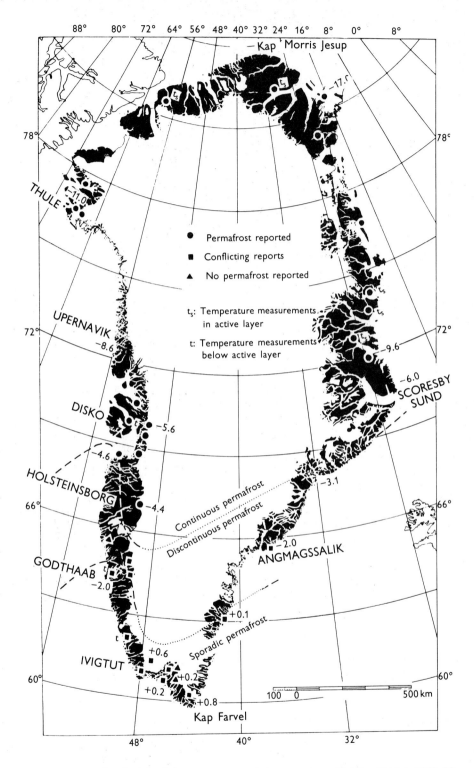

3.8 Permafrost in Greenland in relation to mean annual air temperatures, °C (*after Weidick, 1968, 73, Figure 25*)

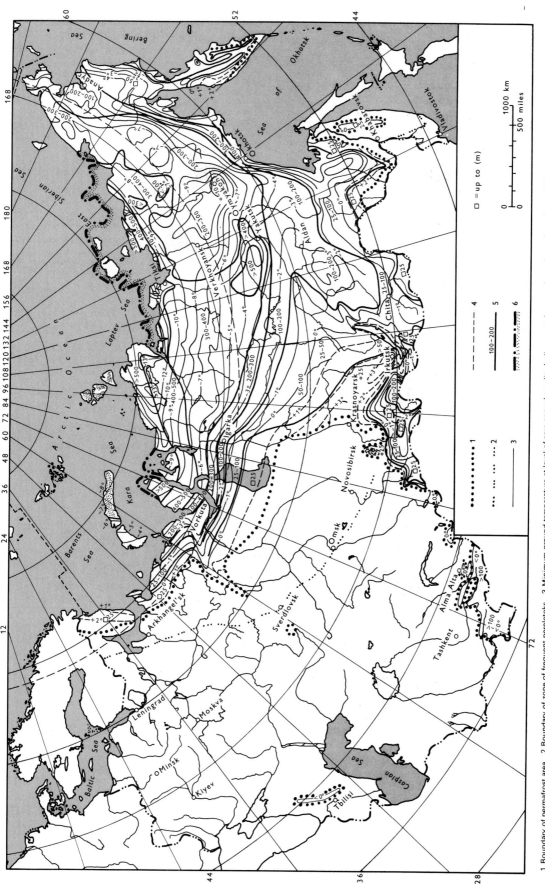

3.9 Permafrost in the Soviet Union (*after Baranov, 1959, Figure 24, opposite 201; cf. Baranov, 1964, 831, Figure 24*)

1 Boundary of permafrost area   2 Boundary of zone of frequent pereletoks   3 Minimum ground temperature at level of zero annual amplitude (in mountainous regions, shown for valleys)   4 Soil isotherms at 1–2 m depth under natural conditions   5 Maximum thickness of permafrost (m)   6 Permafrost zone under Arctic Ocean.   *Note* Temperatures in °C.

□ = up to (m)

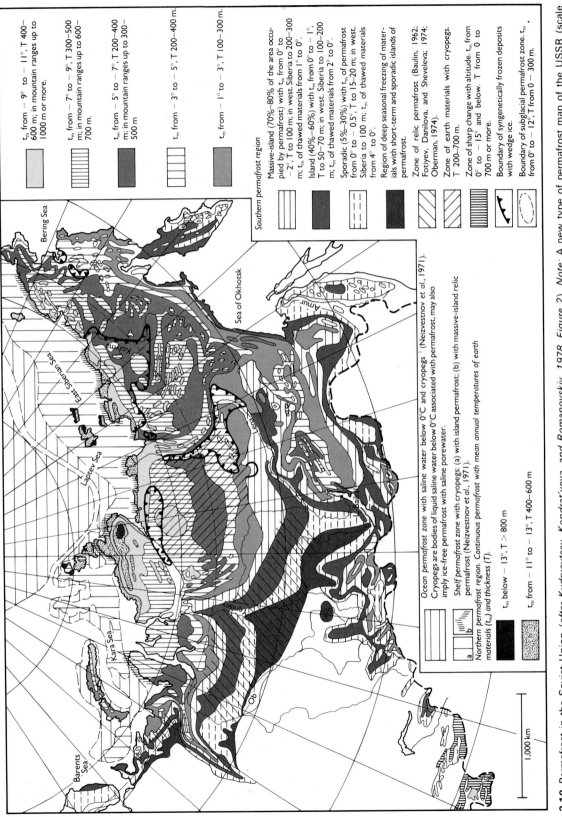

3.10 Permafrost in the Soviet Union (*after Kudryavtsev, Kondrat'yeva, and Romanovskiy, 1978, Figure 2*). *Note:* A new type of permafrost map of the USSR (scale 1:4 000 000) has been compiled by Popov, Rozenbaum, and Vostokova (1978) but was not received in time to include in this volume. It is a cryolithologic map; that is, it shows the distribution, structure, and origin (type of cryolithogenesis) of various kinds of ice-bearing permafrost.

Legend:

t_m from $-9°$ to $-11°$, T 400—600 m; in mountain ranges up to 1000 m or more.

t_m from $-7°$ to $-9°$, T 300—500 m; in mountain ranges up to 600 m.

t_m from $-5°$ to $-7°$, T 200—400 m; in mountain ranges up to 300—500 m

t_m from $-3°$ to $-5°$, T 200—400 m.

t_m from $-1°$ to $-3°$, T 100—300 m.

Southern permafrost region

Massive-island (70%—80% of the area occupied by permafrost) with t_m from 0° to $-2°$, T to 100 m; in west. Siberia to 200—300 m; t_m of thawed materials from 1° to 0°.

Island (40%—60%) with t_m from 0° to $-1°$, T to 50—70 m; in west. Siberia to 100—200 m; t_m of thawed materials from 2° to 0°.

Sporadic (5%—30%) with t_m of permafrost from 0° to $-0.5°$, T to 15—20 m; in west. Siberia to 100 m; t_m of thawed materials from 4° to 0°.

Region of deep seasonal freezing of materials with short-term and sporadic islands of permafrost.

Zone of relic permafrost (Baulin, 1962; Fotiyev, Danilova, and Sheveleva; 1974; Oberman, 1974).

Zone of earth materials with cryopegs. T 200—700 m.

Zone of sharp change with altitude. t_m from 0° to $-15°$ and below. T from 0 to 700 m or more.

Boundary of syngenetically frozen deposits with wedge ice.

Boundary of subglacial permafrost zone. t_m from 0° to $-12°$, T from 0 — 300 m.

Ocean permafrost zone with saline water below 0°C and cryopegs[1] (Neizvestnov et al., 1971). Cryopegs are bodies of liquid saline water below 0°C associated with permafrost, may also imply ice-free permafrost with saline porewater.

Shelf permafrost zone with cryopegs: (a) with island permafrost; (b) with massive-island relic permafrost (Neizvestnov et al., 1971).

Northern permafrost region. Continuous permafrost with mean annual temperatures of earth materials (t_m) and thickness (T).

t_m below $-13°$, T > 800 m

t_m from $-11°$ to $-13°$, T 400—600 m

Bering Sea

Sea of Okhotsk

Amur

Kolyma

East Siberian Sea

Laptev Sea

Lena

Kara Sea

Ob

Irtysh

Barents Sea

1,000 km

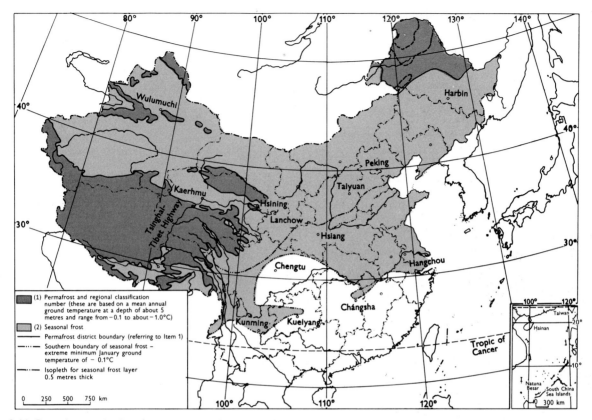

3.11 Permafrost in the Peoples Republic of China (*after Academia Sinica, 1975, 20–1*)

permafrost in North America coincides roughly with the $-1°$ isotherm for the mean annual air temperature (R. J. E. Brown, 1967*a*; 1970, 21; 1975, 30; Gold and Lachenbruch, 1973, 5); according to others it approximates the 0° isotherm (Péwé, 1975, 47). Some variation is expectable because of regional differences in topography, vegetation, snow cover, and other factors. Not only the amount of snow but also its timing relative to freeze-up is important (R. J. E. Brown, 1975, 29). Discontinuous permafrost in Canada is in the belt of greatest peatland concentration, and the presence of permafrost here is controlled in part by the insulating qualities of the peat (Figure 3.14).

Permafrost in highlands may extend far south of the limit as usually mapped. In the Canadian Cordillera where the southern limit is mapped at valley-bottom levels the lower altitudinal limit is estimated to rise from 1220 m (4000 ft) at lat. 54°30′N to about 2134 m (7000 ft) at lat. 49°N (R. J. E. Brown, 1970, 12). It is present at altitudes of 2655 m and 2691 m on Lookout Mountain on the Continental Divide between Alberta and British

Columbia (Scotter, 1975). In the southern Coast Mountains of British Columbia, permafrost has been reported at lat. 49°58′N near Garibaldi Lake (Mathews, 1955, 96), but it is probably relic, or the result of the 'Balch' effect (cf. Balch, 1900, 147–9, 159–61) involving cold air settling where air circulation is restricted, inasmuch as the ground here, consisting of loose cinders, has barely frozen to a depth of 20 cm in winter since 1958 (J. R. Mackay, personal communication, 1978). In the USSR both continuous and discontinuous permafrost, excluding highland occurrences, fails to reach as far south as in North America (Figure 3.3). That discontinuous permafrost in North America may be in disequilibrium was noted above; the same imbalance probably exists in the USSR (cf. Embleton and King, 1975, 32).

Significant amounts of alpine permafrost exist in various parts of the world. The area covered in the middle and low latitudes of the Northern Hemisphere as estimated by Gorbunov (1978, 283) is less than 160×10^4 km^2, excluding uplands of the Far East and eastern Siberia. As compiled by Fujii and

Table 3.3. Area of alpine permafrost in the middle and low latitudes of the Northern Hemisphere (*after Fujii and Higuchi, 1978, Table 2*)

	Location	Area	Total area
ASIA	Tibetan Plateau Karakoram Mountains	157.8×10^4 km²	
	Himalaya Mountains	10.0×10^4 km²	
	Tien Shan Mountains Pamir	19.1×10^4 km²	
			186.9×10^4 km²
EUROPE	Alps Mountains	0.5×10^4 km²	0.5×10^4 km²
NORTH AMERICA	Rocky Mountains (Canada) Coast Mountains (Canada)	27.8×10^4 km²	
	Rocky Mountains (USA) Sierra Nevada (USA)	17.4×10^4 km²	
			45.2×10^4 km²
NORTHERN HEMISPHERE			232.6×10^4 km²

Higuchi 1978) and shown in Table 3.3, the amount is considerably greater – 232.6×10^4 km². Of this amount 80.4 per cent is in Asia, 19.4 per cent in North America, and only 0.2 per cent in Europe (although in the Swiss Alps permafrost ice is estimated to have a volume comparable to 9 per cent of the glaciers – Barsch, 1978); the lower limit of various occurrences is shown in Table 3.4. Some Canadian examples were noted above. In the United States permafrost has been reported in peat at an altitude of 2957 m (9700 ft) in the Beartooth Mountains of northwestern Wyoming (Pierce,

Table 3.4. Altitude of lower limit of alpine permafrost in the middle and low latitudes of the Northern Hemisphere (*after Fujii and Higuchi, 1978, Table 1*)

Area	Locality	Latitude	Longitude	Lower limit (m)	References
ROCKY Mts	Near Cassiar	59°17′N	129°48′W	1370	R. J. E. Brown (1969)
and	Beartooth Mts.	44°53′N	109°30′W	<2960	Pierce (1961)
COAST Mts	Niwot Ridge	40° N	106° W	3500	J. D. Ives and Fahey (1971)
	Tesuque Peak	35°47′N	105°47′W	3720	Retzer (1965)
	Mt Elbert	39°07′N	106°27′W	4000	Baranov (1959)
	Mt Whitney	36°35′N	118°17′W	(4420)	Retzer (1965)
MEXICO	Citlaltepetl	18°30′N	97°50′W	4600	Lorenzo (1969)
E. CANADA	Mt Jacques Cartier	49° N	66° W	1270[1]	R. J. E. Brown (1967a)
E. USA	Mt Washington	44°15′N	71°20′W		Antevs (1932)
ANDES	Central Chile Andes	33° S	70° W	4000	Lliboutry (1956)
ALPS	Corvatsch Mt	46°25′N	9°50′E	2700	Barsch (1969b)
TIEN SHAN	Tien Shan	42° N	78° E	2700	Gorbunov (1967)
HIMALAYA	Mukut Himal	28°45′N	83°30′E	5000	Fujii and Higuchi (1976)
	Khumbu Himal	27°55′N	86°50′E	4900~5000	Fujii and Higuchi (1976)
	Near Rongbuk Gl.	28°10′N	86°50′E	4900	Hsieh et al. (1975)
TIBET	Khulun Shan	31°20′N	91°40′E	4500	Chou and Tu (1963)
	Nienching Tangkula Shan	36°20′N	94°50′E	4200	Chou and Tu (1963)
HAWAII	Mauna Kea	19°30′N	155°40′W	4170	Woodcock (1974)
JAPAN	Mt Fuji	35°21′N	138°44′E	2800~2900	Fujii and Higuchi (1972)
	Mt Taisetsu	43°40′N	142°55′E	2150	Fukuda and Kinoshita (1974)

[1] The lower limit on Mount Jacques Cartier is now estimated to be at 1100 m. Extrapolated ground temperatures indicate that permafrost is 50–100 m thick at an altitude of 1280 m, just above treeline (*R. J. E. Brown, personal communication, 1979*).

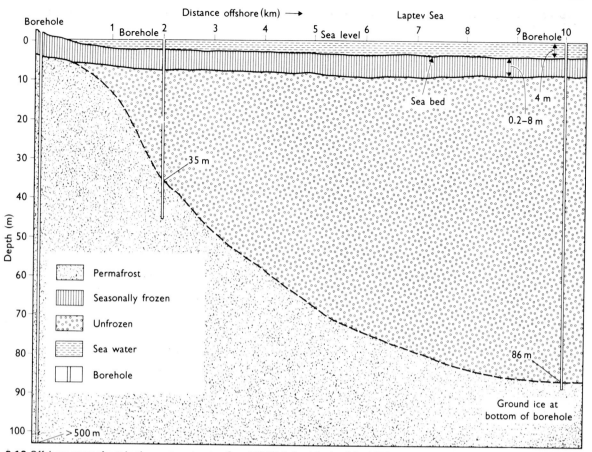

3.12 Offshore permafrost in the eastern Laptev Sea, USSR (*after Vigdorchik, 1977b, Figure 18; cf. Grigor'yev, 1973; 1978; Katasanov and Pudov, 1972; Zhigarev and Plakht, 1974; 1976*)

1961). It occurs in the summit area of Mount Washington (alt. 1917 m) in New Hampshire (R. P. Goldthwait, 1969; John Howe, 1971; Thompson, 1962, 215) where the mean annual temperature is −2.8° (27°F), and it characterizes some areas above treeline in the Rocky Mountains (J. D. Ives, 1973; J. D. Ives and Fahey, 1971). Woodcock (1974) described a curious occurrence at an altitude of 4140 m in the summit area of Mauna Kea (alt. 4206 m), Hawaii. The active layer is about 0.4 m thick and the permafrost extends to a depth of at least 10 m despite a mean annual air temperature of 3.6° at the permafrost altitude. The anomaly is partly explained by a colder micro-climate resulting from radiational cooling and entrapment of air, low incidence angle of sunlight, and dry atmosphere. This could account for the top of the permafrost maintaining itself under the present conditions, although the base may be retreating. The permafrost probably originated under a

colder climate or is a 'Balch' effect (cf. preceding discussion of the Garibaldi Lake, British Columbia, permafrost occurrence. Both occurrences are in loose cinders). Various permafrost occurrences are known at high altitudes in Europe (Barsch, 1969b; 1973; 1977d; Furrer and Fitze, 1970; 1973; Haeberli, 1975; 1978), Japan (Fujii and Higuchi, 1972; Fukuda and Kinoshita, 1974) and elsewhere, including even tropical latitudes given high enough altitudes as at Mauna Kea (discussed above), in Africa where 'fossil ice' (ice-cored glacial deposits) and frozen talus occur on Mount Kenya (alt. 5199 m) (17 058 ft) (Baker, 1967, 68), and in Mexico where relict permafrost occurs within an altitude range of 4600–5000 m on Pico de Orizaba (Heine, 1975).

4 Depth and thickness

To date the maximum reported thickness of permafrost is 1400–1450 m in the vicinity of the upper Markha River in Siberia (Grave, 1968*a*; 1968*b*, 3, diagram opposite 4). Bakker (*in* Büdel, 1960, 43) reported *c.* 1000 m in Siberia, and a calculated thickness of 900 m in the Udokan Range, Siberia (Melnikov, 1963). In North America the maximum reported thickness is about 1000 m near Alert, Ellesmere Island, in the Canadian Arctic, as extrapolated from a temperature of $-16.5°$ at the depth of zero annual amplitude (R. J. E. Brown, personal communication, 1979). The depth is on the order of 700 m at Cameron Island (northwest of Bathurst

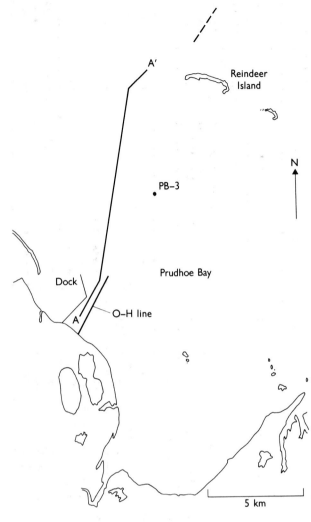

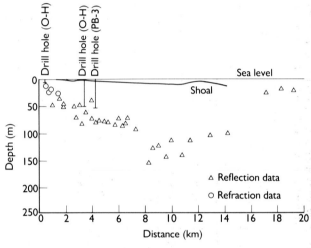

3.13 Upper surface of offshore ice-bonded permafrost along transect A–A′ as indicated by drilling and seismic data, Prudhoe Bay, Alaska (*after Rogers and Morack, 1977a, 18, Figure 4; 20, Figure 5 [Line 15 omitted]; 1977b; cf. 1978, Figure 2 – sparker data omitted*)

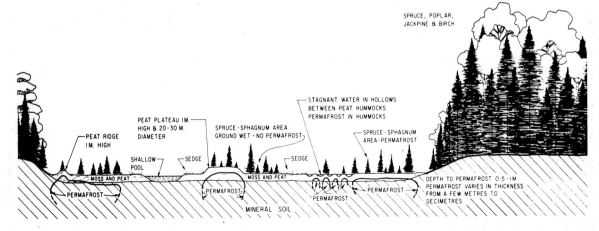

3.14 Profile through typical peatland in southern fringe of discontinuous permafrost zone in Canada, showing relation of permafrost to vegetation and topography (*R. J. E. Brown, 1968, 176, Figure 2*)

Table 3.5. Approximate depth to bottom of permafrost at selected places in Northern Hemisphere. Bottom is generally based on 0°C. Source data originally given in °F or in feet are so indicated; feet are converted to nearest metre. Mean annual temperatures are after authors cited unless otherwise indicated. See also Oberman and Kakunov (1978, 144–5, Table 1)

Location	Depth		Mean annual air temperature		Ground temperature at depth indicated			
	M	Ft	°C	°F	°C	°F	M	Ft
Alaska								
Barrow[6,8]	204 to 405	670 to 1330	-12.0	—	-6.3	—	179	—
Bethel[6]	13 to 184	42 to 603	-1.8[11]	—	—	—	—	—
Cape Simpson[6,8]	250 to 320	820 to 1050	-12.0	—	-9.9	—	42	—
Cape Thompson[6]	305 to 366	1000 to 1200	—	—	—	—	—	—
Fairbanks[6]	30 to 122	100 to 400	-3.5[11]	—	-7.0	—	25	—
Kotzebue[6]	73	238	-6.2[11]	—	—	—	—	—
Nome[6]	37	120	-3.5[11]	—	—	—	—	—
Northway[6]	63	207	—	—	—	—	—	—
Prudhoe Bay[4]	610	2000	—	—	—	—	—	—
Umiat[6,8]	322	1055	-12.0	—	-7.3	—	21	—
Canada								
Aishihik, YT[1,6]	15 to 30	50 to 100	-4.2	24.5	-2.1	28.3	6	20
Alert, NWT	1000[c]	—	-18.0	-0.4	-16.1	3.0	15 to 18	50 to 60
Asbestos Hill, PQ[1]	>274	>900	-8.3	17.0	-7.2 to -6.7	19.0 to 20.0	15 to 61	50 to 200
Churchill, Man.[1]	30 to 61	100 to 200	-7.2	19.0	-2.5 to -1.7	27.5 to 28.9	8 to 16	25 to 54
Churchill, Man.[6]	0 to 42[9]	0 to 140[9]	—	—	—	—	—	—
Coppermine River, NWT[1,2]	180 to 370[c]	—	-11.4	—	-7.0	—	14	—
Dawson, YT[1]	61	200	-4.7	23.6	—	—	—	—
Fort Good Hope, NWT[1,2]	33 to 48[c]	—	-7.6	—	-1.0 to -1.2	—	15	—
Fort McPherson, NWT[1,2]	90 to 150[c]	—	-8.2	—	-2.6 to -4	—	14	—
Fort Simpson, NWT[1]	12	40	-3.9	25.0	0.7 to 1.9	33.2 to 35.4	0 to 2	0 to 5
Fort Simpson, NWT[1,2]	0 to 5	—	-3.9	—	1.4 to 2.0	—	15	—

Table 3.5. (continued)

Location	Depth		Mean annual air temperature		Ground temperature at depth indicated			
	M	Ft	°C	°F	°C	°F	M	Ft
Fort Smith, NWT[1]	(unknown)		-3.2	26.2	≈0.0	≈32.0	5	15
Fort Vermilion, Alta.[1]	0	—	-2.1	28.2	4.3 to 4.4	39.8 to 39.9	0 to 2	0 to 5
Hay River, Alta.[6]	2	5	—	—	—	—	—	—
Inuvik. NWT[1]	>91	>300			-3.3	26.0	8 to 30	25 to 100
Inuvik, NWT[12]	107c	—	-9.6	—	-3.7	—	14	—
Keg River, Alta.[1]	2	5	-0.6	31.0	-0.6 to 0.0	31.0 to 32.0	2	5
Kelsey, Man.[1]	15	50	-3.6	25.5	-0.8 to -0.3	30.5 to 31.5	9	30
Mackenzie Delta, NWT[1]	91	300	-9.1 (Aklavik)	15.6	-4.6 to -3.1	23.8 to 26.5	0 to 30	0 to 100
Mackenzie Delta, NWT[12]	18 to 366c	—	-9.1 to -11.3	—	-0.4 to -6.0	—	14	—
Mary River, Baffin I., NWT[1]	(unknown)		-14.3 (Pond Inlet)	6.3	-12.2	10.0	9	30
Milne Inlet, Baffin I., NWT[1]	(unknown)		-14.3 (Pond Inlet)	6.3	-12.2	10.0	15	50
Norman Wells, NWT[1,6]	46 to 61	150 to 200	-6.2	20.8	-3.3 to -1.9	26.0 to 28.5	15 to 30	50 to 100
Port Radium, NWT[1]	107	350	-7.1	19.2	—	—	—	—
Rankin Inlet, NWT[1]	305	1000	-11.6 (Chesterfield Inlet)	11.2	-9.4 to -8.3	15.0 to 17.0	30	100
Resolute, Cornwallis I., NWT[1,8]	396c	1300	-16.2	2.8	-4.1 / -9.9 / -5.6	24.6 / 25.0 / 21.9	16 / 30 / 135	52 / 98 / 443
Schefferville, PO[16]	>76	>250	-4.5	23.9	-1.1 to -0.3	30.0 to 31.5	8 to 58	25 to 190
Thompson, Man.[1]	15	50	-3.9	24.9	-0.6 to 0.0	31.0 to 32.0	8	25

Table 3.5 (continued)

Location									
Tuktoyaktuk and coast, NWT[1,2]	140 c	—	−11.9 to −12.8	—	−8.0	—	14	—	
Tundra Mines Ltd. NWT[1]	274	900	−8.3	17.0	−1.7	29	99	325	
United Keno Hill Mines Ltd., YT[1]	137	450	−4.3 (Elsa)	24.2	−2.2 to −1.5	28.0 to 29.3	30 to 61	100 to 200	
Uranium City, Sask.[1,6]	9	30	−4.4	24.0	−0.6 to 0.0	31.0 to 32.0	9	30	
Winter Harbour, Melville I., NWT[5]	557	—	−16.0[2]	—	—	—	—	—	
Yellowknife, NWT[1,6]	61 to 91	200 to 300	−5.4	22.2	−0.3	31.4	12	40	
Yellowknife, NWT[1,2]	0 to 80	—	−5.4 to −6.2	—	1.8 to −0.5	—	15	—	
Greenland Thule[6]	518[1] o	1700[1] o	—	—	—	—	—	—	
Spitsbergen ....[6]	241 to 305	790 to 1000	—	—	—	—	—	—	
Braganza Bay, Lowe Sound[7]	320	—	—	—	—	—	—	—	
USSR Amderma[8]	500 c	—	—	—	−4.6 −5.3 −3.8	—	20 100 270	—	—
Bykhanay[8]	650 m	—	−12.0	—	−3.8 −0.2 10.4	—	50 500 1000	—	—
Dzhebariki-Khaya[6]	416 c	—	−11.0	—	−5.5 −3.7	—	30 170	—	—
Kozhevrikova[8]	600 m	—	−13.0	—	−12.5 −3.6	—	15 400	—	—
Magan[8]	450 c	—	−10.0	—	−3.0 −0.8 −0.5	—	20 300 400	—	—
Markha River, upper reaches[3]	1450 to 1500	—	—	—	—	—	—	—	
Mirny[8]	550 m,c	—	−9.0	—	−2.7 −1.8 0.0	—	15 300 550	—	—

Table 3.5. (continued)

Location	Depth		Mean annual air temperature		Ground temperature at depth indicated			
	M	Ft	°C	°F	°C	°F	M	Ft
Namtsy[8]	560[c]	—	-10.0	—	-2.5 / -2.2 / -1.6	— / — / —	20 / 200 / 500	— / — / —
Nordvik[6]	610	2000	—	—	—	—	—	—
Noril'sk[8]	325[m] / 400[c]	—	-8.0	—	-7.5 / -0.3 / 0.2	— / — / —	50 / 320 / 330	— / — / —
Salekhard[8]	350[m] / 380[c]	—	-7.0	—	-4.0	—	20	—
Taimyr[6]	305 to 610	1000 to 2000	—	—	—	—	—	—
Tiksi[8]	630[m]	—	-14.0	—	-11.1 / -8.8 / -3.3	— / — / —	20 / 200 / 500	— / — / —
Ukodan Range[8]	900[c]	—	-12.0	—	-7.5 / -5.0 / -3.3	— / — / —	50 / 300 / 500	— / — / —
Ust'port[8]	425[m] / 500[c]	—	-11.0	—	-3.0 / -1.2	— / —	50 / 330	— / —
Var'yegan[8]	>315	—	—	—	—	—	—	—
Vilyuy River, mouth[8]	420[c]	—	-10.0	—	-2.1 / -1.0	— / —	20 / 240	— / —
Vorkuta[6]	131	430	—	—	—	—	—	—
Yakutsk[6]	198 to 250	650 to 820	-10.13[11]	—	—	—	—	—

[1] R. J. E. Brown (1970, 10, Table 1).
[2] Embleton and King (1975, 30, Table 2.1).
[3] Grave (1968a; 1968b, 3, diagram opp. 4).
[4] Howitt (1971); Robert Stonely in Lachenbruch (1970b, J2–J4).
[5] A. Jessop in R. J. E. Brown (1972, 116).
[6] Stearns (1966, 21, Table 1).
[7] Werenskiold (1953, 197).
[8] Yefimov and Dukhin (1966, 94–5; 1968).

[9] Varies with distance from Hudson Bay and Churchill River.
[10] Estimated from temperature gradient and from temperature at 306 m (1005 ft) depth.
[11] US Department of Commerce (1975).
[12] Judge (1973b, 125, Table 35; 129–31, Table 36; 138, Figure 17).
[13] R. J. E. Brown (personal communication, 1979). See text.
[c] Calculated.
[m] Measured.

Island), in the Mackenzie Delta area, and in northern Quebec (A. E. Taylor and Judge, 1977, 11–12, Table 2), although climatic considerations indicate thicknesses may be considerably greater in the interior of Baffin Island as well as Ellesmere Island (R. J. E. Brown, 1972, 116–17). In Alaska the greatest reported thickness is at Prudhoe Bay where 610 m (2000 ft) has been cited (cf. Howitt, 1971; Robert Stoneley *in* Lachenbruch, 1970*b*, J2–J4) and 650 m is given in Figure 3.7. Table 3.5 shows approximate thicknesses in the Northern Hemisphere. Regional variations in thickness in Alaska, Canada, and the USSR are shown in Figures 3.7, 3.5, and 3.9, respectively.

In addition to other methods, a number of geophysical approaches, such as seismic surveying (including hammer seismic techniques – cf. Lorenz King, 1977), gravity profiling (for massive ice – Rampton and Walcott, 1974), electrical resistivity (J. Henderson and Hoekstra, 1977; Seguin, 1977), and remote-sensing techniques are being used or tested for delineating the distribution of permafrost and massive ice within permafrost. The determination of permafrost thickness and character is difficult but is being actively attacked (J. L. Davis *et al.*, 1976; Ferrians and Hobson, 1973, 481–90; Mackay, 1975*f*), promising approaches being magnetotelluric sounding (Koziar and Strangway, 1978), and electrical properties (Frolov, 1976) including drawing of isothermal curves from electrical logging (Seguin, 1977). The application of geophysics to permafrost regions continues to expand as reviewed by W. J. Scott, Sellmann, and Hunter (1979).

Permafrost is not limited to the Earth. Martian surface temperatures indicate its presence, and perhaps some of it is ice rich (D. M. Anderson, Gatto, and Ugolini, 1973; Carr and Schaber, 1977; Kuz'min, 1977; Sharp, 1974), although this has been questioned (Black, 1978*a*). Possibly some Martian ground ice occurs as a hydrate of CO_2 (Milton, 1974).

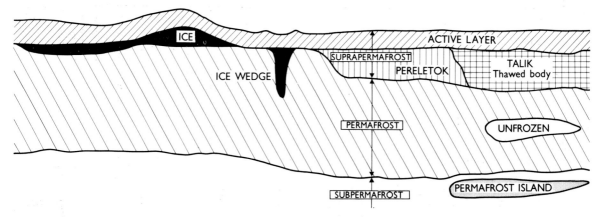

In the discontinuous zone, the permafrost breaks (both horizontally and vertically) into islands and separate masses. The unfrozen islands become more continuous

Active layer extends to continuous permafrost

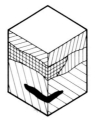

Transition layer with talik

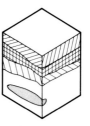

Discontinuous zone Talik, layers, island

3.15 Some structural features of permafrost (*after Stearns, 1966, 17, Figure 7*)

Table 3.6. Classification of frozen ground (*after Pihlainen and Johnston, 1963, Table 1, 12–17*)

Ice not visible by naked eye
  Poorly bonded
  Well bonded
Ice visible by naked eye but 2.5 cm (1 in) or less thick
  Individual ice crystals or inclusions
  Ice coatings on particles
  Random or irregularly oriented ice formations
  Stratified or distinctly oriented ice formations
Ice greater than 2.5 cm (1 in) thick
  Ice with soil inclusions
  Ice without soil inclusions

5 Structure

The basic types of frozen ground in a Canadian (Pihlainen and Johnston, 1963) and CRREL classification (Linell and Kaplar, 1966; cf. Andersland and Anderson, 1978, 29–36; R. F. Scott, 1969. 17) are shown in Table 3.6.

Undersaturated, saturated, and supersaturated permafrost (or frozen ground) refer, respectively, to ground in which ice fails to fill all pores, about fills all pores, and exceeds all the pore space (Black, 1953, 127–8). Pore ice (or interstitial ice) is confined to pore spaces, whereas massive ice (large mass of supersaturated ice) commonly weighs 1000 per cent more than any contained soil (Mackay and Black, 1973, 186–7). However, the ice content of peat, even in the absence of ice lenses, can be as much as 2000 per cent of the dry weight (Kinosita *et al.*, 1979, 20, 41). Segregated (or segregation) ice refers primarily to origin rather than distribution of ice in frozen ground, and designates ice masses formed as the result of water being brought to the freezing plane by the freezing process itself (Taber, 1929, 430. Cf. Freezing process in chapter on General frost-action processes). In the CRREL classification above, the term segregated ice rather than ice was used in the original. It was employed in a nongenetic sense and has been omitted by the present writer to avoid confusion, since the genetic usage is dominant and is followed here.

The thickness and orientation of ice accumulations are especially critical aspects of the structure. The amount of ice tends to control the behaviour of permafrost upon thawing, and the amount and orientation of the accumulations may indicate the origin of the ice.

Some characteristic structures of permafrost as determined by certain forms of ice, including ice lenses and ice wedges described below, are shown in Figure 3.15, which also shows certain features described later, including thaw areas (taliks), temporarily frozen areas simulating permafrost (pereletoks), the upper surface of permafrost (permafrost table), the ground above (suprapermafrost layer), and the layer of seasonal freezing and thawing (active layer). The suprafrost layer is not necessarily the same as the active layer because a pereletok or talik may lie between the permafrost table and the active layer.

6 Forms of ice

a General Ground ice, defined as ice in the ground regardless of amount or configuration (National Academy of Sciences, 1974, 31),[3] tends to be concentrated in the upper levels of permafrost (Figures 3.15–3.16), but massive ice has been encountered to depths of 45 m or more (Figure 3.17). Compared with ice crystals in seasonally frozen ground, ice crystals in permafrost have been described as being 14–25 times larger except where shearing of the permafrost has caused much smaller crystals by recrystallization (Tsytovich, 1975, 104–5). Ground ice may take many forms, some of the more common being pingo ice, ice lenses and massive ice beds, ice veins, and ice wedges, and more rarely buried glacier ice and buried icings. Overviews include the discussion by Mackay (1972a), Mackay and Black (1973), Mackay, Konishchev, and Popov (1979), and the monographs (in Russian) by Vtyurin (1975) and Vtyurina (1974), the last being devoted to seasonally frozen ground. Both the growth and the thawing of ground ice are the direct cause of many characteristic periglacial landforms as stressed by numerous workers (cf. R. J. E. Brown, 1973a; Rampton, 1973). A detailed classification of ground ice, based on the origin of the water and on the transfer process involved, is shown in Figure 3.18. Gas hydrates may resemble ice and possibly constitute a gas resource in permafrost and also beneath it (Davidson *et al.*, 1978; Judge, 1977, 103–4; Katz, 1971). It is often difficult to distinguish between different forms of ground ice by the nature of the ice alone, independent of its form and occurrence, but structure, petrofabrics, and geochemistry of the ice can be

---

[3] There are also other usages. A widely divergent one (Alekseyev, 1973; 1974) includes all ice at air-ground interface and excludes most ground ice as the term is generally employed.

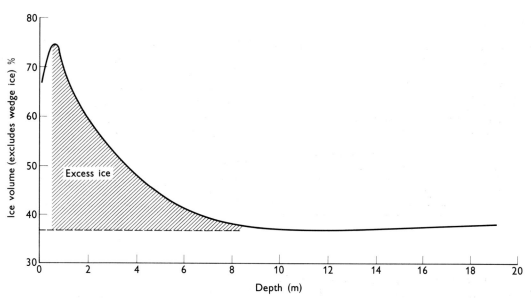

3.16 Ice volume versus depth based on auguring and coring, Barrow area, Alaska (*after Jerry Brown and Sellmann, 1973, 36, Figure 3*)

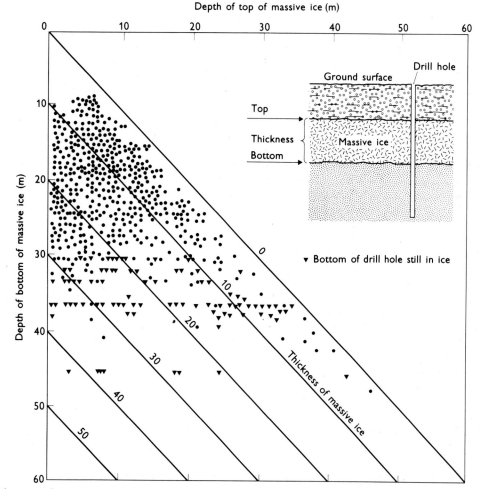

3.17 Thicknesses of massive ice, plotted from 560 beds encountered in drill hole logs in Canadian Western Arctic (*after Mackay, 1973a, 225, Figure 4*)

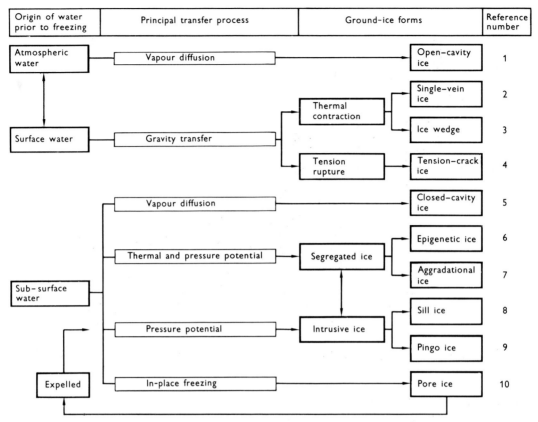

3.18 Classification of ground-ice forms (*after Mackay, 1972a, 4, Figure 2. Courtesy, Assoc. American Geographers*)

diagnostic in places. The fact that elongation of air bubbles in ice develops normal to the cooling surface and follows the heat-flow direction is helpful (Gell, 1974a).

b Glacier ice　Glacier ice may become buried under insulating debris and preserved as ground ice. The structure of glacier ice is rarely horizontal over large distances; rather it tends to be folded and faulted as a result of glacier movement. The nature of glacier ice and movement is reviewed in glaciological texts, including Paterson (1969) and Shumskiy (1959; 1964a), and further discussion here is omitted as beyond the scope of the present volume.

c Icings　An icing (sometimes called Aufeis or naled) is '... a mass of surface ice formed during the winter by successive freezing of sheets of water that may seep from the ground, from a river, or from a spring' (S. W. Muller, 1947, 218) (Figures 3.19–3.21).[4] Like glacier ice, icings may become buried

and preserved as ground ice, as reported, for example, near the head of an outwash plain in Spitsbergen (Cegła and Kozarski, 1977). Although descriptions of burial and preservation appear to be rare, icings *per se* are widespread periglacial features as discussed in Chapter 9.

The structure of icings commonly parallels the underlying surface (Shumskiy, 1964b, 192–5), and unlike glacier ice it is rarely if ever deformed. The ice ('overflow ice') of river icings is reported to have a higher average extinction coefficient (0.43 cm^{-1}) than normal (clear) river ice (0.092 cm^{-1}) (Wendler, 1970). The fact that most large river icings contain little dirt and overlie clean gravel in valley bottoms helps to distinguish buried icings from beds of massive segregated ice, discussed below, which commonly contain silt bands and occur in silty soil.

d Pingo ice　Pingo ice, like buried icings and

[4] It has been suggested that icing be used to designate the process only, and naled the deposit (Harden, Barnes, and Reimnitz, 1977, 28–9) but the need for this is not persuasive. Although often used synonymously, Aufeis as defined in the American Geological Institute's *Glossary of geology* (Gary, McAfee, and Wolf, 1972, 46) refers to river icings only. This would make icing the more general term but it should be noted that the original use of Aufeis was equally general (Middendorff, 1853; 1861, 439–53).

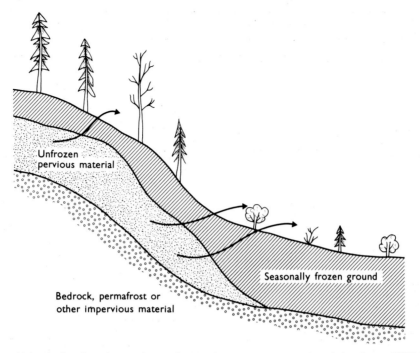

3.19 Development of icing by freezing of groundwater flowing from seepages or springs (*after Carey, 1973, 28, Figure 16a*)

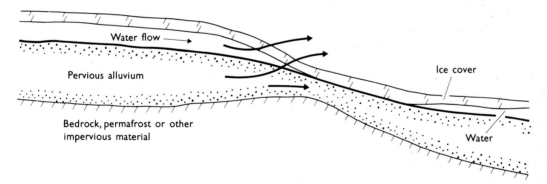

3.20 Development of icing by a river breaking through its ice cover during freeze-up (*after Carey, 1973, 27, Figure 15b*)

glacier ice, is commonly a massive body of fresh-water ice. It is usually formed by (1) progressive, all-sided freezing of a water body or water-rich sediments or (2) injection of groundwater under artesian pressure into permafrost in the manner of a laccolith. In both cases, ice forms the core of a mound and tends to assume a planoconvex shape. Pingos are described in the chapter on Some periglacial forms.

The structure of pingo ice is not well known.

Drilling of a pingo in the Mackenzie Delta area of northern Canada revealed ice with no apparent structure; the ice, some 9 m (30 ft) thick, was milky from small air bubbles (Pihlainen, Brown, and Legget, 1956, 1122). An exposure of pingo ice in the same area showed many crystals with diameters of 2.5–5.0 cm (1–2 in), the maximum crystal size noted being 20 cm (8 in); the optic axes had a preferred orientation towards the centre of the pingo's ice core (Mackay and Stager, 1966, 367).[5] Ice cores are also

[5] See also Gell (1978*b*), published while the present volume was in press.

3.21 Icing, Anaktuvuk Pass area, Alaska. Note smooth stone pavement between rougher stream bed and eroded Aufeis. Photo by S. C. Porter

described as having almost vertical banding and (in another case) irregular shear zones and fractures, perhaps reflecting different modes of growth (French, 1975a, 459; Pissart and French, 1976, 941–4, Figures 6–7, 9, 11–12).

e Ice lenses and massive ice beds An ice lens is a dominantly horizontal layer of ice. It may be less than a millimetre to tens of metres thick and a few millimetres to hundreds of metres in extent (Figures 3.22–3.24). However, very large lenses (say over 2 m thick and 10 m in shortest diameter and having an ice content of at least 250 per cent on an ice-to-dry-soil weight basis – Mackay, 1973a, 223) are better termed massive ice beds (Figure 3.24), or massive icy beds if the amount of contained soil, commonly silt, is large. Some massive ice beds are up to 40 m or more thick (Figure 3.17) and up to 2 km or more in horizontal extent (Mackay, 1973a), and some are responsible for prominent topographic rises as in the peculiar 'involuted hills' of the Mackenzie Delta area

of northern Canada (French, 1976a, 83; Mackay, 1963a, 138) and in pingos.

The terms are purely descriptive but many ice lenses and beds of massive ice form *in situ* by segregation of ice during freezing, as discussed under Freezing process in the chapter on General frost-action processes. If the ice is of segregation origin, its structure tends to be parallel to the freezing surface (i.e. usually dominantly horizontal), with any bubbles tending to be elongated and aligned normal to the horizontal layering. Soil fragments broken by freezing show a vertical separation (Mackay, 1971b, 411). It should be noted that according to recent Soviet research some frozen

3.22 Small ice lenses in clayey silt, northwest Siberia, USSR. Scale in centimetres. Photo by A. P. Tyrtikov

3.23 Large ice lens, right bank Lena River, 90 km north of Yakutsk, USSR. Lenses here are in bedded sand and up to 3 m thick

ground textures (fabrics) can change into others with changes in temperature, pressure, and other conditions, including the demonstration that ice lenses can develop from pore ice as a soil warms and thaws (Mackay, Konishchev, and Popov, 1979).

f Ice veins An ice vein, as the term is used by the present author, is a tabular, wedgelike, or irregular body of ice, hair thin to 5 cm (2 in) thick, whose longest dimensions lie in a dominantly vertical plane and may be several metres or more long. The term is purely descriptive. This usage follows that of T. L. Péwé (personal communication, 1970) in stressing the thinness of veins as opposed to ice wedges, discussed below. However, these terms tend to be used interchangeably in translations from the European literature. It should be noted that the term ice vein has also been used for thin (1–10 cm thick) tabular ice masses without regard to orientation (Mackay, 1974a, 230–1).

g Ice wedges An ice wedge is '... a narrow crack or fissure of the ground filled with ice which may extend below the permafrost table' (S. W. Muller, 1947, 218). As defined, the term is purely descriptive, and this usage is followed here, but to many workers the term has come to imply an origin by frost cracking, discussed later. Ice wedges start as ice veins.

Ice wedges tend to be V-shaped (Figure 3.25) and to have a characteristic, predominantly vertical structure (as opposed to the predominantly horizontal structure of undeformed ice lenses and massive ice beds), imparted by dirt and air bubbles oriented parallel to the wedge edges (Figure 3.26) as a result of the frost-cracking process. The dirt particles and the bubbles tend to be vertically elongated (Black, 1974a, 257, 263). The ice has a milky appearance because of many small bubbles less than 0.3 mm in diameter (Gell, 1974b). According to Shumskiy (1964b, 196–205), the ice crystals tend to be columnar with C axis vertical in the upper part of a wedge and horizontal at depth, but a cataclastic texture may develop as an ice wedge grows.

Ice wedges studied by Black (1954, 844–6; cf. 1963, 265–8; 1974a, 251–64) at Barrow, Alaska, had variously shaped ice crystals – equidimensional, prismatic, and irregular – with straight to sutured boundaries. Grain sizes ranged from 0.1 to 100 mm,

and Gell's (1974b) studies showed that crystal size tends to increase outward from central cracks of wedges. In Black's studies, long crystals were commonly vertical whereas short ones were generally normal to the horizontal axis of a wedge and either horizontal or inclined normal to a fracture or shear plane. The optic axis directions were either predominantly vertical, horizontal and normal to the axial plane of a wedge, or normal to the horizontal axis and inclined to one or both sides. According to Black (1974a, 264), the vertical optic axis direction is presumably in response to the vertical temperature gradient, whereas the inclined or horizontal lineations seem to result from rotation of crystal lattices during shearing, but it has also been suggested that freezing of water trickling into a wedge crack may be in response to horizontal temperature gradients (J. R. Mackay, personal communication, 1978). Sampling of surface and buried ice wedges near Fairbanks, Alaska, revealed fabrics similar to those in the Barrow area. The surface wedges had the better developed lineations of the optic axis and the more complicated fabric, whereas recrystallization in the buried wedges caused a more equigranular texture as well as offsetting of silt layers and a tendency for air bubbles to lie along grain boundaries (Black, 1978b). From studies of Tyndall figures in ground ice, Péwé (1978) concluded that the optic axis directions in the buried wedge ice he examined were essentially random except along veins and shear zones. In such studies the possibility of recrystallization as a factor needs to be considered if there have been significant temperature changes over a long enough time.

Shape and structure provide unambiguous criteria for well-developed ice wedges, and the ice texture and fabric, and the composition of included gases may assist in more doubtful cases.[6]

7 Thermal regime of permafrost

a General The thermal regime of permafrost (Figure 3.27) is dependent on the quantity of heat affecting the permafrost and the overlying layer that freezes and thaws annually – the active layer, described later. The quantity of heat can be expressed as

$$Q_h = Ak_h i_g$$

where Q_h = the quantity of heat flowing through

3.24 Massive segregated ice, Stanton, Northwest Territories, Canada. Section is about 7 m high. Photo by J. R. Mackay (*cf. Mackay, 1973a, 224, Figure 1*)

[6] See also Gell (1978c), published while the present volume was in press.

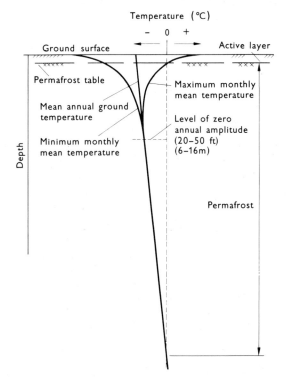

Temperature (°C)

Ground surface

Active layer

Permafrost table

Maximum monthly mean temperature

Mean annual ground temperature

Level of zero annual amplitude
(20–50 ft)
(6–16m)

Minimum monthly mean temperature

Depth

Permafrost

3.27 Typical temperature regime in permafrost (*after R. J. E. Brown, 1970, 11, Figure 6*)

an area at right angles to direction of flow per unit of time (cal s⁻¹), A = area (cm²), k_h = thermal conductivity (cal cm⁻¹ s⁻¹ °C⁻¹), and i_g = geothermal gradient (°C cm⁻¹). The ultimate sources of heat are the earth's interior and the sun (the latter's effect on the earth's surface being some 6000 times the greater – Judge, 1973*a*, 37), and the geothermal gradient expresses their combined effect. These ultimate heat sources are quite stable but the distribution of heat is affected by so many variables that thermal conditions vary widely.

It is convenient to discuss first the relatively stable part of the permafrost, affected by mean annual air temperatures and long-term temperature trends, before discussing the less stable part above the zone of zero annual amplitude where seasonal temperature changes dominate the thermal regime. The many factors affecting the regime here, including the interaction of snow cover, soil thermal conductivity, and latent heat can introduce non-linear effects that

3.25 Ice wedge near Brakes Bottom, Seward Peninsula, Alaska
3.26 Ice wedge at Shamanskiy Bereg, left bank Lena River, 120 km north of Yakutsk, Yakutia, USSR. Scale given by tape with major intervals in decimetres

can appreciably influence mean annual ground-surface temperatures and the shape of regimen curves such as illustrated in Figure 3.27 (Goodrich, 1978). In many respects it is the regime of the active layer and the upper part of the permafrost that is the most critical for frost-action processes.

b Geothermal heat flow The geothermal heat flow inhibits freezing at the base of stable permafrost, and thaws the base of degrading permafrost. To illustrate (Terzaghi, 1952, 30–1): The rate of geothermal heat flow (q_i) varies somewhat from place to place but approximates 40 cal m⁻² yr⁻¹. Considering the heat of fusion of ice (80 cal g⁻¹ or *c.* 70 cal cm⁻³), and assuming an ice content of 30 per cent, the heat (q_f) required to melt frozen ground at 0° would be $q_f = 0.3 \times 70$ cal cm⁻³ = 21 cal cm⁻³. The maximum rate of basal thawing (q_i/q_f) would be

$$q_i/q_f = \frac{40 \text{ cal}}{\text{cm}^2 \text{ yr}} \bigg/ \frac{21 \text{ cal}}{\text{cm}^3} = 2 \text{ cm yr}^{-1}$$

On this basis

$$t = \frac{H \text{ (cm)}}{2 \text{ (cm yr}^{-1}\text{)}}$$

where t = time required for thawing from below, and H = thickness of permafrost. If $H = 200$ m, $t = 10\,000$ years. According to comparative calculations (Terzaghi, 1952, 31), given this situation, a sudden rise in mean annual air temperature of 2° from −1° to 1° would result in less than 100 years in thawing 15 m of permafrost from the surface down while thawing from the base up would amount to only about 2 m. Near-surface heat flow can be significantly perturbed by climatic changes (Sharbatyan, 1974*b*; Sharbatyan and Shumskiy, 1974). Corrections required because of Pleistocene glaciation have been calculated for Canada and generalized in a map by Jessop (1971, 164, Figure 1).

c Geothermal gradient Where the geothermal heat flow is constant, the geothermal gradient is inversely proportional to conductivity. Thus if the mean surface temperature at two places is the same, permafrost extends deeper where conductivity is higher. The effect is startlingly illustrated by Robert Stoneley's observation that permafrost at Prudhoe Bay, Alaska, is 50 per cent thicker than at Barrow and 100 per cent thicker than at Cape Simpson, despite similarity of mean surface temperatures; the increasing thickness was ascribed to increasingly higher proportions of silicious sediments and consequently higher conductivity as between Cape Simpson, Barrow, and Prudhoe Bay (Lachenbruch, 1970*b*, J1).

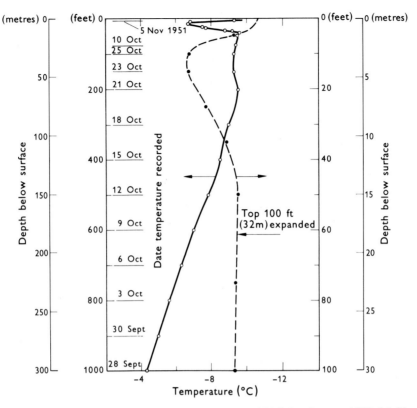

3.28 Ground temperatures at Thule, Greenland, September–November 1951 (*after Stearns, 1966, 34, Figure 25*)

Geothermal gradients in permafrost range from about $1°$ 22 m^{-1} ($1°$F 40 ft^{-1}) to $1°$ 60 m^{-1} ($1°$F 110 ft^{-1}) (Stearns, 1966, 33, Table v). The gradient at Thule, Greenland, which is about $1°$ 54 m^{-1} ($1°$F 100 ft^{-1}) is illustrated in Figure 3.28.

The term steep gradient commonly refers to a high °C m^{-1} ratio but gradients are also reported in m °C^{-1}, and depending on the axes chosen in plotting temperature, a high gradient may have a gentle slope in a graph.

Given sufficient time, the thickness of permafrost is related to temperature and geothermal gradient by the formula (Terzaghi, 1952, 27)

$$H_p = \frac{T}{-i_g}$$

where H_p = thickness (m), T = mean annual ground-surface temperature (°C), and i_g = geothermal gradient (°C m^{-1}).

The geothermal gradient can provide proof of climatic change and indicate its amount rather precisely. For instance, Lachenbruch and Marshall

(1969) and Gold and Lachenbruch (1973, 10–15) described temperature profiles from boreholes along the arctic coast of Alaska that show similar pronounced curvatures (Figure 3.29). The curvature is due to a climatic warming following establishment of thermal equilibrium represented by the linear gradient and its extrapolation to the surface. The extrapolated parts of the Barrow and Cape Simpson curves indicate a former mean annual temperature of about $-12°$. The change to the present mean annual temperature (*c.* $-9°$) is about $+3°$. Because repeated observations at the Barrow borehole provided information on the rate of temperature change as a function of depth, Lachenbruch and Marshall were able to calculate that the mean annual *ground–surface* temperature at Barrow must have risen about $4°$ since the middle of the nineteenth century. They concluded that the one degree difference from the $3°$ change with respect to the present surface temperature indicated an approximate $1°$ cooling that had been underway for no longer than a decade or so. Thermal profiles at Cape Simpson and Prudhoe Bay show a similar trend of climatic change.

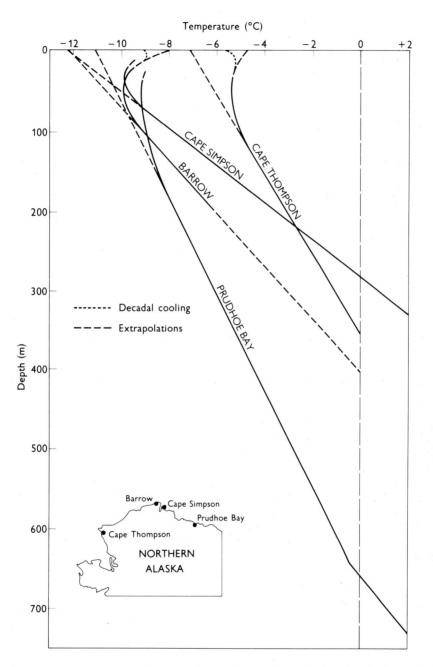

3.29 Geothermal gradients in four boreholes in Arctic Alaska. See text (*after Gold and Lachenbruch, 1973, 10, Figure 1, and Lachenbruch and Marshall, 1969, 302, Figure 2*)

Although thermal profiles from boreholes in Spitsbergen also reveal similar curvatures at depths roughly comparable to those in Alaska, the tentative inference was a warm period between 1920 and 1960 rather than one starting earlier (Liestøl, 1977, 11, Figure 3; 12). For the northern USSR, Balobayev (1978) calculated that temperature profiles in disequilibrium permafrost, thawing from below, showed that temperatures 20 000 years ago were 10°–13° colder than now, with permafrost thicknesses reaching 600–800 m. This calculation rests on a number of assumptions.

  d *Depth of zero annual amplitude*   The depth of zero annual amplitude – the depth to which seasonal change of temperature can propagate into the ground, when no phase change is involved – increases in direct proportion according to the equation (Terzaghi, 1952, 22)

$$z_1 = \sqrt{12at_1}$$

where z_1 = depth, a = diffusivity, and t_1 = time. Diffusivity (cf. Terzaghi, 1952, 11) is defined as

$$a = \frac{k_h}{C_h \times W_t}$$

where a = diffusivity (cm² s⁻¹), k_h = thermal conductivity (cal cm⁻¹ s⁻¹ °C⁻¹), C_h = heat capacity (cal g⁻¹ °C⁻¹), and W_t = unit weight (g cm⁻³).

  Temperatures at the depth of zero annual amplitude can be calculated according to the equation (cf. Grave, 1967)

$$t_{TA} = t_b + \frac{A_{cm}}{2}\left(1 - \frac{1}{f}\right)$$

where t_{TA} = temperature (°C) at depth of zero annual amplitude ('bottom of thermoactive layer'), t_b = mean annual air temperature, A_{cm} = annual amplitude of mean monthly air temperatures, and

$$f = e^{+z\sqrt{\frac{\pi}{kT}}}$$

where z = thickness of snow cover in metres,[7] k = thermal diffusivity[7] of snow (in m² h⁻¹), and T = period of oscillations equal to one year measured in hours.

  The temperature at the depth of zero annual amplitude approximates the mean annual ground-surface temperature (cf. National Academy of Sciences, 1974, 40), and the mean annual temperature at the top of permafrost. Also it is reported

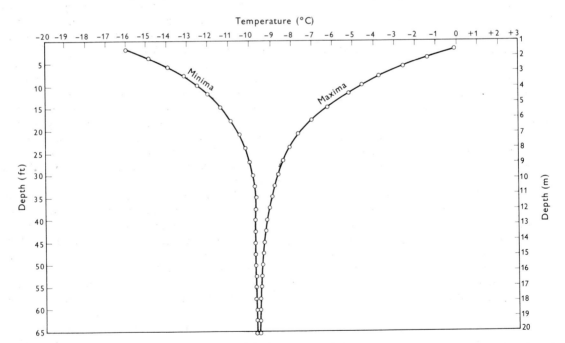

3.30 Amplitude of seasonal temperature changes at Barrow, Alaska, 1949–50 (*after MacCarthy, 1952, 591, Figure 3*)

---

[7] The metres and the diffusivity (instead of conductivity) are emendations to Grave (1967) (N. A. Grave, personal communication, 1977).

to be generally about 3° (6°F) warmer than the mean annual air temperature (R. J. E. Brown, 1967*a*; 1970, 20). For 38 sites in Canada (of which 20 involved permafrost), the average was 3.3° with a standard deviation of 1.5° (Judge, 1973*b*, 124). However, there is considerable regional variation (1.5°–9.3°) in figures from various parts of the world as compiled by Grave (1967, 1341, Table 1).

The convergence of minimum and maximum temperatures to the level of zero annual amplitude is illustrated in Figure 3.30.

Above the depth of zero annual amplitude, the permafrost is subjected to seasonal fluctuations of temperature. Nevertheless, a mean annual surface temperature can be deduced by upward extrapolation of the geothermal gradient from the level of zero annual amplitude, provided the measured gradient has achieved equilibrium and there are no recent climatic changes. However, to depths of 15 m at any one general location, variations in terrain, snow-cover, and vegetation can lead to variations of 2° in mean annual ground temperature (R. J. E. Brown, 1973*b*, 31). As previously noted, it is this upper part of the permafrost and the overlying active layer that are especially critical for frost-action processes and effects.

8   Aggradation and degradation of permafrost

a General   There are numerous ways in which permafrost can build up (aggrade), and thaw (degrade), all of which are controlled by the thermal regime. Changes that can lead to aggradation or degradation are climatic, geomorphic, and vegetational, and the ways in which they can affect permafrost are shown in Figures 3.31–3.32, following Mackay (1971*a*). In addition to the geomorphic changes specifically listed in Figures 3.31–3.32, changes induced by shifting shorelines can be very important. Permafrost on the land thins seaward and can become offshore permafrost (Lachenbruch, 1957). Many lakes and rivers that are too deep to freeze to the bottom have extensive taliks beneath them, some piercing through the entire thickness of permafrost (Lachenbruch *et al.*, 1962, 795–9; M. W. Smith, 1976). Consequently, permafrost thickness commonly increases with distance from the water bodies until their influence disappears. In all these situations a shift in the shoreline of the ocean, lake, or river would introduce a thermal anomaly affecting the distribution of permafrost. Detailed studies pertaining to rivers include those by J. R. Williams (1970, 19–20; 26–52) in Alaska, and by M. W. Smith (1976) in the Mackenzie Delta in northern Canada.

b Syngenetic and epigenetic permafrost   Aggradational permafrost may be syngenetic or epigenetic depending on whether concurrent sedimentation is present or absent (Baranov and Kudryavtsev, 1966, 100–1). Thus, except for the ice, the rocks and soil of epigenetic permafrost are older than the freezing, whereas those of syngenetic permafrost must be about the same age as the freezing. Epigenetic permafrost may be characterized by a concentration of well-developed segregated ice in the upper horizons because of the water that is drawn upward from considerable depth during freezing, as described under Freezing process in the chapter on General frost-action processes. In syngenetic permafrost the concurrent sedimentation and freezing favour accretion of thin layers near the top of the rising permafrost table. Although this is commonly an ice-rich zone (Mackay, 1972*a*, 10; Pissart, 1975), a large amount of water and generally lower freezing rates in developing epigenetic permafrost permit much thicker layers of segregated ice.

c Thermokarst   Thermokarst comprises '... karst-like topographic features produced by the melting of ground-ice and the subsequent settling or caving of the ground' (S. W. Muller, 1947, 223). The term was introduced by Ermolaev in 1932 (French, 1976*a*, 104). Cryokarst (Kryokarst – cf. Schunke, 1977*c*, 45, footnote 1) is synonymous but less common. Thermokarst records degradation of permafrost. The disturbance of the thermal regime may be local such as disruption of insulating vegetation, or more general such as climatic change. There are many cases of well-developed thermokarst features, due to local influences only, in the continuous permafrost zone. Thermokarst resulting from climatic change is more likely to be associated with the discontinuous zone where the thermal balance of permafrost is more delicate, with changes being apparent within 20 years (Thie, 1974); however, by the same token, local disturbances in this zone are also more likely to cause thawing of permafrost. As discussed later the origin of thermokarst must be interpreted with care.

d Taliks   A talik is '... a layer of unfrozen ground between the seasonal frozen ground (active layer) and the permafrost. Also applies to an unfrozen layer within the permafrost as well as to the unfrozen ground beneath the permafrost' (S. W. Muller, 1947, 223).[8]

---

[8] This definition is subject to the criticism cited in the footnote to the definition of permafrost table in the following section.

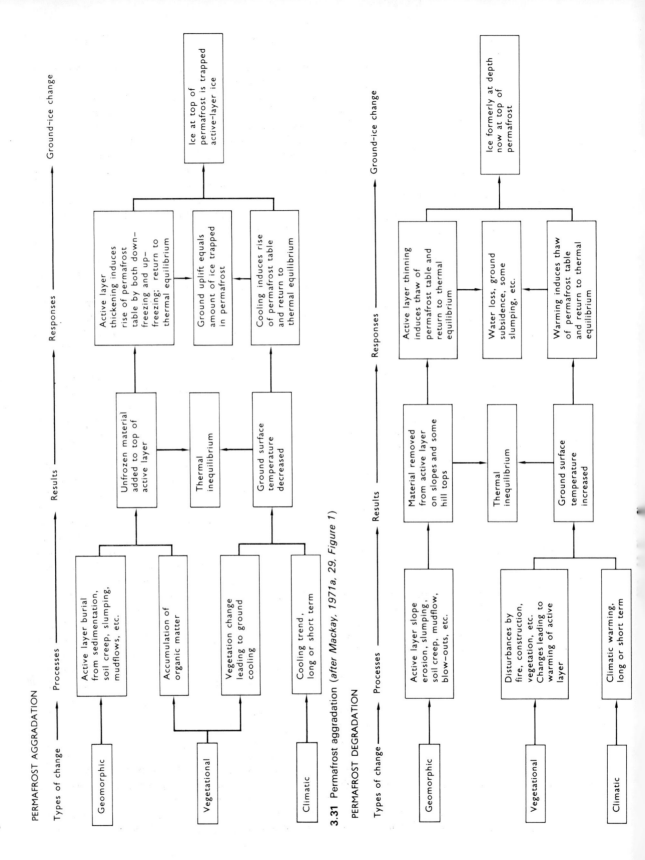

3.31 Permafrost aggradation (*after Mackay, 1971a, 29, Figure 1*)

PERMAFROST AGGRADATION

| Types of change | → | Processes | → | Results | → | Responses | → | Ground-ice change |

Geomorphic

Active layer burial from sedimentation, soil creep, slumping, mudflows, etc.

Accumulation of organic matter

Vegetational

Vegetation change leading to ground cooling

Climatic

Cooling trend, long or short term

Unfrozen material added to top of active layer

Thermal inequilibrium

Ground surface temperature decreased

Active layer thickening induces rise of permafrost table by both down-freezing and up-freezing; return to thermal equilibrium

Ground uplift equals amount of ice trapped in permafrost

Cooling induces rise of permafrost table and return to thermal equilibrium

Ice at top of permafrost is trapped active-layer ice

PERMAFROST DEGRADATION

| Types of change | → | Processes | → | Results | → | Responses | → | Ground-ice change |

Geomorphic

Active layer slope erosion, slumping, soil creep, mudflow, blow-outs, etc.

Vegetational

Disturbances by fire, construction, vegetation, etc. Changes leading to warming of active layer

Climatic

Climatic warming, long or short term

Material removed from active layer on slopes and some hill tops

Thermal inequilibrium

Ground surface temperature increased

Active layer thinning induces thaw of permafrost table and return to thermal equilibrium

Water loss, ground subsidence, some slumping, etc.

Warming induces thaw of permafrost table and return to thermal equilibrium

Ice formerly at depth now at top of permafrost

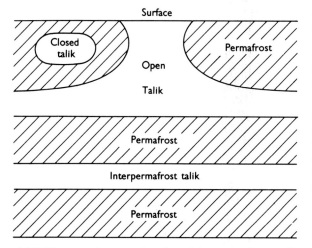

3.33 Diagram of open, closed, and interpermafrost talik (*courtesy, Jaromír Demek, 1969*)

Soviet investigators recognize three kinds of talik but contrary to Muller they do not commonly designate the ground below the base of permafrost as talik. The three kinds are open (skvoznoy), closed (zamknuty), and interpermafrost (mezhmerzlotny) talik (Figure 3.33). Taliks within permafrost, even in the continuous zone, are a common feature of the thermal regime and may or may not indicate degradation of permafrost. Some are due to local heat sources, including groundwater circulation, and may be associated with quasi-equilibrium conditions. Other taliks result from a marked change of thermal regime and thus indicate permafrost degradation, in which case thermokarst is likely to be present. Groundwater flowing in open taliks in alluvium beneath frozen rivers can provide the major water supply in some areas of continuous permafrost (cf. Sherman, 1973).

9   Thermal regime of active layer

a General   The active layer is the 'layer of ground above the permafrost which thaws in the summer and freezes again in the winter' (S. W. Muller, 1947, 213). Association with permafrost is inherent in this definition but some investigators (cf. Péwé, Church, and Andresen, 1969, 7) also apply the term to the layer of ground that freezes and thaws in a non-permafrost environment and a few others have used it in a still different sense (cf. Corte, 1969a, 130). However, the term is widely used as originally defined, and the present writer follows this usage.

A monograph on the active layer has been authored by Vtyurina (1974).

The thermal regime of the active layer is controlled by the same environmental factors that control the development of frozen ground as discussed earlier. They include the basic factors of climate, topography, material, and time, and the dependent factors of snow and ice cover, moisture, and vegetation. These interact in a complex fashion and determine such regime aspects of the active layer as its thickness, depth to the underlying permafrost table, upward freezing from the permafrost table, the zero-curtain effect, and the structure of the active layer. Environmental factors can cause considerable variation in the thickness of the active layer. A general range would be 15 cm to 5 m (Ferrians and Hobson, 1973, 479) but a range of 3–12 m has been reported from an area as small as 1 km², the thickest occurrences being in valley sites where thawing is accelerated by movement of groundwater (Nicholson and Thom, 1973, 161–2). Thickness variations within short distances are common across some beaches (Owens and Harper, 1977). Normally, the active layer is thicker in sands and gravels than in fine-grained soils, other conditions being equal. Nevertheless, the reverse has been reported where the coarse material has permitted meltwater to penetrate rapidly to depth and refreeze, thereby building up an icy layer that delays thawing because of its latent heat of fusion (Semmel, 1969, 42, 43 – footnote 10). The thickness of the active layer in the same spot can vary appreciably from year to year – by up to 71 cm in cases reported by R. J. E. Brown (1978, Table III). A year-to-year variation amounting to 25 per cent of the total depth of the active layer was reported as normal by Nicholson (1978, 430), based on extensive observations in an area of discontinuous permafrost.

Mathematical models of thermal regimes, based on quantitative parameters representative of the controlling factors, are becoming increasingly sophisticated. Simulations derived from a model, utilizing values for a test site at Barrow, Alaska, have shown a remarkable similarity with the observed regime and help to evaluate the relative importance of selected factors (Nakano and Brown, 1972).

b Permafrost table   The permafrost table is the 'more or less irregular surface which represents the upper limit of permafrost' (S. W. Muller, 1947, 219). As such it is the dividing surface between the permafrost and the active layer.

Any frozen surface in the active layer as it thaws downward towards the permafrost table is called a

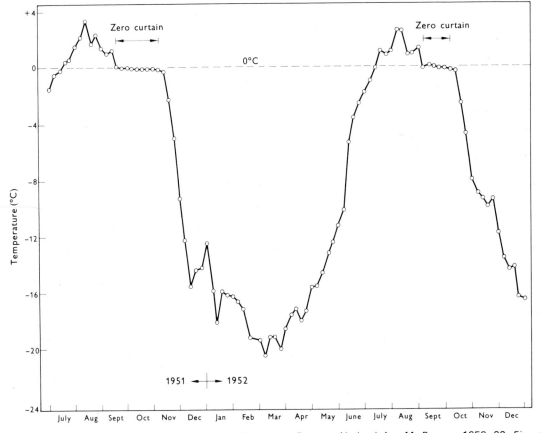

3.34 Time-temperature curve at a depth of 25 cm (10 in) near Barrow, Alaska (*after M. Brewer, 1958, 22, Figure 4*)

frost table and must not be confused with permafrost or the permafrost table. The term frost table is also applicable to a frozen surface in seasonally frozen ground in a non-permafrost environment.[9]

Other conditions remaining constant, the thermal regime controls the depth to the permafrost table which in the zone of continuous permafrost usually coincides with the thickness of the active layer. However, the position of the permafrost table represents an average condition, and because of short-term fluctuations of regime, such as an unusually cold winter, warm summer, or thick snow cover, the permafrost table need not coincide with the maximum depth of thaw in a given year. Consequently there may be short-term anomalies that can cause a temporary talik at the average position, or a pereletok, defined as '. . . a frozen layer at the base of the active layer which remains unthawed for one or

two summers (Russian term meaning "survives over the summer"). Pereletok may easily be mistaken for permafrost' (S. W. Muller, 1947, 219). R. J. E. Brown and Péwé (1973, 72) regarded the distinction as artificial and included pereletok with permafrost even though most definitions of permafrost would exclude a temporarily frozen layer that survived for one summer only.

The permafrost table is a crucial boundary in several ways. It is commonly a rigid surface capable of bearing considerable loads without deforming. It also prevents moisture from seeping downward and it thereby favours a high moisture environment and the formation of ice lenses.

c Zero curtain The zero curtain can be defined as the zone immediately above the permafrost table '. . . where zero temperature (0 °C) lasts a considerable period of time (as long as 115 days a year) during

[9] Frost table is synonymous with tjaele, a frequently used term in the European literature. However, as pointed out by Bryan (1951), tjaele has also been erroneously applied to permafrost.

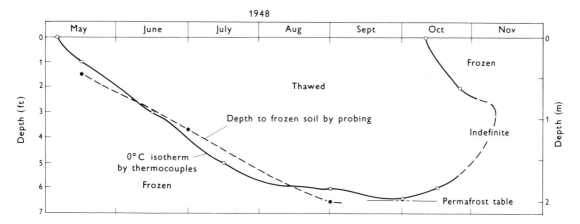

3.35 Freeze-thaw cycle in active layer at Fairbanks, Alaska. Note upward freezing from permafrost table in October (*after Stearns, 1966, 24, Figure 13*)

freezing and thawing of overlying ground' (S. W. Muller, 1947, 224).[10]

The zero curtain is caused by the latent heat of fusion of ice (80 cal g^{-1} water), which delays freezing and thawing. The higher the moisture content near the permafrost table, the greater the delay. The zero curtain (Figure 3.34) plays a major role in some frost-action processes and is a notable aspect of the thermal regime of the active layer (cf. Nakano and Brown, 1972, 31–6).

d Upward freezing Because of the thermal regime of permafrost, freezing of the active layer may proceed upward from the permafrost table as well as downward from the ground surface (Black, 1974*a*, 253, Figure 4; Drew *et al.*, 1958; Mackay, 1973*d*; 1974*d*; cf. Meinardus, 1930, 40; Schmertmann and Taylor, 1965, 50; US Army Corps of Engineers, *in* Washburn, 1956*b*, footnote 19, 842; Viereck, 1973, 61). Figures 3.35–3.36 illustrate upward freezing from the permafrost table.[11] According to Mackay's (1973*d*, 392) observations in the Mackenzie Delta area of northern Canada, the greatest upfreezing is where the active layer is thickest; the amount of up-freezing found by Mackay ranged from 2 to 13 cm. In many places upward freezing is shown by bubble patterns in the ice of the active layer (Gell, 1974*a*), and by distinctive ice layering parallel to the bottom of the active layer (Vtyurina, 1974, 33–4), which

record the upward-moving as well as the downward-moving freezing front.

Upward freezing of a frost table in a non-permafrost environment is unlikely. Details as to the prevalence and significance of the process in a permafrost environment are meagre. However, it is probably an important factor in some frost-action processes as discussed later.

e Effect of structure on thermal regime The thermal regime of frozen ground, whether seasonally frozen or permafrost, is strongly affected by structure of permafrost, especially the amount and distribution of the ice. As noted above in discussing the zero curtain, heat flow is influenced by the latent heat of fusion of ice; it is also influenced by the attitude of ice masses (Bakulin, Savel'yev, and Zhukov, 1957, 73–4; 1972, 3). Thus ice lenses and massive ice beds, being predominantly horizontal, delay thawing of underlying ground compared with adjacent ground lacking such ice masses. On the other hand, the predominantly vertical attitude of ice veins and ice wedges promotes vertical heat flow because of the greater thermal conductivity of ice than soil.

f Effect of thermal regime on structure Inversely, the thermal regime also influences the structure of frozen ground. For one thing, there are generally smaller ice masses in the active layer than in permafrost. Also the active layer tends to have a threefold

---

[10] In this definition, the phrase 'zone immediately above the permafrost table' has been substituted for Muller's (1947, 224) phrase 'layer of ground between active layer and permafrost', because such a layer is inconsistent with Muller's definitions of permafrost table and active layer. The same caveat can be applied to his definition of talik.

[11] As illustrated by Figure 3.35, the depth to frozen soil by probing can differ from the position of the 0° isotherm, especially in fine-grained soil in which probing may penetrate below the isotherm where the soil is not hard frozen due to unfrozen water, discussed later. A depth discrepancy of several decimetres is quite expectable (Mackay, 1977*c*).

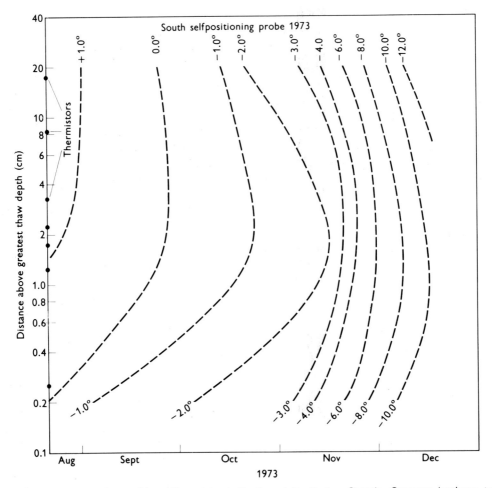

3.36 Upfreezing from permafrost table at Garry Island, Northwest Territories, Canada. Compare isotherm trends with similar trend of 0° isotherm for freeze-up period in Figure 3.35 (*after Mackay, 1974d, 251, Figure 3*)

structure consisting of (1) an upper zone with small ice inclusions; (2) a middle zone that is desiccated because of withdrawal of moisture towards freezing fronts, one moving down from the surface, another up from the permafrost table (upward freezing); and (3) a lower zone with a mixture of thin and thicker ice lenses immediately above the permafrost table (Zhestkova *et al.*, 1961, 45; 1969, 5–6).

10   Origin of permafrost

Most permafrost may have originated during the Pleistocene. The evidence for this is that (1) Tissues of woolly mammoths (*Mamuthus primigenius*) and other Pleistocene animals have been preserved in permafrost, indicating presence of permafrost at time of death (Gerasimov and Markov, 1968, 11; Vereshchagin, 1974), and there is no reliable evidence that the woolly mammoth survived into the Holocene (Farrand, 1961), although it appears to have lived as late as the Alleröd interstadial in northern Siberia (Heintz and Garutt, 1965, 76–7). (2) The upper boundary of some permafrost is considerably deeper than the present depth of winter freezing (Gerasimov and Markov, 1968, 12). This evidence merely indicates that some permafrost is not of present-day origin. (3) In places the temperature of permafrost decreases with depth, indicating residual cold (Gerasimov and Markov, 1968, 12). Again, this evidence does not necessarily prove that the residual cold demonstrates Pleistocene permafrost. (4) The thickest permafrost is commonly in areas that remained unglaciated, and therefore not insulated by ice, during the Pleistocene (Gerasimov and Markov, 1968, 14). (5) In the Soviet Union in the area covered by the early Pleistocene Kara Sea transgression, there are two layers of permafrost, the lower one dating from before the transgression (Grave, 1968a, 52–3; 1968b, i–ii, 6–8). (6) Permafrost in the Central Yakutian Lowland of Siberia has been reported as existing continuously since at least Middle Pleistocene time (Katasonov and Ivanov, 1973, 10–11) and in northern Yakutia since the Lower Pleistocene (Katasonov, 1977). 'According to present-day concepts, the formation of the permafrost in the Soviet Northeast dates from the first half of the Pleistocene and has continued without interruption right up to the present time' (Grigor'yev, 1978, 166). The present writer is not aware of the detailed evidence for continuous permafrost, although the presence of permafrost in the Soviet Union at least as far back as the early Pleistocene is indicated by syngenetic ice-wedge casts in the Olerskaya Formation of the Kolyma Lowland (Arkhangelov and Sher, 1973; 1978). (7) Some permafrost in the Mackenzie Delta area of northern Canada is believed to be of early Wisconsin age or older, since it is glacially deformed but occurs where there is no evidence of glaciation during the last 40 000 ^{14}C years (Mackay, Rampton, and Fyles, 1972). The presence in these sediments, beneath till, of ice-wedge casts with relict wedge ice still present at the bottom is confirming evidence (Mackay, 1976a).

In formerly glaciated areas, much permafrost is post-glacial. The argument for this is that: (1) Permafrost would have thawed beneath the thickest Pleistocene ice sheets (Büdel, 1959, 305). The argument is supported by the finding of water beneath the Antarctic Ice Sheet (Gow, Ueda, and Garfield, 1968) but it need not apply to all parts of an ice sheet (Grave, 1968a, 51–2; 1968b, 6). (2) In polar regions permafrost is forming today in many areas where retreat of glaciers or recent emergence has exposed unfrozen material. Also, taken together, evidence from Alaska (Jerry Brown, 1965, 39–47; Hopkins, MacNeil, and Leopold, 1960, 55; Péwé, 1958; Sellmann, 1967, 16–19) and from Siberia (Jahn, 1975, 29 – citing A. Kudryavtsev) argues for widespread development of postglacial permafrost, some (but not all) of it probably following the Hypsithermal warm interval dated at between about 2500 and 9000 years BP (Flint, 1971, 524).

Clearly the freezing and thawing of permafrost at depth can be consequent on past as well as present environmental conditions. Ground temperatures change particularly slowly where the latent heat of fusion of ice is involved, and permafrost boundaries are correspondingly slow to shift as emphasized by Jessop (1973). Nevertheless, ground-temperature profiles and meteorological records suggest that the southern boundary of continuous permafrost in Canada's Mackenzie Valley moved north as much as 320 km (200 mi) from the late 1800s to the 1940s following an approximate 3° rise in mean annual temperature, then began shifting south in response to a temperature lowering of about 1° (Mackay, 1975e). A recent increase in permafrost has also been reported in the USSR (Belopukhova, 1973; 1978).

In summary, (1) the upper part of most continuous permafrost is in balance with the present climate but the base may be slowly thawing, stable, or aggrading, depending on the past and present climate and the geothermal heat flow, (2) most discontinuous permafrost, both at its surface and base, is either out of balance with the present climate or in such delicate equilibrium that the slightest climatic or surface change will have drastic disequilibrium effects, although a number of years may be required for major changes in permafrost boundaries.

4 General frost-action processes

I Introduction

Frost action, in the *Glossary of geology* (Gary, McAfee, and Wolf, 1972), is defined as

(a) The mechanical weathering process caused by alternate or repeated cycles of freezing and thawing of water in pores, cracks, and other openings, usually at the surface.... (b) The resulting effects of frost action on materials and structures. Syn. *freeze-and-thaw action; freeze-thaw action.*

The weathering aspect is often secondary and to include results leads to confusion between cause and effect. The best definition would appear to be the simplest – the synonym *freeze-and-thaw action.* It is important to note that some water in fine-grained soils freezes and thaws at temperatures well below 0° and that subfreezing temperature variations cause volume changes in materials, especially ice and soil in the present context as discussed later. Such subfreezing changes are logically part of frost action regarded as a very general process that subsumes many subsidiary processes, including a number discussed in this volume.

Parts of this chapter are based on previous discussions by the present writer (Washburn, 1956*b*; 1967; 1969*a*; 1969*b*), and some statements are verbatim or only slightly altered from one or another of these earlier discussions.

II Freezing process

1 General

The processes discussed below are general in the sense of operating independently of mass-wasting and other slope processes discussed later.

The freezing of soil is a complicated thermodynamic process (Tyutyunov, 1964). Much remains to be learned about the process, including the movement of moisture towards a freezing plane in fine-grained soils, the expulsion of soil particles from developing ice and the consequent *in situ* segregation of clear ice masses in frozen ground, and many other facets of soil freezing (cf. R. D. Miller, 1966; 1972; 1978; Wissa and Martin, 1968, 12–39; Yong and Osler, 1971). A generally accepted theory of frost heaving, including ice segregation, remains to be established (cf. D. M. Anderson, 1977; D. M. Anderson and Morgenstern, 1973, 273–4; Penner, 1977; Radd and Oertle, 1973, 377). Ice segregation is discussed below in the section on Critical conditions for ice segregation. In addition to forming ice lenses, the process is believed to account for some pingos (discussed in the next chapter) and for some occurrences of massive, horizontally layered ground ice (Figure 3.24).

The nature of phase-boundary water in freezing soils is far from well established (D. M. Anderson,

1968; 1970; 1977). There is some agreement that an electric double layer adjacent to a mineral surface, and the dipole nature of water are basic elements in the movement of water to a freezing plane. However, as reviewed by R. F. Scott (1969, 1–10), there are two primary hypotheses regarding the growth of an ice lens. According to Jackson and Chalmers (1957), the ice molecules are at a lower energy state than the water of the electric double layer, which in turn is at a lower energy level than the free water. Systems tend towards a lower energy level, so the free water flows towards the developing ice lens. The energy released by the freezing of supercooled water in fine-grained soils supplies the energy for frost heaving. This theory was developed and an experimental verification was attempted by Chalmers and Jackson (1970). On the other hand, according to Cass and Miller (1959) osmotic pressure, resulting from the concentration of cations in the electric double layer, causes a flow of relatively pure water to the ice-water interface. Supercooling is not necessarily involved because the ions in the double layer cause a reduction in the equilibrium temperature of the water there. Thus some of the ice in frozen ground can thaw at temperatures below 0°.

Whatever its complete explanation, the tendency for water to flow to a freezing front is influenced by a number of factors. Their combined effect can be regarded as a suction potential, or water potential ψ, in a freezing system (D. M. Anderson, 1971, 2–4). Table 4.1 shows ψ in various units and their relation to capillarity and freezing-point depression.

2   Effect of closed *v.* open systems

The classic work on soil freezing was carried out by Taber (1929, 1930*a*; 1930*b*) who experimented with systems that were either closed or open with respect to water. He demonstrated that water is drawn towards a freezing front to form lenses of segregated ice parallel to the front and that in a closed system the build up of ice lenses tends to desiccate the still unfrozen soil. He found that heaving pressures are in the direction of crystal growth, normal to the freezing plane, and that heaving in a closed system is limited to the volume change of water to ice but in an open system is dependent on the amount of water drawn into it from outside. As a result heaving

can be far greater than in a closed system; in fact heaving can be caused by freezing of liquids that contract rather than expand with the phase change. According to Radd and Oertle (1973, 378, 383), the process leading to growth of ice lenses is probably a general phenomenon of crystal growth in finely divided solids in the presence of abundant moisture and a heat sink.

It should be noted that the amount of thaw settling following heaving in a closed system can considerably exceed the heaving; experiments indicate that during the first few cycles it can be as much as 20 per cent of the original volume prior to heaving (McRoberts and Nixon, 1975*a*, 162). This is significant in considering the behaviour of clay soils that because of low permeability react as if in a closed system.

3   Effect of moisture

Moisture is the *sine qua non* of frost action. This seems obvious but is specifically cited to stress its critical importance. Without moisture frost action is impossible,[1] and with it the amount can determine the results. In many places moisture differences strongly influence the occurrence of patterned ground, as noted for instance by Hastenrath (1973, 176–8) in comparing differences in the periglacial zonation of Mounts Kenya and Kilimanjaro in East Africa (cf. also Spönemann, 1977, 313–8), and by Kelletat (1977*b*, 218–19) in the Apennines. The amount of moisture can also determine the ease with which it moves under freezing conditions. Thus experiments by Volkova (1973*a*; 1973*b*; 1978), involving closed-system freezing of loam at −5°, showed a lack of significant water migration at a moisture content less than the plastic limit (lower plasticity limit) but appreciable migration at higher moisture contents. Although moisture is essential for frost action, it has been suggested that a very wet soil may impede certain aspects (patterned ground) (Hövermann and Kuhle, 1978, 324, 328).

4   Effect of temperature

To the extent that an open system prevails, it can favour build-up of ice lenses and much heaving, depending on the temperature. Very rapid freezing

[1] A speculative exception would be the breakdown of dry rocks by fatigue from long-continued fluctuations of temperature extremes but moisture is rarely absent.

Table 4.1. Energy state of soil water as expressed by the water potential ψ (*D. M. Anderson, 1971, 3, Table II*)

Appearance of soil	Soil water types and so called soil water constants	Water potential, ψ (all values are negative) cm of water	Atmospheres (approx.)	ergs/g ×10^8	Cal/g	Relative humidity at 25°C %	Freezing point depression °C	Equivalent capillary diameter mm	Common methods of estimating ψ
	Oven dry	10^7	10^4	98000	235·2				
Dry	Hygroscopic water (unavailable to terrestrial plants and most micro-organisms)	10^6	10^3	9800	23·5	50	90		
						75			
		10^5	10^2	980	2·35	93		Colloidal	A
	Hygroscopic coefficient	31623	30·6			98			
	Wilting point	14125	13·6			99	1·12	0·0002	
		10^4	10	98	0·24		0·4	Coarse clay	
Moist	Capillary water (Available to terrestrial plants and micro-organisms)						0·2		
							0·1	0·002	
		10^3	1	9·8	0·024				
	Field capacity	501	0·5				0·04	Silt	
		10^2	10^{-1}	0·98	0·002		0·01	0·02	
	Aeration porosity limit	50	0·05					Fine sand	G
Wet	Gravitational water (subject to drainage) (available to higher organisms)						0·2		
		10	10^{-2}	0·098	0·0002			Coarse sand	
								2·0	
	Saturation	1	10^{-3}	0·0098	0·00002				

A Psychrometry and water vapour pressure
B Freezing point
C Pressure membrane
D Bouyoucos resistance blocks
E Thermal units
F Centrifuge
G Tension meters
H Neutron thermalization and gamma ray attenuation

reduces migration of moisture so that pore water tends to freeze *in situ*. However, if penetration of freezing is in equilibrium with movement of moisture towards the plane so the plane is stabilized, thick ice lenses can form as a result of the continued contribution of moisture. Much depends on pressure as well as temperature (Low, Hoekstra, and Anderson, 1967) and grain size. Reasons why a series of ice lenses usually form instead of a single lens have been discussed by Martin (1959) and Palmer (1967), among others. Hallet (1978) has suggested that cyclic precipitation of solutes may be a factor in places.

The efficiency of ice segregation and its relation to thermal gradient and conductivity can be evaluated by the equation (Arakawa, 1966; Penner, 1972)

$$E = \frac{\sigma L}{K_1(\partial T_1/\partial_x) - K_2(\partial T_2/\partial_x)}$$

where E = segregation efficiency, σ = rate of ice segregation (as mass of ice per unit area per unit time at freezing front), L = latent heat of fusion, K_1 = temperature of frozen soil, K_2 = temperature of unfrozen soil, T_1 = thermal gradient in frozen soil, and T_2 = thermal gradient in unfrozen soil.

$E = 1$ for 'perfect' segregation, $0 > E < 1$ for 'imperfect' segregation, and $E = 0$ for no segregation since $\sigma = 0$ in a nonfrost-susceptible soil.

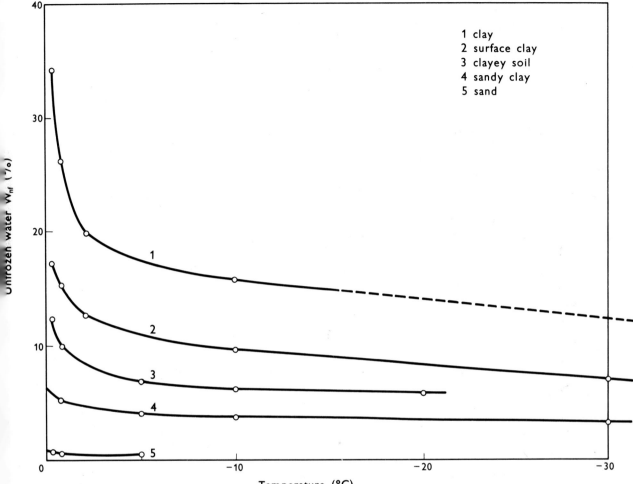

4.1 Unfrozen water content of frozen soils with changes of temperature (*after Tsytovich, 1957, 116, Figure 1*)

Table 4.2. Per cent of unfrozen water as related to temperature and grain size (*Tsytovich, 1958*)

Temperature	Per cent of unfrozen water		
	Pure sand	Sandy clay	Clay
−0.2° (31.5°F)	0.3	18.0	42.0
−10° (14°F)	0.15	9.0	> 20.0

Temperature also controls the amount of unfrozen water in fine-grained soils (Figure 4.1; Tables 4.1–4.2; cf. also Tsytovich *et al.*, 1959, 111–12, Tables 2–3, Figure 5; 1964, 7, 96–9, Tables II–III; 105, Figure 5). The unfrozen water, which surrounds mineral grains as a film 3 to 50 Å or more thick, is readily measured to − 10° and retains high proton mobility to below − 50° (D. M. Anderson and Tice, 1970, 1). Unfrozen water content, especially at temperatures below − 5°, can be estimated from the equation (D. M. Anderson and Tice, 1972, 13; 1973, 119–22)

$$W_u = a\theta_f\beta$$

where W_u = unfrozen water (as per cent of dry soil weight), θ_f = temperature below freezing in degrees C, and a and β = parameters derived from the specific surface area of the soil. Subsequent work suggests that total ice content as well as temperature is a variable that can affect the results and lead to somewhat larger unfrozen water contents than previously believed if much ice is present (Tice, Burrous, and Anderson, 1978). Atterberg liquid limit determinations also have a predictive capacity with respect to unfrozen water contents (Tice, Anderson, and Banin, 1973; 1976).

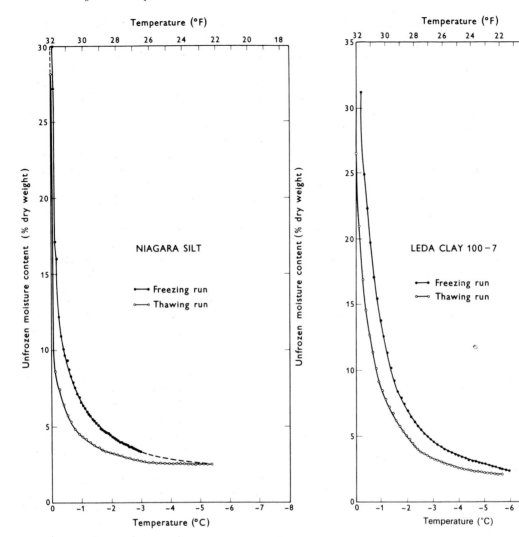

4.2 Unfrozen water content during freezing and thawing (*after P. J. Williams, 1963, 121, Figure 5*)

4.3 Unfrozen water content during freezing and thawing (*after P. J. Williams, 1963, 122, Figure 6*)

Experiments led Dirksen and Miller (1966, 172) '...to conclude that water moves rather freely within the frozen soil providing the temperature is not too low.' Other experiments show that the diffusion coefficient for sodium ions in frozen bentonite at $-15°$ is only an order of magnitude less than that expected at $25°$ (Murrmann, 1973, 358), and even substantial ice lensing need not lead to total impermeability (cf. P. J. Williams, 1977, 49). One result is that weathering of silicates may be a continuing process in permafrost (Ugolini and Anderson, 1973).

It should be noted that the temperature at which the ice in a soil thaws may differ from that at which it freezes (Figures 4.2–4.3). This hysteresis is partly due to differences in apparent specific heat (combined result of latent heat and specific heat) resulting from compaction (= consolidation in engineering terminology) during freezing, whose effect in this respect is similar to that of drying (P. J. Williams, 1967, 1–10).

5   Effect of pressure

Pressure lowers the freezing point of water by a small amount ($0.0073°C$ per kg cm^{-2}). Except in the case of large pressures, the effect is probably negligible compared to other factors affecting freezing. However, as noted under Permafrost in the chapter on Frozen ground, the pressure beneath the Antarctic Ice Sheet at the site of a drill hole reduced the freezing point to an estimated − 1.6° and inhibited freezing.

Pressure also tends to counter the build up of ice lenses. As a result, other conditions being equal, ice lenses normally decrease in size with depth. Similarly, frost heaving can be artificially reduced by applying a large pressure to the top of freezing soil. The pressure required to stop the flow of water to the freezing front, and hence heaving resulting from ice segregation and growth of ice lenses, has been termed the shut-off pressure, defined as '... the effective stress at the frost front which will cause neither flow of water into or away from the freezing front' (McRoberts and Nixon, 1975*b*, 42). It should be emphasized that the shut-off pressure does not control heaving that results from the volume expansion of freezing water. Critical conditions for ice segregation as related to shut-off pressure are discussed under Effect of grain size.

6 Effect of mineralogy

The mineralogy of the fines can influence the freezing process with respect to migration of water and heaving. Clays with expandable structure are able to hold more water but the water is relatively immobile compared with non-expandable clays. Consequently, strong frost heaving is more likely to be associated with kaolinite than bentonite or montmorillonite (Dücker, 1940; Grim, 1952; Linell and Kaplar, 1959, 92–9). The nature of exchangeable cations affects the permeability (which is reflected in the liquid limit) and consequently the heave rate of clayey soils, but the effect varies with the type of clay mineral (Lambe, Kaplar, and Lambie, 1969, Table II, 6; 19). Salts such as NaCl and CaCl$_2$ influence freezing by depressing the freezing point significantly (Yoder, 1955) but other additives can be more effective in reducing the frost susceptibility of soils (Brandt, 1972; Lambe and Kaplar, 1971; Lambe, Kaplar, and Lambie, 1971). The effect of salinity in Antarctic soils was noted in the chapter on Environmental factors.

7 Effect of grain size

Pore size and therefore grain size strongly affect the growth and form of ice in the soil by influencing

(1) the freezing temperature of phase-boundary water and the water in fine capillaries, and (2) the movement of water to the freezing front.

It is well known that the amount of unfrozen water varies inversely with grain size, other conditions remaining equal (Figure 4.1). This accounts for the fact that silt and particularly clay freeze solid at a lower temperature than coarser material (Beskow, 1935, 13, 31–42; 1947, 5, 14–21). Some aspects of this are discussed below and others under Frost heaving and frost thrusting and under Mass displacement later in this chapter.

The basic factor accounting for the different freezing temperatures is the greatly increased contact area between solids and water as grain size is reduced, and the resulting increased tendency for surface effects to keep capillary and phase-boundary water from crystallizing. Thus in a closed system there would be less ice, the finer the grain size and the higher the freezing temperature. However, in an open system where water is continuously available for movement to the freezing plane, the water in material of fine grain size (i.e. with small pores) may remain mobile longer and hence build up larger ice lenses. In fact in incompletely frozen clays where moisture movement is still possible, ice lensing can continue behind the freezing plane and, given constant volume, pressures can continue to increase with decreasing temperature (Hoekstra, 1969). Much depends on the rate of freezing as discussed previously.

Grain size influences the movement of water to the freezing front because the potential for drawing water to the freezing front increases with decrease in grain size. Laboratory experiments by Adler-Vignes and Dijkema (1975) convinced them that the driving force is the difference in chemical potentials of unfrozen adsorbed water and capillary water; the results of subsequent work on the influence of pressure were reported as being in accord with thermodynamic theory (Biermans, Dijkema, and De Vries, 1976). Whatever the exact cause, this suction potential, or water potential ψ (Table 4.1), is absent in coarse sands and gravels lacking fines, so that the segregation of ice lenses as opposed to development of interstitial ice is inhibited. On the other hand, soils may be so fine grained as to become essentially impermeable, and consequently the movement of water to the freezing front to form ice lenses and cause heaving is at a maximum where the suction potential and permeability combine most effectively (Figure 4.4). As a result silt is particularly prone to heaving, since it permits relatively easy

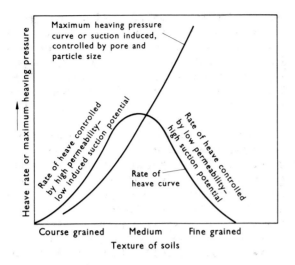

4.4 Diagram of relation between heaving rate, heaving pressure, and grain size (*after Penner, 1968, 22, Figure 1*)

migration of moisture during freezing, and in an open system this factor favours a greater ice content in silt than in clay (other conditions remaining equal), as recognized long ago by Johansson (1914, 84, 93–4) and investigated in detail by Beskow (1935; 1947) and others. Linell and Kaplar (1959, 86, Figure 3; 88) found that silt and lean (slightly plastic) clays generally exhibit higher heave rates than fat (plastic) clays. ('Clays' in the engineering terminology they used includes all grain sizes <0.074 mm in diameter that produce plasticity – US Army Waterways Experiment Station, 1953, 3–4.) However, heaving pressures, which are also controlled by grain size, continue to increase beyond the point where heaving rates decline with decrease in grain size (Figure 4.4). According to experiments by Jumikis with a gravelly silty sand, 'The porosity [which in soils is strongly influenced by grain-size distribution and arrangement] is the primary factor controlling the amount of heat flow, moisture transfer, and all related phenomena...' (Jumikis, 1973, 308). For the soil used, the maximum frost penetration and ice accumulation occurred at about 42 per cent porosity. In general, grain size determines thermal conductivity, other conditions being equal. Thus disregarding water content, conductivities would normally increase in the order fine-grained to coarse-grained soils to bedrock (Ingersoll, Zobel, and Ingersoll, 1954, 288, Table A1; Kersten, 1949, 59–61).

As noted in the chapter on Environmental factors, the role of grain size has been stressed in the follow-

ing 'rules of thumb'. According to these, a soil is frost susceptible (i.e. subject to build up of ice lenses and severe heaving) if it has (1) several per cent of particles <0.07 mm in diameter (passing through 200-mesh screen – *c.* upper limit of silt) (Terzaghi, 1952, 14), or (2) (the Casagrande criterion) contains more than 3 to 10 per cent particles <0.02 mm in diameter (*c.* middle of silt range), depending on the uniformity of the soil (Casagrande, 1932, 169; US Army Arctic Construction and Frost Effects Laboratory, 1958, 28–30). These 'rules of thumb' are subject to many factors (Dücker, 1956; 1958), and some soils heave that would be considered nonfrost susceptible by these standards (Corte, 1961*a*, 10; cf. Corte, 1962*d*, 20). For instance, Kaplar (1971, 1) noted that sandy soils from Greenland showed substantial heaving when the content of particles finer than 0.02 mm was about 1 per cent, whereas the content in some other sandy materials could be up to 20 per cent before there was much heaving. More reliable criteria based on heaving rates are now being used by the US Army Corps of Engineers, and improved techniques are being investigated (Kaplar, 1971; 1974). Nevertheless, the influence of the smallest grain sizes in controlling ice-lens heaving pressure supports the view that grain size is a valid criterion in measuring frost susceptibility (Penner, 1973; Sherif, Ishibashi, and Ding, 1977), even though completely satisfactory frost susceptibility criteria remain to be established (Jessberger, 1973).

8   Critical conditions for ice segregation

a Primary heaving   The foregoing effects are inherently combined in the following expression specifying the critical conditions for growth of segregated ice as cited by Mackay (1971*b*, 411; following D. H. Everett, 1961; cf. Aguirre-Puente, Vignes, and Viaud, 1973; 1978; Andersland and Anderson, 1978, 85–9; Penner, 1977; P. J. Williams, 1967, 99; 1972)

$$p_i \rightleftharpoons p_w = \frac{2\sigma}{r_{iw}} < \frac{2\sigma}{r} \qquad (1)$$

where p_i = pressure office, p_w = pressure of water, σ = surface tension ice-water, r_{iw} = radius of ice-water interface, r = radius of [largest] continuous pore openings. This model of ice lensing, sometimes known as the capillary model, has been termed primary heaving by R. D. Miller (1972, 6–7) for reasons discussed in the next section.

Combining (1) with an air-intrusion value (P. J. Williams, 1968, 1383–4)

$$p_a - p_w = \frac{2\sigma_{aw}}{r} \qquad (2)$$

where p_a = pressure of air necessary to displace water, and σ_{aw} = the surface tension air-water, gives

$$p_i - p_w = \frac{\sigma_{iw}}{\sigma_{aw}}(p_a - p_w) \qquad (3)$$

Thus the air-intrusion value $p_a - p_w$ is a measure of $p_i - p_w$, thereby under certain conditions providing an estimate of frost susceptibility as discussed by Williams. The inequality $p_i - p_w < \frac{2\sigma}{r}$ (1) can be expressed as

$$p_i < \frac{2\sigma}{r} + p_w \qquad (4)$$

and for ice lenses to grow

$$p_w > p_i - \frac{2\sigma}{r} \qquad (5)$$

(5) is similar to the form illustrated later in connection with pingo growth (Figure 5.51), with p_i being analagous to the weight plus bending resistance of the overlying material (overburden) being lifted and deformed by a growing pingo.

Another way of looking at these relationships is in terms of effective stress regarded (cf. Mackay, 1976c; McRoberts and Morgenstern, 1975, 132–3) as

$$\sigma - u = C \qquad (6)$$

where σ is accepted as equivalent to p_i, $u = p_w$ (porewater pressure) as above, and $C = \frac{2\sigma_{iw}}{r}$ is a constant for a given soil (soil constant). Because it is the pressure at which water will flow neither to nor from the freezing front it is theoretically equivalent to the shut-off pressure cited previously under Effect of pressure (cf. McRoberts and Nixon, 1975b, 46). The constant C has a reported range (in kg cm^{-2}) of 0–0.075 for sands, 0.075–2.0 for silts to silty clays, and >2.0 for clays (P. J. Williams, 1967, 101). Accordingly frost-heaving clays should exert more stress and lift a greater overburden than coarser-grained soils. The depth to which ice lenses can form in clays is probably at least 200 m (Mackay, 1976c, 61).

Shut-off pressure as a practical design criterion is questionable. Recent research indicates that initial water expulsion from laboratory soil specimens under pressures as high as 4 kg cm^{-2} was followed by water intake and further, although slight, heaving, and that it may not be feasible to control heaving caused by ice segregation (Penner and Ueda, 1977).

It should be stressed that application of the capillary theory to ice-water systems is not universally accepted. For one thing, the theory does not consider possible complications introduced by soil-particle surface chemistry (Higashi, 1977, 24). Also, although experimental data support the theory as applied to an ice-water meniscus with stationary molecules, the theory is not proved for a freezing meniscus where molecules are being replaced (Takagi, 1977, 59; 1978, 1). As a result Takagi espoused a concept of ice segregation involving coupled heat and water flow and the suction of water to the freezing front to replenish a thin water layer of constant thickness. He proposed the concept as a new freezing mechanism that he called segregation freezing (Takagi, 1977, 59, 61; 1978, 1–2). Yet, however new some of the formulations, the growth of segregated ice as opposed to pore ice is a long-standing concept whatever the exact mechanism, and in the present state of knowledge the term segregation freezing should not be limited to any one theory.

b Secondary heaving The formation of ice lenses may be considerably more complicated than suggested by the concept of primary heaving, whose ramifications are in themselves far from simple. Thus according to R. D. Miller (1972), ice lensing and secondary heaving occur behind the freezing front at what R. D. Miller (1978) has termed the lensing front.

Transport of water in frozen soils is probably not restricted to fluid phases. In a frozen permeameter a series-parallel mode of transport should exist in which the ice phase may move with uniform translational velocity in a stationary pore system formed by particles having adsorbed films of unfrozen water. (R. D. Miller, Loch, and Bresler, 1975, 1029.)

According to this concept, development of a fringe of pore ice beneath an ice lens does not prevent the pore ice from moving, because the '... movement can be accommodated by continuous transformation of ice to film water at the "windward" ice/film interfaces while water is continuously transformed to ice at "leeward" interfaces' (R. D. Miller, Loch, and Bresler, 1975, 1029–30. Cf. R. D. Miller, 1977). The coupled heat and mass transport considerations posed by the secondary heaving may be amenable to computer simulation (R. D. Miller, 1978).

The concept of ice lensing occurring behind the freezing front has been independently arrived at in a different form by Soviet researchers (E. D. Ershov *in* R. D. Miller, 1977; Fel'dman, 1967; 1972). Also from another viewpoint, freezing of unfrozen water with decrease of temperature behind a freezing front is well known as discussed above under Effect of grain size.

c Porewater expulsion   Regardless of the exact ice-segregation mechanism, porewater expulsion from an advancing freezing front can supply ample water for segregation freezing. Even massive ice beds (Figure 3.24) can grow as segregated ice if porewater pressures are adequate to replenish groundwater that is transformed to ice (Mackay, 1971*b*; 1973*a*; 1976*c*). As stressed by Mackay, such pressures would be generated by aggrading permafrost encroaching on permeable water-saturated sediments underlying fines. This situation applies to many massive ice beds underlain by sand or gravel that can serve both as an aquifer and as a reservoir for water expelled from the freezing of adjacent coarse sediments. Water expulsion in advance of a freezing front under certain conditions has been experimentally determined (Balduzzi, 1959; 1960, 25–6, 31; McRoberts and Morgenstern, 1975). Given sufficient porewater pressure, ice segregation might proceed in medium sands, and in such a system might constitute a transition to injection ice as in some pingos (Mackay, 1971*b*, 411–12, 419; 1973*b*, 999–1001, Figure 23) (Figure 5.51). On the other hand, clayey soils are more likely to develop a small-scale, reticulate structure of ice lenses and ice veins because moisture movement and therefore supply are severely restricted and must come from immediately adjacent areas only, as described by Mackay (1974*a*, 234–5) who discounted several other explanations for the reticulate structure. However, the possible importance of hydraulic fracturing as a cause remains to be determined (Mackay, 1975*d*; McRoberts and Nixon, 1975*a*).

Among other things the nature of the segregation process can determine the ratio of ground heaving to depth of permafrost (with respect to an independent datum) in an area newly exposed to freezing (Figure 4.5).

III   Estimates of freezing and thawing

1   Freezing and thawing indexes

Freezing and thawing indexes are measures of the heat balance at the ground surface (surface index) or at a height of 1.5 or 1.8 m (5 or 6 ft) above it (air index), usually given in degree days of freezing or thawing over an unbroken freezing or thawing period (Sanger, 1966, 253; cf. Andersland and Anderson, 1978, 126–30). For °C

$$I = \int_0^t T dt$$

where I = index, T = mean temperature for a day as represented by (maximum + minimum temperature)/2, and t = the period.

For °F, which has been the most commonly used unit for freezing and thawing indexes

$$I = \int_0^t (T - T_0) dt$$

where $T_0 = 32°$ F.

Except in change-over months characterized by both freezing and thawing, the monthly index is given by (Boyd, 1976, 177)

$$I = NX$$

where I = index, N = number of days in a given month, and X = average temperature for that month, less 32° if °F.

In autumn and spring change-over months, the monthly index can be approximated by

$$Y^2 - NXT = N^2k^2$$

where k = a constant. The solution gives two values for Y: a positive for the thawing index, and a negative for the freezing index.

Freezing and thawing indexes provide a measure of the severity of climate. They lend themselves well to mapping (Figure 4.6), are useful in projecting depths of freezing and thawing (A. L. Are and Demchenko, 1976; Fahey, 1974; Sanger, 1966), and the thawing index in combination with mean annual temperature provides an interesting approach to projecting the presence or absence of permafrost and whether it is continuous or discontinuous (Pihlainen, 1962). The Stefan equation, presented in the section on Seasonally frozen ground in the preceding chapter, has the freezing or thawing index as one of its terms, and if the other terms are

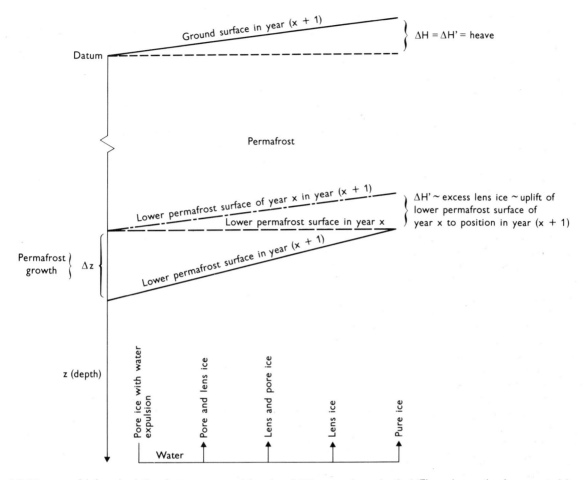

4.5 Diagram of inferred relation between ground heaving (ΔH), permafrost depth (ΔZ), and growth of segregated ice (ΔH') (*after Mackay, 1975b, 472, Figure 1*)

known, the appropriate index can be obtained (approximately). In some situations this permits comparison of former with present indexes, and conclusions regarding the former presence or absence of permafrost as illustrated by Maarleveld (1976, 65). On the other hand these indexes do not necessarily correlate well with mean annual temperature because of differences with respect to summer and winter temperatures in continental and maritime climates (P. J. Williams, 1961, 341, Table 1; 342, Figure 2), and for the same reason they give no information as to frequency of freezing and thawing.

2 Freeze-thaw cycles

The frequency of freeze-thaw cycles is an important control in the effectiveness of various kinds of frost action, including frost wedging. However, the purely climatic factor of the number of times the air temperature passes through the freezing point is not, in itself, an adequate measure of the effectiveness. Frost wedging on north slopes in the Swiss Alps appears to be especially great (Barsch, 1977c, 155), although temperature measurements on the Jungfrau suggest that south slopes have many more days (196:22 on the Jungfrau) with air-temperature cycles through the freezing point (Mathys, 1974, 57). The temperatures themselves and their duration, the rapid attenuation of temperature fluctuations with depth, the insulating effect of snow and vegetation, and the nature of the bedrock or soil must all be taken into account in evaluating the frequency and effectiveness of freeze-thaw cycles.

Large discrepancies between air and ground-surface temperatures are possible as a result of

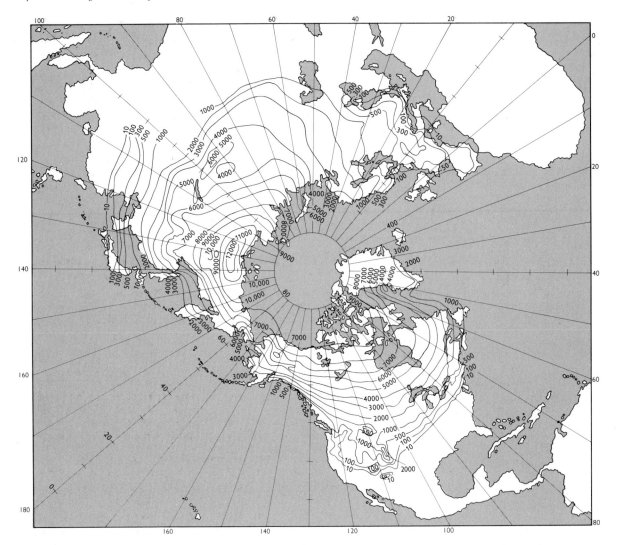

4.6 Map of mean annual air freezing indexes (°F) for Northern Hemisphere. Units explained in text (*after Corte, 1969a, Plate 6 opposite 161*)

insolation on dark surfaces. A difference of over 30° has been observed in Northeast Greenland, and similar occurrences have been noted elsewhere (cf. Washburn, 1969a, 45–6). In the Antarctic thawing of snow or ice adjacent to rock has been reported at air temperatures as low as −16.5° (G. Taylor, 1922, 47) and −20° (Souchez, 1967, 295), and freeze-thaw cycles in bedrock joints during negative air temperatures have also been observed there (Andersen, 1963, B142; Aughenbaugh, 1958, 168). On the other hand, because of outgoing radiation, it is also possible for ground freezing to occur at positive air temperatures as reported from Iceland

(Steche, 1933, 222, Figure 8). Records of the Astrakhanskaya Hydrometeorological Observatory, USSR, show that decreases of temperature through 0° are consistently more frequent at the soil surface than in the air (Chigir, 1972, 96–7, Tables 1–2; 1974, 96–7, Tables 1–2).

Thus the frequency of freeze-thaw cycles at and beneath the ground surface can vary widely from the frequency in the air above. A climatic regime having many air cycles of small temperature range favours extreme examples; Meinardus (1923, 433; 1930, 50–1) reported that at Kerguelen Island in the Subantarctic no frost was observed at a depth of 5 cm in 1902–3, despite 203 air freeze-thaw cycles and 238 days with freeze-thaw cycles at the ground surface. At Cornwallis Island in the Canadian Arctic where the air cycles have a larger temperature range, Cook and Raiche (1962, 67, Table 1) found, from 1 May to 30 September 1960, 23 ground cycles in the range $-2.2°$ to $0°$ (28° to 32°F) at the surface but only one cycle at a depth of 2.5 cm and none at depths of 10 cm and 20 cm. By contrast with the 23 ground-surface cycles there were only 15 air cycles. Observations in Arctic Canada convinced Cook (1966, 129) that

... there are no cycles, apart from the annual cycle, at depths below a few centimetres, and it follows that assumptions still widely held among some geographers and geologists (that frost cycles are a vigorous process producing frost splitting at depth in arctic countries today) are not valid.

Air freeze-thaw cycles become more common farther south in Canada (Fraser, 1959, 44–5), and the trend continues into the United States (Llewelyn Williams, 1964). Under such conditions frost wedging is probably more frequent. Whether it is also more effective with respect to size and quantity of resulting debris is less certain. In fact French (1976a, 19) questioned the importance of short-term cycles because of their shallow penetration and he suggested that the annual cycle may be more effective in frost wedging than was once thought. It is certainly the most important cycle in promoting frost-action effects at depth.

There are several methods for estimating the frequency of ground freeze-thaw cycles on the basis of air cycles. One method is to select a temperature range for an air cycle that supposedly also represents a ground cycle. There are many inherent uncertainties (cf. Chambers, 1966b, 77–80; Matthews, 1962), and several workers employing the method have done so with reservations (Cook and Raiche, 1962; Fraser, 1959). Fahey (1973) has demonstrated its unreliability. One safeguard is to directly compare ground cycles with air cycles during a control period, and to work out a correlation from these data that permits estimates of ground cycles over a much longer interval (cf. Washburn, 1967, 133–8).

IV Frost wedging

1 General

Frost wedging is the prying apart of materials, commonly rock, by expansion of water upon freez-

ing. It is synonymous with congelifraction (Bryan, 1946, 627, 640), gelifraction, frost riving, frost shattering, and frost splitting. Frost wedging was described from northern Canada as early as 1742 (Middleton, 1743, 158–9). The prying force is not necessarily confined to the 9 per cent volume expansion accompanying the freezing of water but in porous material may be due, in even greater measure, to the directional growth of ice crystals as first emphasized by Taber (1929; 1930a; 1930b). Frost wedging is of widespread interest and importance to engineers and geologists, and despite the large amount of research that has been carried out, much still remains to be done (cf. Aguirre-Puente, 1978).

Some porous rocks may behave like certain soils in absorbing moisture, and frost wedging, although facilitated by foci of weakness, does not necessarily require pre-existing fractures (Taber, 1950; Willman, 1944). Based on his experiments at the US Army Cold Regions Research and Engineering Laboratory, Mellor (1970, 42–3) reported

... that the observed freezing strains come close to the strains for complete tensile failure of the rocks, while the potential freezing strains for completely saturated rock frozen rapidly far exceed the tensile failure strain for sandstone and limestone. It should also be noted that internal cracking of these rocks almost certainly occurs at tensile strains well below the ultimate tensile fracture strain.

In general, however, freezing of pore water strengthens rock, the strength increasing with lowering of temperature to $-120°$ (Mellor, 1971).

The maximum pressure generated by the phase change of water to ice is 2115 kg cm^{-2} under ideal laboratory conditions (Bridgman, 1912, 473–84). Because of pressure melting, the pressure generated by freezing is temperature dependent in that it increases with lowering temperature to $-22°$ at which temperature the maximum theoretical pressure is attained and Ice III with less volume replaces the pre-existing common Ice I. As stressed by Grawe (1936), the maximum theoretical pressure is probably never attained in nature. For one thing, the maximum tensile strength of rock is much less, being on the order of 250 kg cm^{-2}.

2 Frost wedging *v.* hydration shattering

It has recently been argued that hydration shattering may be the most important process in producing the effects commonly ascribed to frost wedging. Dunn and Hudec (1965, 115–38; 1966) reported that ordering and disordering of water molecules may simulate 'frost deterioration' and they '... concluded that sorptive interactions with water vapor or liquid water are far more destructive of shales, siltstones, and argillaceous carbonate rocks than is freezing and thawing' (Dunn and Hudec, 1972, 65). According to Hudec (1974, 313), in

> ... fine-grained sedimentary rocks, the frost phenomenon is of minor importance. The rocks fail due to the expansive action of the rigid, adsorbed, non-freezable water contained in the *sorption sensitive* rocks. The failure is by fatigue due to expansion and contraction on sorption and desorption, and is enhanced by large temperature fluctuations and cycling in the freezing and thawing range. True frost action is active only in the coarser grained rocks, and only in those that saturate critically by *absorption of bulk* water.

Based on his experimental data, Hudec (1974, 325, 327) reported that

> Those rocks that are known to be frost sensitive, i.e., those that break up under repeated freezing and thawing, adsorb large proportions of their total saturation water, and have the least amount of this water in the frozen state. Also, addition of salt, which normally accelerates the frost action on rock (and concrete) has the effect of reducing the amount of ice forming in the rock. The obvious conclusion must be that ice action is not responsible for rock breakdown, and is not a major force in mechanical weathering of the rocks that make up the major part of the earth's surface.

Many factors are involved and the conclusion is not all that obvious to the present writer. The correlation that rocks normally regarded as frost susceptible by engineering standards have the most unfrozen water and least ice during freezing experiments could perhaps also be interpreted as indicating that little ice is required for their breakdown. It would be useful to carry out further hydration experiments restricted to temperatures above 0° as included in experiments conducted by Konishchev (1973; 1978).

That chemical effects and hydration of minerals can be important in causing volume changes leading to distintegration of rocks has long been recognized and is not at issue.[2] Also the possibility that ordering and disordering of water molecules may play a role, including an enhanced effect in clays (Allan Falconer, 1969), is not denied. Konishchev (1973; 1978) has argued for the importance of 'hydrational' and 'cryohydrational' weathering and concluded,

subject to further study, that they characteristically produce finer grain sizes than frost wedging, the latter being responsible for coarse sand and larger particles under favorable conditions. At issue is whether these effects can really account for the angular fragments of sound rock such as those commonly ascribed to frost wedging, or in places to salt wedging, or more rarely to purely thermal changes (cf. I. S. Evans, 1971; W. F. Hume, 1925, 24–8; Wilhelmy, 1977, 181–3). Even if ordering and disordering of water is capable of producing such fragmentation, additional evidence would be required to prove that frost wedging is unimportant. S. E. White (1976a) in a provocative review comparing frost action and hydration shattering suggested the possibility that even block fields and block slopes (Figure 4.7), discussed later in some detail in the chapter on Mass-wasting processes and forms, may be hydration phenomena (cf. S. E. White, 1976b, 92, 94). However, the fact that such accumulations of large angular rock debris are characteristic of low-temperature alpine and polar environments argues strongly for an origin by frost wedging and clearly puts the burden of proof on those who would claim otherwise. The importance of frost wedging is also supported by the experience of alpinists that rockfall in high mountains is usually consequent on thawing and is most common where frost action is diurnal. Pending further evidence, the present writer accepts in the following pages the traditional view of the effectiveness of frost wedging but some of the conclusions may well be subject to revision as further badly needed research on frost action is carried out. New ideas keep appearing such as the untested suggestion that rocks containing undercooled water in pore spaces may fail because of shock waves produced by cavitation-induced nucleation of ice (Hodder, 1976).

3 Factors

Not only is the presence of moisture mandatory but the amount available can be critical, since the bulk freezing strain of a rock increases with increasing water content (Mellor, 1970, 50). Laboratory experiments consistently demonstrate that the effectiveness of frost wedging depends on the amount of water available (cf. Martini, 1967; 1973). For instance, limestone cubes originally 10 cm on a side

---

[2] Whether such hydration is as important in splitting rock into large angular fragments as apparently held by Wilhelmy (1977, 182) seems less certain to the present writer. If jointing is present various processes could operate.

break down much more readily in water 4 cm deep than 1 cm deep (Guillien and Lautridou, 1970), and rocks half immersed generally disintegrate more rapidly than those that were saturated but surrounded by only a thin film of water (Potts, 1970, 112–13). Frost action has been reported to be particularly intense in the neighbourhood of thawing snow patches, also on shores where rocks are frequently wetted (Mackay, 1963a, 57; Taber, 1950).

Given adequate moisture, the nature of a rock is probably the most important factor determining its susceptibility to frost wedging. Sedimentary rocks such as siltstone and shale containing mica or illitic clays display, by virtue of the horizontal orientation of the micaceous minerals, planes of fissility through which water migrates preferentially. Thus shales break down more readily than igneous rocks (Potts, 1970, 114–22), although crystalline rocks rich in biotite or other mica may be also vulnerable to frost wedging. The process may become more critical if both biotite and muscovite are altered and the interlayer potassium has been replaced by hydrated calcium, magnesium, or sodium. Different behaviour may be anticipated in different minerals as the result of variations in hydration and expansion. Porosity as a factor in unfractured rocks was emphasized by Lautridou (1975), who reported that carbonate rocks having pore sizes in the range 0.0001 to 0.003 or 0.004 mm were prone to disintegration under the prevailing experimental regime, their susceptibility increasing with increasing pore size in this range.

Some experimental data indicate the primary importance of freeze-thaw cycles for a given rock type (Coutard *et al.*, 1970, 37; Lautridou, 1971, 69, 79; Potts, 1970, 113–14; Wiman, 1963, 116). Legget, Brown, and Johnston (1966, 25–6) reported that 400 cycles resulted in a 10 to 20 per cent increase in shale and sandstone fragments having a size range of 0.1 to 10 mm.

Experimental evidence of rock breakdown supports the view that the intensity of freezing as well as the number and length of freeze-thaw cycles[3] is important (Tricart, 1956, 295–7). As noted by Schunke (1977c, 48–9), widespread angular rock debris in periglacial environments appears to depend more on low temperatures than on frequency of freezing and thawing.

Battle (1960, 92–3) found that uncracked porous rocks in the laboratory did not fail above about − 10°, and he also concluded that a temperature drop of about 0.1° per minute is necessary. Depending on the freezing rate, two different rocks can reverse their susceptibility to breakdown (Thomas, 1938, 63–8, 94). Mellor (1970, 31–43, 50–1) found that the bulk freezing strain of a rock increases not only with increasing water content but also, for any given water content, with increasing freezing rate, thereby lessening the opportunity for redistribution or extrusion of porewater, and he suggested that at water contents > 50 per cent saturation, the freezing strain may be enough to cause internal cracking. Aguirre-Puente (1975, 9–11), on a theoretical basis, also argued that rapid (brutal) freezing would seal off fine pores that might otherwise conduct water to the surface and thereby dissipate capillary water having a lower freezing point than water in larger pores (cf. Bertouille, 1975, 8–10). Thus, rapid freezing of saturated rocks should favour frost wedging by creating a closed system that would promote pressure effects. The importance of closed-system freezing in rock splitting (éclatement), as opposed to more superficial disintegration (écaillage superficiel), was strongly supported by Lautridou and Ragot (1977, 4–5), whose laboratory experiments showed that under such conditions fine-grained limestone was highly susceptible. Coarse-grained rocks, they believed, were much less so because of porewater expulsion.

Where cracks exist, frost wedging would be furthered by water freezing from the surface down and creating a solid plug of ice (Battle, 1960, 93–4; Pissart, 1970a, 44–5).

On the other hand it has been suggested that slow freezing may promote frost wedging of fine-grained uncracked rocks on wet soil by permitting flow of water to the freezing front to build up disruptive ice crystals, whereas rapid freezing would inhibit the flow of water (Taber, 1950). The apparently contradictory viewpoints are probably readily reconcilable if the controlling factors were exactly specified for a given case.

Finally, intimately related to the time factor and frequency of freezing and thawing is the fatigue factor. As noted by Bertouille (1975, 10), it can explain why many Quaternary deposits contain abundant frost-wedged rocks, some of which to fracture in the laboratory require thermal shocks far

---

[3] As previously discussed in the chapter on Environmental factors, and above under Estimates of freezing and thawing, large discrepancies can occur between air and ground temperatures. Consequently the number of shifts of air temperature through the freezing point, which is sometimes cited as an environmental measure of frost action, is by no means synonymous with the number of freeze-thaw cycles in rocks and soil.

exceeding climatic influences. The extent to which fatigue can precondition natural specimens and influence laboratory results is not always known.

The multiplicity of factors affecting frost wedging – rock type, moisture, temperature, time – complicate the interpretation of the resulting products as is frequently stressed (cf. Guillien and Lautridou, 1970, 40–5; Martini, 1967). Also as previously discussed, and noted by Pitty (1971, 185) in citing Wiman's (1963) observations, hydration shattering may be a factor. All these matters need further study.

4   Products

Frost wedging characteristically produces angular fragments that can be of widely varying size, ranging from huge, house-size blocks to fine particles.

'Frost weathering' has been cited as the mechanism by which unconsolidated sediments and shales disintegrate to granule-size particles by growth of segregated ice and subsequent thawing (Harrison, 1970), but the predominant size to which rocks can be ultimately reduced by frost action is much smaller and usually held to be silt (cf. Hopkins and Sigafoos, 1951, 59; Sørensen, 1935, 24–5; Taber, 1953, 330), although as stressed by Hopkins and Sigafoos the parent rock can exert a critical influence on the size of its products during distintegration. Laboratory experiments by Guillien and Lautridou (1970) showed that frost wedging of certain kinds of limestone can produce particles as fine as clay. Experiments reported by B. Meyer (*in* Semmel, 1969, 51) apparently resulted in production of clay (chlorite), as did experiments by Leshchikov and Ryashchenko (1973; 1978), who reported kaolinite and montmorillonite as new components. To some extent chemical as well as physical effects would seem to be involved. Experiments by McDowall (1960) suggest that even clays may be further comminuted by frost action. Konishchev, Rogov, and Shchurina (1975) reported that clay minerals (hydromica, kaolinite, montmorillonite) subjected to 50 freeze-thaw cycles underwent changes, the most marked being destruction of the hydromica crystal lattice. However, Guillien and Lautridou (1974) stressed that very fine-grained limestones furnished little sediment as small as silt or fine- or medium-grained sand, and that comminution essentially ceased after 200–300 freeze-thaw cycles. Similarly, little material finer than 0.06 mm was produced in experiments carried out by Potts, who con-

cluded that 'processes other than frost shattering produce silt and clay particles which are found in solifluction deposits' (Potts, 1970, 120). It would also seem that the significance of frost wedging in producing cold-climate loess remains to be fully established; probably there is considerable variability in view of other silt-producing processes such as glacial erosion.

Frost wedging resulting from repeated temperature cycles at negative temperatures appears possible (cf. Barsch, 1977c, 155; Fukuda, 1972), since water in minute rock openings should freeze and thaw at such temperatures like some of the water in fine-grained soils (Figures 4.1–4.3, Tables 4.1–4.2), as previously discussed in this chapter under Freezing process.

Finally it should be noted that Mackay, Konishchev, and Popov (1979), citing Konishchev, reported that disintegration by freeze-thaw cycles leads to a weathering sequence of common rock-forming minerals in which the ultimate size reduction of quartz (0.05–0.01 mm) is smaller than for feldspar (0.1–0.05), a reversal of the usual sequence in the absence of freezing and thawing.

5   Environmental aspects

a General   Interpretation of shattered bedrock as a result of frost wedging under a former cold climate dates back at least to Conrad (1839, 243), who reported

Occasionally I have seen the upper portions of limestones and sandstones broken up, a distance of several feet from the surface, but the fragments remain *in situ*. ... Indeed, I think it impossible to account for this breaking up of the rocks to a distance of many feet below the surface, except by the agency of intense cold, freezing the water which filled the fissures, and thus forcing the rocks into tabular fragments, and disturbing their position by the lateral and upward pressure.

In periglacial environments, thawing of snow or ice adjacent to dark rocks warmed by insolation is common at subfreezing air temperatures and must be a potent factor in frost wedging when meltwater seeps into joints and refreezes. Since subfreezing temperatures occur below the thawed layer throughout the year in a permafrost environment, refreezing of meltwater in lower-lying jointed bedrock or in cracks in unconsolidated material is not confined to the spring or autumn and to freeze-thaw cycles induced by changes of surface temperature. Nevertheless, the maximum effect of frost wedging as

4.7 Block slope, Hesteskoen, Mesters Vig, Northeast Greenland (*cf. Washburn, 1969a, 34, Figure 18*)

gauged by release of rock fragments probably occurs in the spring as shown by the frequency of rockfalls then. Products of rockfall accumulate on the snow during the spring thaw in many places, and the process itself has been observed while the sun was thawing a cliff face, so it seems certain that the rockfall in such places is triggered by thawing and is therefore primarily the result of frost wedging. Exclusive of earthquakes, frost wedging is commonly regarded as probably the most effective cause of rockfall in periglacial environments, other conditions such as rock structure being equal. Both daily and seasonal observations support this view (Luckman, 1976; Rapp, 1960a, 104–9; 1960b, 17–23; Washburn, 1969a, 35). It has been estimated that 'frost weathering' over a 50-year period caused steep rock faces in Longyeardalen, Spitsbergen, to retreat at a rate of 0.3 mm yr^{-1} (Jahn, 1976a, 122), and that frost wedging in the Swiss Alps during the Holocene has caused some valley walls to retreat 10–25 m at an average rate of about 2.5 mm yr^{-1}

(Barsch, 1977c, 148, 155). In many places, the frost wedging may be as much in response to an annual freeze-thaw cycle as to shorter-term cycles.

Coarse and angular, fresh rock debris masswasting from bedrock attests to the importance of frost wedging in polar and alpine regions. Some slopes without, or with only low, cliffs at their heads are formed mainly of material much like talus rubble. Where such coarse, angular debris is derived from the underlying rock without benefit of rockfall, frost wedging must be the predominant process responsible for detaching the fragments and forming block fields (on nearly horizontal surfaces) and block slopes (Figure 4.7), about which much remains to be learned as discussed later in the chapter on Masswasting processes and forms.

Büdel (1969; 1977, 60–1, 79–81) suggested that frost wedging of bedrock in stream courses that dry up in the autumn accounts for especially rapid valley deepening in a permafrost environment. According to this view (Büdel, 1977, 60–1), bedrock in the

upper part of the permafrost is subject to frost cracking with frost wedging occurring as the result of air
penetrating and depositing 'needle ice' as a sub-
limate, thus forming an 'ice rind'. Then recurring
floods every few years remove the bedload, more or
less completely, leading to thawing and removal of
the broken bedrock. Continous exposure of more
bedrock to such effects accelerates stream erosion.
However, the thickness and thermal regime of the
bedload deposit of periglacial streams subject to
drying is not well known, and the erosion mechanism
described by Büdel is not well established. No ice-
rind effect of the kind described was observed in
Iceland by Schunke (1975a, 188, 229), in Greenland
by Stäblein (1977, 30–1), or in Spitsbergen by Bibus
(1975, 115) or Semmel (1976, 398; cf. Bibus, Nagel,
and Semmel, 1976, 39–41). Although Semmel observed an ice rind in perennially frozen sediments of
some stream beds, the sediment thickness was such as
to protect the underlying bedrock from erosion, and
^{14}C dating of buried moss layers, representing former
surface horizons, showed this condition had persisted
for some 1800 years or longer. He concluded that
only the glacier-fed streams are capable of eroding
the bedrock (Semmel, 1977, 36–7).

The stress on frost wedging in periglacial environments should not obscure the fact that chemical
weathering can be significant, whether in alpine
regions (cf. Boch, 1946, 213; Caine, 1974, 729–30;
Rapp, 1960a, 165–8, 184–5; Reynolds, 1971; Reynolds and Johnson, 1972; Slaymaker, 1974, 321–4),
the Arctic (Cogley, 1972; Isherwood, 1975; D. I.
Smith, 1972; Washburn, 1969a, 43–4), or the Antarctic (cf. Boyer, 1975; Campbell and Claridge, 1975;
Ugolini and Anderson, 1973; Ugolini, Bockheim,
and Anderson, 1973). Generally it is a question of
relative importance with the dominance of physical
weathering tending to mask chemical effects. In
places chemical and physical effects combine as in salt
wedging (Bradley, Hutton, and Twidale, 1978;
I. Evans, 1970; Goudie, 1974; Johnston, 1973;
Mortensen, 1933; Wellman and Wilson, 1965).
Like frost wedging, this process can break up rocks
and produce silt-size particles, and at least in chalk
and sandstone, especially in the case of sulphates,
it can be much more effective than frost wedging
under the laboratory conditions investigated by
Goudie (1975, 8–10). Both frost wedging and salt
wedging can operate in cold–arid environments, and
their combined effect on chalk was dramatic in
Goudie's experiments. The fact that salts and ice can
cause many convergent phenomena has been stressed
by Kaiser (1970, 171–5) and Tricart (1970) among

others, and their effects in cold–arid environments
may be difficult to separate. It is striking how many
desert features are similar whether in periglacial or
warm environments (Cameron, 1969; Kaiser, 1970;
Meckelein, 1965; Mortensen, 1930), and in many
places the processes are the same, whether related
to low precipitation or edaphic aridity (Meckelein,
1974).

b Periglacial tors   Tors have generated much
controversy. As defined by Linton (1955, 476)

A *tor* is a residual mass of bedrock produced below the
surface level by a phase of profound rock rotting effected
by groundwater and guided by joint systems, followed by
a phase of mechanical stripping of the incoherent products
of chemical action.

Accordingly a tor would be a two-stage form. This
genetic definition was criticized by Pullan (1959)
who argued that one-stage forms were also present
and he therefore redefined the term.

A *tor* is an exposure of rock *in situ*, upstanding on all sides
from the surrounding slopes and it is formed by the
differential weathering of a rock bed and the removal of
the debris by mass movement. (Pullan, 1959, 54.)

This definition is adopted here.

Tors occur in a variety of lithologies and tend to
be located where jointing is more widely spaced than
in the surrounding bedrock. In height tors range
from a few metres to tens of metres, and they occur
primarily in highlands, both in summit areas and
on side slopes. They have been described from polar,
subpolar, and temperate environments; however,
some tors are associated with deeply weathered bedrock requiring a warm-temperate or tropical
environment – a circumstance giving rise to the two-
stage hypothesis.

The origin of tors has been hotly debated (cf.
Bunting, 1961; Cunningham, 1969; R. Dahl, 1966a,
78–84; Demek, 1964c; Dyke, 1976; Isherwood, 1975,
121–42; Lester King, 1958; Linton, 1955; 1958; 1964;
Palmer and Radley, 1961; Raeside, 1949; Wilhelmy,
1958, 55–62), the suggested processes including frost
action and mass-wasting, wind action, and subsurface weathering and later exhumation. The
question is complex but there is now reasonable
agreement that tors can have multiple origins. The
evidence appears convincing that many if not most
tors occurring in the Arctic (French, 1976a, 154–5;
Washburn, 1969, 39–42) and Antarctic (Derbyshire,
1972; Selby, 1972) are true periglacial tors produced
mainly by frost wedging and mass-wasting, aided
in places by wind (and in the Antarctic also by salt

wedging— Selby, 1972), which leave the most resistant rocks standing in relief. The importance of chemical weathering in development of tors on Baffin Island in Arctic Canada was stressed by Isherwood (1975, 141–2), who although supporting a one-stage process concluded that frost wedging merely aided a final disintegration phase. However, the tors studied occurred in an area (zone 1) '... considered to represent weathering throughout the Quaternary, and possibly even longer' (Isherwood, 1975, 12), so that multiple climatic changes may have affected the nature of the weathering. The question is open in that some evidence suggests the tors may have survived glaciation and that tors do not necessarily indicate its absence, especially if the ice was cold based – i.e. at a pressure-temperature below freezing (Sugden and Watts, 1977).

Periglacial tors are one kind but, despite contrary suggestions (cf. Worsley, 1977, 205), many tors are probably of different origin so that tors *per se* can not be accepted as diagnostic of a former periglacial environment (Embleton and King, 1975, 167, 174–5; French, 1976a, 232–4). The continuing problem is the extent to which some specific occurrences can or cannot be related to former periglacial influences, as in England (Worsley, 1977, 209–10).

V Frost heaving and frost thrusting

1 General

'The pressures generated by freezing water are exerted in all directions, but they are expressed in soil movements only upward and horizontally. The vertical expression has been termed heave; the horizontal, thrust' (Eakin, 1916, 76). Although Eakin's statement is open to the objections indicated below, the distinction between heave and thrust is important (cf. Hopkins and Sigafoos, 1954). Frost heaving is the predominantly upward, frost thrusting the predominantly lateral movement of mineral soil during freezing.

Taber (1929, 447–50; 1930a; 1930b, 116–18) demonstrated that the pressure generated by the growth of ice crystals is at right angles to the freezing isotherm and not necessarily in all directions; and because freezing extends downward from the ground surface he stressed the role of frost heaving as opposed to thrusting (cf. Taber, 1943, 1458–9; 1952). However, the complexities involved in the freezing of heterogeneous material can result in pressures in various directions as pointed out by Hamberg (1915, 600–3). Various investigators have argued that varying conductivities in heterogeneous material influence the orientation of cooling surfaces and introduce lateral movement in places (cf. Cook, 1956, 17; Corte, 1962c, 14–17; 1962e, 58; Schmid, 1955, 92–5, 122, 130). Resistance to expansion is also an important factor controlling direction of expansion during freezing (Beskow, 1935, 59; 1947, 30–1);[4] although the effect would normally be upward, this is not necessarily the case and frost thrusting cannot be neglected in considering frost action. Where thrusting as well as heaving may be involved, frost expansion is a useful term covering either direction or both.

Thermal expansion of ice upon temperature increase also causes lateral stresses referred to as thrusting (Laba, 1970). However, this process does not conform to Eakin's definition of *frost* thrusting, which specifies the pressure of freezing water and is a separate process.

The general process of frost heaving is determined by the thermodynamics of the freezing process and the growth of pore ice and ice lenses, and is another way of looking at this complex and the controlling factors discussed in this chapter under Freezing process. Where permafrost is aggrading the amount of surface heaving depends on the nature of ice accumulation at the permafrost base (Mackay, 1975a), but elsewhere frost heaving is associated with the active layer above permafrost or with seasonally frozen ground. Because of its practical significance in foundations for buildings, roads, and airfields in cold environments, frost heaving has been studied particularly by civil engineers, who have made many valuable contributions. A recent summary from the engineering viewpoint has been presented by Kaplar (1970). Methods of measuring frost heaving in the field have been reviewed by James (1971). The practical implications and problems, especially in permafrost regions, are covered in numerous publications cited in the chapter on Frozen ground. The emphasis in the following is on certain specific frost-heaving processes that help to characterise the periglacial environment.

2 Heaving of joint blocks

One of the striking results of frost action is heaving

[4] Freezing may also cause contraction by desiccation, as discussed in the next chapter under Patterned ground, and by withdrawing water from small pores to freeze in air-filled larger pores (A. B. Hamilton, 1966; cf. Yong and Osler, 1971, 278–9).

4.8 Frost-heaved bedrock, Mesters Vig, Northeast Greenland (*cf. Washburn, 1969a, 51, Figure 25*)

of joint blocks. Blocks, frost wedged from bedrock along joints, are raised well above the general surface in places, although the blocks are still tightly held by the surrounding bedrock (Figure 4.8).

Yardley (1951) has described such occurrences as 'frost thrust blocks', but this term is inappropriate because frost thrusting applies to lateral rather than vertical movements. Frost-heaved blocks are common in permafrost regions. For instance, they have been observed on Spitsbergen (Bertil Högbom, 1910, 41–2; 1914, 274–7), in the Canadian Arctic (Yardley, 1951), Northeast Greenland (Washburn, 1969a, 51–2) and elsewhere. Such heaving could endanger pipelines in jointed bedrock.

3 Upfreezing of objects

a Stones Ejection of stones from fines by upfreezing is commonly accepted because of prominent edgewise projecting stones in periglacial environments (Figure 4.9), the often-reported but rarely documented appearance of stones on previously cleared farmers' fields, the heaving of posts and other artifical structures, and other similar field occurrences. However, detailed observations and measurements relating to stones are conspicuously few. Morawetz (1932, 39) observed upward displacements of 4–7 mm for pea- to nut-size stones as a result of three to four freeze-thaw cycles in an alpine environment. Field experiments in Germany by Schmid (1955, 88–9, 130) were negative with respect to upheaving of stones not on edge; edgewise stones apparently heaved like posts but details were not reported.

In general wherever frost heaving is prominent any tabular stones within the soil tend to be on edge. In central Iceland, Schunke (1974a, 21) found that 60–70 of every 100 stones he counted had their long

4.9 Frost-heaved block in diamicton, Mesters Vig, Northeast Greenland. Scale given by trench shovel (*cf. Washburn, 1969a, 53, Figure 28*)

axis vertical. In the high plateau area of northwest Iceland, the count was 55–60 per cent (Schunke, 1975a, 62). In the Silvrettagruppe of the Swiss Alps, Vorndrang (1972, 12–13, 19–21, 44–5, 47, 51) found that a site whose material had been excavated, thoroughly mixed, and replaced showed, a year and 46 freeze-thaw cycles later (93 passes through 0° at a depth of 10 cm), sorting effects that included upfreezing and edgewise orienting of stones. Of the surface stones, 38 per cent were edgewise, almost approaching the average 41 per cent of edgewise stones found to a depth of about 50 cm during excavation. Interestingly, in view of the much greater number of freeze-thaw cycles nearer the surface, over 90 per cent of the stones were found on edge at depths of about 60 to 120 cm in a horizon where long axes were in the direction of the slope, the conclusion being that this long-axis orientation was due to solifluction. The possibility should be considered that such slope movements may also cause stones to be on edge, but freezing progressing upward from a permafrost table (which in this case lay at a depth of about 1 m) is probably critical in setting stones on edge at depth (cf. discussion of

frost push under Mechanisms later in this section).

Laboratory experiments carried out and summarised by Corte (1966b) prove a vertical sorting involving an upward displacement of coarse soil particles relative to fine as the result of some types of frost action. Upfreezing of stones that lay immediately beneath the surface occurred within 7 freeze-thaw cycles in an experiment set up by the writer in the Periglacial Laboratory of the University of Washington's Quaternary Research Center, and subsequent experiments with glass objects of various sizes, shapes, and orientations demonstrated that some vertically oriented lath shapes that had been emplaced at a depth of 1.4 cm reached the surface after 2 cycles, whereas objects with other orientations moved up more slowly (Burrous, 1977).

b Targets Kokkonen (1930), concerned with the upfreezing of plants, investigated the upfreezing of wood pegs of various lengths and inserted to various depths in southern Finland. He found that the longer the peg (i.e. the deeper its insertion), the greater was its heave, provided the peg insertion did not exceed depth of freezing.[5] Schmid (1955, 80–1) working in Germany found that the deeper the peg

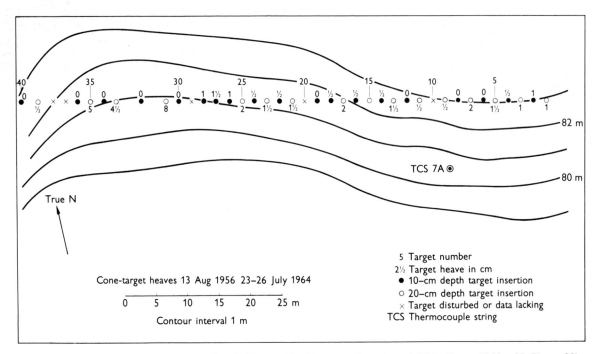

4.10 Cone-target heaves, Experimental Site 7, Mesters Vig, Northeast Greenland (*cf Washburn, 1969a, 65, Figure 38*)

[5] Some pegs showed a heave increase greater than that of the ground surface as of a given date. Kokkonen (1930, 97–8) regarded this as reflecting progressive freezing of unfrozen water with increasingly low temperature but the cause was probably short-term freeze-thaw cycles between observation dates and perhaps upfreezing by ice segregation at the base of the pegs pushing them up (cf. discussion of frost push under Mechanisms later in this section).

4.11 Experimental Site 11, Mesters Vig, Northeast Greenland. View west, 27 August 1956 when dowels installed (*cf. Washburn, 1969a, 69, Figure 43*)

insertion, the less the heave but his observations related to relatively shallow frost penetration. Czeppe (1959; 1960) working in Spitsbergen found that wood pegs inserted to a depth of 15 cm were heaved out within a year, whereas the annual heave of pegs inserted to a depth of 35 cm was about 5 cm and of those at a depth of some 60 (or 65) cm was about 10 cm. Czeppe (1966, 70–1) also reported that targets inserted to depths of 20–35 cm heaved 5 cm in a year whereas those reaching depths of 40–50 cm heaved 10–11 cm (results from another site were inconsistent). Czeppe (1959; 1960) concluded that only the shallowest targets had been appreciably heaved by autumn freeze-thaw cycles and that the heave of the longer targets by the annual cycle was proportional to their depth of insertion.[6] However, the effective change

of level between some targets and the ground may have been consequent on spring thawing and collapse of mineral soil around the target pegs. Somewhat similar experiments were carried out at Signy Island in the Antarctic by Chambers (1967, 18–19), who used a reference frame and rods that were sleeved so that they did not move until the freezing front reached their depth of insertion. Among other things, this experiment showed that some of the deeply seated rods moved down rather than up during freezing, as if positive soil movements in some places were being compensated by negative movements in others (Chambers, 1967, 12–15). Further work is required to confirm and explain such behaviour; conceivably the negative movements resulted from desiccation by withdrawal of soil water to form ice lenses in the positive areas

[6] Although Jahn (1961a, 9) reported that these and related experiments showed that depth of insertion had no essential effect on amount of target heaving, the apparent discrepancy is explained by the fact that Jahn was comparing the shallowest targets, which were affected by spring freeze-thaw cycles, with the deeper targets subject to the annual cycle only.

Table 4.3. Target heaves, Experimental Site 7, Mesters Vig, Northeast Greenland (*cf. Washburn, 1969a, Table E IV*)

Target no. and dry or wet	13 Aug 1956 Depth cm	1957 Heave (noted)	8 Aug 1958 Heave cm (estimated)	21 Aug 1959 Heave cm	25 Aug 1960 Heave cm	21 Aug 1961 Heave cm	23–26 July 1964 Heave cm	Remarks
1 D	20		3.0	1.0	1.5	2.0	1.0	
2 D	10		2.5	1.5	1.0	1.0	1.0	
3 D	20		0.5	1.0?	2.0?	2.0?	0.5	
4 D	10	x	1.0	0.5	0.5	1.0	1.5	0.5-cm void at base
5 D	20	x	1.0	0.5	1.0	0.5?	3.5	
6 D	10		0.0	1.0?	1.5	1.0	0.0	
7 D	20		1.0	0.5	1.0	1.0	2.0	
8 D	10		0.5	0.0	0.0	0.5	0.0	
9 D	20		0.5	0.5?	0.5?	1.0?	0.5	Disturbance noted July '64
10 D	10		1.5	0.5?	0.5?	1.0		
11 D	20		0.0	0.5?	0.5?	0.5?	0.5	
12 D	10		1.5	1.0?	2.0?	1.5?	0.0	
13 D	20	x	3.0	1.5	1.0	1.5	1.5	0.5-cm void at base
14 D	10		0.5	0.5	1.0	1.0	0.5	
15 D	20		0.5	0.5	0.5	0.5	1.0	
16 D	10		0.0	0.5?	0.0	0.0	0.5	
17 D	20	x	1.5	0.5	1.5	1.5	2.0	Small pebble at base
18 D	10		1.0	0.5	0.5	0.5	0.5	
19 D	10		0.0	?	0.0	0.0	0.0	1.9-cm void at base
20 D	10		0.0	0.0	0.0	0.0		Disturbance noted July '64
21 D	20		0.5	0.5	0.5	1.0	1.5	
22 D	10		1.0	0.5	0.5	0.5	0.5	
23 D	20	x	2.0	2.0	2.0	1.5	1.5	
24 D	10		1.5	0.5	0.5	1.0	0.5	
25 D	20		0.5	0.5	0.5	1.0	2.0	
26 D	10		1.0	0.5	0.5	1.0	1.0	
27 D	10		0.5	0.5	0.5	0.5	1.5	
28 D	10		0.5	0.0	0.0	0.5	1.0	
30 W	10		0.0	?	?	?	0.0	Large pebble at base
31 W	20	x	3.0	1.5	0.0	2.5	8.0	6-cm void at base
32 W	10		1.5	1.0	1.5?	1.0	0.0	
33 W	20		0.0	1.0?	?	1.0?	4.5	
34 W	10		2.0	0.5?	0.0	0.0	0.0	Well vegetated
35 W	20	x	3.0	3.0	3.0	4.0	5.0	Well vegetated
36 W	10		0.0	0.5?	0.0	0.5	0.0	Disturbance noted July '64
38 D	10		0.0	?	0.0?	?		Well vegetated
39 D	20		0.0	?	?	?	0.5	
40 D	10		0.0	?	?	0.0?	0.0	
Mean (with standard error)								
Questioned occurrences excl.								
10-cm targets								
Dry			0.7 (0.2)	0.5 (0.1)	0.5 (0.1)	0.6 (0.1)	0.6 (0.1)	
Wet			0.9 (0.5)	1.0 (0.0)	0.0 (0.0)	0.5 (0.3)	0.0 (0.0)	
20-cm targets								
Dry			1.1 (0.3)	0.8 (0.2)	1.1 (0.2)	1.3 (0.1)	1.4 (0.2)	
Wet			2.0 (1.0)	2.3 (0.8)	1.5 (1.5)	3.3 (0.8)	5.8 (1.1)	

4.12 Experimental Site 11, Mesters Vig, Northeast Greenland. View west, 20 July 1960 (*cf. Washburn, 1969a, 69, Figure 44*)

as suggested by J. R. Mackay (personal communication, 1978). From field observations in Illinois, D. L. Johnson and Hansen (1974, 91) concluded that cylindrical objects oriented normal to the freezing front move upward faster than spherical objects.

Experiments in Northeast Greenland (Raup, 1969, 21–6, 83–93, 122–3, 143–6; Washburn, 1969a, 58–81) utilized targets of two types – cone targets and dowels. The cone targets were mounted on wood pegs 1.5 cm in diameter. They were designed for theodolite observations of mass-wasting and were generally aligned across a slope at intervals of 2 m, and inserted in the ground to alternate depths of 10 and 20 cm. The dowels were short sticks 0.3 cm in diameter. They were commonly spaced at intervals of 10 or 20 cm and inserted to depths of 5 or 10 cm in lines or grid patterns to record frost-action and mass-wasting effects as measured with reference to strings or wires.

Several of the experimental sites and the heave data from them are illustrated in Figures 4.10–4.12 and Table 4.3.

The data clearly show that moisture content of the mineral soil, the vegetation, and depth of target insertion were critical variables controlling target heaving. Temperature conditions, although obviously important, were sufficiently similar at any one experimental site, and between most sites, to be considered constant. Grain size at most sites was also sufficiently similar to be considered constant. Exceptions were several sites that were poor in fines and also characteristically dry so that the relative importance of grain-size and moisture conditions at these sites could not be well determined.

The critical importance of moisture was strikingly apparent in that the greatest target heaves were confined to characteristically wet places. However, vegetation was a controlling factor where moisture conditions were similar, since target heaves were consistently greater in thin tundra than in more richly vegetated areas.

Depth of target insertion influenced heaving in that, almost without exception, 10-cm targets heaved more than adjacent 5-cm targets, and 20-cm targets heaved more than adjacent 10-cm targets. Not all target heaves were progressive from year to year; in fact some targets tended to drop back slightly in some years, probably due to the weight of the target and/or that of overlying snow while the ground was thawed.

Most of the heaving probably occurred in the autumn as indicated by the following facts: (1) The greatest heaving was commonly in places that tended to remain wet throughout the summer because of lingering snowdrifts and were therefore particularly favoured with moisture during the autumn freeze-up; (2) most of these places tended to be protected from spring freeze-thaw cycles by lingering snow; (3) even places that were commonly snow free and wet in the early spring showed relatively little heaving; (4) low temperatures following thawing were more characteristic of autumn than spring.

The field experiments of Chambers, Czeppe, and Kokkonen, and those from Northeast Greenland show that given sufficiently deep frost penetration for peg heaving to overcome retarding friction, progressively greater depths of peg insertion correlate with greater heaving, other factors remaining equal. The field observations are also supported by laboratory experiments with wood parallelepipeds (D. L. Johnson, Muhs, and Barnhardt, 1977, 135-9, 144-5) and with glass laths, discs, and spheres. Laboratory observations by Burrous (1977) confirm Hamberg's (1915, 609) conclusion that the maximum up-freezing an object undergoes during a single freeze-thaw cycle is proportional to the object's 'effective height', which is the vertical dimension of the buried portion frozen to, and therefore heaved

with, the adjacent material. In areas of relatively shallow freezing, long dowels may be sufficiently restricted from heaving so that shorter dowels heave more, as observed by D. L. Johnson and Hansen (1974, 83-5) but their data also show that dowels 12-17 cm long tended to heave more than shorter dowels. Dowel diameter seemed to have little effect. Of course, if objects extend from the active layer into permafrost and are securely anchored there or are anchored by upfreezing from the permafrost table, they will not heave. On this principle, houses and other structures, such as pipeline supports, in permafrost environments are often built on deep-seated piles, and to assure good anchorage in 'warm' permafrost the piles are sometimes refrigerated.

c Other objects   Many kinds of objects are subject to upfreezing. In far northern graveyards it is not unusual for coffins to break the ground and become exposed, as the writer has observed on Herschel Island in Arctic Canada. Upfreezing of artifacts can assist the archaeologist in finding them but can also be seriously misleading by changing their stratigraphic position (D. L. Johnson and Hansen, 1974, 90; D. L. Johnson, Muhs, and Barnhardt, 1977, 145-6).

d Mechanisms   The mechanics of upfreezing is poorly known, as was recognized by Schmid (1955, 23, 86-91). Bertil Högbom (1910, 53-4) suggested that when ground expanded during freezing it carried stones with it but, in contracting during thawing, fines adhered to each other and left the stones behind. Repetition of the process would thus lead to ejection of stones from fines, although Bertil Högbom (1910, 49) apparently did not specifically apply the hypothesis to upfreezing of stones as opposed to lateral sorting in patterned ground. Hamberg (1915, 603-10) argued that this explanation for stones not returning to their original position was inadequate. He elaborated the hypothesis that stones were pulled vertically with expansion of fines during freezing and did not return all the way during thawing because thawed material collapsed around them while their bases remained frozen. Beskow (1930, 626-7), adopting Hamberg's explanation, stressed the importance of growth of ice lenses in the process; he added the suggestion that stones, in addition to being hindered from dropping back by slumping in of material during thawing, would also be hindered because cavities left by stones as they were heaved would tend to be narrowed by frost thrusting during freezing (Figure 4.13). Vilborg (1955) emphasized slumping in of sand and gravel during the heaving. Hamberg failed to

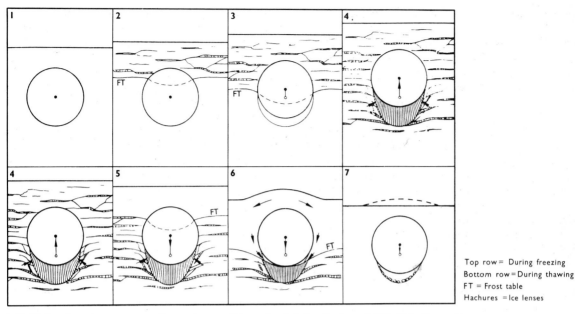

4.13 Upfreezing of stones according to frost-pull hypothesis (*after Beskow, 1930, 627, Figure 4*)

discuss the formation and filling of such cavities, although his hypothesis logically entails these consequences; in any event the addenda cited strengthen this aspect of what can be called the frost-pull hypothesis. Time-lapse photography has demonstrated several stages of the frost-pull mechanism in the laboratory (Figure 4.14). Based on this mechanism the maximum distance a stone would be heaved in one cycle of freezing would be approximately (Kaplar, 1969, 36)

$$D = \frac{H_R L}{R_f}$$

where D = vertical distance (mm), H_R = heave rate (mm day^{-1}), L = vertical height of stone below its greatest horizontal diameter (mm), R_f = rate of frost penetration (mm day^{-1}). Thus large stones (with large L) would tend to heave more than small stones, a circumstance consistent with Pissart's (1977a, 144) inference from field experiments that large stones move upward more rapidly than small stones.

Cailleux and Taylor (1954, 32), Grawe (1936, 177), Bertil Högbom (1914, 305), and Nansen (1922, 117–18) argued that upfreezing is explained by the greater heat conductivity of stones than fines, whereby ice would form around stones (Högbom) or at their base (Cailleux and Taylor, Grawe, Nansen) and force them up, and they thought that

seeping in of fines during thawing would prevent stones returning to their original position. The above concept can be called the frost-push hypothesis; it was proposed again, apparently independently, by Bowley and Burghardt (1971). Their laboratory experiments showed a correlation between rate of upfreezing and number of freeze-thaw cycles, but they did not discuss the possibility that the frost-pull mechanism caused the heaving, as it demonstrably did in Kaplar's (1965; 1970) experiments.

Streiff-Becker (1946, 154–5) argued that stones are kept from sinking back by preservation of ice at their base as the result of their poor conductivity, but Schmid, like Bertil Högbom and Nansen, pointed out that stones are better conductors than finer material; similarly, it has been stressed that diffusivity, which takes account of heat capacity as well as conductivity, also leads to stones being better conductors (cf. Bowley and Burghardt, 1971; Washburn, 1956a, 808; 1956b, 855–6). Only to the extent that a sizeable accumulation of ice would delay thawing because of its latent heat of fusion would there be a tendency for such accumulations to persist. The result pictured by Streiff-Becker is therefore misleading. Ice and frozen ground may support a stone at its base while thawing proceeds near its top, but this would be despite, not because of, differences in conductivity between stones and finer material.

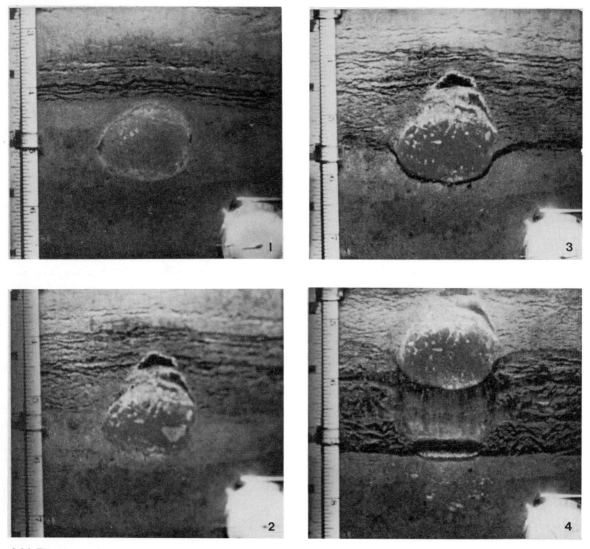

4.14 Time-lapse photography of upfreezing stone. Scale in inches (*after Kaplar, 1965, 1520, Figures 1–4*)

Following lifting of a stone by the frost-pull mechanism, thawing of soil above the bases of stones could lead to collapse of the ground around their upper portions and thus contribute to their upfreezing, and if very near the surface would leave the upper portions of stones projecting. However, where edgewise stones project in an area where the tops of many non-edgewise stones are at the surface without projecting markedly, such an explanation is dubious.

The shape and orientation of stones probably influences their behaviour significantly (Figure 4.15). For instance, frost-heaved, wedge-shaped stones with their narrow end down would not readily sink back upon thawing of the ground. Because tabular stones that are partially emergent at the ground surface are commonly nearly vertical (Figure 4.9), it is believed that upfreezing characteristically causes such stones to become oriented with their long axes normal to the freezing surface. Completely buried dowels are known to tend to rotate towards the vertical (D. L. Johnson, Muhs, and Barnhardt, 1977, 139–45). Suggested mechanisms call upon application of Stokes Law (modified), lateral compression during thawing (Cailleux and Taylor, 1954, 33, 54), or force couples set up during freezing (Figure 4.16) (Pissart, 1969; 1973*a*, 102–5;

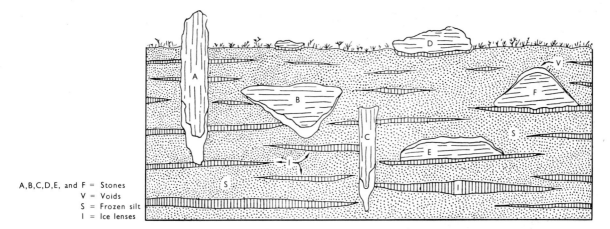

A,B,C,D,E, and F = Stones
V = Voids
S = Frozen silt
I = Ice lenses

4.15 Diagram of upfreezing stones (*after Taber, 1943, 1453, Figure 3*)

Schmid, 1955, 98). That completely buried objects can also rotate in the opposite direction is shown by Burrous' (1977) laboratory experiments. In these experiments with glass objects, all discs and lath shapes that were buried at a 45° angle at a depth of 9.1 cm rotated towards the horizontal while moving upward and (except for a broken lath fragment) failed to emerge at the surface within 32 freeze-thaw cycles when the experiment was terminated, whereas all inclined objects except small discs that were originally buried at a depth of 1.4 cm were extruded with little or no change in angle within 10 cycles. All of the more deeply buried, long (6 cm) glass laths broke at about a third of the distance from their upper end, presumably in response to an imposed force couple. All objects that were initially horizontal moved but slightly and failed to emerge within the 32 cycles.

Schmid (1955, 90–1) referred to an upward pressing of stones into still unfrozen ground by formation of ice at their base but omitted details as to how the freezing isotherm could reach the base of a stone without having already passed the level of the top and frozen the adjacent material. Because Taber (1943, 1455) believed that stones cannot be forced through overlying frozen soil, he rejected the view that growth of ice beneath stones could explain their upfreezing. However, the possibility of a buried stone being forced into still unfrozen fines is favoured by (1) the fact that freezing is partly a function of grain size so that silt and clay freeze solid at a lower temperature than coarser material (cf. Table 4.2 and

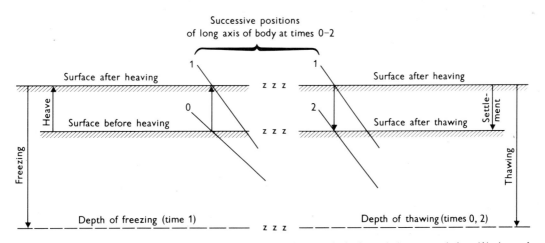

4.16 Frost heaving tending to rotate a longitudinal body towards a vertical plane. It is assumed that (1) the unfrozen soil permits rotation of the body, and (2) soil infiltration beneath it during thawing prevents body from settling back to original position

discussion under Freezing process in this chapter), and (2) upward freezing from the permafrost table. Ice lenses in soft clay have been observed in laboratory experiments (Taber, 1929, 432; 1930*a*, 311), and similar occurrences in the field have been reported in material with temperatures of −4° to −5° (Holmquist, 1898, 418) and as low as −15° (Terzaghi, 1952, 14–15). In frozen clay there can be 15 to 20 per cent unfrozen water at −10° (Tsytovich *et al.*, 1959, 111–12, Tables 2–3, Figure 5; 1964, 7; 98–9, Tables II–III; 105, Figure 5; cf. Tsytovich, 1958). In general, unfrozen water in fine-grained mineral soil at temperature ranges to −25° can cause appreciable variations of strength with temperature (Lovell, 1957), although the relationship is complex (Yong, 1966). Partial thawing as well as freezing is frequently cited in the Soviet literature (cf. Savel'yev, 1960, 163–6). It may be significant that frost-heaved stones are commonly in materials having an appreciable content of fines. To the extent that freezing takes place from a permafrost table upward, as well as from the surface downward, it would also promote frost heaving of stones into unfrozen material. Upward freezing of the frost table occurs in some permafrost areas as reported long ago by Domrachev (1913, cited in Shvetsov, 1959, 81–2; 1964, 8–9), and it is supported by recent observations as reviewed elsewhere (Washburn, 1967, 138–40), but the magnitude of the phenomenon is not well known. Fritz Müller (1954, 130–1) reported freezing at depth to be more rapid than at the surface at Ella Ø, Northeast Greenland. That a stone can be lifted by bottom-upward freezing in the laboratory has been demonstrated by Mackay who reported (J. R. Mackay, personal communication, 1977)

In my upward-freezing experiments [Figure 4.17], I have used about 40 different rocks and carried out over 50 experiments. In upward freezing, the rock surface, if minutely rough, serves as a soil surface. The ice is unable to propagate into the small depressions, just as ice has difficulty propagating into a small soil pore. As long as there is film flow from the sides, some ice freezes and the stone is uplifted. The stone never rests freely on the ice surface but is bonded more and more strongly to it until eventually uplift is impossible. I have loaded small tile specimens (20 cm²) with weights of 2.1 kg/cm², and have not prevented uplift.

Even if material immediately above or adjacent to a stone is frozen, it would not necessarily prevent heaving of the stone, since the force of ice crystallization beneath the stone might lift overlying frozen material along with the stone. Also, if the stone were

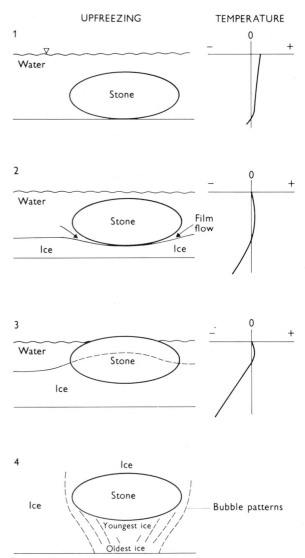

4.17 Uplift of stone by bottom-upward freezing. Laboratory experiment by *J. R. Mackay, 1977*

near the surface, localized shearing could lead to relative displacement of the stone with respect to the surrounding material. However, even without this shearing of frozen material the stone would work its way to the surface if the ice beneath it became replaced by thawed material. The effect then would be the same as the infilling of a void left by a stone that had been lifted by frost heaving of overlying material as depicted in the frost-pull hypothesis.

An effective argument against the frost-push hypothesis at one site was presented by Chambers (1967, 18) who showed that wood stakes with comparatively low thermal conductivity, and squared as well as pointed bases, heaved the same amount as elongate stones. Nevertheless, other considerations are consistent with or, in some instance, favour the frost-push hypothesis. Vegetation- and silt-capped stones (Washburn, 1969a, 55, Figures 30–1) can indicate a rather rapid forcing up of individual stones. Measurements show that the average upward movement in places can be as little as 0.6 cm ($\frac{1}{4}$ in) due to autumn and winter freezing but that the total annual amount may be up to 5.1 cm (2 in) for individual stones as the result of subsequent movement in the spring or early summer (L. W. Price, 1970, 107–8). Price suggested this later movement was the result of thawing soil settling around a stone while the base was still frozen or of upfreezing by frost push because of freeze-thaw cycles at the base. In any event, a movement that breaks the vegetation cover in the way described above suggests appreciable accretions of ice at the base of the stone and supports the frost-push concept. That ice does develop preferentially adjacent to stones in material containing fines, and especially beneath them as argued by Nansen (1922, 117), has been frequently observed (Fritz Müller, 1954, 35, Figure 76; 128, Figure 52; Washburn, 1950, 41). In line with the frost-push hypothesis, this should occur because the better diffusivity of stones than fines would lead to earlier freezing of fines in contact with stones than of fines at the same level in the ground farther away, and consequently moisture should be preferentially drawn to stones during freezing, particularly to their base from below where the supply of moisture would tend to be uninterrupted. Also stones provide discontinuities that of themselves would favour localization of ice in heterogeneous material (Beskow, 1935, 41–2; 1947, 21). Thawing as well as freezing would first occur adjacent to stones and result in a tendency for them to drop back, but material thawed around their upper portions while their bases were still frozen would tend to fill in around the stones and prevent return to their original position.

In the writer's opinion, the frost-pull mechanism is the most common. However, where there is evidence of rapid upfreezing, as in the case of some vegetation- and silt-capped stones, the frost-push mechanism may be the explanation. In any event, as shown by Mackay, the frost-push mechanism can be demonstrated in the laboratory when freezing is from the bottom upward.

As reviewed by Corte (1966b), his detailed laboratory research demonstrates that repeated freezing from the surface downward can cause an upward displacement of coarse soil particles relative to fine. However, the exact mechanism is not clear and the experiments are hardly applicable to the forceful and rapid upfreezing of some stones and to the accompanying soil shearing (Washburn, 1969a, 55, Figures 30–1).

The above analysis is focused on the upfreezing of stones from depths below the influence of needle ice, which is essentially a surface or near-surface phenomenon as discussed under Work of needle ice in this chapter. It is well known that needle ice can lift sizeable stones, as illustrated by Troll (1944, 582, Figure 10; 1958, 29, Figure 10), and it may be a factor in the upfreezing of stones whose bases are below the surface but near it. Schmid (1955, 87) maintained that the only stones affected by needle ice would be those lying on the surface.

The discussion has dealt primarily with frost heaving and the upfreezing of individual stones but frost thrusting could operate in a similar manner. This could happen wherever freezing isotherms are appropriately oriented in heterogeneous material or adjacent to contraction or dilation cracks.

It should also be noted that objects may move upward through a shallow layer as the result of cycles of wetting and drying (Cooke, 1970, 571–8; Howard, Cowen, and Inouye, 1977; Jessup, 1960, 194; Springer, 1958, 64–5). How significant and widespread the process is does not appear to be known but in an Arctic environment it is unlikely to approach the importance of upfreezing.

4 Local differential heaving

Local differential heaving is frost heaving that is significantly greater in one spot than in the area around it because of less insulation or other factors. Taber (1943, 1458–9) cited local differential heaving as an important cause of some forms of patterned ground. Under differential heaving, Taber (1943, 1452–6) included various aspects of frost heaving such as upfreezing of stones. The formation of dilation cracks by doming or irregular arching of the ground is another aspect (Benedict, 1970a). Thus local differential heaving does not specify the exact heaving process but emphasizes its local effect. Although usually in soil, local doming can also occur in fissured bedrock (Gordon Thom, 1978).

5 Work of needle ice

a General   Needle ice (Taber, 1918, 262), also known

4.18 Needle ice, Santa Maria near Curitiba, Brazil. Scale in millimetres. Note gravel lifted by needles

as Kammeis, pipkrake, and shimobashira, is an accumulation of slender, bristle-like ice crystals (needles) practically at, or immediately beneath, the surface of the ground (Figure 4.18). Elongation of the needles is perpendicular to the cooling surface (i.e. parallel to the temperature gradient), commonly producing a nearly vertical structure on a horizontal surface except for long needles that tend to curve. Anomalously, the optic axis of needles tends to be perpendicular rather than parallel to the temperature gradient (Corte, 1969a, 131; Steinemann, 1953, 502–5; 1955, 4–6).

Needles can range in length from a few millimetres to as much as 35–40 cm (Krumme, 1935, 39), depending on temperature, moisture, and soil conditions. Needles 0.5–3 cm long are quite common. There may be several tiers of needle ice separated by a thin layer of mineral soil, each tier sometimes representing a different freezing regime. In many cases tiers represent diurnal cycles but shorter or longer periods can be represented (Beskow, 1947, 7). Two tiers have been formed in the laboratory during a single freeze (Soons and Greenland, 1970, 585–7, 591–2). Needle ice can lift stones as large as cobbles and weighing as much as 15 kg (Mackay and Mathews, 1974*b*, 355), and thawing and collapse of needles can be a significant factor in sorting (Fahey, 1973, 277–8; Gradwell, 1957; Hay, 1936; Troll, 1944, 586–8, 669; 1958, 32–4, 92), in frost creep (Ellenberg, 1974*a*, 3, Figure 2), and in differential downslope movement of fine and coarse material (Mackay and Mathews, 1974*b*).

It is noteworthy that needle ice often occurs as if raked. Troll (1944, 585–6; 1958, 30–2, 36–7) reported that the direction of 'raking' is parallel to the wind. On the other hand, Mackay and Mathews (1974*a*; 1975) on the basis of experiments and repeated observations since 1960 in Canada (British Columbia) and New Zealand concluded that it is parallel to the late-morning sun and is a shadow effect developed during thawing rather than being a freezing effect as indicated by Troll (cf. Outcalt, 1970, 88). Observations in Africa support the sun hypothesis but with respect to local sunrise rather than late-morning sun (Furrer and Freund, 1973, 199, Photo 9; Hastenrath, 1973, 164–5; Hastenrath and Wilkinson, 1973, 159), and it is also supported by observations in Mexico (Heine, 1977*a*, 62) and the Peruvian Andes (Hastenrath, 1977, 358). Some of Schubert's (1973) observations in the Venezuelan Andes would seem to be consistent with both the wind hypothesis (which he favoured) and the sun hypothesis, but his arguments in support of the former are hardly convincing (Lliboutry, 1974). However, the fact that the raking effect can occur on the equator (Mt Kenya) where Troll observed a northwest-southeast orientation coinciding with nightly winds, and observations by Beaty (1974) indicative of the influence of wind rather than sun suggest that the wind hypothesis may apply in some situations even though the sun hypothesis is more generally applicable.

Grain-size analyses of mineral soil in which needle ice has been seen appear to be rather few. Gradwell (1954, 245, Table 3; 249–52, 256–7) found that needle ice was best developed in loamy soil where fines exceeded 30 per cent, although other factors were also critical. Observations in Northeast Greenland (Washburn, 1969*a*, 82–5) support the views of Beskow (1947, 6) and of several earlier investigators that a very high percent of fines may inhibit needle ice under certain conditions. Depending on the rate of freezing, water movement through a 'tight' soil can be too slow for needle growth, in the same way that restricted moisture migration can inhibit other forms of frost heaving (cf. Washburn, 1967, 102–3). Loose soil can favour growth of needle ice, and repeated growth itself loosens the soil. In places in Japan where this has led to soil deflation, a protective measure is to inhibit needle ice by compacting the soil (Ellenberg, 1974*a*, 1).

According to Outcalt (1971), three conditions must be fulfilled for needle ice to form: (1) A surface equilibrium temperature at least as low as $-2°$ for ice nucleation;[7] (2) a soil-water tension, $\gamma < \dfrac{2\sigma}{r}$, where σ = interfacial energy ice-water and r = effective radius of soil pores; and (3) the rate of heat flow from the freezing plane is balanced by the latent heat of fusion of the soil water supplying the growing ice crystal.

Needle ice is probably important in the origin of some forms of patterned ground, especially small sorted forms (cf. Hay, 1936; Heine, 1977*a*, 71; Pissart, 1977*a*, 149; Troll, 1944, 586–8, 669; 1958, 32–4, 92) but its role here is not completely clear (Brockie, 1968). Occurrences of needle ice in Greenland (Boyé, 1950, 132; Washburn, 1969*a*, 82–5), in Arctic Alaska (K. R. Everett, 1963*a*, 50–1, 135–7), and in the Canadian Arctic (Washburn, 1956*b*, 847, footnote 23) show that it is fairly widespread in Arctic regions. It is probably much more widespread elsewhere (cf. Troll, 1944, 575–92; 1958, 24–37), and its disruptive work, including the effect on vegetation (Brink *et al.*, 1967; Schramm, 1958), seems to be far more common than suggested by the relatively few observations of needle ice itself, the difference presumably resulting from the fact that the effects long outlive the process. In fact the disrupting effect on vegetation is believed to be responsible for exposing soil to wind action and deflation of turf in a variety of periglacial environments (Troll, 1973). In many places an initial absence of vegetation favours needle ice, including its development in many non-periglacial environments, even warm-arid ones (J. Hagedorn, 1975, 37). Nubbins, discussed below, are probably also due to needle ice.

b Nubbins   A nubbin, as used here, is 'a small round-to-elongate earth lump, one to several centimetres in diameter' (Washburn, 1969*a*, 85–6). Nubbins include Feinerdeknospen (fine-earth buds)

---

[7] The undercooling does not imply an air temperature $< -2°$ but can result from evaporation and thermal radiation; it is a momentary phenomenon, since once ice nucleation starts the temperature of the freezing water rises to 0° because of the latent heat of fusion. In nature, freezing usually begins close to 0° (cf. P. J. Williams, 1967, 93).

4.19 Nubbins, Mesters Vig, Northeast Greenland. Scale given by 17-cm rule (*cf. Washburn, 1969a, 87, Figure 62*)

amid stony accumulations (Furrer, 1954, 233, Figures 15–16; 1955, 149; Höllermann, 1964, 101) but they are not restricted to such features.

Nubbins observed in Northeast Greenland (Washburn, 1969a, 85–8) tended to be elongate downslope, were commonly 1–2 cm wide and 2–6 cm long, and gave the surface a broken and very irregularly wrinkled aspect (Figure 4.19). Some nubbins were bare of vegetation, others were covered with black organic crust; in several places only the crust itself had the nubbin shape. The mineral soil of a representative nubbin was gravelly$_9$–clayey$_{18}$–silt$_{36}$–sand$_{37}$ in the top 5 cm and showed no significant difference

at the 5–10 cm depth. There were a few stones at the surface but the nubbins occurred whether or not stones lay in inter-nubbin depressions. Frozen nubbins were quite porous and contained small granular ice masses. Although needle ice was not seen in the nubbins, some nubbins with black organic crust were disrupted as if by needle ice. Moreover, needle ice was observed in the same central areas that had nubbins but not in areas where nubbins were largely lacking.

Nubbins or forms very similar to them were ascribed to the work of needle ice by Furrer (1954, 233–4; Figures 15–16; 1955, 149), Heine (1977a, 62;

4.20 Gaps around stones, Mesters Vig, Northeast Greenland. Scale given by 17-cm rule (*cf. Washburn, 1969a, 89, Figure 63*)

64, Figure 4; 68, Figure 11; 73), Mohaupt (1932, 32), Fritz Müller (1954, 135, 198), and Troll (1944, 612–13; 1958, 52–3). Furrer, Heine, and Mohaupt observed needle ice in the forms but Müller and Troll did not. Although it seems probable that most nubbins owe their origin to needle ice, Stingl (1971, 28–9) reported other types.

6   Other surface effects

a Gaps around stones   Narrow gaps encircling stones at the surface of the ground are another manifestation of frost heaving (Figure 4.20). They resemble features discussed by Behr (1918, 102–10), Bertil Högbom (1914, 302–3), and Schmid (1955, 89–90). Investigations in Northeast Greenland (Washburn, 1969a, 88–90) confirm Behr's and Schmid's suggestion that these gaps can be explained by the heaving and lifting of the immediate surface material away from the stones, since freshly frozen crusts were observed to have done just that. No ice was present in the resulting gaps, so that ice wedging between the stones and the adjacent mineral soil was not a factor in these instances. The top 3 cm of the soil

had become so porous after freezing and heaving that irregular voids extended right through it. The ice responsible for the heaving was in two forms – small spots of needle ice with needles up to 0.5 cm long and, more prevalently, numerous tiny irregular lenses up to 0.1 cm thick. The frost-heave explanation for the gaps is probably generally valid with respect to stones that are large enough and of such shape that they would escape being pushed up by needle ice or pulled up by adjacent soil at the very start of freezing, particularly since partially buried stones because of their greater thermal conductivity than the adjacent soil would tend to become anchored by freezing at their base.

Many of the stones had shallow encircling gaps that were 0.3–1.0 cm wide and had a washed appearance; some of these gaps contained granules and small pebbles. Very probably these gaps originated as described and were then modified by meltwater runoff and rainwash.

b Surface veinlets   Surface veinlets are '. . . sliver-like ice crystals that grow in a thin surface layer of the ground and intersect it vertically or at large angles so that their exposed sections appear needlelike' (Washburn, 1969a, 90). Individual veinlets gener-

4.21 Slits left by thawing of surface veinlets, Mesters Vig, Northeast Greenland. Scale given by 17-cm rule (*cf. Washburn, 1969a, 89, Figure 64*)

ally taper towards both exposed ends but feathery forms occur, and branching at angles of 30° and 60° is described as common. Descriptions include those of Behr (1918, 110–12; cf. 115–17), T. M. Hughes (1884, 184), and Marbut and Woodworth (1896, 992). Hughes, and also Marbut and Woodworth, referred to needles (among other designations) but the term surface veinlets avoids confusion with needle ice.

Surface veinlets observed in Northeast Greenland (Washburn, 1969a, 90) were in clayey silt and in silty sand. Upon thawing they left characteristic slits commonly a fraction of a millimetre up to 0.3 cm wide, 1–9 cm long, and 0.5–1.0 cm deep (Figure 4.21). Drying may contribute to their width with the result that the slits become desiccation cracks as described by Marbut and Woodworth. Although the needle-like aspect is commonly emphasized because of the surface appearance, the slits left by thawing are considerably deeper than wide. J. R. Mackay (personal communication, 1977) has sug-

gested that surface veinlets result from supercooling of the surface soil, analagous to the steeply dipping plates with horizontal c axis that form in supercooled water (cf. P. V. Hobbs, 1974, 575–7).

VI Mass displacement

1 General

The term mass displacement is defined '... as the *en-masse* local transfer of mobile mineral soil from one place to another within the soil as the result of frost action' (Washburn, 1969a, 90). It will be used mainly for upward and/or downward movements of one kind of mineral soil into another but includes lateral movements. Among possible causes of mass displacement are artesian pressure, changes in density and intergranular pressure, cryostatic pressure, differential volume changes in frozen soil, thawing pressure (the Mackay effect), and irregular upward freezing from the permafrost table.

2   Artesian pressure

By definition, artesian pressure arises from differences in hydrostatic head. It differs from cryostatic pressure and changes in density and intergranular pressure in being slope induced, although it may aid these processes where they operate on slopes. Artesian pressure of water-saturated soil beneath a frozen or dried surface is perhaps a more common mass-displacement process than is usually recognized. In general, saturated taliks favour the possibility of liquefaction (cf. Finn, Yong, and Lee, 1978).

Selzer (1959) observed the growth of isolated earth mounds up to 60 cm high, consisting of unfrozen stony mineral soil that was apparently injected upward from beneath a frozen surface layer some 50 cm thick. Eye-witness accounts of such soil movements are very rare, and detailed field and laboratory data are largely lacking. Shilts (1973, 9–11) observed that a number of nonsorted circles (mudboils) in till in the Hudson Bay region, northern Canada, showed upward injections (diapirs), which he interpreted as having formed by artesian pressure beneath a rigid surface layer of vegetation and desiccated soil, perhaps aided by cryostatic pressure. He also showed that the soil had low liquid limits (commonly 10–20 per cent moisture) and was very prone to liquification (Shilts, 1974). He subsequently stressed even more strongly the role of artesian or hydrostatic stress in the origin of both nonsorted and sorted circles in central Keewatin (Shilts, 1978). There is abundant evidence that artesian flow of water beneath frozen ground is a very common process. Its role in growth of pingos is discussed later.

3   Changes in density and intergranular pressure

In soil mechanics it is well known that

The stresses that act within a saturated mass of soil or rock may be divided into two kinds; those that are transmitted directly from grain to grain of the solid constituents, and those that act within the fluid that fills the voids. The former are called *intergranular pressures* or *effective stresses*, and the latter *porewater pressures* or *neutral stresses* ... only the intergranular pressures can induce changes in the volume of a soil mass. Likewise, only intergranular pressures can produce frictional resistance in soils or rocks. (Peck, Hanson, and Thornburn, 1953, 58.)

The concept of effective stress is also discussed under Freezing process earlier in this chapter.

Where rock particles in frozen ground are separated by ice lenses and do not touch, density relationships are changed and melting of the ice can radically alter the pressure relationships. Whereas formerly only intergranular pressures existed, significant porewater pressures can be generated (cf. Chamberlain and Blouin, 1976, 2–7). If more meltwater is produced than can be accommodated in pore spaces or escape rapidly enough to effect normal thaw consolidation (in soil-engineering terms), porewater pressures will keep some of the soil particles separated, reduce shear resistance caused by friction, and can cause liquefaction of the soil that was associated with the ice lenses (cf. R. F. Scott, 1969, 58). Theoretically this effect can be evaluated by means of the thaw-consolidation ratio (McRoberts and Morgenstern, 1974a, 456–8; Morgenstern and Nixon, 1971)

$$R = \frac{a}{2\sqrt{c_v}}$$

where R = thaw-consolidation ratio, $a = \dfrac{d}{\sqrt{t}}$ with

d = depth of thaw and t = time (s), and c_v = coefficient of consolidation (cm^2 s^{-1}). To apply the model, the assumption is made that a is governed by the Stefan equation (cf. discussion of Seasonally frozen ground in chaper on Frozen ground) in the form (Nixon and McRoberts, 1973, 443)

$$d = \sqrt{\frac{2\,k_u\,T_s}{L}}\,\sqrt{t}$$

where d = as above, k_u = thermal conductivity of thawed soil (cal cm^{-1} s^{-1} $^\circ$C^{-1}), T_s = step temperature causing thaw ($^\circ$C), L = volumetric latent heat of frozen soil (cal cm^{-3}), and t = time, so that

$$a = \sqrt{2\,k_u\,T_s\,/\,L}$$

The thaw-consolidation ratio indicates the relative influence of the rate of thaw-water production at the thaw interface and the rate of thaw-water expulsion from the overlying thawed soil. Under these circumstances the overlying soil if sufficiently heavy will tend to sink down and displace any underlying, liquified and less–dense soil (especially fines, and possibly small stones), which would tend to rise towards the surface. Whether the overlying heavier soil would act as a mass or as individual particles would depend on conditions but large stones would most probably act as individual particles.

The process would be repetitive year after year as ice lenses redeveloped and thawed again, and the upward movement of material could be by small increments. In addition to melting of ice lenses, other less general changes such as slumping and drainage modifications could also lead to increased porewater pressures with similar effects. How delicate the equilibrium can be between stones and underlying fines is illustrated by stones sinking into mud 5 m from where a person was disturbing the ground by walking (Dybeck, 1957, 144). Disturbance of the ground by walking has also led to upwelling of mud that formed a nonsorted circle, a variety of patterned ground discussed in the next chapter.

I heard a sound behind me. When I looked I observed mud spurting up from one spot along the path we had just taken. The mud continued to pour out for at least 15 minutes until it covered an area about 30 cm in diameter, forming a nonsorted circle essentially identical to the others nearby. (J. R. Reid, 1970*a*.)

On a slope such hydrostatic pressures would be aided by a hard desiccated surface and artesian effects, as suggested by Reid in reporting the above case. A similar upwelling caused by disturbance was reported by French (1976*a*, 43; 44, Figure 3.12).

Mortensen (1932, 421–2) was among the first to suggest the hypothesis that moisture-controlled density differences are important in the origin of patterned ground but he thought of the process as convection. Related significant contributions include those of Jahn (1948*a*, 36–41, 89–100; 1948*b*, 52, 54, 57–8) and Sørensen (1935, 32–53). The present concept (cf. Washburn, 1956*b*, 855) is consistent with (1) observations of plugs of finer material, some characterized by vertically oriented elements, intruding coarser material (Bunting and Jackson, 1970, 200–7; Cook, 1956; Corte, 1962*b*, 12, 16, 25; Figures 100*a*, 100*b*; 1963*c*, 18, 20–1, 36; Figure 11; Mackay, 1953, 34–6; Washburn, 1956*b*, 844); (2) experiments by Anketell, Cegła, and Dżułyński (1970), Cegła and Dżułyński (1970), and Dżułyński (1963; 1966, 16–21), the last outlining an essentially similar concept but with less emphasis on melting of ice; and (3) experiments by Jahn and Czerwiński (1965), who stressed equilibrium disturbances introduced by freezing and thawing. As described by Dżułyński and his co-authors, differing viscosities as well as densities of the materials

constitute an important factor. The hypothesis has the advantage over the cryostatic concept in that intrusion is into unfrozen material, and it accounts for displacement of fines into coarse material such as beach gravels that tend to remain free of ice and unconsolidated, however low the temperature. On the other hand it would hardly explain the upward displacement of coarse material along the margins of sorted circles to provide the sorted aspect as depicted by Fedoroff (1966*a*, 141).

The hypothesis is particularly applicable to some types of patterned ground (cf. Chambers, 1967, 20; Rohdenburg and Meyer, 1969) and to interpenetrating beds (involutions), discussed in the next chapter. Field testing of the hypothesis presents major difficulties, and the ideas may be more amenable to detailed laboratory investigation. Recent experiments at the Periglacial Laboratory of the University of Washington's Quaternary Research Center support the hypothesis as applied to involutions (Washburn, Burrous, and Rein, 1978). Although primarily considered in relation to the active layer, the concept has also been invoked for diapiric features attributed to thawing of permafrost (Eissmann, 1978).

Some investigators, for instance Kostyayev (1969), stressing the role of a high-moisture environment whether or not it is ice-induced, have argued that density-controlled mass displacements explain a variety of patterned ground forms and that therefore such forms are azonal and have no necessary periglacial significance. This argument in an extreme form can be countered by pointing out the lack of typical periglacial-type forms of patterned ground in other environments. On the other hand, pending further research, differences of opinion will continue as to the significance of density-controlled mass displacements as a periglacial process.

4   Cryostatic pressure

Cryostatic pressure was originally described as the hydrostatic pressure set up in pockets of unfrozen material trapped between the downward-freezing active layer and the permafrost table when the active layer becomes irregularly anchored to the permafrost table by freezing to it in some places sooner than others (cf. Washburn, 1956*b*, 842–5). An impermeable and resistant surface like an upward freezing layer, or bedrock, could have the same effect as the permafrost table. Quite generally considered, cryostatic pressure as a concept can be extended to apply not only to the active layer but also to pressures that may be generated during freezing of thicker horizons (cf. discussion of closed-system pingos in next chapter).

Cryostatic pressure is a cause of heaving as noted by S. W. Muller (1947, 21, 68) and is obvious in small ponds where water trapped beneath the downward-freezing ice and the permafrost table freezes and bulges the ice into mounds up to 1–2 m high, as reported by Black (1973b, 19) from the Antarctic and as the present writer has also observed there. Cryostatic pressure can probably cause thrusting, and was favoured long ago by Kessler (1925, 96) as a cause of periglacial involutions, discussed later. It has been invoked by Philberth (1960; 1964, 142–7, 169–90) to explain the regularity of some forms of patterned ground, and by Corte (1967a) to explain the formation, in the laboratory, of small soil mounds and intrusive phenomena. However, subsequent reports by Corte (1969b; Higashi and Corte, 1972), extending the same experiments, emphasized the role of frost heaving rather than cryostatic pressure.

The significance of cryostatic pressure as a cause of mass displacement in patterned ground remains to be determined. Direct field evidence is meagre. Field work by Shilts and Dean (1975) led them to suggest that cryostatic pressure causes subaqueous 'rib-and-trough' patterns, which are irregular to stripe-like patterns, both sorted and nonsorted, in silty till having low liquid limits, but Shilts (1978, 1054) subsequently related these features to circles (mud-boils) for which he advocated a different origin. Cryostatic pressure also appears to be the explanation invoked by Crampton (1977a, 646–7) for sorted and nonsorted nets in which displacement of material is indicated by their microfabric.

However, the ice lenses formed by the downward-moving freezing front from the surface and the upward-moving front from the permafrost table tend to desiccate the intermediate soil as noted by Vtyurina (1974, 70). Moreover, although field measurement of pressures in freezing and thawing soils is difficult, work by Mackay and MacKay (1976) showed that the desiccation during freezing inhibits plasticity of the unfrozen material and that pressures in clayey soils appear to be much less during freezing than thawing. In fact the pressure readings during thawing were so high that Mackay and MacKay (1976, 895) regarded them as suspect but perhaps caused by swelling of clay in line with variations in the coefficient of thermal expansion of fine-grained soils as reported by Russian investi-

gators and discussed below in the section on Differential volume changes in frozen soil. Later work by Mackay (1979) supported high pressures during thawing.

On the other hand, in places where there is water expulsion during freezing (as reviewed earlier in this chapter in discussing critical conditions for ice segregation), the water content of the unfrozen material could be increased, leading to its increased mobility under pressure. The variable freezing temperatures resulting from variations in grain size and moisture content, and the presence or absence of water expulsion from adjacent sediments (which is itself dependent on grain size and other factors) are among the critical variables that would control cryostatic pressure. The above view was advanced by J. R. Mackay (personal communication, 1977) in suggesting that cryostatic pressures are therefore confined to coarse-grained soils where porewater expulsion would be expectable.

It has been argued by Cholnoky (1911, 130–1), Eakin (1916, 80–1), and most recently in a more detailed way by Nicholson (1976, 333–9) that irregular freezing fronts in certain, *already-established*, forms of patterned ground can cause a pressure in unfrozen soil, even without invoking anchoring of the active layer to the permafrost table, and that a type of circulatory movement can result whereby the unfrozen soil is forced down and laterally from under the border of the forms towards the central areas and then upward. This concept would accord with evidence of lateral movement from beneath borders towards centres as reported from permafrost areas by Crampton (1977a, 643–4) and Fedoroff (1966a, 137, Figure 5; 153, Figure 7; 161, Figure 9; 1966b, 91, 94) among others. That an upward movement occurs in the central area of some forms is well established as noted in the next chapter in discussing mass displacement in the origin of patterned ground. However, the validity of the process as outlined above is not proved and suffers from the difficulties that (1) in the absence of anchoring of the active layer, pressure on unfrozen mobile material would tend to lift the frozen overburden as a unit (cf. Washburn, 1956b, 845); and (2) any buckling of the active layer would tend to cause stress relief by shearing along the contact between frozen and unfrozen soil as in folding of competent and incompetent strata (a point advanced by J. R. Mackay, personal communication, 1977).

A view somewhat similar to Nicholson's was advanced by Schunke (1977c, 282) in partial explanation of hummocks (thúfur) in Iceland, in which he stressed the effect of lateral pressure arising from freezing of interhummock areas. He referred to this lateral thrusting as cryostatic but lateral advancement of freezing fronts could conceivably achieve a buckling without involving a hydrostatic effect.

The concept that irregular freezing fronts without permafrost could lead to cryostatic pressures and mass displacements has also been advanced in explanation of bulbous pockets of sandy clayey silt 1–8 m wide and some 2 m deep penetrating downward into gravel (Conant, Black, and Hosterman, 1976). Although former permafrost and other alternative explanations were considered unlikely by these investigators, it is difficult to understand how mass displacement of the kind described would relieve stress more easily than simple differential heaving, and the present writer regards the origin of the features as unresolved.

Laboratory experiments concerning cryostatic pressure are also rather inconclusive. Pissart (1970a, 40–4) demonstrated that pressures set up within a freezing mass, presumably as the result of some areas within the mass remaining unfrozen longer than others, are up to 4 kg cm^{-2}. These pressures may be basically cryostatic in being caused by confined freezing in a closed system but the system can be local as part of a larger open system. Thus, as stressed by Pissart, pressure effects of this restricted kind are to be expected in seasonal freezing in a non-permafrost environment as well as in a permafrost environment.

5　Differential volume changes in frozen soil

The water in fine-grained soils may freeze through an appreciable subfreezing temperature range as discussed at the beginning of the present chapter, with concurrent effects on volume changes in frozen soils (cf. Elkhoraibi, 1975; P. J. Williams, 1976). In clayey soils such variations may help to account for apparently anomalous variations in the coefficient of thermal expansion of frozen soils as reported by Soviet investigators, inasmuch as some of the values greatly exceed that of pure ice (c. 45 × 10^{-6} °C^{-1} at −40°) (P. V. Hobbs, 1974, 347, Figure 5.1; 349, Figure 5.2). According to Votyakov (1966, 14, Table 1), the coefficient can be as much as 345 × 10^{-6} °C^{-1} for a clay with a moisture content of 40 per cent undergoing a temperature change from −0.4° to −1.6° (reported as an expansion in going from

−1.6° to −0.4°). According to subsequent reports, the coefficient can exceed 2000 × 10^{-6} °C^{-1} for clay and be 100–400 × 10^{-6} °C^{-1} for loams and sandy loams (cf. Grechishchev, 1973a, 27; 1973b, 29; 1978, 229). In frozen clayey soil the linear coefficient of thermal contraction increases with temperature and fineness ('disperseness') of soil particles, and has maximum value at a moisture content corresponding to the plastic limit (lower plasticity limit) or maximum molecular moisture capacity for the thawed state (Votyakov, 1973a, 80; 1973b, 102; 1978, 266). Similarly, Shusherina, Rachevskiy, and Otroshchenko (1970, 278–9, Tables 1–2; 282) reported that contraction in Nikol'skii loam (grain size: 5.4% > 0.1 mm; 67.7% 0.1–0.005 mm; 26.9% <0.005 mm) decreases with increasing ice content, rather than increasing as in sand; and at 0.42 per cent saturation it is 372 × 10^{-6} °C^{-1} (cf. Romanovskiy, 1973b, 251). This is seven times or more that of pure ice at −40°. More recently Shusherina and Zaitsev (1976), examining fine sandy to silty permafrost soils with a clay content ranging from about 4 to 17 per cent, found that both expansion and contraction could occur upon cooling from −1° to −6°. In this temperature range, the critical difference appeared to depend on texture, with soils characterized by pore ice showing expansion (35 to 605 × 10^{-6} °C^{-1}) and those dominated by ice lenses exhibiting slight contraction or no detectable expansion (Shusherina and Zaitsev, 1976, 195–6). It was concluded that contraction of the ice lenses upon cooling compensated for the expansion of the soil. On the other hand in the temperature range of about −6° to −60°, all the permafrost samples contracted (up to 75 × 10^{-6} °C^{-1}), but considerably less than in the case of the artificially prepared soils studied by previous investigators. Unpublished laboratory work by P. S. Marshall (personal communication, 1977) with unsaturated frozen soils (pure Kaolin and Kaolin mixed with Fairbanks silt) supported large negative coefficients near 0°, positive values up to about twice that of ice for the silty clay but less than that of ice for the pure clay near −3°, and about half that of ice for both soils at lower temperatures of −10° to −20°.

In summarizing some of the regularities revealed by the Soviet investigations, Grechishchev (1976, 20; cf. Grechishchev, 1970, 22; 1975, 20) referred to Votyakov's work and stated

Figure 4 shows a large variety of "stabilized deformation – temperature" curves.[8] Firstly, they are distinctly non-

---

[8] Grechishchev had earlier noted (p. 19) that frozen soil continued to deform for an appreciable time after a constant temperature had been established in a sample (cf. Votyakov, 1973a, 81, Figure 2g; 82; 1973b, 101, Figure 2d; 103; 1978, 266, Figure 2d).

SOIL TYPES

Name	Clay Content % by Wt.	Minimum Diam. of Roll in Plastic Limit Procedure	Liquid Limit	Plasticity Index
Glina (clay)	>30	<1	53–68	23–24 and up
Suglinok	30 to 10	1 to 3	32–38	12–16
Supes	10 to 3	>3	22–38	1–14
Pesok (sand)	<3	N. P.	—	—

It will be remembered that 3 mm is the size for the plastic limit test.

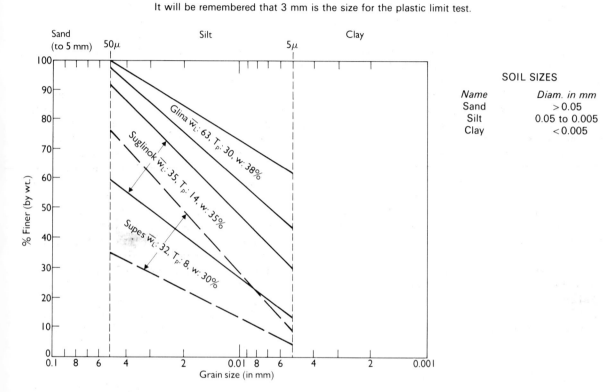

SOIL SIZES

Name	Diam. in mm
Sand	>0.05
Silt	0.05 to 0.005
Clay	<0.005

linear. Fairly steep curves are characteristic of temperatures close to 0°C. The maximums occur within the first few degrees below 0°C, and the curves level out as the temperature falls. Secondly, the sign of thermal deformation (contraction or expansion) depends on the water content and the grain-size composition of the soil. For example, kaolin clay containing 22% and 29% water undergoes considerable contraction in volume as the temperature drops from 0° to −2°C (positive thermal deformation). The same clay containing 37% water expands within the same temperature range. Supes[9] expands within any temperature range (if the initial temperature is 0°C), while a mixture of clay and sand contracts, etc. Thirdly, the sign of thermal deformation for the same soil but at different final temperatures may differ. For example, sand containing 6.5% water expands up to −3°C[10] and contracts at lower temperatures.

Votyakov's (1966) explanation for his surprising findings was that structures formed by clays and other fine particles become dehydrated when the content of unfrozen water is reduced by freezing at increasingly lower temperatures, leading to volume reduction of the soil; with rising temperatures the process is reversed. Shusherina, Rachevskiy, and Otroshchenko (1970, 283) called upon unknown processes in the freezing of clays.

Apparently independently of Votyakov, Pissart (1976c, 278–9; cf. Pissart, 1972b) indicated that vari-

[9] Supes and suglinok are Russian terms, common in translations because of lack of specific English equivalents. Both supes and suglinok are silts containing more clay than sand (i.e. a kind of loam) but supes has less clay than suglinok. The following characteristics apply (National Academy of Sciences, 1978, xi).
[10] The minus sign is erroneously omitted in Grechishchev (1975, 20).

ations in the unfrozen water content of soil could explain some volume changes during freezing and thawing (*c.* 0.35 mm progressive increase in length of a soil sample 10.5 cm long, during temperature lowering from 0° to *c.* −22° in the case cited). Furthermore he suggested that such variations could account for pressure effects and the origin of involutions and a variety of patterned-ground types, and that contrary to the commonly accepted volume expansion of frozen ground with rising temperature, unfrozen-water effects would tend to counteract the expansion and even cause fissuring of soil. Whether such effects are quantitatively adequate for the indicated results remains to be demonstrated.

6   Thawing pressure (the Mackay effect)

Earlier, in discussing cryostatic pressure as a cause of mass displacement, it was noted that desiccation of the central portion of the active layer tends to occur as water is withdrawn to form ice lenses near the surface and near the permafrost table, and that Mackay had observed higher pressures during thawing than freezing. This has led him to suggest that heaving near the surface and the permafrost table during freezing, followed by wetting and expansion of the desiccated layer and collapse of the heaved layers during thawing, lead to a pump-like circulatory movement and to soil displacements that may help to explain some hummocks (Mackay, 1979) and perhaps some circles and polygons (J. R. Mackay, personal communication, 1978). The forms might start as nonsorted circles that originated by differential frost heaving (cf. section on Patterned ground in next chapter). An important requirement is that the frost table beneath forms becomes bowl shaped as thawing approaches the permafrost table. In places this condition accords with observations by Mackay and others but its generality remains to be established. Observations must be made late enough in the year to assure maximum thaw depth, since greater thaw beneath borders than centres has been frequently reported earlier in the thawing period. That thaw depths may reverse during this period was noted by Sharp (1942a, 288–90).

7   Irregular upward freezing from the permafrost table

Conceivably upward freezing from the permafrost table can be sufficiently irregular to cause displace-

ments within the soil. In discussing involutions, Gravis and Lisun (1974, 80; cf. Gravis, 1971[11]; 1974, 139–40) mention 'freezing of the active layer from its floor' as a cause of soil injections but without details. Very little is known about the effects of upfreezing from the permafrost table, and much more research will be required before the various mass-displacement processes can be fully evaluated or, more generally, before the behaviour of the active layer is fully understood.

VII   Frost cracking

1   General

Frost cracking is fracturing by thermal contraction at subfreezing temperatures. This definition conforms to general usage in excluding dilation cracking by differential heaving, although other usages occur (cf. Benedict, 1970a; Jahn, 1975, 33). The term frost crack, as used by the present writer, carries no implication except origin, whereas in Europe it is often additionally restricted to very narrow features (*c.* < 0.2 cm, often fossil) that have not evolved into wedge forms (discussed later). The writer is grateful to Dr Johannes Karte for this terminological clarification.

Frost cracking was suggested as early as 1823 by Figurin (1823, 275–6) in explanation of large polygonal cracks in Siberia. However, not all cracking is polygonal (Figure 4.22). Following other pioneer investigators in Siberia (cf. Shumskiy, 1959, 276–9; 1964a, 5–9), especially Middendorff (1861, 505–6), the process was elaborated in considerable detail by Bunge (1902, 205–9) and Leffingwell (1915; 1919, 205–14) on the basis of their respective observations in Siberia and Arctic Alaska. Although attacked by several later workers (Pissart, 1964a; Schenk, 1955a, 64–8, 75–6; 1955b, 177–8; Taber, 1943, 1519–21), the process is accepted by most investigators including Pissart (1970a, 10–19). Taber's criticism was effectively answered by Black (1963; cf. Washburn, 1956b, 851), who measured seasonal changes in crack widths, and by the work of Berg and Black (1966) and Mackay (1972c; 1974c), who demonstrated repeated cracking and the growth of ice wedges. A sound theoretical basis for the process was elaborated by Lachenbruch (1961; 1962; 1966) and supported by field observations of Kerfoot (1972). As discussed below, frost cracking is not restricted

---

[11] Gravis (1971) was not obtained in time to review.

4.22 Linear frost crack, Mount Pelly, Victoria Island, Northwest Territories, Canada (*cf. Washburn, 1947, Figure 2, Plate 31*)

to a permafrost environment but it rarely forms well-defined and persistent features elsewhere. Frost cracking, and wedge structures described later in this chapter, have been comprehensively reviewed by Dylik (1966), Dylik and Maarleveld (1967), and Jahn (1975, 31–84) among others.

Even though frost cracking is a valid process, other cracking processes such as desiccation and dilation leave features simulating fossil frost-crack phenomena (Benedict, 1970a Dżulyński, 1965; Johnsson, 1959; Wright, 1961, 941–2). That important unresolved problems exist and much remains

to be learned about the growth of ice wedges has been emphasized by Popov (1978, 165–7).

2   Characteristics

Ice is the critical material in frozen ground. Pure ice has a coefficient of linear contraction of about 45×10^{-6} °C^{-1} at $-40°$, increasing somewhat towards $0°$, with some variability depending on crystal orientation (P. V. Hobbs, 1974, 347, Figure 5.1; 349, Figure 5.2). If the ice has a salt content, the coefficient also varies with the salinity (D. L. Anderson, 1960, 310–15). However, contraction calculations based on a simple elastic model do not conform in either time of cracking, depth or spacing of cracks, or crack closing, to field observations by Mackay (1975c, 1672–4), who concluded that the question was still open as to whether a modified elastic or a visco-elastic model would be required. According to Glen (1975, 23; cf. Glen, 1955) deformation of ice can be approximated by a power law of the form

$$\epsilon = A\sigma^n$$

where ϵ = creep rate (time^{-1}), σ = stress (bars, 1 bar = 10^6 dynes cm$^{-2} \approx$ 14.5 pounds in^{-2}), A = a constant (stress^{-n} time^{-1}) at a given temperature[12], and $n = 2 - 4$ (although n may be slightly less at stresses < 1 bar – but see Weertman, 1973, 331–2).

Aside from the type of deformation there are other complications in evaluating the effect of thermal stresses, including the fact that the ice content of frozen ground is highly variable, and that according to recent Soviet work the linear coefficient of thermal contraction can be much greater for frozen ground than for pure ice as discussed in the preceding section.

As indicated by Lachenbruch (1966, 65–6), frost cracking is more dependent on the rate of temperature drop than on the actual subfreezing temperature at time of cracking. The insulating effect of snow is an important factor here as shown by a high inverse correlation between snow depth and cracking frequency (Dylik and Maarleveld, 1967, 11; Mackay, 1974c, 1376), a sparse snow cover in places being due more to wind than to low annual precipitation. Whether edaphic aridity caused by good drainage in porous soils promotes frost cracking, as suggested by Schunke (1974b, 164), is questionable. Fine-grained, moisture-rich soils are probably highly susceptible to frost cracking, although not as favourable as gravel for preservation of ice-wedge casts as noted later. The first time cracking occurs in a permafrost environment it starts at the surface and extends to a depth of 3 m (10 ft) or more (Lachenbruch, 1970b, J1) in a pattern and spacing determined by temperature conditions and the rheological behaviour of the frozen ground, and open cracks as deep as 4 m have been observed by Mackay (1972c; 1974c, 1370). The crack spacing may be on the order of two to three times the crack depth (cf. Lachenbruch, 1962, 35–9). Most frost

cracks are in unconsolidated material, and in such material in a permafrost environment the crack pattern is primarily imprinted in the permafrost rather than in the active layer where thawing tends to destroy the pattern. Therefore subsequent cracking often starts in ice wedges at the permafrost table and is propagated upward as well as downward (Lachenbruch, 1966, 65–6; Mackay, 1974c, 1374–5). That cracking is not necessarily upward has been demonstrated by Mackay (personal communication, 1975) whose observations with electronic 'fissure' detectors at Garry Island in the Canadian Arctic showed both directions (16 down, 11 up). Where frost cracking occurs in bedrock, as reported for some nonsorted polygons (Arthur *in* Corte, 1969a, 140), or takes place in unconsolidated material where permafrost is lacking, Lachenbruch's analysis would not apply.

In a model of frost cracking presented by Romanovskiy (1973a; 1973b, 237–50),[13] based on the mean annual temperature at the base of the active layer and the annual amplitude of surface temperature, cracking occurs progressively under the influence of the maximum thermal gradient. Although a simple elastic model is assumed, an ice-soil mixture may behave somewhat differently as noted above. Lachenbruch's concept of upward cracking from the permafrost table is not included, rather the model assumes that repeated cracking from the surface downward determines the crack pattern in the permafrost, with some cracks terminating sooner than others, resulting in a two-storied crack structure. Also, the effect of unstable fast fracturing that extends to depths greater than the depth of ambient tension (Lachenbruch, 1966, 66–7) is not considered. That fracturing can be very rapid is suggested by reports of sharp explosive-like noises (Bunge, 1884, 446; 1902, 205; Leffingwell, 1919, 205; Stefansson, 1921, 282; Thoroddsen, 1913, 253; 1914, 259), but other observations indicate it can be slower (measured in seconds) than is commonly believed (Knight, 1971, 394). The quantitative effect of these factors on the model remains to be determined.

The actual temperature at which frost cracking can start, given an adequate rate of temperature drop, depends on many factors and only approximations have been suggested. With respect to air temperatures, Spethmann (1912, 246) reported observing new cracks in Iceland following a temperature minimum of about − 10°. In Finland

[12] The exp − 1 in time was inadvertently omitted in the reference cited.

[13] Further discussion, not considered in the following, appears in a monograph (Romanovskiy, 1977 – in Russian), which was received through the kindness of Professor Romanovskiy while the present volume was in press.

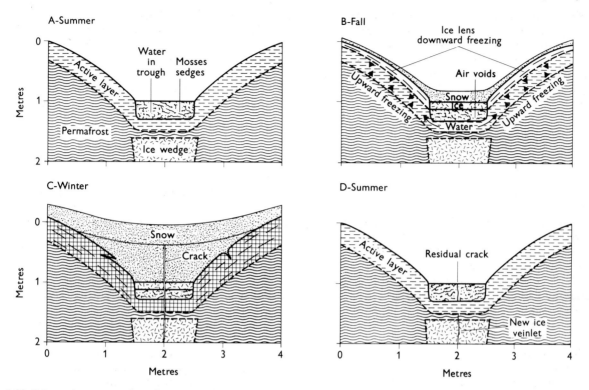

4.23 Schematic cross section of an ice wedge and overlying trough showing changes over one year, including cracking and addition of a new vein to an older ice wedge (*after Mackay, 1974c, 1378, Figure 15*)

Aartolahti (1970, 21–3) found fresh cracks in conjunction with sudden temperature drops from $-7.9°$ to $-27.7°$, $-5.9°$ to $-23.4°$, and $-0.4°$ to $-17.2°$.

The relation of frost cracking to mean annual air or ground temperatures is of great interest, but since it is especially pertinent to ice-wedge polygons discussion is deferred to the section on Nonsorted polygons in the next chapter.

Frost cracking in a permafrost environment is commonly accompanied by growth of ice wedges or sand wedges. Ice wedges, whose ice characteristics are described under Permafrost in the chapter on Frozen ground, grow as the result of surface water, ground water, or water vapour entering a fissure in permafrost and freezing. Ice wedges start as ice veins, which are the focus of subsequent cracking, except in massive ice where they do not normally become a plane of weakness for subsequent cracking as they do in soil (Gell, 1974b; 1978a). The width of ice added in any one year may be very small (Figure 4.23) – 1 to 20 mm (Shumskiy, 1964b, 198) and probably 20 per cent or less of the mid-winter's crack width in some Arctic areas (Mackay, 1975c, 1673–4) – but repeated cracking at the same place over a period

of years leads to well-developed ice wedges (Figures 4.24–4.26), some being massive in suitable material. Because of the lateral growth of the wedge, adjacent sediments commonly develop an upward bend (Figures 4.25–4.26) as the result of expansion during warming. Whether the expansion is directed outward from the wedge, as usually thought, or from the adjacent sediments (cf. Jahn, 1975, 55), is not fully established, the question reflecting the recent Soviet investigations of the coefficient of thermal contraction and expansion of frozen ground as previously discussed. Exposures parallel to the trend of an ice wedge can show continuous ice and give the impression of a continuous ice layer (Figure 4.27). Except for sections exactly at right angles to the trend of a wedge, the true width of a wedge is always less than the apparent width. Other things remaining equal, large ice wedges tend to develop in fines or in peaty soil rather than sandy material, the width of wedges in pure sand being limited to 0.1–0.5 m (Shumskiy, 1964b, 197). It has been reported that very large wedges extend to depths of 60 m along a bluff of the Yana River in Siberia (Jahn, 1975, 35), and can grow so large laterally, squeezing and

4.25 Ice wedge at Kurankh, right bank Aldan River, 170 km above junction with Lena River, Yakutia, USSR

deforming the intervening soil into columns, that the volume of ice exceeds the volume of soil (Shumskiy, 1964*b*, 199); also that adjacent wedges may even form essentially continuous stratiform masses (Figure 4.28), which may be up to 80 m thick in the New Siberian Islands and along the Arctic coast of Siberia (Gerasimov and Markov, 1968, 14; Grave,

4.24 Ice wedge, Ostrov Chagen, Lena River near Shamanskiy Bereg, 130 km north of Yakutsk, Yakutia, USSR. 15-cm rule as scale

1968*a*, 53; 1968*b*, 9). However, some stratiform ice masses described as of ice-wedge origin may well be massive beds of segregated ice such as those described by Mackay (1971*b*; 1972*a*, 15–19); 1973*a*; and Mackay and Black, 1973); this would be particularly true if the ice had a predominantly horizontal layering as opposed to vertical foliation[14]. According to Black (1976*a*, 6–7), ice wedges rarely widen to more than 5–6 m, apparently because the flow of squeezed material is entirely within a wedge

[14] An old view that stratiform ice masses are buried glacier ice can be shown to be erroneous in many areas but may apply in some places (French, 1976*a*, 82; Mackay, Konishchev, and Popov, 1979).

4.26 Ice wedges, left bank, lower reaches Yana River, USSR. Photo by A. I. Popov. Scale given by *c.* 2.5 m-high tree, top left. Grain size of sediments 60–70 per cent 0.01–0.05 mm

when the slope of its sides approaches 45°. Gasanov (1973; 1978) suggested a different limiting mechanism involving access of moisture to the wedge but his concept requires clarification. In any event the extent to which lateral growth of an ice wedge is self-limiting deserves more study. As noted by Black, presumably many wedges stop growing before reaching a limit and other wedges may start, due to changes in the thermal regime such as might be caused by variations in distribution of snowdrifts and moisture.

A point may be reached at which the thermal regime of an initial wedge is changed – the upper part of the

permafrost cooling more slowly in early winter than adjacent areas. The wedge is subsequently abandoned as a locus of thermal contraction and a new crack is initiated elsewhere. (Black, 1976*b*, 79, 81.)

Ice wedges may be syngenetic or epigenetic in the same sense that permafrost may be one or the other – that is, depending on the presence or absence of con-

4.27 Continuous ice exposure parallel to an ice wedge, left bank Aldan River, 130 km above junction with Lena River, Yakutia, USSR

4.28 Stratiform ice mass interspersed with baydjarakhs, Novosibirskiye Ostrova, USSR (*Toll, 1895, Plate III*)

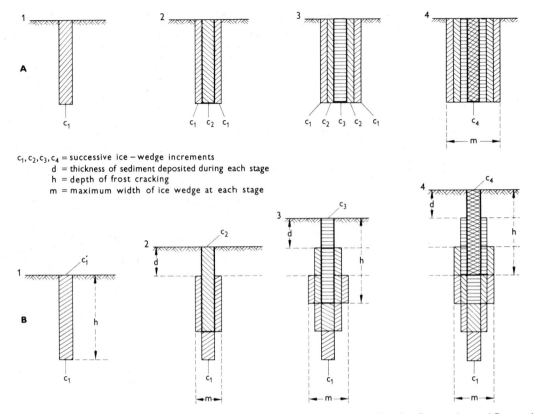

4.29 Diagram of growth of an epigenetic ice wedge (A) and syngenetic ice wedge (B) (*after Dostovalov and Popov, 1966, 103, Figure 2*)

tinued sedimentation. Wedges growing in the absence of sedimentation are epigenetic. They increase in thickness but not in the vertical dimension (Figure 4.29[A]). Growth is limited by the compressive stresses generated by repeated cracking and crack filling. Wedges growing concurrently with sedimentation are syngenetic and increase in both thickness and vertical dimension (Figure 4.29 [B]). Assuming a constant depth of cracking, the width of syngenetic wedges is limited by the expression (Dostovalov and Kudryavtsev, 1967, 203–4; Dostovalov and Popov, 1966, 103)

$$\bar{m} = \frac{hc}{\bar{d}}$$

where $\bar{m}$ = total (mean) horizontal thickness of ice wedge, $\bar{h}$ = mean depth of frost crack, $\bar{c}$ = mean width of crack at each stage, taken as half the surface width,[15] and $\bar{d}$ = mean thickness of sediment

deposited during each stage, or rate of sedimentation. Thus the maximum width of ice wedges is inversely proportional to rate of sedimentation and is attained after h/d stages, or 3 stages in Figure 4.29(B). Because of concurrent development of ice lenses, syngenetic ice wedges are more likely than most to intersect lenses in a honeycomb-like arrangement.

Sand wedges, formed by repeated frost cracking and infilling with dry sand or loam, occur in the arid parts of the Antarctic (Berg and Black, 1966, 70–3; Péwé, 1959; Ugolini, Bockheim, and Anderson, 1973), and intermediate, composite ice and sand forms also occur. Original sand wedges imply readily moveable sand and therefore in some situations a degree of aridity. It has also been argued that they imply exceptionally low temperatures because the surrounding soil would have had a low coefficient of contraction by virtue of low ice content (French,

---

[15] Dostovalov and Popov (1966, 103) do not give the reason for taking $\bar{c}$ = c/2, but as explained by Dostovalov and Kudryavtsev (1967, 203) this allows for the wedge shape of the ice, since this shape is omitted from the idealized diagram (Figure 4.29[B]).

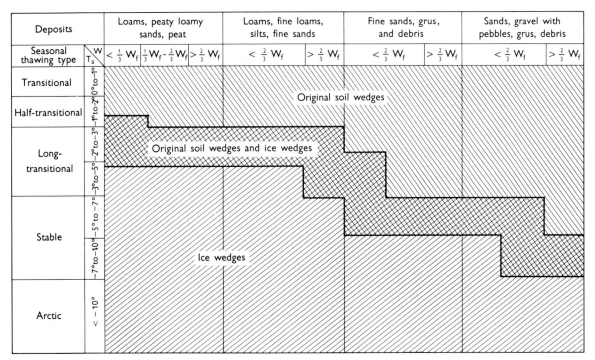

Deposits		Loams, peaty loamy sands, peat			Loams, fine loams, silts, fine sands		Fine sands, grus, and debris		Sands, gravel with pebbles, grus, debris	
Seasonal thawing type	W T_a	$< \frac{1}{3} W_f$	$\frac{1}{3} W_f - \frac{2}{3} W_f$	$> \frac{2}{3} W_f$	$< \frac{2}{3} W_f$	$> \frac{2}{3} W_f$	$< \frac{2}{3} W_f$	$> \frac{2}{3} W_f$	$< \frac{2}{3} W_f$	$> \frac{2}{3} W_f$
Transitional	0° to −1°									
Half-transitional	−1° to −2°					Original soil wedges				
Long-transitional	−2° to −3° −3° to −5°		Original soil wedges and ice wedges							
Stable	−5° to −7° −7° to −10°		Ice wedges							
Arctic	< −10°									

4.30 Theoretical correlation of ice wedges and soil wedges as related to some controlling variables. T_a = temperature at level of zero annual amplitude, W_f = frozen water in proportion to total water (*after Romanovskiy, 1976, 290, Figure 1*)

1976a, 25, 236; cf. Berg and Black, 1966, 75), but that such a low ice content originally existed could be difficult to prove. Figure 4.30 shows an interesting effort by Romanovskiy (1976) to correlate (1) temperatures at depth of zero annual amplitude (T_a), (2) frozen soil water (W_f) in relation to total water, and (3) soil type, with (4) development of ice wedges and original soil wedges.

Attempts have been made to measure growth rates of ice wedges at Point Barrow, Alaska (Black, 1963, 261–4, 268–9; 1974a, 264–74) and Garry Island, N.W.T., Canada (Mackay, 1972c; 1973c; 1974c; 1976d), and along with sand and composite wedges, in Victoria Land, Antarctica (Berg and Black, 1966; Black, 1973a; Black and Berg, 1964). The wedges form large nonsorted polygons (described in the next chapter). Not all wedges crack annually (only about 40 per cent at Garry Island, 50 per cent at Barrow), some grow narrower rather than wider in any one year, and there is considerable range in individual growth rates. On an average, the suggested rate at Point Barrow was 1 mm yr^{-1}, at Garry Island less than 1 mm yr^{-1} for wedges that cracked annually over a 10-year period (Mackay, 1976d), and in Victoria Land the rate ranged from −0.4 to 1.6 mm yr^{-1} depending on the

site (Black, 1973a, 202, Table XVI). Attempts have been made to use the width of wedges in Victoria Land as an age indicator of the surface in which they occur but there are many variables, and the method must be used very cautiously (Calkin, 1971, 386). Growth rates of sand wedges can be non-linear (Ugolini, Bockheim, and Anderson, 1973, 253), and Mackay (1974c, 1374; 1379, Figure 16; 1381) found that cracking frequency of ice wedges is a function of width, with medium-size wedges cracking the most often. Most importantly, recent and still unpublished field data by Mackay strongly suggest that reported rates of wedge growth, including his own earlier reports, '...do not reflect growth rates but only differential movements of the active layer and the top of the permafrost' (J. R. Mackay, personal communication, 1978).

Ice wedges require certain temperature conditions for growth and preservation. In Alaska, for instance, the southern limit of presently active wedges roughly coincides with a mean annual air temperature of −6° to −8° (Péwé, 1966a, 78; 1966b, 68; 1975, 52). Further environmental data are cited in the discussion of nonsorted polygons. Upon thawing of permafrost, the site of an ice wedge is filled in by collapsed material that becomes a cast of the wedge.

4.32 Ice-wedge casts just west of Eistedfo Gurig, Wales, Great Britain. Note nonsorted polygons at surface beyond fence, and upturned beds adjacent to collapse structure of ice-wedge casts

Such ice-wedge casts (Figures 4.31–4.32) are among the few acceptable criteria for former permafrost. Although fine-grained, moisture-rich material favours growth of ice wedges, it does not favour development of ice-wedge casts because such material would tend to flow rather than maintain steep faces when thawed (Black, 1965a, 220; 1969b, 229; Romanovskiy, 1973b, 262). Consequently, although ice wedges grow preferentially in fine-grained material they are best preserved in angular gravel and are most common in gravel deposits (Black, 1976a, 10). As repeatedly stressed in other

4.31 Ice-wedge cast, Zakrecie, Poland

contexts, thawing can be caused by purely local factors rather than climatic change. For instance, shifting stream channels or encroachment of other water bodies may lead to thawing of ice wedges and their replacement by subaqueous deposits (Danilov, 1973a).

The infilling of a thawed ice wedge may be sand and form a so-called sand wedge. However, as noted above, sand wedges can form instead of ice wedges under arid conditions such as those in the Antarctic dry valleys today. Thus as features of former permafrost, fossil sand wedges may record such an origin or be an ice-wedge cast associated with a moister climate. The difference in origin might be hard to establish unless the infilling is clearly eolian and contemporaneous with the frost cracking (cf. Cailleux, 1973b, 51–2, 58–9). Péwé (1959) suggested that some of the loess or loam wedges of Europe may be similar to Antarctic sand wedges rather than being ice-wedge casts formed by the thawing of ice wedges and collapse of the adjacent or overlying material.

Much ambiguity exists in the use of the terms soil wedge and soil vein,[16] inasmuch as some investigators, especially in the USSR, restrict the use of soil wedge and soil vein (and their equivalents, ground wedge and ground vein) to crack fillings that are not in permafrost (cf. Jahn, 1975, 37–9, 63). In view of the general nature of the term soil wedge and the difficulty, in places, of distinguishing between (1) active-layer soil wedges, (2) soil occupying cracks in seasonally frozen ground (i.e. in non-permafrost environment), (3) soil as original fillings in permafrost cracks, and (4) soil replacing ice wedges (ice-wedge casts), it seems best to use the term soil wedge (or vein) as the non-ice counterpart to ice wedge (or vein). Contemporary occurrences could then be unambiguously distinguished as permafrost active-layer, or seasonally frozen-ground soil wedges (or veins), and fossil occurrences could be simply referred to as such. Sand, loam, and loess wedges would then become specific kinds of soil wedges. Permafrost wedges consisting of both ice and soil (including sand) are known as composite wedges (Berg and Black, 1966, 75; Black and Berg, 1964, 111; Jahn, 1975, 74–8) and readily fit into the scheme (Table 4.4). The scheme is not confined to wedges originating in frost cracks; the constitution of a wedge is not always an indicator of the cracking process. Nevertheless, unless otherwise stated, the term ice wedge or soil wedge in periglacial literature normally implies a frost-crack filling. A more detailed classification of wedge structures is presented by Jahn (1975, 51–84).

Permafrost soil wedges formed in dry polar climates differ from ice-wedge casts in being (1) thinner, (2) vertically foliated, and (3) limited to material small enough to enter narrow cracks up to 1–2 cm wide. New cracks tend to form between older ones and with time result in the development of many vertical narrow wedges. Ice-wedge casts on the other hand, as the result of collapse of material into voids left by melting ice, tend to (1) be broader, (2) have

[16] By analogy with ice wedges and ice veins (cf. section on Forms of ice in chapter on Frozen ground), the width of soil veins is here taken to be <5 cm, that of soil wedges, ⩾5 cm.

Table 4.4. Wedge (and vein) terminology

	Contemporary wedges (veins)						Fossil wedges (veins)					
	Permafrost		Active layer or seasonally frozen ground				Former permafrost			Former active layer or seasonally frozen ground		
Ice	*Soil*	*Ice and soil*	*Ice*	*Soil*	*Ice and soil*		*Ice*	*Soil*	*Ice and soil*	*Ice*	*Soil*	*Ice and soil*
Permafrost ice wedge	Permafrost soil wedge	Permafrost composite wedge	Active-layer ice vein[1]	Active-layer soil wedge	Active-layer composite wedge		Ice-wedge cast	Fossil permafrost soil wedge, or fossil permafrost sand wedge (if sand predominates)[2]	Fossil composite wedge[3]	(Probably not identifiable as such)	Fossil active-layer soil wedge	(Probably not identifiable as such)
			Seasonally frozen-ground ice vein[1]	Seasonally frozen-ground soil wedge	Seasonally frozen-ground composite wedge						Fossil seasonally frozen-ground soil wedge (probably rare)	

[1] Ice veins (<5 cm thick) are much more probable than ice wedges (⩾5 cm thick) in seasonally frozen ground and the active layer.
[2] Although use of the term sand-wedge cast (French, 1976a, 239) or soil-wedge cast by analogy with ice-wedge cast removes the ambiguity as to the contemporary or fossil nature of a feature, it introduces another problem in that cast implies replacement, which is inappropriate except for ice. Use of fossil is well established and unambiguous in the periglacial literature.
[3] Composite wedges are contemporary, not fossil, features as noted in the discussion, and the use of composite wedges for the latter without the modifier fossil should be avoided even if the context is clear (cf. French, 1976a, 239; Goździk, 1973, 112).

foliation that is less distinctly vertical, and (3) contain larger stones if the overlying or adjacent material is stony (Black, 1969*b*, 229).

Frost cracking of seasonally frozen ground in a non-permafrost environment is known to occur (cf. Black, 1976*a*, 18–19; Washburn, Smith, and Goddard, 1963), as is frost cracking confined to the active layer above permafrost (cf. Pissart, 1968, 177–9; 1970*b*, 15–16). However, frost cracks demonstrably formed in a non-permafrost environment rather than occurring in one formerly characterized by permafrost appear to be either temporary or formed under sufficiently intense freezing to at least approach, or be consistent with, discontinuous permafrost.

The present writer questions whether it has been demonstrated that well-developed soil wedges in materials not subject to large volume changes by wetting and drying form in a non-permafrost environment. The situation described by Danilova (1956; 1963) and Bobov (1960) and cited by Pissart (1970*a*, 17–18) as a demonstration is not convincing because, even if present cracking occurs, the wedges may not have formed under present conditions. In fact Danilova (1956, 117–18; 1963, 94) thought they had formed under former colder conditions, and based the hypothesis that the cracks were confined to a seasonally frozen layer on an interpretation of wedge features. Certainly, some wedges are limited to the active layer (Katasonov, 1972; 1973*b*; 1975; Katasonov and Ivanov, 1973, 8–10, 32–5; Katasonov and Solovyev, 1969, 12–18; Romanovskiy, 1973*a*; 1973*b*). Yet, almost all the soil wedges and related structures illustrated by Katasonov (1973*b*; 1975) were continuous with underlying ice veins or ice wedges in permafrost – a situation quite consistent with the view that large, well-developed soil wedges generally indicate a permafrost environment. Seasonal frost cracking would be favoured by a strongly continental climate with intense winter freezing, as in parts of the USSR. Soil-wedge polygons of such origin are reported to form at a mean annual ground temperature as high as 2°–3° but they are of small diameter and limited distribution (N. N. Romanovskiy, personal communication, 1978). Mean annual air temperatures are normally several degrees lower than ground temperatures, so even such limited occurrences would still be strongly suggestive of mean annual air temperatures approaching 0° or lower and, in many environments, of discontinuous permafrost. By the same token, large-diameter, well-developed soil-wedge polygons would be strongly suggestive of continuous permafrost. More data are needed on the entire subject of soil wedges.[16]

Pissart (1970*a*, 18–19) reviewed features of active-layer wedges as follows. (1) The cracks are developed in sands; (2) the wedge bedding is more or less vertical, and if a range of grain sizes is present there is sorting with the largest grain sizes at top; (3) the sediments enclosing the wedge tend to be down-warped along the contact, whereas in ice-wedge casts they are upwarped; (4) the structures are shallow, being confined to the active layer or, in the absence of permafrost, to seasonally frozen ground; (5) when fully developed the wedges may be very broad, approaching equilateral triangles. Pissart pointed out that the first three characteristics also apply to some wedges where permafrost occurs but that the great vertical extent of permafrost cracks (> 2 m) can be used as evidence of permafrost.

The following criteria for recognition of ice-wedge casts were listed by Black (1976*a*, 11–12)

1. Supporting evidence of permafrost, such as a [former] permafrost table (as recorded by change in fabric of the soil and by secondary deposits)....
2. Type of host [enclosing material] that ... [would have been] supersaturated in normal permafrost conditions where winter ground temperatures are not extremely cold.
3. Evidence to suggest little snow cover or thin active layer, and cloudy, wet, cool summers.
4. Multiple wedges forming polygons whose diameters and configurations are consistent with space filling requirements and supposed coefficients of thermal expan-

---

[16] In a publication received while the present volume was in press, Dionne (1978*b*, 196–7) described some nonsorted polygons on the east side of James Bay in subarctic Canada as [seasonal] frost-crack polygons. They were well delineated, 1–15 m in diameter, and the wedges were 15–30 cm wide at the surface and extended to depths of 20–50 cm, terminating in a commonly oxidized, narrow crack. The polygons occurred in sandy deposits in areas believed to be free of permafrost, although ground temperatures were largely lacking. Mean annual air temperatures in the region range from −2.5° to −4.5°. The sites, some barren of vegetation, were exposed to deflation and usually had little if any snow cover. Other, somewhat larger soil-wedge polygons at higher altitude were associated with permafrost, and the entire region is characterised by scattered patches of permafrost, mostly in palsas.

Although Dionne regarded the polygons as probably contemporary forms, he noted that several Holocene periods of lower temperature had probably affected the region, especially the post-Hypsithermal cooling and the Little Ice Age (whose respective dates Dionne cited as *c.* 2500 yr BP and 1430–1850 AD) when temperatures may have been 2°–4° lower than at present (Dionne, 1978*b*, 209, 213, 221). Thus the soil-wedge polygons may have originated when permafrost was more widespread.

4.33 Small, secondary ice wedge at right, subdividing low-centred pond deposits formed by large primary wedges like one at left. Small wedge is about 1.5 m high (*Black, 1976a, 8, Figure 6*)

sion of the host. Uniformly small polygons (less than 3–4 m in diam.) with uniformly sized wedges do not occur in permafrost today. Initial polygons are typically 10–40 m in diameter. On subdivision into smaller polygons, wedges of different sizes are formed .... [Figure 4.33.]

5. Pressure effects from widening wedges should be preserved locally adjacent to some wedges in a group. Upturned strata and realigned clasts are typical.

6. Slump fabrics should show stratification arcuate downward across wedges ... [Figure 4.34] in contrast to the vertical fabric of primary sand wedges.... [Figure 4.35.] In some wedge casts in a group, portions of the host material or stones many centimeters across should be present to show that filling was not accomplished in contraction cracks less than a centimeter wide, but in the full width of the wedge.

Although this last aspect would support the ice-wedge cast interpretation, its absence would not be inconsistent with a former frost crack lacking a well-developed ice wedge.

Neither the presence nor absence of any one of these criteria is critical. Certainly independent

evidence of permafrost (point 1) supports an ice-wedge-cast interpretation but in many places ice-wedge casts may be the principal evidence of former permafrost. The supersaturated nature of the enclosing material (point 2) while suggestive is weakened by the recent work of Votyakov (1966; 1973a; 1973b; 1978), Shusherina, Rachevskiy, and Otroshchenko (1970), and others as previously discussed in the section on Mass displacement. The problem arising from point 1 also applies to point 3; in addition the evidence of little snow cover and cool summers is of little value taken alone. The criterion that ice-wedge polygons less than 3–4 m in diameter with uniformly sized wedges do not occur in permafrost today (point 4) would be very valuable if confirmed; however, confirmation is needed in the opinion of the present writer. Pressure effects in the sediments enclosing a wedge, and realigned clasts (point 5) are certainly very commonly associated with ice-wedge casts but not exclusively so. As noted by Black (1976a, 11), 'Soil wedges with expandable material or ground of expandable material [in the

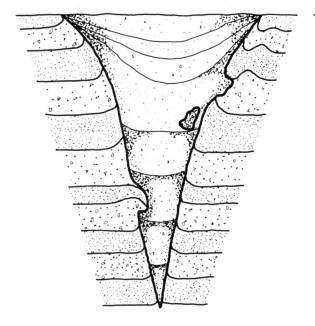

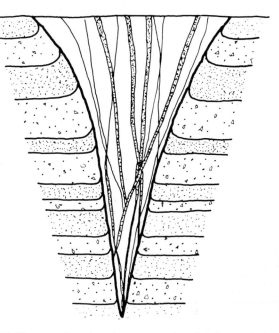

4.34 Diagram of ice-wedge-cast cross section showing slump structures in cast and upturned beds in enclosing material (*Black, 1976a, 12, Figure 8*)

4.35 Diagram of sand-wedge cross section showing vertical fabric in wedge and upturned beds in enclosing material (*Black, 1976a, 12, Figure 9*)

presence of wetting and drying] in which wedges of nonexpandable materials occur are immediately suspect as being of nonpermafrost origin.' Also, although upturned beds adjacent to ice-wedge casts are common (Figure 4.34), down-turned beds are expectable in enclosing materials subject to slumping (Figure 4.31). Reorientation of clasts can occur in many different situations. The vertical fabric of primary sand wedges (point 6) is good evidence against an ice-wedge origin but supporting evidence is needed that a vertical fabric in a wedgelike feature is proof of former permafrost. According to French (1976a, 239), a characteristic of fossil sand wedges in central Poland is that they tend to be much closer together (3–5 m apart) than ice-wedge casts in the same region. (He also suggested this might indicate that some fossil sand wedges resulted from desiccation cracking or from intense cold commensurate with thermal contraction cracking in an arid, well-drained site.) Clearly no single criterion is adequate to demonstrate the origin of an ice-wedge cast or of a fossil soil wedge. It is the weight of all the evidence that is critical. 'Any single wedge is untrustworthy.... we need to find evidence that proves or at least reasonably documents casts of former ice wedges. We then have a powerful tool

at our disposal for interpreting past climatic changes' (Black, 1976a, 3, 23).

VIII Sorting by frost action

A number of frost-action processes contribute to size sorting of mineral soil. Included are upfreezing of stones, work of needle ice, mass displacement due to moisture-controlled changes in intergranular pressure, and probably cryostatic pressure, all of which have been discussed above.

Among the sorting processes is the often-discussed gravity movement of stones into polygonal cracks from dome-shaped central areas (cf. Washburn, 1956b, 851). The fact that some plants have taproots in the central areas of sorted polygons but their above-ground portions at the margins supports such a radial movement (Lötschert, 1972, 7). Richmond (1949, 150–1) stressed that melting of ice wedges would be conducive to radial movement. Because stones have a higher diffusivity than fines, thawing occurs first adjacent to stones (cf. Washburn, 1956a), as noted in the field by Schunke (1975a, 63), who repeatedly observed stones sliding to stony borders. Slopewash running into cracks of nonsorted patterned ground in stony soil can presumably transform such patterns into sorted ones (Czeppe, 1961). These and other processes relating to the origin of sorted forms of patterned ground have been reviewed elsewhere (Washburn, 1956b, 838–860).

The work of Corte (1961a; 1962a–1962e; 1963a; 1963b; 1966a; 1966b) on sorting accompanying movement of a freezing front provided new data. Laboratory work showed that under certain conditions there is a general tendency for finer material to migrate ahead of the freezing front whatever its direction of movement, while coarser material is more readily entrapped by freezing and left behind. The factors are complex and include the orientation of the freezing front, rate of freezing, moisture content of the mineral soil, and shape, size, and perhaps density of particles. All these enter into what Corte termed horizontal and vertical sorting. In addition he reported a mechanical sorting involving downward movement of fines in an unfrozen upper layer under the influence of a freezing front that simulated upward freezing from a permafrost table (Corte, 1962c, 8–9; 1962e, 54, 65; 1966b, 193–7, 232). That sorting in nature occurs by migration of fines ahead of a freezing front has been challenged by Inglis (1965) but supported by Jackson and Uhlmann (1966) and Pissart (1966a). Kaplar (1969, 35–6) argued that Corte's vertical sorting would not accompany a downward-moving freezing front because of the difficulty of pushing fines into the underlying soil. The upfreezing of individual stones, described in the present chapter under Frost heaving and frost thrusting, is probably related to some aspects of the sorting investigated by Corte but it is distinct in involving, in places, a dynamic heaving through fines. It has been shown that particles in ice can migrate under a temperature gradient (Hoekstra and Miller, 1965, 6–8; Radd and Oertle, 1973, 379, 382–3; Römkens, 1969; Römkens and Miller, 1973). This movement of fines is towards the warm side, as in upward freezing from a permafrost table, but whether it has significance for sorting in nature is not known.

In fact all the sorting processes related to frost action are poorly understood and much remains to be learned from both field and laboratory work. The possibility that rise of capillary moisture, or movement of moisture to loci of ice development, may carry clay-size particles and contribute to sorting merits investigation (cf. Bakulin, 1958 – cited *in* Kachurin, 1959, 374; 1964, 10; Cook, 1956, 16–17; Johansson, 1914, 19–20, 22; Schunke, 1975a, 133–4; Thoroddsen, 1913, 253–4; 1914, 260; Washburn, 1956b, 841–2). Laboratory work of R. Brewer and Haldane (1957, 303, 308) supports the possibility.

The rapidity of sorting in some situations is demonstrated by an upward increase in grain size of material (< 20 mm in diameter) after as few as 12–20 freeze-thaw cycles (Strömquist, 1973, 88–90, 145). Microsorting effects, involving particles 0.02–0.05 mm in diameter, have been reported as occurring after 28 freeze-thaw cycles (Konishchev, Faustova, and Rogov, 1973). Marked sorting consisting of coarse carbonate debris (up to 11 cm long), surrounding finer debris containing only small (but unspecified) amounts of sand, silt, and clay, was obtained by Kuhle after 387 cycles in the laboratory. There was some evidence, only partly related to the experiment, that excessive moisture inhibited rather than promoted sorting (Hövermann and Kuhle, 1978, 324–8).

Sorting effects can be achieved by differential weathering where moisture concentrations or other favourable conditions occur. Such effects, although not the result of a sorting process, may be important in the origin of some forms of patterned ground as held by Nansen (1922a, 111–20) and others and discussed in the section on Patterned ground (Origin – cracking non-essential) in the next chapter.

Finally as noted in the sections on Frost wedging and on Frost heaving and frost thrusting in the present chapter, Frost heaving and frost thrusting, processes other than frost action can result in convergent effects, including sorting.

5 Some periglacial forms

I Introduction

Patterned ground, periglacial involutions, string bogs, palsas, and pingos are characteristic periglacial forms resulting from some of the frost-action processes discussed in the previous chapter. Stone pavements are less typical but also of periglacial interest. Except for some varieties of patterned ground (especially stripes) and involutions the above forms are confined to horizontal surfaces or gentle slopes. Periglacial phenomena that are primarily allied to mass-wasting, nivation, and slopewash are discussed in subsequent chapters. Although block fields are low-gradient features, they are so closely related to block slopes and block streams that they are included with them in the chapter on Mass-wasting processes and forms.

II Patterned ground

1 General

'Patterned ground is a group term for the more or less symmetrical forms, such as circles, polygons, nets, steps, and stripes, that are characteristic of, but not necessarily confined to, mantle subject to intensive frost action' (Washburn, 1956*b*, 824). Much of the following discussion is based on the publication cited.

Although various active forms of patterned ground occur in a variety of environments as reviewed by Troll (1944), periglacial forms are the most widespread. Some cartographic syntheses relate patterned ground to environment (Figure 5.1), but there are so many forms of patterned ground to consider that much detailed work will be required for accurate syntheses. For instance, Troll (1944, 661) contended that the lower limit of contemporary patterned ground rises toward the continental interior away from sources of moisture, whereas Hövermann (1960; 1962), stressing the role of temperature, argued that the trend is opposite to treeline and snowline and is lower in Eurasia than in European mountains at comparable latitudes. Kaiser (1965, 139–47) reported similar divergent trends of snowline and of lower limit of periglacial features with increasing continentality in the mountains of Lebanon and Syria, both at present and during Würm time, the main cause being the primary dependence of snowline on moisture and of periglacial features on low winter temperature (Kaiser, 1969, 31). More recently Hövermann and Kuhle (1978) presented field and laboratory evidence suggesting that a combination of moderate moisture and frequent freeze-thaw cycles, such as

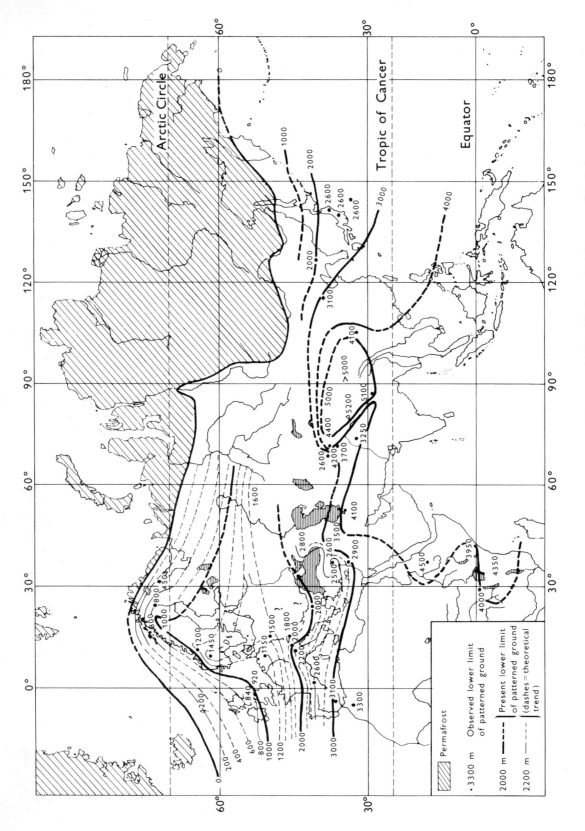

5.1 Lower limit (m) of patterned ground in the mountains of Eurasia and Africa. The lower limit in eastern Siberia lies much farther north than shown (*Troll, 1969, 233*) (*Troll, 1947, 164, Figure 1; key translated*)

found in arid to semi-arid mountains of the Near East (cf. Kuhle, 1978*c*) and northern Africa, represents optimal conditions for many alpine periglacial features. This conclusion would account for the lower limit of such features descending from humid or oceanic, towards arid or continental environments. The presence of low-lying patterned ground in the mountains of certain arid to semi-arid regions is now well established but interpretations of the climatic significance of the features remain much less certain (cf. Höllermann and Poser, 1977, 341–2; Mensching, 1977, 294–6). The complexities for the Tropics and Subtropics were reviewed by Heine (1977*b*).

Comparing periglacial features alone, the lower limit of sorted forms of patterned ground can be either higher or lower than the lower limit of gelifluction, depending on various factors. In addition to moisture and temperature, these include the susceptibility of the soil to frost sorting and gelifluction (Schunke, 1975*a*, 226) and the presence or absence of vegetation (Höllermann, 1972, 238–45). According to Höllermann, a vegetation cover deters frost sorting more than gelifluction and in many places controls the lower limit of sorted patterns. His discussion presents a thoughtful, climatically oriented overview of the interaction of these various factors. There is an upper as well as lower altitudinal limit to patterned ground and gelifluction, the upper limit being determined not only by perennial snow and ice but also by lack of thawing (Kuhle, 1978*a*, 290, 296; 1978*b*). The survey by Fritz (1976) for the eastern Alps, the review by Heine (1977*b*, 161–6) for the Tropics and Subtropics, a symposium overview by Höllermann and Poser (1977), and another held under the auspices of the Comité National Français de Géographie and the Union Géographique Internationale (Association Géographique d'Alsace, 1978) illustrate the difficulties introduced by the controlling environmental factors and the variety of forms and processes to be considered in establishing an altitudinal zonation of contemporary periglacial features. Such problems and the dating of forms are magnified in interpreting the environmental significance of fossil forms and warn against temptation to over interpret the evidence.

The kind of detailed regional studies needed for a world survey include those of Garleff (1970), Jan Lundqvist (1962), and Rudberg (1977) for Scandinavia; Schunke (1975*a*) for Iceland; Kelletat (1970*a*) for Scotland; Höllermann (1967) for the Pyrenees, the eastern Alps (cf. also Fritz, 1976; Stingl, 1969; 1971), and the Apennines (cf. also Kelletat, 1969; 1977*b*); and Brosche (1977; 1978)

for the Iberian Peninsula, and Furrer's (1965*a*) comparison of the altitudinal ranges of various kinds of patterned ground in the Swiss Alps and in the Karakoram Range of northern India; Graf's (1971) similar comparison between the Alps and the Bolivian and Peruvian Andes; and Kuhle's (1978*a*; 1978*b*) study in the Dhaulagiri-Himalaya (Nepal). Also helpful are more local inventories such as the one for the mountains of Snowdonia, North Wales (Ball and Goodier, 1970), and for the east side of James Bay in subarctic eastern Canada (Dionne, 1978*b*), as well as detailed local studies such as those of Rudberg (1970; 1972). Figure 5.2 shows the distribution of some studies of periglacial features in the mountains of Europe, Great Britain, and Iceland since 1950. A valuable symposium volume (Poser, 1977) summarizes some of the results to date and, besides illustrating some of the difficulties involved in such studies as noted above, shows the many advances that have been made.

The upper limit of patterned ground is generally determined by snow, ice, or steep slopes but Kuhle (1978*a*) reported an independent upper limit in the Dhaulagiri-Himalaya, which he implied was similar to the lower limit in being controlled by decreasing frequency of freeze-thaw cycles (Kuhle, 1978*a*, 290, 296, 300).

Excluding patterned ground controlled by local factors adjacent to glacier fronts, the sequence from lower to higher altitude as reported by Furrer (1965*a*, 72) for the Swiss Alps and the Karakoram Range comprises the following zones: (1) Zone of ploughing blocks (Wanderblöcke) – i.e. stones sliding downslope with a ploughing action, leaving a linear depression upslope and forming a frontal ridge downslope. Most earth hummocks and turf hummocks (Erdbülten) lie in this zone; (2) zone of garlands (Girlanden) – i.e. terrace-like features; (3) zone of earthflows (Erdströme); (4) zone of miniature patterned-ground; (5) zone of sorted or stony lobate forms (Steinzungen); (6) zone of large sorted patterned ground. These zones overlap and the average altitudes of zones 5 and 6 are practically identical. Furrer (1965*a*, 72) cited Jäckli (1957, 21) for the view that zone 6 tends to be characterized by permafrost. Graf (1971, 62–6, 110) reported a similar zonation in the eastern Alps and the Bolivian and Peruvian Andes. By comparison, zones are poorly developed in the Japanese Alps, where 'All kinds of [periglacial] forms can occur together in quite different altitudes' (Ellenberg, 1974*b*, 56). Among the number of environmental differences from Europe, vegetation is reported to be one of the most important (Ellenberg, 1974*b*, 63), and in many places a thick snow cover inhibits the development of periglacial features (Ellenberg, 1976*a*, 150–1). Nevertheless, in Japan and far-eastern USSR a well-defined, temperature-related sequence of patterned-ground and other periglacial forms has been reported that is similar to such sequences elsewhere (Figure 5.3).

2   Classification

Many patterned-ground forms are known and many different terms have been employed for them. A purely descriptive terminology in common use is based on geometric form and presence or absence of prominent sorting[1] between stones and finer material as determined by the presence or absence

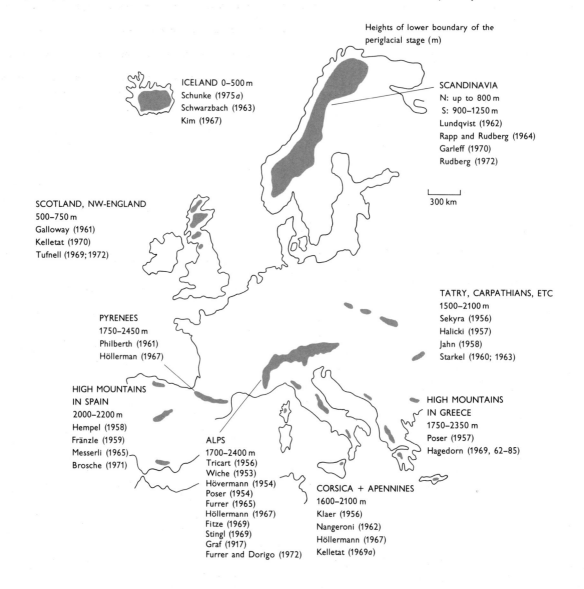

Heights of lower boundary of the periglacial stage (m)

ICELAND 0–500 m
Schunke (1975*a*)
Schwarzbach (1963)
Kim (1967)

SCANDINAVIA
N: up to 800 m
S: 900–1250 m
Lundqvist (1962)
Rapp and Rudberg (1964)
Garleff (1970)
Rudberg (1972)

300 km

SCOTLAND, NW-ENGLAND
500–750 m
Galloway (1961)
Kelletat (1970)
Tufnell (1969; 1972)

TATRY, CARPATHIANS, ETC
1500–2100 m
Sekyra (1956)
Halicki (1957)
Jahn (1958)
Starkel (1960; 1963)

PYRENEES
1750–2450 m
Philberth (1961)
Höllerman (1967)

HIGH MOUNTAINS
IN SPAIN
2000–2200 m
Hempel (1958)
Fränzle (1959)
Messerli (1965)
Brosche (1971)

HIGH MOUNTAINS
IN GREECE
1750–2350 m
Poser (1957)
Hagedorn (1969, 62–85)

ALPS
1700–2400 m
Tricart (1956)
Wiche (1953)
Hövermann (1954)
Poser (1954)
Furrer (1965)
Höllermann (1967)
Fitze (1969)
Stingl (1969)
Graf (1917)
Furrer and Dorigo (1972)

CORSICA + APENNINES
1600–2100 m
Klaer (1956)
Nangeroni (1962)
Höllermann (1967)
Kelletat (1969*a*)

5.2 Some studies of periglacial features in the mountains of Europe, Great Britain, and Iceland since 1950. Many more could be added. Altitudes refer to lower boundary of periglacial realm (*after Ellenberg, 1974b, 55, Figure 1 – updated*)

[1] The terms well sorted and poorly sorted are used in the geologic sense to indicate, respectively, well-developed and poorly developed uniformity in grain size. The engineering usage is the opposite with well sorted indicating a wide range of grain sizes.

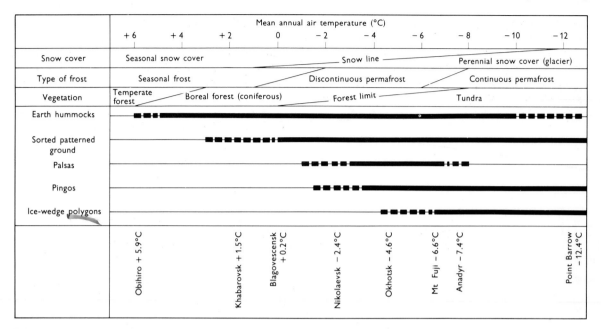

	Mean annual air temperature (°C)									
	+6 +4 +2 0 −2 −4 −6 −8 −10 −12									
Snow cover	Seasonal snow cover			Snow line			Perennial snow cover (glacier)			
Type of frost	Seasonal frost			Discontinuous permafrost			Continuous permafrost			
Vegetation	Temperate forest	Boreal forest (coniferous)		Forest limit			Tundra			
Earth hummocks										
Sorted patterned ground										
Palsas										
Pingos										
Ice-wedge polygons										

Obihiro + 5.9°C Khabarovsk + 1.5°C Blagovescensk + 0.2°C Nikolaevsk − 2.4°C Okhotsk − 4.6°C Mt Fuji − 6.6°C Anadyr − 7.4°C Point Barrow − 12.4°C

5.3 Zonal distribution of patterned ground and other periglacial features in Japan and far eastern USSR (*after Koaze, Nogami, and Iwata, 1974a, 183, Table 1*)

of stony borders (Washburn, 1950, 8–13; 1956*b*, 826–38). However, it should be recognized that less obvious sorting as revealed by excavation or grain-size analysis is not necessarily excluded (Figures 5.6–5.7) (cf. Fitze, 1975, 79–81) and that some patterned ground is gradational in both pattern and sorting as pointed out by Black (1952, 125) and others. The separation of patterned ground into sorted and nonsorted forms has been criticized by Nicholson (1976, 331–2) as being unsatisfactory for descriptive purposes because he regarded the term sorted as having major genetic implications; furthermore he believed such implications were unwarranted in places. The sedimentological aspect also concerned Shilts (1978, 1056) who regarded its application to patterned ground as confusing, since both nonsorted and sorted forms could have central areas with variable sorting in detail. The objections would seem to be largely negated by the common use of sorted, bedded, etc. to describe a condition, not the process responsible for it, and by the condition being specifically defined as prominent sorting between borders and central areas (or between adjacent stripes in striped forms of patterned ground). Thus defined, the distinction between sorted and nonsorted forms of patterned ground has proved to be a useful and widely adopted descriptive parameter.

The following forms of patterned ground are recognized:

Circles	{ nonsorted sorted
Polygons	{ nonsorted sorted
Nets	{ nonsorted sorted
Steps	{ nonsorted sorted
Stripes	{ nonsorted sorted

Circles, polygons, and nets are equiforms, defined as '... patterns with a surface form that is equidimensional in several directions' (Nicholson, 1976, 330). Use of this term should not obscure the importance of clearly distinguishing between circles and polygons, since their origin can be very different. The unit component of equiforms is a cell or mesh, and especially in the case of polygons it is desirable to differentiate small from large forms, because here too their origin can be quite different (Washburn, 1969*a*, 123–49). The critical diameter separating small from large forms remains to be standardized. One metre is used here, also by Jahn (1975, 43), but other usages occur; for instance, Schunke (1975*a*, 119) adopted 0.5 m, and Stingl and Herrmann

5.4 Nonsorted circle (Experimental Site 3), Mesters Vig, Northeast Greenland (*cf. Washburn, 1969a, 108, Figure 69*)

(1976, 212), 0.8 m for sorted forms in Iceland; and Katasonov (1973a, 74), 3 m for nonsorted forms in Siberia. Troll (1944, 562) regarded 10–25 cm as the general diameter of small (miniature) sorted forms.

Circles, polygons, and nets characteristically occur on nearly horizontal surfaces and become elongated ('elongates' according to Nicholson, 1976, 330) and tend to merge into stripes over a transition gradient that commonly ranges from about 2° to 7°, depending on conditions (cf. Hussey, 1962; Washburn, 1956b, 836–7). Although all stripes are confined to slopes, the sequence of slope relationships is less well established for nonsorted stripes than sorted, and both nonsorted and sorted stripes can occur without merging upslope or downslope into 'elongates'. Both moisture and grain size can strongly influence transition gradients. For instance, in the Bolivian Andes the transition gradient

5.5 Nonsorted circles joined by narrow neck, Mesters Vig, Northeast Greenland (*cf. Washburn, 1969a, 109, Figure 70*)

between small polygons and stripes can range from 4° to 20° depending on moisture (Graf, 1976, 438–9).

The various kinds of patterned ground can be associated with specific major soil types and can strongly influence their detailed distribution (Drew and Tedrow, 1962; Tedrow, 1977, 250–66).

Some kinds of patterned ground occur subaqueously (cf. Dionne, 1974a; Shilts, 1974, 231, 234; Shilts and Dean, 1975) but are commonly assumed to have originated subaerially. That this may not always be the case is suggested by the observations of Shilts and Dean in the Canadian Arctic, and by the present writer's observations there of presumed frost cracks beneath near-shore sea ice that was breaking up but had probably been frozen to the bottom.

Inactive and fossil patterned ground can be of great value in environmental reconstructions and paleoclimatic studies. Inactive (also called relic or relict) forms have features similar to active forms (except for evidence of inactivity) and might become reactivated through some environmental change. Fossil forms are commonly less obvious, may be revealed in cross-section only, and are more clearly irreversible in activity. In periglacial literature, the term fossil does not imply former organic life.

As with many other features, there is a striking convergence of patterned-ground forms in cold climates and warm arid to semi-arid climates. Superficially similar forms include all the main geometric varieties, both nonsorted and sorted, that are listed above. Forms known as gilgai, characterized by a

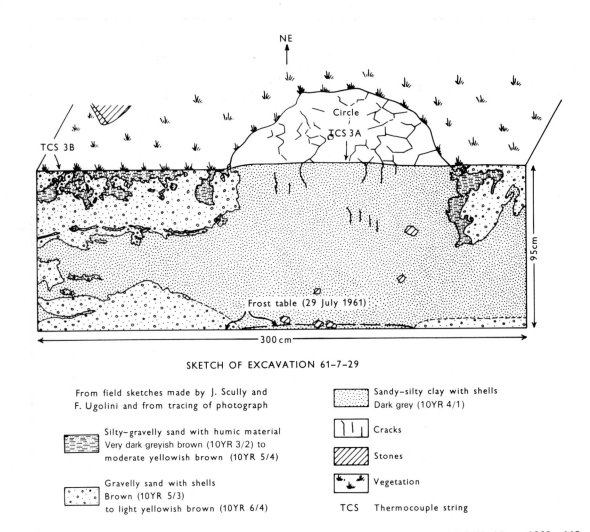

SKETCH OF EXCAVATION 61-7-29

From field sketches made by J. Scully and
F. Ugolini and from tracing of photograph

 Silty–gravelly sand with humic material
Very dark greyish brown (10YR 3/2) to
moderate yellowish brown (10YR 5/4)

 Gravelly sand with shells
Brown (10YR 5/3)
to light yellowish brown (10YR 6/4)

 Sandy–silty clay with shells
Dark grey (10YR 4/1)

Cracks

Stones

Vegetation

TCS   Thermocouple string

5.6 Cross-section of nonsorted circle (Experimental Site 3), Mesters Vig, Northeast Greenland (*cf. Washburn, 1969a, 112, Figure 72*)

Key for cross sections A – A' to D – D'

Gravelly sand with humic material
Dusky brown

Cracks

Gravelly sand with shell fragments
Moderate yellowish brown (10YR 5/4)

SKETCHES OF EXCAVATION 58-7-19 12 m EAST OF EXPERIMENTAL SITE 3

Stones

Gravelly–clayey sand–silt to sandy–silty clay with shell fragments
Light to dark grey with brownish to greenish tints in places

Vegetation

5.7 Parallel cross-sections of nonsorted circles joined by narrow neck, Mesters Vig, Northeast Greenland (*cf. Washburn, 1969a, 114–15, Figure 74*)

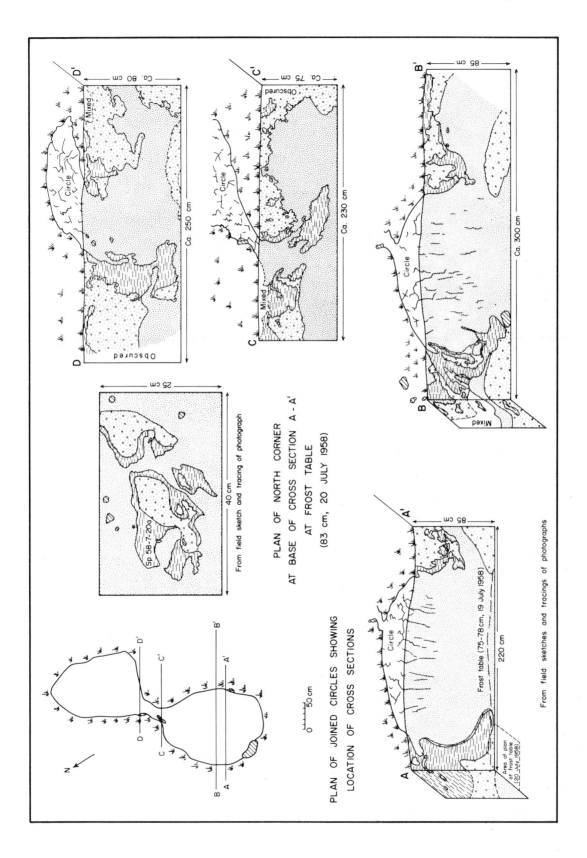

PLAN OF NORTH CORNER
AT BASE OF CROSS SECTION A - A'
AT FROST TABLE
(83 cm, 20 JULY 1958)

From field sketch and tracing of photograph

PLAN OF JOINED CIRCLES SHOWING
LOCATION OF CROSS SECTIONS

From field sketches and tracings of photographs

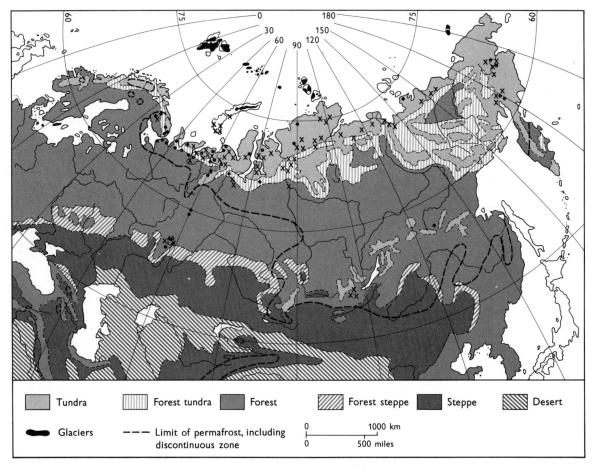

5.8 Distribution of active nonsorted circles and related forms with flat (●) and domed (×) surfaces in northern Eurasia (*after Frenzel, 1959, 1011, Figure 8*)

significant content of swelling clays, occur in Australia (Cooke and Warren, 1973, 129–35; Costin, 1955; Hallsworth, Robertson, and Gibbons, 1955; Mabbutt, 1977, 54–5, 130–4) and elsewhere (cf. Bremer, 1965, 227), and may be more widespread than is commonly recognized (cf. E. M. White and Agnew, 1968). Nonsorted and sorted forms in the salt-rich environments of warm deserts are also widespread (cf. Hunt and Washburn, 1966; Meckelein, 1959, Figures 53–54, 58, opp. 102; 122; Figure 69, opp. 126).

3   Circles

a Nonsorted   Nonsorted circles are patterned ground whose mesh is dominantly circular and lacks a border of stones (Figures 5.4–5.5). The term mud-

boil is frequently used in Canada for both nonsorted and sorted circles. Nonsorted circles are characteristically margined by vegetation, and occur singly or in groups. Common diameters are 0.5–3 m. The central areas tend to be slightly dome-shaped and cracked into small nonsorted polygons. In places the long axis of stones and sand-size particles tends towards the vertical (Fahey, 1975, 158–60).

The mineral soil normally has a high content of fines and may or may not contain stones. Cross-sections indicate that the material of the central areas in some forms has risen from depth (Figures 5.6–5.7) (cf. Graf, 1971, 128–9; Pissart, 1976c, 280–1; Shilts, 1978, 1062).

Nonsorted circles are prominent in polar, subpolar, and highland environments. Their distribution in northern Eurasia is shown in Figure 5.8.

Recognition of cold-environment, inactive non-

5.9 Sorted circles, Hornsundifiord, Vest Spitsbergen. Photo by Alfred Jahn

sorted circles that are out of balance with their present surroundings is dependent on proof of inactivity. Surface features such as lichen-covered stones and vegetation-covered central areas provide evidence. Fossil forms in stratigraphic sections would be hard to recognize but may be represented by some involutions (cf. Hopkins and Sigafoos, 1951, 98; Poser, 1947a, 12; 1947b, 233).

  b Sorted 'Sorted circles are patterned ground whose mesh is dominantly circular and has a sorted appearance commonly due to a border of stones surrounding finer material' (Washburn, 1956b, 827) (Figures 5.9–5.12). The qualification 'commonly' was introduced because of stone pits, discussed below, which are relatively rare, isolated, circular accumulations of stones that can be regarded as stone-centred sorted circles (Figure 5.13) in contrast to the more common stone-bordered forms. In the following the term sorted circles will apply to the stone-bordered forms unless otherwise indicated, in which case the term stone pits will be used. Debris islands are sorted circles amid blocks or boulders (Figures 5.10–5.12). In many places cross-sections show that the finer material was derived from depth (Schunke, 1975a, 119, 221; Abb. 11) (Figure 5.12), although in some other situations the finer material appears to have no such connection. Some occurrences of debris islands are transitional to sorted circles (Schunke, 1975a, 119, Abb. 9). Like nonsorted circles, sorted ones occur singly or grouped, and the size range is also similar. In many places the size of sorted circles increases with depth of frost action and decreases where the soil is thinner than this depth (Fritz, 1976, 259–60). Sorted circles are common on nearly horizontal surfaces but debris islands may occur on gradients as steep as 30°. Such

5.10 Sorted circle (debris island), Mesters Vig, Northeast Greenland. Scale given by 16-cm rule (*cf. Washburn, 1969a, 155, Figure 96*)

slope forms tend to have central areas that are considerably less steep than the general gradient but can be as high as 25°.

The central areas have a concentration of fines, either with or without stones. The stones of the borders surrounding the central areas tend to increase in size with the size of the circles. Tabular stones tend to be on edge with their long axis in the vertical plane parallel to the border, the next most common long-axis orientation being at right angles to it.[2] This fabric is reported to characterize the central areas as well as the borders, with the former

[2] As used here 'on edge' does not necessarily imply that the long axis is dominantly horizontal in the vertical plane. The prevailing orientation of *a* and *b* axes in the vertical plane in the bordering stones of sorted forms of patterned ground still remains to be established (Watson and Watson, 1971, 113–14).

5.11 Sorted circle (debris island) on 31° gradient, Hesteskoen, Mesters Vig, Northeast Greenland. Scale given by hammer at centre (*cf. Washburn, 1969a, 157, Figure 98*)

having most of the stones dipping at angles >45° (Furrer, 1968; Furrer and Bachmann, 1968, 9–12).

Like stone-bordered sorted circles, the stone-centred sorted circles known as stone pits (G. Lundqvist, 1949, 336) have a variable appearance. Some consist of a central stone surrounded by smaller stones (Dionne, 1974a, 333–4; Jan Lundqvist, 1962, 79, Figure 43; 80) or by slabby edgewise fragments, and in places the latter variety has been ascribed not to frost action but to wave action (Dionne and Laverdière, 1967; Dionne, 1971b) or 'hydrodynamic' action (Ball, 1976; Kostyayev,

1973, 350–2). Other varieties consist of numerous stones amid vegetation (Beskow, 1930, 634–5, Figures 9–10; Tedrow, 1977, 261, Figure 14–12) or on bare ground (Figure 5.13). Some forms are up to 2.6 m in diameter and occur in the central area of sorted nets (Dionne, 1974a, 334, Figure 10). Stone pits described by Schunke (1975a, 62–3, 118) and Stingl (1974, 257–9) were so closely associated with the apparent break-up of sorted circles, nets, and stripes that it was concluded they were derived from them, the mesh intersections of some sorted nets, for instance, becoming a protostone pit.

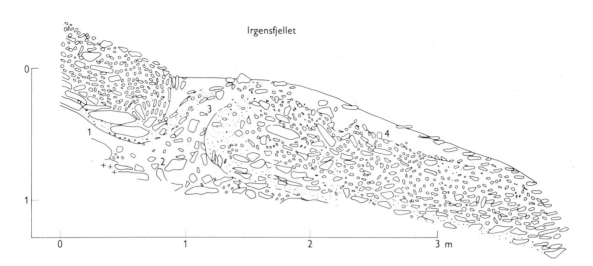

Irgensfjellet

5.12 Cross-section of sorted circle (debris island) on 15°–22° gradient, Irgensfjellet, Blomstrandhalvöya, Kongsfjorden, Vest-Spitsbergen. Relatively fine material, gray brown (1–2) at depth to brown (3–4) near surface, intruding coarser debris and then extending over it on downslope side. Small crosses indicate ice between fragments of the underlying limestone bedrock (*after Herz and Andreas, 1966, Tafel 42, E*)

5.13 Stone pits in basalt detritus, Bardaströnd, northwest Iceland. Scale length 80 cm (*Schunke, 1975a, Abb. 7, cf. p. 74–5*)

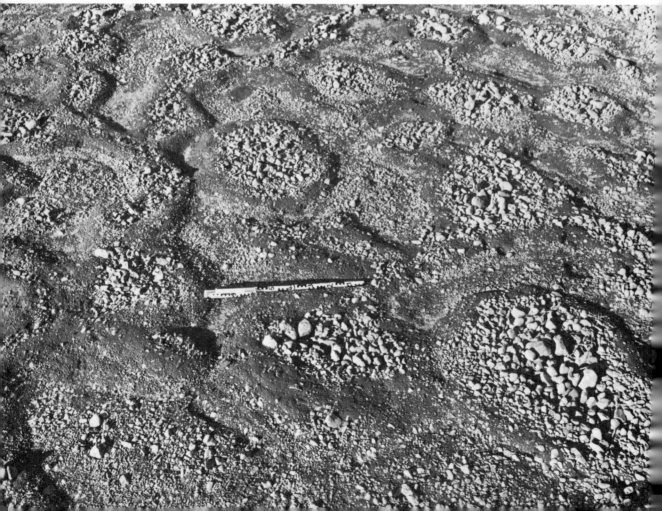

Sorted circles are characteristic of polar environments. However, they also occur in subpolar and highland regions. Although best developed where there is permafrost, they are present in Iceland in places where it is lacking (Steche, 1933, 209), and apparently active forms up to 3 m in diameter occur on the subantarctic island of South Georgia (R. A. S. Clayton, 1977, 94) whose mean annual temperature in 1973 and 1974 was 2.2°, with a minimum temperature of − 14.2° during that period (Limbert, 1977a, 133; 1977b, 141). It would be interesting to establish whether or not extrazonal factors influence the occurrences. A model has been suggested in which, given a threshold freeze-thaw frequency, circles become inactive as stony borders become well developed, leading to stability of borders and revegetation. Thus inactivity would depend on circle evolution rather than on any independent environmental change (Thorn, 1976a, 22–3). However, applicability of this somewhat speculative model is limited to already existing threshold conditions.

Inactive sorted circles, identified as such by lichen-covered stones in the borders and central areas and by the vegetated nature of the central areas, are more easily recognized than inactive nonsorted circles. Fossil forms in stratigraphic section should also be easier to identify but well-documented reports are rare for all sorted varieties of patterned ground. In Jahn's (1975, 134) opinion this may be because (1) in Europe, where the extensive research might have been expected to find such records, the periglacial region was largely tundra rather than barren soil, and the vegetation tended to minimize frost sorting; and (2) thawing and collapse tended to make sorted forms unrecognizable.

4 Polygons

a Nonsorted Nonsorted polygons are patterned ground whose mesh is dominantly polygonal and lacks a border of stones.[3] As noted at the start of this chapter, it is convenient to classify polygons into small forms (diameter < 1 m) and large forms (diameter > 1 m). Meshes of small forms (Figures 5.14–5.15) measure as little as 5 cm across, those of large forms (Figures 5.16–5.17) can exceed 100 m (Black, 1952, 130). Nonsorted polygons are group

forms whose mesh (or borders between polygons) is commonly but not always delineated by a furrow and a crack. Where vegetation is sparse, the vegetation is generally concentrated along the furrow and emphasizes the pattern. Nonsorted polygons are most frequent on nearly horizontal surfaces but are by no means confined to them. Small forms have been noted on gradients as high as 27°; large ones are known to occur on slopes as steep as 31° in polar regions (Büdel, 1977, 62, footnote 13). Extraordinarily well-preserved large fossil forms occur on slopes of 5°–30° in the Italian Alps (Kelletat and Gassert, 1974).

The mineral soil of nonsorted polygons can be well-sorted fines, sand (including dune sand – Aartolahti, 1972) or gravel, or it can be a diamicton.[4] According to Schunke's (1975a, 128–9) observations in Iceland, large nonsorted polygons there require good soil drainage. However, in many coastal lowlands at least, this requirement is inapplicable to large polygons of the ice-wedge variety such as those in Figure 5.16.

Ice-wedge polygons have an ice wedge coincident with the borders and their borders tend to be raised or depressed with respect to the central areas. A raised border has the furrow noted above but the furrow has a marginal bulge on each side reflecting deformation caused by growth of the underlying ice wedge and subsequent expansion of the permafrost upon rise of temperature (Black, 1976a, 6) or, in Jahn's (1972a, 287–8; 1975, 44–53) view, by intrusion and freezing of water expelled from the furrow by cryostatic pressure. Raised borders are responsible for low-centre polygons. Depressed borders are caused by thawing and give rise to high-centre polygons if the thawing degrades ice wedges enough to leave the central areas standing in relief. Furrows overlying ice wedges tend to reflect the width of ice wedges but not necessarily their depth. Furrows can be absent (Root, 1975), or misleading in that some overlie sand wedges and only thin ice veins (Mackay, 1974b) or occur where thawing is affecting only cracks and ice veins in the active layer. During the thaw season, the low-centred polygons often contain ponds in the central areas, whereas the high-centred ones hold water in the bordering depressions (Hussey and Michelson, 1966, 165–70). Some ice-wedge polygons enclose

---

[3] Some fossil polygons that appear nonsorted at the surface have bordering stones that are covered with finer sediments and vegetation, but many of these polygons could have originated as nonsorted forms. In the case of former ice-wedge polygons, the stones could have accumulated in troughs left by melting wedges (Benedict, 1979).

[4] A diamicton is a nonsorted terrigeneous sediment containing a wide range of grain sizes, regardless of origin (cf. Flint, 1971, 152–4).

5.14 Small nonsorted polygons, DeSalis Bay, Banks Island, Northwest Territories, Canada (*cf. Washburn, 1950, 26, Figure 1, Plate 6*)

5.15 Small nonsorted polygons on a 27° gradient, Mesters Vig, Northeast Greenland. Scale given by 35-mm camera case at centre (*cf. Washburn, 1969a, 133, Figure 84*)

small pingo-like, ice-cored mounds (French, 1971a). Nonsorted polygons can have a pronounced microrelief. In the vicinity of Barrow, Alaska, 10.7 per cent of a 100–km² area had a microrelief of 0.5–1.5 m because of them (Sellmann *et al.*, 1972, vii, 17–27). Analogously to ice-wedge polygons, sand-wedge polygons have sand wedges instead of ice wedges; they are especially prominent in the Antarctic. Some nonsorted polygons occur essentially in bedrock (Arthur *in* Corte, 1969a, 140; Berg and Black, 1966,

69–70; W. E. Davies, 1961a; Walters, 1978; Washburn, 1950, 47–9). Topographic and compositional differences between borders and central areas of polygons commonly cause differences in vegetation. This can be true even in fossil forms; for instance, some presumed ice-wedge casts, because of a higher moisture retention, are known to be more favourable for crop growth than the central areas (Christensen, 1974). Nonsorted polygons in cross-section are illustrated in Figures 5.18–5.19.

Reported rates of ice-wedge and sand-wedge development are mentioned under Frost cracking in the preceding chapter. Some large nonsorted polygons with furrows 60–70 cm wide and 15–25 cm deep overlying sand wedges with thin ice veins can originate within 10 years as known from their presence on a recently drained lake bottom

5.16 Large nonsorted polygons, Arctic Coastal Plain, Alaska. Smaller polygons on west formed as river channel (beyond photo) migrated northwestward from low escarpment separating polygon types. Photo by H. J. Walker

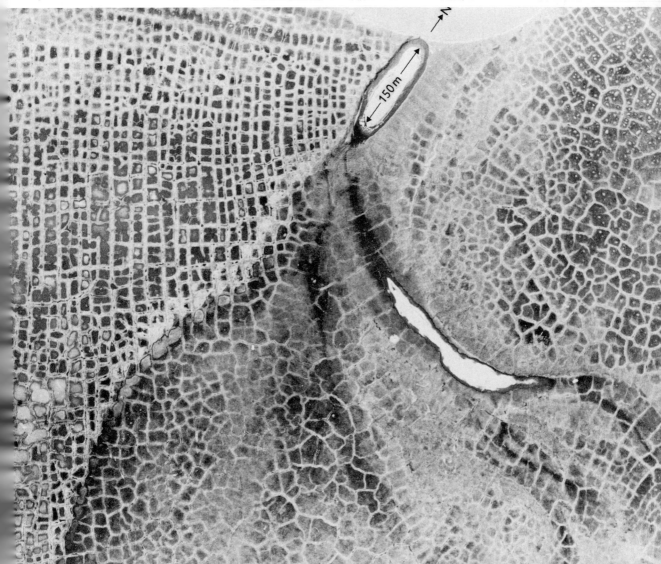

5.17 Large nonsorted polygons, Mesters Vig, Northeast Greenland. Polygon marked by ice axe is 3 m in diameter (*cf. Washburn, 1969a, 148, Figure 92*)

(Mackay, 1974*b*); some large nonsorted polygons probably require very much longer to become so obvious.

Small nonsorted polygons due to desiccation are ubiquitous. Large forms occur mainly in two contrasting arid environments – cold and warm. The large, cold-environment forms are characteristically polar, and ice-wedge polygons and probably most sand-wedge polygons are necessarily associated with permafrost. The distribution of ice-wedge polygons and related forms in northern Eurasia is shown in Figure 5.20. In Alaska the association of active forms with continuous permafrost implies, for their growth, a mean annual *air* temperature of about −6° to −8° in the south to −12° in the north, and a mean annual degree-days (°C) freezing range of 2800 to 5400 (Péwé, 1966*a*, 78; 1966*b*, 68; 1975, 52) (Figures 5.21–5.23). Minimum temperatures at the top of the permafrost here range from some −11° to −30°. Winter temperatures of −15° or colder at

the top of the permafrost have been suggested as necessary for active ice-wedge polygons (R. J. E. Brown and Péwé, 1973, 82; Lachenbruch *in* Péwé, 1966*a*, 78; 1975, 52). Annual precipitation is about 20 cm as rain and less than 140 cm as snow. Where inactive ice wedges are associated with discontinuous permafrost in Alaska, the mean annual air temperature ranges from about −2° in the south to −6° to −8° in the north, and the mean annual degree-days freezing range is 1700 to 4000. Precipitation is greater in the west than in the east, and the snowfall is 100 to 200 cm.

Subsequently Péwé (1974, 40) indicated that −7° or −8° could be applied quite generally as a maximum mean annual air temperature for the growth of ice-wedge polygons, but extending the Alaskan observations to other regions where different conditions may prevail is stretching the evidence. Péwé's −6° upper limit is more conservative, and −5° is even safer (cf. chapter on Environmental recon-

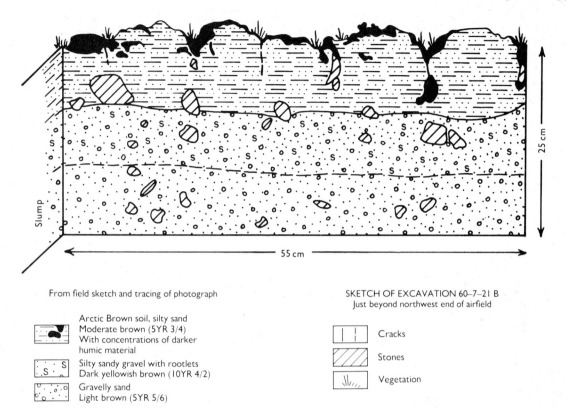

25 cm

Slump

55 cm

From field sketch and tracing of photograph

Arctic Brown soil, silty sand
Moderate brown (5YR 3/4)
With concentrations of darker
humic material

Silty sandy gravel with rootlets
Dark yellowish brown (10YR 4/2)

Gravelly sand
Light brown (5YR 5/6)

SKETCH OF EXCAVATION 60–7–21 B
Just beyond northwest end of airfield

Cracks

Stones

Vegetation

5.18 Cross-section of small nonsorted polygons, Mesters Vig, Northeast Greenland (*cf. Washburn, 1969a, 135, Figure 86*)

structions). The subject is of critical importance in re-constructing paleo temperatures based on fossil frost-crack features. In northern Finland, Aartolahti (1972; 131) reported infrequent cracking at a mean annual air temperature of − 1.5°. With respect to the USSR, Jahn (1975, 31) cited Baulin *et al.* (1967) to the effect that a mean annual air temperature of − 2° was adequate for frost cracking and development of ice wedges but the evidence is not known to the present writer. In Spitsbergen active ice-wedge polygons have been reported at Longyearbyen where the mean annual air temperature has been variously cited as − 4.9° (Svensson, 1973, 174), and − 5.5° (1957–1965) to − 6.2° (1957–1968) (Steffensen, 1969, 273, 332). Researchers at the USSR Academy of Sciences' Institute of Geography in Moscow told the present writer in 1976 that a mean annual air temperature of − 5° to − 7° was adequate for permafrost cracking and development of new ice

wedges in northern Siberia and that ice wedges are present (but not newly developing) at a mean temperature of − 3° to − 4°. Also, small ice-wedge polygons (1.5 × 1.5 m, 1 × 2 m) occur along the Angara River at mean annual air temperatures of about − 2° to − 3°, and active wedges are found in more easterly regions at − 4° (N. N. Romanovskiy, personal communication, 1978). This puts in doubt R. B. G. Williams' (1975, 99–100) suggestion on the basis of Soviet occurrences that − 8° to − 10° would be a good approximation for ice-wedge polygons, especially in sand and gravel. With respect to *ground* temperatures,[5] Baulin, Dubikov, and Uvarkin (1973, 13; 1978, 65) reported that permafrost crack-ing in western Siberia can occur at mean annual ground temperatures of about − 4° to − 5°. Based on calculated values rather than direct observation, Kudryavtsev (1974, 323–4; Kudryavtsev *et al.*, 1977, 376) stated that frost cracks reach the base of the

[5] As noted in the chapter on Frozen ground, the mean annual ground temperature, whether at the ground surface or at the top of permafrost, is approximated by the temperature at the depth of zero annual amplitude, which is reported to be generally about 3° warmer than the mean annual air temperature but with a considerable regional variation.

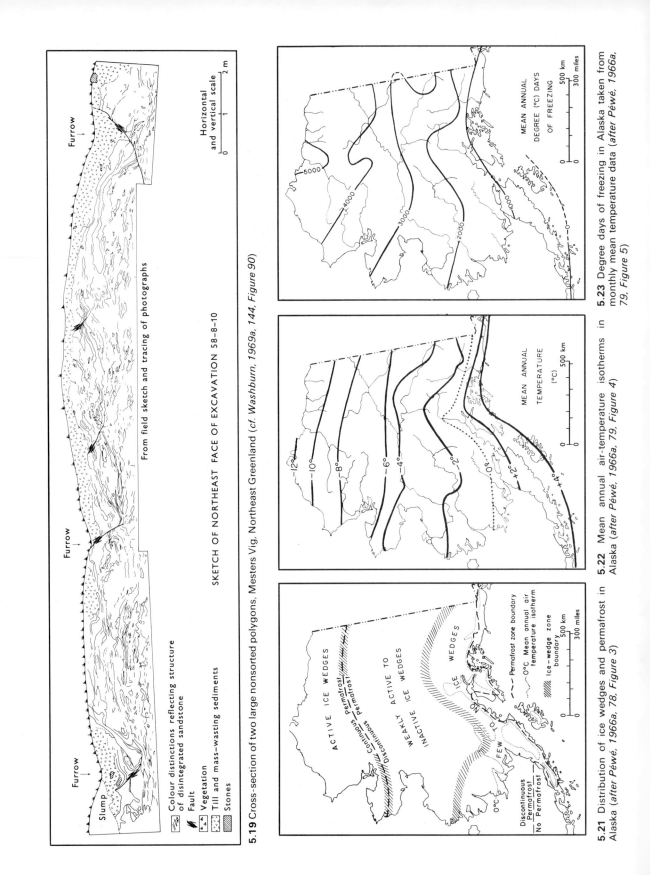

5.19 Cross-section of two large nonsorted polygons, Mesters Vig, Northeast Greenland (*cf. Washburn, 1969a, 144, Figure 90*)

Colour distinctions reflecting structure of disintegrated sandstone

Fault

Vegetation

Till and mass-wasting sediments

Stones

SKETCH OF NORTHEAST FACE OF EXCAVATION 58–8–10

From field sketch and tracing of photographs

Furrow   Furrow   Furrow   Furrow   Slump

Horizontal and vertical scale

0   1   2 m

5.23 Degree days of freezing in Alaska taken from monthly mean temperature data (*after Péwé, 1966a, 79, Figure 5*)

MEAN ANNUAL DEGREE (°C) DAYS OF FREEZING

5000   4000   3000   2000   1000   0

500 km

300 miles

5.22 Mean annual air-temperature isotherms in Alaska (*after Péwé, 1966a, 79, Figure 4*)

MEAN ANNUAL TEMPERATURE (°C)

−12°   −10°   −8°   −6°   −4°   −2°   0°   +2°   +4°

500 km

5.21 Distribution of ice wedges and permafrost in Alaska (*after Péwé, 1966a, 78, Figure 3*)

ACTIVE ICE WEDGES

Continuous Permafrost

Discontinuous Permafrost

WEAKLY ACTIVE TO INACTIVE ICE WEDGES

NO ICE WEDGES

FEW ICE WEDGES

Discontinuous Permafrost

No Permafrost

0°C

Permafrost zone boundary

0°C Mean annual air temperature isotherm

Ice–wedge zone boundary

500 km

300 miles

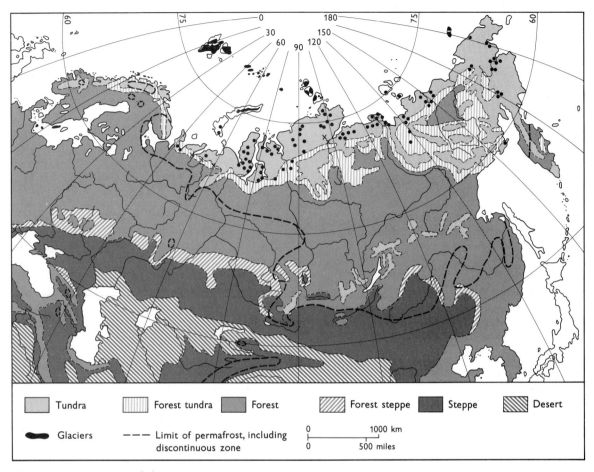

5.20 Distribution of active (●) and subfossil (x) ice-wedge polygons and related forms and oriented lakes in northern Eurasia (*after Frenzel, 1959, 1010, Figure 7*)

active layer when the ground temperature (presumably mean annual) is −2° to −3° in sandy loams and peats, or to −4° and −6° in sands, pebbles, and grus, and that cracks extend into permafrost at lower temperatures. A similar approach was taken by Romanovskiy (1973b, 270–1; cf. 245) who reported that the following mean annual ground temperatures would suffice for permafrost cracking and build up of ice wedges: −2° to −4° for 'dusty' sandy or sandy-clayey deposits, −5° to −6° for fine-grained sands, and −7° to −8° for clayey sandy or gravelly deposits; he stressed that there is no direct connection, independent of the nature of the soil, between air temperature and ground temperature and that it can be very misleading to deduce air temperatures from ice-wedge features Romanovskiy, 1973b, 275) (cf. Figure 4.30). In Canada

Mackay (1975c, 1668) reported that infrequent cracking of permafrost may occur at a mean annual ground temperature just below 0°. A comparison of the distribution of ice-wedge polygons and ground temperatures in the Mackenzie Delta area (Mackay, 1963a, 70, Figure 28; 1974e, 390, Figure 2) shows that both sparse and numerous polygons occupy the same temperature field of −3° to −4°. Also,

(1) at Garry Island, Pelly Island, and the Tuk area where we have many direct measurements of ground temperature and ice-wedge cracking, ground temperature of −8°C ± 1°C produces only about 40 to 50 per cent cracking; (2) at the outer (distal) end of the Delta, with mean ground temperatures of −1° to −2°C, ice wedges are now growing; (3) at Inuvik, with mean ground temperature of about −4° to −5°C, ice wedges are inactive. I feel

that temperature is only "half" of the reason behind crack-ing. In recently drained lakes, where the temperature gradient is steep, because permafrost is very thin, cracking may be frequent the first few years, and then slow down. (J. R. Mackay, personal communication, 1978.)

The ground-temperature drop needed for crack-ing varies with conditions. Based largely on his studies near Barrow, Alaska, Black (1963, 264; 1969a, 228) cited a rapid temperature drop of 4° as being adequate to crack frozen ground having a thermal coefficient of contraction approaching that of ice; a change of about 2° would be adequate to propagate the crack downward. Where the co-efficient is near that of rock (*c.* 1/5–1/6 that of ice), field data suggest that a temperature drop of about 10° is required to start the crack, and a change of some 4° is needed to propagate it downward. Black (1976a, 9) emphasized that it is the amount and rapidity of temperature change in the top 5–10 m of permafrost that is critical, not the mean annual air temperature except as colder temperatures promote more cracking. As noted above he reported that at Barrow a sudden change of only 4° caused cracking in ice-rich permafrost, yet changes of 8°–10° were required where the permafrost was not supersaturated or changes occurred less rapidly (Black, 1976a, 8–9). Lachenbruch (personal com-munication *in* Péwé, 1975, 52) suggested that crack-ing of ice wedges is expectable when the tem-perature at the *top of the permafrost* falls below − 15° to − 20°, and Péwé indicated that the sparse thermal data available are consistent with such winter temperatures in the continuous permafrost zone of Alaska where ice wedges are most active.

The effects of snow cover and vegetation on ground temperature are especially critical factors requiring consideration as noted earlier. They seriously inhibit attempts to infer mean annual air temperatures from periglacial features and to trans-fer conclusions from one region to another without due regard for environmental differences.

As discussed under Frost cracking in the preceding chapter, large seasonal frost-crack polygons are reported that simulate some ice-wedge polygons in surface appearance. Thus subsurface investigation may be required to prove the presence or absence of ice wedges or ice-wedge casts. The critical problem is to avoid misinterpretations leading to inferences that temperatures were low enough for continuous permafrost and the origin of ice-wedge polygons (cf. Schunke, 1975a, 129). For instance, the report of ice-wedge polygons in the loess soils

of interior Iceland (Thorarinsson, 1964, 330–5) is misleading inasmuch as no ice wedges were described, although fresh, polygonal frost cracks were observed. Seasonal frost cracking has been reported in the high interior of Iceland (Friedman *et al.*, 1971; Schunke, 1974b; 1975a, 128–9; Stingl, 1974, 253–5; Svensson, 1977, 69–70), and appears reasonable in view of the area's cold winters and the discontinuous permafrost in palsas (Priesnitz and Schunke, 1978), but such observations do not show that present-day conditions in Iceland are suit-able for forming ice-wedge polygons as has been claimed (Embleton and King, 1975, 85). On the other hand, in the opinion of the present writer, most large seasonal frost-crack polygons probably originate in an environment that has sufficiently intense freezing to at least approach, or be consistent with, *discontinuous* permafrost, and that their poten-tial significance from this viewpoint should not be overlooked if occurrences are widespread.

The influence of grain size on the distribution of large nonsorted polygons is illustrated by observa-tions in Iceland that those developed in silty, highly frost-susceptible soil occur at lower altitudes (by 200–250 m) than those in sandy soil (Schunke, 1975a, 127, 217). The influence of moisture on development of large polygons is not well established. The occur-rence on the Arctic Coastal Plain of Alaska and Canada of well-developed, low centre ice-wedge polygons with water at their surface (Figure 5.16) appears inconsistent with observations that large nonsorted polygons in Iceland are absent where there is a high water table (Schunke, 1975a, 127–8, 217). However, different polygon types may be involved. According to Schunke the Icelandic poly-gons presumably originated as seasonal frost-crack polygons that did not require permafrost, but the question is complicated by the fact that permafrost was formerly more widespread and the polygons could be largely relic, with some being recently reactivated (Schunke, 1975a, 127–9, 228, 233). Conceivably, the inhibiting role of high moisture in Iceland could reflect the marginal temperature conditions for permafrost and the lack of it where freezing would have to overcome the retarding effect of the latent heat of fusion.

Small nonsorted polygons are likely to be passing features following an environmental change, and inactive forms and fossil forms are rarely reported. However, large inactive and fossil nonsorted poly-gons are among the most reliable evidences of environmental change, although inactive forms may be difficult to distinguish from weakly active forms that crack rarely. In the case of ice-wedge polygons, ice wedges can be present in both the inactive and the weakly active forms and both occur in the discontinuous permafrost zone of Alaska. Truncated inactive ice wedges can occur below a younger generation of ice wedges as the result of a thawing episode followed by renewed growth of permafrost, and thus be evidence of environmental change. Such cases are known from both Alaska (Sellmann, 1967, 1972) and Canada (Mackay, 1975a). How old inactive ice wedges can be is uncertain but some in the Mackenzie Delta region of northern Canada are older than 40 000 ¹⁴C years (Mackay, 1976a), but no ice wedges are demonstrably older than Wisconsin according to Pewé (1975, 53).

In places, fossil ice-wedge polygons can be recognized from their surface appearance as suggested by E. P. Henderson (1959a, 48–57), although the cases he cited are troublesome because of their large size and central depressions, some of the depressions being suggestive of collapsed pingos. Generally independent evidence to eliminate alternative possibilities (such as desiccation polygons) is required. For instance, the grain-size analyses cited by Svensson (1976a, 11, Figure 4) in confirmation of air photographs of patterns in northern Germany, which are recognizable on the ground, strongly support the fossil ice-wedge interpretation because the deposits are too coarse grained to make desiccation cracking plausible. Most fossil ice-wedge polygons commonly manifest themselves in section through ice-wedge casts, formed by the melting out of the former wedges and the slumping in of adjacent or overlying sediments. As seen in section, the spacing between ice-wedge casts is commonly less than the true polygon diameters, but in some cases greater, depending on the orientation of the exposure. Some of the variations that may occur during infilling were discussed by Brüning (1964). If the infill is sand, fossil ice-wedge polygons might be difficult to distinguish from fossil sand-wedge polygons (Péwé, 1959). However, as previously noted, original sand wedges are likely to be thinner, vertically foliated, and devoid of stony material (Black, 1969b, 229).

In each case the forms would be incontrovertible evidence of former permafrost if correctly identified. Fossil ice-wedge polygons and sand-wedge polygons are widespread (cf. review by Dylik and Maarleveld, 1967, 14–21) and constitute the basis for many paleoclimatic reconstructions of periglacial environments. The crux of the problem is to distinguish the fossil forms from other features that may closely resemble them, especially in cross section. Examples

are (1) differential solution in calcareous deposits (Byrne and Trenhaile, 1977; Grubb and Bunting, 1976; Yehle, 1954), (2) decay of tree roots (C. S. Denny and Walter Lyford, personal communication, 1956), (3) dikes resulting from glacial drag (cf. Dionne and Shilts, 1974), (4) large-scale desiccation cracks in warm arid climates (Knechtel, 1951; Lang, 1943; Neal, 1965; Neal, Langer, and Kerr, 1968; Willden and Mabey, 1961) and in temperate climates (cf. E. M. White, 1972; E. M. White and Agnew, 1968; E. M. White and Bonesteel, 1960), and (5) perhaps seasonal frost cracks in non-permafrost areas.

Some of the problems were discussed by Dżułyński (1965), Johnsson (1959), and Wright (1961, 941–2). Among critical criteria are (1) the polygonal character of the crack pattern in plan, (2) the depth of the cracks, and (3) the nature, structure, and organic content of the surrounding and infilling sediment (cf. section on Frost cracking in preceding chapter). The presence of polygonally oriented casts would eliminate non-cracking processes, deep cracks if of frost-crack origin would indicate either permafrost or very deep seasonal frost approaching permafrost conditions, and the nature and organic content of the sediment might differentiate between frost cracking and large-scale desiccation cracking in an arid environment. The problem of differentiating frost-cracking from other cracking processes is discussed further in considering the origin of patterned ground.

Nonsorted polygons have been described as emerging from beneath a small ice cap on Baffin Island in the Canadian Arctic (G. Falconer, 1966), and nonsorted polygons and stripes that may also be fossil in the sense of being exhumed from beneath glacier ice have been reported from the Antarctic (Stephenson, 1961).

Lithified large nonsorted polygons of pre-Pleistocene age that have been interpreted as permafrost-crack polygons are known from strata as old as Precambrian (A. M. Spencer, 1971, 40–5, 49–53; G. M. Young and Long, 1976) but their periglacial origin is usually difficult to prove (cf. discussion by Deynoux and Trompette, 1976, 1308–9; and reply by Schermerhorn, 1976, 1317–18). However, the association with glacial deposits and reported pingos in the Ordovician of the Sahara strengthens the evidence for this occurrence (Beuf *et al.*, 1971, 259–335).

b Sorted 'Sorted polygons are patterned ground whose mesh is dominantly polygonal and has a sorted appearance commonly due to a border of

5.24 Small sorted polygons, Mesters Vig, Northeast Greenland. Scale given by 16-cm rule (*cf. Washburn, 1969a*, 162, *Figure 103*)

stones surrounding finer material' (Washburn, 1956*b*, 831) (Figures 5.24–5.25).[6] The minimum mesh size of sorted polygons is probably on the order of 10 cm across and thus slightly larger than that of nonsorted forms, but otherwise there is no size distinction in the small forms. However, the maximum reported mesh size of large sorted forms is some 10 m and thus an order of magnitude smaller than in large nonsorted polygons. Both small and large forms are most frequent on nearly horizontal surfaces and they do not appear to occur on as steep slopes as do some nonsorted polygons. In places small sorted forms occur in the central bare areas of larger sorted forms.

The central areas are similar to those of non-sorted forms in having a concentration of fines, either with or without stones. The bordering stones tend to increase in size with the size of the polygons but to decrease in size with depth whatever the polygon dimensions. According to R. P. Goldthwait (1976*a*, 33), the size of the largest common clasts (10–50 per cent by weight) may in fact control the diameter of sorted forms as follows, with the ratio of average size of large common clasts to the diameter between centres of stony borders ranging from 1:5 to 1:10.

Clasts	Diameter
2–10 cm chips mixed	20–50 cm
6–30 cm stones mixed	50–200 cm
20–150 cm stones mixed	2–5 m
0.5–5 m blocks mixed	5–20 m

This general relation, noted long ago by Meinardus (1912*b*, 25), seems also to be correlated with the

5.25 Large sorted polygons, Kjölur plateau, southwest Iceland. Scale length 2 m (*Schunke, 1975a, Abb. 3, cf. p. 80*)

5.26 Excavation across stony border of sorted net. Scale given by 16-cm rule (*cf. Washburn, 1969a, 167, Figure 107*)

[6] Sorted polygons with 'reverse' sorting – i.e. with coarser centres than borders – have been reported from southeast Libya. Maximum diameters average 2.2 m, and the borders are 10–15 cm wide. These polygons may be related to a crack pattern in underlying bedrock but their exact nature has not been established (Underwood, 1974).

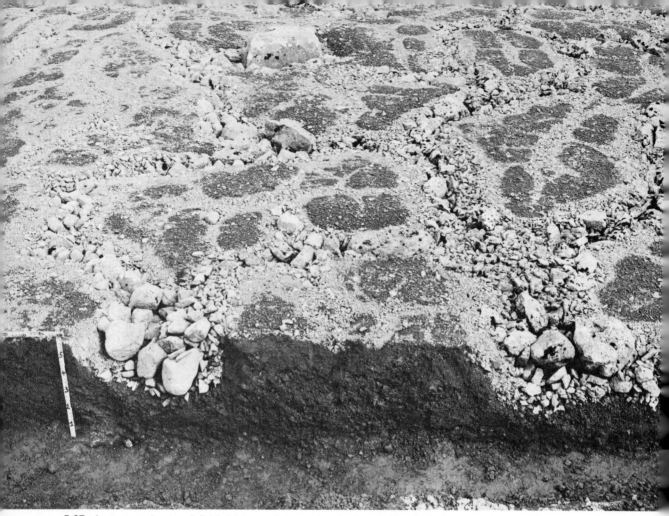

5.27 Large sorted nets subdivided into smaller nets, basalt plateau, Thorskafjardarheidi, northwest Iceland. Scale figures in decimetres; diameter of large nets, 2.5 m. Photo by Ekkehard Schunke (*cf. Schunke, 1975b, 51, Figure 2*)

observation that the diameter of a sorted polygon increases with increasing thickness of the soil subject to polygon development (Bout, 1953, 77–8; Schunke, 1975a, 41–3, 62, 91–2, 95, 105, 120; Stingl and Herrmann, 1976, 221–5), an observation that also applies to some nonsorted polygons (Segerstrom, 1950, 115). However, in places large sorted polygons having central areas up to about 1.4 m across occur in strongly frost-heaving soil only 0.5 m thick (Schunke, 1975a, 121, 198, 269, Table 3, Obs. 47). Tabular stones tend to be on edge and oriented parallel to the border (Figure 5.26),[7] suggesting a lateral thrust. This orientation developed in some field experiments by Pissart (1977a, 146–7), who as a cause favoured frost thrusting from the border toward the centre or *vice versa*. Expansion of the soil in the central area following desiccation and wetting is another possibility as indicated by

Prechtl (1965, 106). It was rejected by Pissart because pronounced deformation was so rapid (within a year) but it is not clear to what extent the preparation of his experiments influenced the results. There is a tendency for the largest stones in a border to be concentrated along the axis (Figure 5.27), a fact Prechtl interpreted as reflecting more rapid frost heaving and ejection of large stones than small (cf. section on Upfreezing of objects under Frost heaving and frost thrusting in the preceding chapter).

Stones in the borders tend to decrease in size with depth. In some cases at least, this may be because the stones there, including those immediately adjacent to the central fines, remain moist longer than stones near the surface and therefore weather more rapidly into smaller sizes, thus providing a secondary sorting effect. The borders themselves

---

[7] See footnote to section on Sorted circles in this chapter.

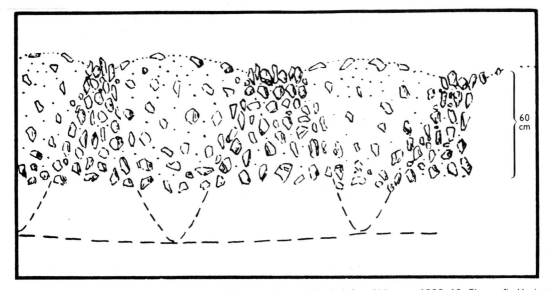

60 cm

5.28 Stony border of sorted polygons merging with stony layer at depth (*after Ahlmann, 1936, 10, Figure 4*). Horizontal dashed line shows approximate lower limit of active layer

narrow downward and terminate in some forms (Figures 5.26–5.27), whereas in other forms they broaden with depth and merge into a stony layer (Figure 5.28). In some forms the downward-narrowing borders are coincident with a crack pattern, and in places polygons are transitional from nonsorted to sorted (Schunke, 1975a, 42, 44).

The grain-size contrast between borders and centres leads to major differences in thermal regime (Rydquist, 1960, 66, Figure 16; Sharp, 1942a, 288–90).

Small sorted polygons occur in many different environments. Cold-environment forms are known from polar regions to temperate highlands. They can form within a year or two where permafrost is absent and seasonal freezing extends to depths of 40–50 cm or more, the probable mean annual temperatures being 5° to −2° (R. P. Goldthwait, 1976a, 34). Superficially very similar features also occur in warm-arid regions, including Death Valley, California, where forms up to 2.7 m (9 ft) across are associated with a crack pattern in rock salt and in gypsiferous and stony silt (Hunt and Washburn, 1966, B118).

Large sorted polygons are best developed in permafrost environments. On the basis of the literature and a lifetime of observation in many different places, R. P. Goldthwait (1976a, 34) concluded that regional occurrences of sorted forms exceeding 2 m in diameter imply ice-rich permafrost and that this in turn requires adequate moisture and mean annual

air temperatures of −4° to −6° or colder for tens to hundreds of years. Clearly extrazonal forms whose presence can be explained by special conditions must be excluded. The importance of edaphic conditions was stressed by Stingl and Herrmann (1976) who concluded, on the basis of multivariate analysis of voluminous observations by Stingl and Schunke, that such conditions were more important than climate in explaining the distribution of large and small sorted polygons and nets in Iceland. Even where climate can be shown to be paramount, it remains to be demonstrated that sorted forms such as those cited by Goldthwait normally require permafrost for their development, however probable this seems. Active sorted polygons 1.8–3.7 m (6–12 ft) in diameter are reported from the Avalon Peninsula of Newfoundland where permafrost is absent but an impermeable till could be regarded as a substitute (E. P. Henderson, 1968). Because large-sorted polygons are clearly best developed in permafrost environments, fossil forms (if not in saliferous material and of the warm-arid type) are commonly regarded as reasonable evidence of former permafrost. At the same time, pending further study, they should not be accepted as proof. The relic or inactive nature of sorted forms is reasonably established by evidence of immobility as determined by (1) lichens covering the tops but not bottoms of stones; (2) precise measurements over a period of years; (3) rephotographing after many (50–100) years; (4) development of a thick, con-

5.29 Hummocks, Mesters Vig, Northeast Greenland

tinuous vegetation mat over the central area; (5) well-developed soil horizons, especially non-polar types (R. P. Goldthwait, 1976*a*, 34). Like nonsorted polygons, some ancient sorted patterns are appearing from beneath receding glaciers, both in the Arctic (where the writer has seen evidence of this near Thule, Greenland) and in the Antarctic (Chambers, 1970, 97; Yoshida and Moriwaki, 1977). As noted in the discussion of sorted circles, fossil forms of any sorted patterns are rarely recognized in stratigraphic section.

5   Nets

Nets are patterned ground whose mesh is neither dominantly circular nor polygonal. In most other respects such intermediate forms are similar to impinging circles and polygons, and the same nonsorted and sorted terminology applies. The size range and slope relation of most nets are similar to those of circles and polygons, although no nets are known to be as large as the largest nonsorted polygons.

The constitution of most nets parallels that of circles and polygons except for special nonsorted forms known as hummocks (thúfur in Iceland), which are characterized by a knob-like shape and vegetation (Figures 5.29–5.31). The Icelandic forms can be hump shaped, plateau shaped, shield shaped, and ridge shaped (Schunke, 1977a, 5–8, 40). Some forms (earth hummocks) have a core of mineral soil; others (turf hummocks) may or may not have such a core or may have a core stone (Raup, 1965). Well-developed hummocks are normally up to 50 cm high and 1–2 m in diameter, although hummocks up to 78 cm high and 3.2 m in average diameter have been cited by Tarnocai and Zoltai (1978, 584). The total silt and clay content of the 173 hummocks they examined ranged from 58 to 99 per cent, the clay varying from 17 to 57 per cent (Tarnocai and Zoltai, 1978, 583, 587). Hummocks can occur on slopes as steep as 15° (Raup, 1965, 2; Schunke, 1975a, 132) or 20° (Sharp, 1942a, 282–3). Observations in Iceland indicate that a water table near the ground surface tends to inhibit hummock development (Schunke, 1975a, 132, 216; 1977a, 34, 38; Thorarinsson, 1951, 148), but other observations, for instance in Greenland (Raup, 1965, 33–77) as well as Iceland (Rutten, 1951, 166), show that this is not always the case. In places much would depend on the time of year. Icelandic thúfur are reported to re-form within 1–2 decades after having been levelled in farmers' fields (Gruner, 1912, 71).

Hummocks are prominent in some polar areas such as Ellesmere Island, Arctic Canada (Habrich, 1968), Northeast Greenland (Raup, 1965), and northern Eurasia (Figure 5.32), but they are usually considered to be most common in subpolar and alpine environments. Jahn (1975, 14) reported that hummocks in the latter, in both his and Sekyra's (1960) experience, occur at lower altitudes than solifluction or (at still higher levels) than sorted patterned ground. Clearly hummocks can develop where permafrost is lacking as in the coastal region of Iceland (Schunke, 1977a, 23, 46; Steche, 1933, 215–16, 221; Thorarinsson, 1951, 148) (Figure 3.3). In Japan the southern limit of active earth hummocks approximates the 6° mean annual isotherm (Koaze, Nogami, and Iwata, 1974a, 177–80). In fact Gerlach (1972, 72) concluded that hummocks should not necessarily be accepted as a typical indicator of a periglacial environment. Nevertheless the periglacial environment is where they are most prevalent.

In the Mackenzie Valley of northern Canada, they are particularly common in the northern half,

where the average ice content of the associated soils was found to be highest, being up to 185 per cent by weight according to Zoltai and Tarnocai (1974, 15, 33), who found no evidence of present or former ice wedges in the borders; however, a crack pattern in similar features has been observed by J. R. Mackay (personal communication, 1977). That many hummocks were active rather than relic was suggested by outward tilting of living trees on their sides, and soil profiles through the hummocks (Zoltai and Pettapiece, 1974), and reaction–wood crescents due to compressed tree rings indicated alternating periods of stability and instability that were perhaps climatically related (Zoltai, 1975). Active hummocks appeared to be closely related to the presence of ice-rich permafrost, and on the basis of soil profiles Zoltai and Tarnocai (1974, 9, 36) concluded that the hummocks they observed were relic where permafrost was lacking or had a low ice content. Extensive radiocarbon dating of hummocks studied by Zoltai, Tarnocai, and Pettapiece (1978) showed that ages differed by as much as 1480 ^{14}C years within the active layer of a single hummock (Hummock 16) because of involutions (cryoturbations). There was no consistent age difference between the margin and the centre of a hummock, such as might result from a circulatory movement of soil. A concentration of dates suggested that there was intensive development of hummocks some 4500 ^{14}C years ago. A small difference in average age of hummocks was found along a transect from the Arctic to the Subarctic, which may or may not be significant but suggests the desirability of further patterned-ground age determinations along such transects.

Aside from hummocks, nets have no special distribution distinct from that of the circle or polygon type to which a given net is most closely allied. Some small sorted nets occur in caves (Pulina, 1968). Because of their organic nature, hummocks do not form in vegetation-free areas but are confined to tundra or forest environments. They are not to be equated with knob-like nonsorted polygons described as sols à buttes (cf. Pissart, 1976c, 276).

The recognition of inactive or fossil nets is subject to the same conditions as those cited for analogous types of circles and polygons.

6 Steps

Steps are patterned ground with a step-like form and downslope border of vegetation or stones embanking an area of relatively bare ground upslope. They are

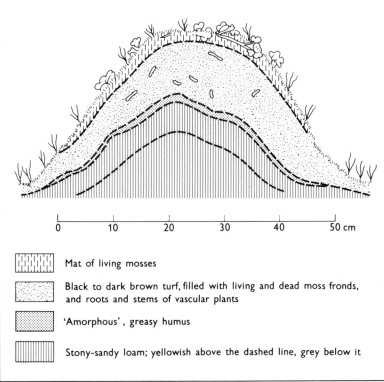

Mat of living mosses

Black to dark brown turf, filled with living and dead moss fronds, and roots and stems of vascular plants

'Amorphous', greasy humus

Stony-sandy loam; yellowish above the dashed line, grey below it

5.30 Cross-section of earth hummock, Mesters Vig, Northeast Greenland (*cf. Raup, 1965, 35, Figure 2*)

5.31 Cross-section of earth hummocks, Thjófadalir, central Iceland. Scale figures in decimetres. The structure is revealed by alternating dark and light beds of volcanic ash and dust and by core concentrations of palagonite detritus. Photo by Ekkehard Schunke (*cf. Schunke, 1975b, 41, Figure 1*)

restricted to slopes, and their downslope border is a low riser fronting a tread whose gradient is less than that of the general slope. Although they are thus terracette-like forms and are probably genetically related to some of the many varieties of terracettes that have been reported (Vincent and Clarke, 1976), steps are derived from circles, polygons, or nets rather than developing independently. Like their parent forms, they can be nonsorted or sorted, depending on whether the riser is characterized by vegetation only (Figure 5.33) or by stones (Figure 5.34). The latter have also been termed stone garlands (Sharp, 1942a, 277–8), but this designation includes other forms as well (cf. Washburn, 1956b, 833).

Probably nonsorted steps are derived mainly from hummocks – perhaps exclusively so. Presumably, sorted steps can be derived from either sorted circles or sorted polygons. They have been reported on gradients of 5°–15° in Alaska (Sharp,

1942a, 278) and on an estimated gradient of 10° in Northeast Greenland (Washburn, 1969a, 150). Strömquist (1973, 114; 116, Figure 34) reported that sorted steps he studied in northern Scandinavia commonly occurred on a steeper gradient than sorted stripes but his illustration suggests he referred to terracette-like forms rather than to steps derived from other patterned ground. The longest dimension of steps tends to be in the direction of steepest gradient. Some sorted steps are clearly an intermediate stage between sorted polygons and sorted stripes (Sharp, 1942a, 277–8). Except for the tendency towards a longer downslope dimension, the size of steps generally conforms to their parent form.

The constitution of steps is similar to that of the upslope forms from which they were derived. Stones in the fronts of sorted steps tend to be imbricated as well as on edge. Sharp (1942a, 295) stated that stones in the forms he studied moved downslope faster than the finer material but the evidence was

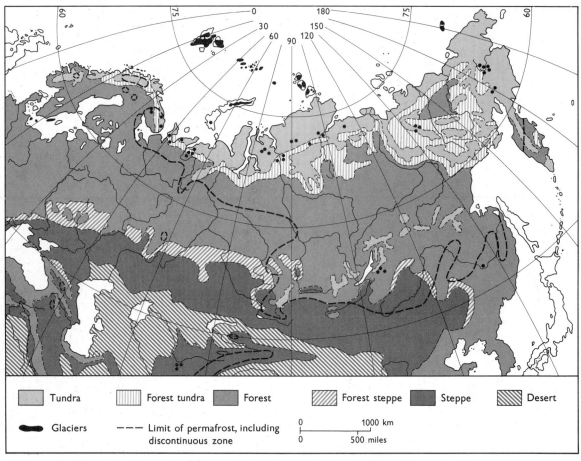

5.32 Distribution of active hummocks (●) in northern Eurasia (*after Frenzel, 1959, 1013, Figure 9*)

Tundra   Forest tundra   Forest   Forest steppe   Steppe   Desert

Glaciers   ——— Limit of permafrost, including discontinuous zone

0   1000 km
0   500 miles

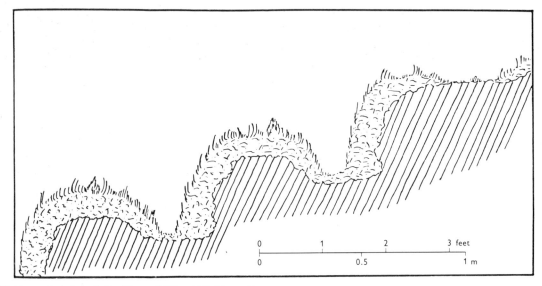

5.33 Nonsorted steps (*after Sharp, 1942a, 283, Figure 7*)

5.34 Sorted steps, Mesters Vig, Northeast Greenland. Scale given by 1-m tape (*cf. Washburn, 1969a, 150, Figure 93*)

not discussed. Where vegetation is present, it can be overridden and buried by the fronts so that it extends back upslope as an organic layer beneath the tread. The environmental association of steps is similar to that of their parent forms. Inactive forms may be identifiable as such but it is doubtful if fossil forms in section can be distinguished from the parent form unless the presence of overridden vegetation is established.

7 Stripes

a Nonsorted 'Nonsorted stripes are patterned ground with a striped pattern and a nonsorted appearance due to parallel lines of vegetation-covered ground and intervening strips of relatively bare ground oriented down the steepest available slope' (Washburn, 1956*b*, 837). Both small forms (Figure 5.35) and large forms (Figure 5.36) occur. The large forms shown in Figure 5.36 were on a gradient of 5°–6°, downslope from nonsorted polygons on a more gentle gradient. In places these vegetated stripes and those without vegetation were about equally wide; in other places the vegetated stripes were 0.3 to 0.6 m (1–2 ft) wide and were spaced some 3.0 to 4.6 m (10–15 ft) apart (Washburn, 1947, 94). Nonsorted

5.35 Small nonsorted stripes, Mesters Vig, Northeast Greenland. Scale given by 16-cm rule at centre (*cf. Washburn, 1969a, 152, Figure 94*)

stripes on slopes of 3°–20° in the Sachs Harbour area of Banks Island, Arctic Canada, were up to several hundred metres long. Here the relatively bare stripes were 1.0–1.5 m wide, alternating with narrower bands of vegetation in depressions 5–15 cm deep (French, 1974a, 73–4). A close similarity in cross section of nonsorted stripes and of small nonsorted polygons on Banks Island was emphasized by Pissart (1976c, 282). Stripes of tussocks (*Poa flabellata*) alternating with depressions containing mosses in nonsorted soil have been described from the Sub-antarctic (Stone, 1975). Hummocks on a slope can also form stripe patterns but appear to be uncommon (Jan Lundqvist, 1962, 58, Figure 30).

The nonsorted stripes in Figure 5.36 consisted of a diamicton. The vegetated stripes and associated organic soil narrowed downward and were thus wedge-shaped in cross-section. The Sachs Harbour stripes in their bare areas contained mostly medium to coarse sand with many pebbles and occasional larger stones.

Except for being slope features, nonsorted stripes have probably somewhat the same distribution as nonsorted circles and some nonsorted polygons but they appear to be less common. Warm-climate forms known as 'wavy gilgai' are the slope analogue of 'normal gilgai' (Costin, 1955; Hallsworth, Robertson, and Gibbons, 1955, 2–7, Plate 4). Large striped forms, manifested by differences in vegetation, are also reported from semiarid environments such as Kenya (Spönemann, 1974).

Fossil nonsorted stripes have been described from

5.36 Large nonsorted stripes, Mount Pelly, Victoria Island, Northwest Territories, Canada (*cf. Washburn, 1947, Figure 2, Plate 27*)

eastern England (R. Evans, 1976; Watt, Perrin, and West, 1966; R. B. G. Williams, 1964), an unusual aspect being that they were on gradients as low as 0°50′ and not exceeding 6°, probably because the associated chalky soil would have been especially prone to solifluction (R. Evans, 1976, 13, 18). Features that were totally vegetation covered and perhaps inactive forms of nonsorted stripes have been reported from North Wales (Ball and Goodier, 1970, 208–9). In Poland, wedge structures of sand, occurring in till and oriented more or less parallel to the gradient, have been interpreted as fossil nonsorted stripes (or possibly greatly elongated polygons) of frost-crack origin (Goździk, 1976). Nonsorted stripes that appear to be exhumed from beneath glacier ice occur in the Antarctic (cf. discussion of nonsorted polygons). The general scarcity of reports suggests that inactive or fossil

nonsorted stripes are rare, difficult to recognize, or both. To the extent that pedogenic horizons might be differentially developed as between vegetated and non-vegetated stripes, the repetitive and more or less regular nature of the differentiation may assist recognition. By contrast, nonsorted circles and small polygons may be less persistent features.

b Sorted 'Sorted stripes are patterned ground with a striped pattern and a sorted appearance due to parallel lines of stones and intervening strips of finer material oriented down the steepest available slope' (Washburn, 1956*b*, 836). Like nonsorted stripes, both small (Figure 5.37) and large (Figure 5.38) sorted stripes occur. Although observed on gradients as low as 1° (A. Pissart, personal communication *in* Collard, 1973, 327), they seldom occur on gradients of less than 2°–3°, and if derived by downslope extension of sorted polygons or sorted nets the transition gradient commonly ranges from about 3° to 7° and may be characterized by sorted steps. However, transition gradients as high as 20° have been reported in dry alpine environments (Graf, 1976, 439).[8] In general, other conditions being equal, low transition gradients probably correlate with high clay, silt, and very fine sand contents, and hence with moisture retention (Figure 5.39). Some stripes occur independently of other sorted forms. Maximum gradients on which sorted stripes occur rarely exceed 30° but can be as steep as 40° (Graf, 1976, 438).[8] The stripes are usually parallel and in places sinuous. The stony stripes may be several centimetres to 1.5 m or more wide, and the intervening finer stripes tend to be several times wider than the coarse. In relief the coarse stripe can be either higher or lower than the finer (Hastenrath, 1973, 177–8). According to Spönemann (1977, 310–11), in places nonsorted stripes are simulated by the temporary frost heaving of fines by needle ice, but upon desiccation of the fines and their removal by wind and water coarse material is left standing in relief and a sorted pattern becomes evident. Such rapid inversions of topography and appearance are unusual. Active sorted stripes as long as 120 m occur in Greenland (Washburn, 1969a, 180), and stripes well over 150 m long but apparently relic are known from elsewhere (Pyrch, 1973, 86, 115).

The stones of the coarse stripes can range from pebbles to boulders, depending on the size of the stripes. The intervening finer material can be stone free or contain stones and be a diamicton. In some cases the coarse material apparently moves faster than the finer (Antevs, 1932, 57–8; Pissart, 1977a, 151, footnote 1; Washburn, 1967, 187–8); in others the opposite is true. Büdel (1977, 70–1) found that fine stripes generally move faster than adjacent coarser material. Measurements by Pissart (1972a, 263) on a 7° gradient showed a rate of 0.2–3 cm yr^{-1} for the coarse stripes as compared with 2–6 cm yr^{-1} for the fine; similarly, measurements on a 6°–9° gradient over a 10-year period showed that the movement along the centre line of the stripes was 0.5–1.8 cm yr^{-1} for the coarse stripes and 4–7 cm yr^{-1} for the fine (Pissart, 1977a, 152). Observations by Benedict (1970b, 203) on a 12°–13° gradient over a 3-year period indicated rates of 1.3–3.9 cm yr^{-1}, with stones in the fine stripe moving downslope about twice as fast as those in the coarse. A similar

difference in rate between coarse and fine stripes was demonstrated by Mackay and Mathews (1974b, 348, 352) for a 10°–15° gradient over a 10-year period during which the average rate in the coarse stripes was about 15 cm yr^{-1} and in the fine stripes about 35 cm yr^{-1} (Figure 5.40). On gradients up to 15°, Chambers (1966a, 23–30) found the fine material moving 15 cm yr^{-1} relative to the coarse. In still other cases there was no significant difference between fine and coarse stripes (Gradwell, 1957, 801; Pissart, 1972a, 263). Where the coarser stripes were the slower moving compared with the finer stripes it has been explained by (1) local obstructions (Mackay and Mathews, 1974b, 354) or, more generally, (2) the stony cover having an insulating effect that retarded gelifluction (Benedict, 1970b, 203), or (3) nocturnal cooling and development of needle ice (Gradwell, 1960, 584; Mackay and Mathews, 1974b, 359). According to Benedict (1970b, 203, 216), faster movement of fine than coarse material is probably characteristic of solifluction rather than frost creep, although the needle-ice observations of Gradwell and of Mackay and Mathews show that frost creep is the critical process in places. It is worth noting that stripes whose average movement was about 10 cm yr^{-1} (with no significant difference between stripe and interstripe areas) were nevertheless judged to be largely relic, based on lichen-covered stones and vegetational aspect (Pyrch, 1973, 90, 94–101, 115–16).

The fabric of sorted stripes tends to be characterized by the long axis of stones being primarily in the vertical plane parallel to the stripe. As a result tabular stones are commonly on edge, especially in the coarse stripes.[9] A secondary long-axis maximum at right angles to the first is reported to be present in the coarse stripes but absent in the intervening finer stripes (Furrer, 1968; Furrer and Bachmann, 1968, 5–8, 11–13). The size of stones generally decreases with depth. Sorting, again depending on size of the forms, extends from a depth of a few centimetres to about a metre, and the stony accumulations normally narrow downward as in the borders of sorted polygons.

Like sorted polygons and nets, sorted stripes occur in many environments, ranging from polar to warm

5.37 Small sorted stripes, east side Glacial Lake, Seward Peninsula, Alaska. View upslope. Scale given by 17-cm rule

---

[8] Graf cited Erdstreifen (earth stripes) and Zellenboden (small polygons), which he distinguished from Steinstreifen (stone stripes) and Steinpolygonen (stone polygons). However, his illustrations and text show that the distinction was based on size, not sorting, and that his discussion of Erdstreifen and Zellenboden included small sorted forms.

[9] See footnote to discussion of sorted circles.

5.38 Large sorted stripes, Hesteskoen, Mesters Vig, Northeast Greenland (*cf. Washburn, 1969a, 179, Figure 113*)

arid. The largest forms are commonly associated with permafrost. Small ones such as those described by Poser (1931, 202–3; 209, Figures 6–7; 211, Figure 9) from the vicinity of Reykjavik, Iceland, can occur where it is absent (Steche, 1933, 221–3). Numerous middle-latitude highlands, many without permafrost, have areas with small sorted stripes that are active as established by direct measurements such as those noted above. In some places stripes formed again within 1–2 years after being destroyed (Michaud and Cailleux, 1950; R. Miller, Common, and Galloway, 1954).

Inactive forms can be recognized by the same criteria that apply to sorted polygons and nets. Fossil forms in stratigraphic sections have been reported by Dücker (1933) and Nørvang (1946, 61–2) but are probably rarely recognized.

8   Origin—general

The genesis of many forms of patterned ground is problematical, not only quantitatively but even qualitatively as emphasized, for instance, by Stingl (1974), who also stressed the problem of multiple origins and illustrated it by careful field observations

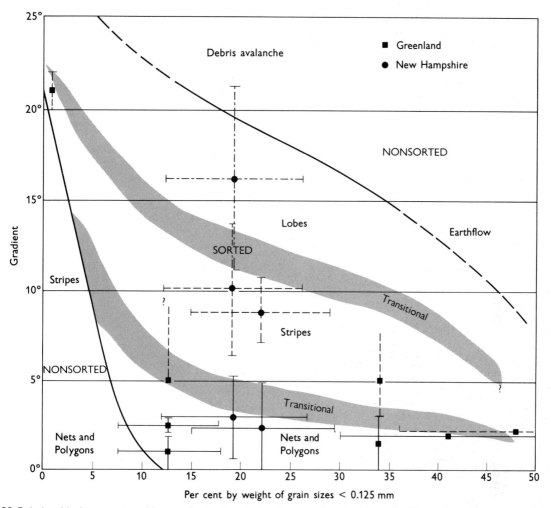

5.39 Relationship between transition gradients of sorted patterns and grain sizes < 0.125 mm, based on observations in Greenland and on Mount Washington, New Hampshire (*after R. P. Goldthwait, 1976a, 32, Figure 4*)

in Iceland and elsewhere. An adequate explanation of patterned ground is a multivariat problem (Caine, 1972a, 56). The following discussion, which reviews various possibilities, is based on Washburn (1956b, 838–60; 1969a; 1970) and the following premises: (1) Patterned ground is of polygenetic origin; (2) similar forms of patterned ground can be due to different genetic processes; (3) some genetic processes may produce dissimilar forms; (4) there are more genetic processes than there are presently recognized terms for associated forms, and (5) it is desirable to maintain a simple and self-evident terminology. Table 5.1 serves both as a summary and as an attempt at a genetic classification of patterned ground; it combines, in matrix form, exist-

ing terms for geometric patterns with existing terms for genetic processes. The resulting combined terms are readily understandable in light of the matrix approach. However, some combinations would be awkwardly long without being shortened and therefore the terms nonsorted and sorted in the combined forms have been abbreviated to N and S.

Only initial processes ('first causes') with respect to origin of the basic geometric pattern (circles, nets, polygons, or stripes) are reviewed. A number of these were considered in the preceding chapter. Once a basic pattern is established, various processes (cf. section on Sorting by frost action in preceding chapter) can modify it to produce sorted forms. For instance, a cracking process in itself provides only

Table 5.1. Genetic classification of patterned ground

PATTERNS		CRACKING ESSENTIAL						CRACKING NON-ESSENTIAL						
		DESICCATION CRACKING	DILATION CRACKING	SALT CRACKING	THERMAL CRACKING — SEASONAL FROST CRACKING	THERMAL CRACKING — PERMAFROST CRACKING	FROST ACTION ALONG BEDROCK JOINTS	PRIMARY FROST SORTING	MASS DISPLACEMENT	DIFFERENTIAL FROST HEAVING	SALT HEAVING	DIFFERENTIAL THAWING AND ELUVIATION	DIFFERENTIAL MASS-WASTING	RILLWORK
CIRCLES	NONSORTED								Mass-displacement N circles	Frost-heave N circles	Salt-heave N circles			
CIRCLES	SORTED						Joint-crack S circles (at crack intersections)	Primary frost-sorted circles, incl.? debris islands	Mass-displacement S circles, incl. debris islands	Frost-heave S circles	Salt-heave S circles			
POLYGONS	NONSORTED	Desiccation N polygons	Dilation N polygons	Salt-crack N polygons	Seasonal frost-crack N polygons	Permafrost-crack N polygons (incl. ice-wedge, permafrost soil-wedge polygons)	Joint-crack N polygons		Mass-displacement N polygons?	Frost-heave N polygons?	Salt-heave N polygons?			
POLYGONS	SORTED	Desiccation S polygons	Dilation S polygons	Salt-crack S polygons	Seasonal frost-crack S polygons	Permafrost-crack S polygons	Joint-crack S polygons	Primary frost-sorted polygons?	Mass-displacement S polygons?	Frost-heave S polygons?	Salt-heave S polygons?	Thaw S polygons?		
NETS	NONSORTED	Desiccation N nets incl.? Earth hummocks	Dilation N nets		Seasonal frost-crack N nets, incl.? Earth hummocks	Permafrost-crack N nets (ice-wedge and ? permafrost soil-wedge nets)			Mass-displacement N nets, incl.? Earth hummocks	Frost-heave N nets, incl. Earth hummocks	Salt-heave N nets			
NETS	SORTED	Desiccation S nets	Dilation S nets		Seasonal frost-crack S nets	Permafrost-crack S nets?		Primary frost-sorted nets	Mass-displacement S nets	Frost-heave S nets	Salt-heave S nets	Thaw S nets		
STEPS	NONSORTED								Mass-displacement N steps	Frost-heave N steps?	Salt-heave N steps?		Mass-wasting N steps	
STEPS	SORTED							Primary frost-sorted steps?	Mass-displacement S steps	Frost-heave S steps	Salt-heave S steps	Thaw S steps?	Mass-wasting S steps	
STRIPES	NONSORTED	Desiccation N stripes	Dilation N stripes?		Seasonal frost-crack- N stripes?	Permafrost-crack N stripes	Joint-crack N stripes?		Mass-displacement N stripes	Frost-heave N stripes	Salt-heave N stripes		Mass-wasting N stripes?	Rillwork N stripes?
STRIPES	SORTED	Desiccation	Dilation		Seasonal frost-crack	Permafrost-crack	Joint-crack S stripes	Primary frost-sorted	Mass-displacement	Frost-heave S stripes	Salt-heave S stripes	Thaw S stripes?	Mass-wasting	Rillwork S stripes

PROCESSES

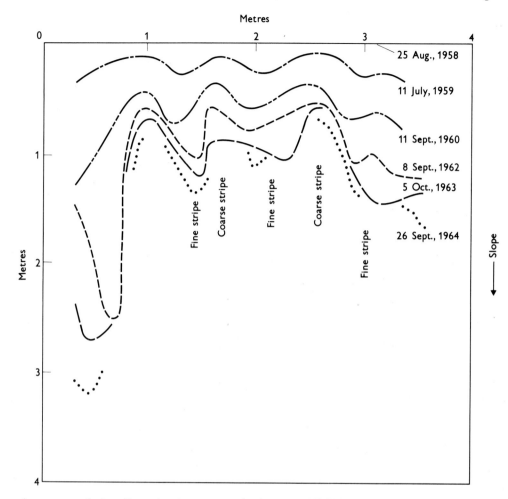

5.40 Annual movement of a lone line painted across sorted stripes on a 10°–15° gradient, Garibaldi Park, British Columbia, Canada (*after Mackay and Mathews, 1974b, 352, Figure 7*)

a nonsorted pattern but a sorting process acting on an initially nonsorted pattern can change it to a sorted pattern. These sorting processes are not specifically included in the table unless they also act as an initial patterning process. Genetic processes in Table 5.1 are divided into (1) processes in which cracking is the essential and immediate cause of the resulting patterned-ground forms, and (2) processes in which cracking is non-essential. Involutions are omitted from the discussion, since they are probably of various origins and associated with many kinds of patterned ground. The table also omits any mention of gelifluction, although this process is present in many forms of patterned ground on slopes. Through 'microgelifluction' it may contribute to sorting, and through 'macrogelifluction' to the

development of nets, steps, and stripes. However, to what extent it is an initial process in forming patterned ground is uncertain.

Many occurrences of patterned ground must reflect the influence of several processes, which can be thought of as end members of a continuum (Washburn, 1956b, 859–60). Probably various combinations are possible. For sorted forms, Stingl (1974, 259–60) suggested cracking, sorting, and turbation movement, the last implying a variety of processes leading to diapir-like movement and essentially corresponding to mass displacement as previously described. A more general continuum would have a patterning process, a sorting process, and a slope process (primarily creep and gelifluction) as end members (Figure 5.41).

The frequency and intensity of freeze-thaw cycles have been stressed by Troll (1944, 559–62, 566–8, 618–19, 673–4; 1958, 10–15, 18–19, 57–8, 95) and others as a genetic factor determining sorting and size of patterned ground. Diurnal cycles and shallow freezing would determine small forms with shallow stony borders, annual cycles with deep freezing would account for large forms with deep-seated borders. Even though small and large forms can occur together, which led Furrer and Freund (1973, 132) to criticize this concept in its extreme form, the concept would explain why there are characteristically small forms at high altitudes in tropical latitudes and widespread large forms in polar latitudes. At the same time, there has been a tendency over the years to oversimplify and interpret, as well as classify, small forms as diurnal-type features and large forms as annual-type features without consideration of intermediate possibilities such as those recognized by Troll (1944, 567; 1958, 19). In studying the distribution of patterned ground in Iceland, Preusser (1973) found it necessary to recognize forms intermediate in size between 'diurnal' and 'annual'. In fact, except where diurnal cycles only are present, there is no reason for regarding small forms as necessarily only diurnal in nature. On the other hand, large polar-type forms are not restricted to polar environments but can occur in low-latitude alpine environments such as the Dhaulagiri-Himalaya (Nepal), given a long-enough freezing period as well as short-term freeze-thaw episodes (Kuhle, 1978a, 294, 296, 299).

It is misleading to emphasize just the end members of a continuous series of events. Given forms demonstrably due to frost action, frequency and intensity of freezing and thawing should be regarded as continuous variables that combine in various degrees, along with other factors, to determine the variety, size, and regional distribution of patterned ground. Recently, Büdel (1977, 51–7) emphasized a complex of frost-related processes acting in various combinations at different times of the year, including the influence of meltwater, upfreezing of stones, cryostatic pressure, frost heaving and thrusting, needle ice, thaw cracking, desiccation cracking, infilling of cracks, eluviation and filtering, frost wedging, and circulatory movements.

In order better to assess the environmental significance of patterned ground, discussion of origin is not limited to frost-action processes. Clearly, environmental interpretations are complicated by the polygenetic character of patterned ground, a difficulty that Stingl (1974, 260) concluded could only

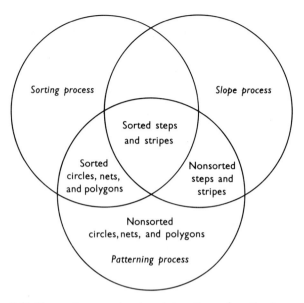

5.41 Venn diagram showing interaction of patterning, sorting, and slope processes in development of patterned ground

be resolved by strengthened genetic investigations combining laboratory and field approaches.

9 Origin—cracking essential

a General Cracking is a very widespread and important process in the initiation of sorted as well as nonsorted patterned ground. Sorted forms initiated by cracking obviously involve the addition of a sorting process. Basic patterns are typically polygonal but gelifluction, possibly ice wedging in cracks, and some volume changes associated with cracking (for instance, in gilgai formation – cf. Cooke and Warren, 1973, 129–35; Hallsworth, Robertson, and Gibbons, 1955) can deform some polygons to nets. Moreover, given favourable slope conditions, it seems probable that small transverse cracks would be narrowed by mass-wasting, whereas cracks paralleling the slope would tend to remain open and determine the location of small stripes (cf. Pissart, 1976a; Washburn, 1969a, 151, 154). Transitions from various-sized, nonsorted and sorted polygons initiated by cracking, to comparable stripes led Stingl (1974, 253) to conclude that sorted as well as nonsorted stripes could develop from crack forms. Study, over a 12-year period, of the evolution of patterned ground at Chambeyron (Basses Alpes) also convinced Pissart (1976a) that sorted stripes (up to 1 m across) could evolve from desiccation cracking combined with sorting processes. Although small, roughly circular desiccation cracks have been observed around concentrations of fines (Chambers, 1967, 4, 7), there is no evidence that normal cracking takes the shape of circles. On the other hand, according to Dostovalov and Kudryavtsev (1967, 220–4) cracking can initiate more or less circular spot tundra (pyatinstye tundry) and medallions (pyatna-medal'ony) (which the present writer equates with nonsorted circles) in that inward freezing from the cracks can squeeze unfrozen material to the surface. They also applied this concept to sorted forms in stony ground. The possibility that cracking can initiate circular forms has also been advanced by Shilts (1978). The present writer has observed bare circular patches in the central areas of presumed former ice-wedge polygons, and he ascribed the circular aspect to frost heaving within polygonal boundaries (Washburn, 1969a, 140–7). If the above views are valid, cracking would be a first cause even though the subsequent movement of soil was responsible for the circular pattern. However, pending more evidence, cracking is not here regarded as a primary cause of most circular patterns.

Abundant evidence indicates that true polygons, as distinct from nets and other patterned-ground forms that are sometimes referred to as polygons, are primarily contraction phenomena. Possible causes of contraction include thawing, synaeresis, partial wetting, drying, and lowering of temperature.

Thawing as a cause of polygons (cf. Washburn, 1956b, 852), and cracking of frozen ground as a result of rising temperatures (Pissart, 1976c, 278–80, 283) lack convincing evidence. Büdel (1977, 55) suggested thawing as a cause of cracking around already established centres of fines, hence not as a first cause. Perhaps a more promising possibility that should be studied is the role that synaeresis cracks (Jüngst, 1934; W. A. White, 1961; 1964) may have in polygon formation. 'Synaeresis cracks are fissures that develop in a suspension where waters are expelled from the clay-water system by internal forces; they may resemble mud cracks in the sediments' (W. A. White, 1961, 561). They form subaqueously in a saline environment where clays are flocculated and they are apparently due to adjustments accompanying realignment of clay particles under the influence of electrochemical forces between the particles (W. A. White, 1961, 569). Small-scale, polygonal tension cracks developed

underwater following formation of polygons by density displacements in laboratory experiments by Dżulyński (1963, 146). According to Kostyayev (1969, 235–6), synaeresis cracks can form polygons with diameters ranging from a few decimetres to tens of metres. Cracks formed by partial wetting in laboratory experiments were reported by Corte and Higashi (1964, 68, 70–1), who suggested surface tension as a possible explanation. That wetting of desiccated soil can cause further shrinkage was noted by J. J. Hamilton (1963, 32, 40–1), who speculated that densification is caused by immersion breaking down crumblike packets of clay particles. Interestingly, S. A. Harris (1964) regarded expansion upon drying of non-clayey soils as the cause of certain gilgai.

These processes remain to be further investigated but to the extent any resulting cracks persist, the cracks probably determine the location of polygons. Nevertheless, pending more data, the above processes are here regarded as probably minor in periglacial patterned ground.

b Desiccation cracking   Desiccation cracking is fissuring due to contraction by drying. It is probably one of the most common and important patterned-ground processes, especially for forms having diameters < 1 m but not excluding larger forms. Desiccation can be due not only to drainage and evaporation but also to withdrawal of moisture to loci of ice formation. Presumably the volume reduction is caused by reorientation of particles, especially clay, as the result of stresses consequent on the moisture transfer. Taber (1929, 457–8) demonstrated that desiccation accompanying freezing could produce tiny polygonal fissures in laboratory specimens. Subsequently Taber (1943, 1522–7) suggested that the same process could produce ice-wedge polygons and sorted polygons, and Schenk (1955a, 64–8, 75–6; 1955b, 177–8) argued that it was important in a wide variety of polygonal forms, including small ones. Laboratory experiments by Pissart (1964a) show that the process can produce small nonsorted polygons up to about 10 cm in diameter, but the polygons were subsurface rather than surface phenomena and were very irregular. That the volume reduction caused by freezing and thawing of clay soils is considerable is well known (cf. Chamberlain and Blouin, 1976; 1978; A. B. Hamilton, 1966) but proof is lacking that it produces more than a small-scale crack pattern. To the extent that desiccation is involved, the fact that polygonal cracks confined to the active layer are best developed at the surface and become rapidly less prominent with depth argues strongly for the predominant effect of subaerial desiccation; in fact Büdel (1960, 42) regarded such evidence as conclusive for forms he studied. Pissart's preliminary report clearly supports the desirability of further work and experimental data relating to Taber's hypothesis, particularly as applied to small polygons. Without further evidence, however, subaerial drying would seem to be the most probable kind of desiccation process responsible for well-defined polygonally patterned ground.

Salt content appreciably influences volume changes associated with wetting and drying as discussed long ago by Wollny (1897, 20–1, 31–2, 51). Sodium chloride, for instance, appears to affect development and doming of desiccation polygons (Dow, 1964; Elton, 1927, 179; Kindle, 1917, 140–4), it is involved in synaeresis cracking (cf. W. A. White, 1961; 1964) as previously described, and it is known to affect frost heaving (Beskow, 1935, 100–1, 232; 1947, 57–8). Also the wide variety of patterned ground in warm deserts (Cooke and Warren, 1973, 129–49), where salt is an important variable in forms related to desiccation as well as in some other forms (Hunt and Washburn, 1966), suggests the desirability of investigating the role of chemical variations in the mineral soil of patterned ground.

A major problem in the origin of some cracks is to distinguish between thermal contraction and desiccation, whatever the latter's cause. This problem as it affects large nonsorted polygons is discussed below under Seasonal frost cracking and Permafrost cracking.

Small nonsorted polygons in West Greenland (Boyé, 1950, 134) and in the arid environment of North Greenland have been ascribed to desiccation (W. E. Davies, 1961b, 49, Figure 2; 50; Tedrow, 1970, 81, Figure 65; 82), as have small polygons in Arctic Canada (D. I. Smith, 1961, 74), in Spitsbergen (Büdel, 1961, 347, 353; Elton, 1927, 176–80, 190–1), and elsewhere in the Arctic, and they are demonstrably due to drying in many environments (Longwell, 1928; Segerstrom, 1950, 115; Shrock, 1948, 188–209; and many others).

The importance of desiccation cracking in the origin of some small sorted polygons and nets has been repeatedly cited, with stones accumulating in cracks by various sorting processes (cf. Chambers, 1967, 11–12; Corte and Higashi, 1964, 60–4; Gradwell, 1957, 796–806; Spönemann, 1977, 309–10; Stäblein, 1977, 21; Washburn, 1956b, 848–9; 1967, 163–4). Field experiments by Pissart (1972a, 251–5;

1974a; 1976a; 1977a, 144–5), involving destruction of an area of small sorted polygons, indicate that cracking was the first stage toward reconstitution of the forms; sorting by upfreezing and movement of stones to the cracks was noticeable in 1–2 years. Desiccation was regarded as the principal cause of cracking (Pissart, 1976a; 1977a, 147–9) but not necessarily the only one (Pissart, 1974a, 246).

In general the conclusion is warranted that most well-developed and regular nonsorted polygons averaging about 1 m or less in diameter are primarily due to desiccation cracking, a conclusion also cited by Tricart (1963, 88; 1969, 91). That desiccation may produce large polygons as well does not affect the argument. For instance, Marnezy (1977b, 378–9) ascribed to desiccation a transitional pattern of nonsorted to sorted polygons in a lake bed that was subject to repeated drying. According to Tricart (1967, 196–7, 201), nonsorted polygons with diameters up to 0.5 m are probably due to air drying, whereas those with diameters of 0.6–3.0 m are more likely to be due to freeze drying. Presumably desiccation cracking and thermal contraction can reinforce each other. Thus Romanovskiy (1977, 69) concluded that some polygons are of dual origin in that desiccation cracks may be widened and deepened by frost cracking. The inverse sequence should also apply.

In addition to small nonsorted polygons, it appears demonstrable, as noted above, that some small sorted polygons are initiated by desiccation cracking. The process has also been applied to some earth hummocks (T. Koaze *in* Ellenberg, 1976b, 181; Zawadski, 1957, 60–4, 81–4), but the absence of cracks is demonstrable in at least some forms studied by Schunke (1975a, 133) and the present writer. Similar-appearing hummocks can probably have different origins (Beschel, 1966).

Desiccation cracking as an initiating cause of some nonsorted stripes and sorted stripes has not been investigated in detail. It is supported by a few field observations (Chambers, 1967, 5–7, 11; Klatka, 1961a, 313–14; 1968, 274–6; Pissart, 1977a, 151–2; Washburn, 1969a, 153–4) and some experimental evidence suggesting that desiccation cracks tend to be emphasized parallel to the gradient (Furrer, 1954, 237–9, 242; Kindle, 1917, 137–9). Desiccation stripes in the above context should not be confused with the 'giant desiccation stripes' described by Neal (1966), which are playa features transverse to the long axes of washes and narrow playas.

c Dilation cracking Dilation cracking is fissuring due to stretching of surficial materials. Although the

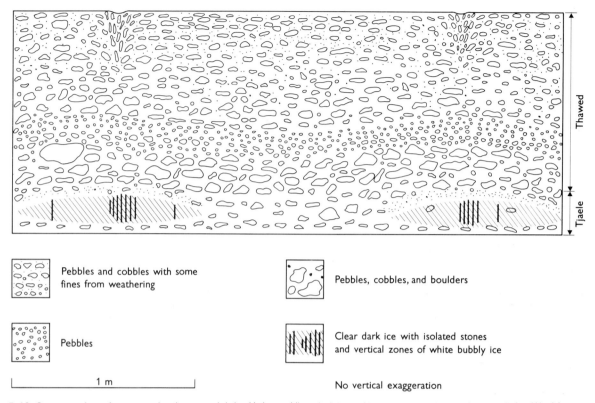

Pebbles and cobbles with some fines from weathering

Pebbles

Pebbles, cobbles, and boulders

Clear dark ice with isolated stones and vertical zones of white bubbly ice

1 m

No vertical exaggeration

5.42 Cross-section of nonsorted polygons, vicinity Holman, Victoria Island, Northwest Territories, Canada (*after Washburn, 1950, 44, Plate 10, Figure 2*)

process has been described as important (Boytsov, 1963, 72–3; Kostyayev, 1965; cf. Jahn, 1975, 34), there is little evidence that it is a widespread cause of patterned ground. Yet it is clearly a genetic process in places where the cracking is consequent on local differential heaving or on sagging of the ground.[10] In this context the terms differential frost heaving and differential thawing as used in Table 5.1 imply processes in which any associated cracking is incidental and non-essential to the origin of patterned ground.

Dilation cracking as opposed to other kinds of cracking may be identified from associated evidence of pronounced heaving or collapse as, for instance, a medial crack along an elongate heaved feature or a radial and concentric crack pattern on a domed one (Benedict, 1970a; Boytsov, 1963, 72–3; Jahn, 1975, 34 and Photo 14; Washburn, 1969a, 109; 110, Figure 71; 125, Figure 78; 136). In addition Benedict (1970a) observed that crack widths exceed

those to be expected from frost cracking.[11] Cracking associated with hydraulic pressure has been cited as the apparent cause of cracking and development of earth hummocks in the Snowy Mountains of Australia where present temperatures are too high to favour frost cracking (Costin and Wimbush, 1973, 119). Some cracks that would normally be ascribed to frost cracking may be dilation features as suggested by arching of beds below the cracks (Figure 5.42). In the case illustrated the cracks were part of a well-developed nonsorted polygon system in beach gravels. Ice wedges in cracks above anticlinal crests in sediments have been described by Soviet investigators (Kostyayev, 1965) as noted by Jahn (1975, 55–6) (although the arching here was attributed to density displacements in the thawed state), and dilation has been cited as a possible process in the origin of large nonsorted polygons in Finland (Aartolahti, 1972, 131). Possibly dilation cracking caused by irregularities of volume change during

[10] Jahn's (1975, 33) usage of compaction (subsidence) cracking as synonymous with dilation cracking emphasizes the sagging effect.
[11] Benedict regarded dilation cracking as a form of frost cracking because of its relation to frost heaving. However, frost cracking as used here and by most authors is restricted to cracks caused by thermal contraction.

freezing and thawing is more common than realized. In places the opposite cause-and-effect relationship should be considered in that moisture seeping down the cracks might localize frost heaving and upwarps (Stege) as suggested by Herz (1964) and Herz and Andreas (1966, 194–5). However, their comprehensive application of this concept to polygons and some other forms is highly speculative and contradicts evidence that heaving is usually minimal in borders compared to central areas. Given the presence of dilation cracking, the process might lead to the various patterned-ground forms associated with cracking, but patterns other than nonsorted polygons are probably very rare, since even nonsorted polygons of this origin appear to be uncommon— unless some small forms that are usually assumed to be desiccation features are in fact initiated by dilation as described by Hopkins and Sigafoos (1951, 67–8).

d Salt cracking Salt cracking describes the fissuring that initiates nonsorted and sorted polygons in hard rock salt and salt crusts of warm deserts. The exact process is not established but may be mainly thermal contraction (Hunt and Washburn, 1966, B120–30; Tucker, 1978, 97). However, desiccation cracking, but patterns other than nonsorted polygons are probably very rare, since even nonsorted polygons of this origin appear to be uncommon – unless sequent to the cracking (cf. Tricart, 1966, 20–1; 1970, 434–5). There is no evidence that salt cracking forms other than polygonal patterns, some of which occur on slopes as steep as 37° without becoming nets.

e Seasonal frost cracking Seasonal frost cracking is fissuring due to thermal contraction of frozen ground in which the fissuring is confined to a seasonally frozen layer. In a permafrost environment this layer is the active layer as discussed under Frost cracking in the preceding chapter (cf. also section on Nonsorted polygons in present chapter). Patterned ground resulting from this process is perhaps widespread and may be difficult to distinguish from forms developed by other processes. Most of the geometric patterns conform to those developed by other cracking processes as modified by slope conditions and presence or absence of superimposed sorting processes.

By definition, frost-crack polygons or, more accurately, seasonal frost-crack polygons, are not necessarily related to permafrost (Hopkins *et al.*, 1955, 139). Such polygons have been described from Alaska (Black, 1952, 131–2; Hopkins *et al.*, 1955, 137–9), Iceland (Friedman *et al.*, 1971; Schunke, 1974*b*; 1975*a*, 128–9; Stingl, 1974, 253–4; Svensson,

1977, 69–70), eastern Alps (Stingl, 1974, 254–5), and elsewhere (Black *in* Corte, 1969*a*, 140; Pataleyev, 1955; J. R. Reid, 1970*b*; Washburn, Smith, and Goddard, 1963), although their occurrence in Alaska has been challenged (R. E. Church, Péwé, and Andresen, 1965, 38; Péwé, Church, and Andresen, 1969, 49).

The Iceland occurrence reported by Stingl consisted of fresh cracks, 1–2 cm wide and 5 cm deep, forming irregular polygons up to 30 cm across, in places apparently beginning as 3 rays from a common centre. The surprising thing was that the cracks appeared following air temperatures no lower than −2°. This indicates the possibility that rather than being true frost cracks, they were the sites of thawed surface veinlets as suggested by the radiating aspect (cf. Figure 4.21), or perhaps re-opened older cracks. Friedman *et al.* (1971) concluded that the large forms they described (up to 100 m across) originated under a former more rigorous climate and have been recently reactivated but that there was no proof that they required permafrost. Their association with involutions led Schunke (1974*b*, 162; 1975*a*, 129) to deduce the former presence of continuous permafrost but the evidence of involutions is suspect as noted later in this chapter. In the case described by Reid, which was on the campus of the University of North Dakota in the winter of 1962–3, cracks developed during several days of −20° to −25° temperatures when snow cover was largely lacking. In places the cracks formed tetragonal polygons and were up to 2 m deep. They remained open at temperatures up to −5° and became filled with wind-blown organic silt as the result of two severe dust storms. However, surface indications of the cracks disappeared by spring. The case described by Washburn, Smith, and Goddard (1963), which was at Hanover, New Hampshire, was similar but the pattern was less well defined and no infill was observed. Although such cracks may be locally common where active, it is doubtful they would normally be preserved as fossil forms in the geologic record. Rather, fossil forms are presumptive evidence of at least such deep seasonal freezing as to approach, or be consistent with, discontinuous permafrost. This and related questions are discussed further in the section on Nonsorted polygons in the present chapter.

Grain size as a criterion in distinguishing between frost cracking and desiccation cracking remains to be investigated in any detail. Frost cracking occurs in many different soils and the influence of grain size on the process, except as it affects distribution of moisture, is not well known. As for desiccation, the process can be eliminated in the cracking of gravels and coarse sands lacking silt and clay. Also, according to E. M. White (1972, 108)

Narrow surface desiccation cracks will form in sandy soils with some clay, but desiccation-caused microrelief does not form unless the soil has more than about 40% clay. Depression soils with 20–40% clay occasionally have 60–90-cm-wide polygons formed by 2–5-cm-wide cracks during droughts.

More information is needed on proportions of clay and silt as they affect desiccation cracking and formation of nonsorted polygons but grain-size restraints will eliminate the process as a factor in some soils and favour alternatives such as frost cracking.

It has been suggested that seasonal frost cracking may initiate some earth hummocks (thúfur) (Thoroddsen, 1913, 253–4; 1914, 258–62), but cracking of any origin remains to be demonstrated as a general explanation for such forms even though cracks occur in places (Drury, 1962, 36–8). With one possible exception, no crack systems responsible for such hummocks were observed by Schunke (1975*a*, 133; 1977*a*, 19–20, 37, 42) in Iceland, nor by Raup (1965, 76) and the present writer in Greenland. Similarly the importance of seasonal frost cracks as a cause of stripes is uncertain, although field occurrences were reported by Furrer and Freund (1973, 193, 197, 200), who also suggested desiccation as a contributing factor.

f Permafrost cracking Permafrost cracking is fissuring due to thermal contraction of perennially frozen ground – i.e. it is frost cracking that extends into the permafrost and is not confined to the active layer. The resulting patterned ground is very widespread in polar environments and is of considerable significance because of its temperature implications and the fact that fossil forms are more subject than most patterned ground to preservation in the geologic record. The literature dealing with active and fossil permafrost cracks and criteria for their recognition is prolific as discussed under Frost cracking in the preceding chapter; further information is provided below. The best-known examples of the process are permafrost-crack nonsorted polygons (following the terminology of Table 5.1) of which two varieties are commonly distinguished – ice-wedge polygons and sand-wedge polygons.[12]

Where cracks bordering polygons can be traced into similar cracks in underlying permafrost, permafrost cracking can be assumed to have originated the polygons.

Confinement of bordering cracks to shallow depth and the active layer in a permafrost environment suggests that the polygons formed by processes other than frost cracking. Development of permafrost requires deep penetration of freezing temperatures, and therefore it can be logically assumed that frost cracking would have affected the permafrost and not be confined to the active layer or, as in places, to just the upper part of it. On the other hand, surface desiccation cracks in a permafrost environment are necessarily confined to the active layer. The 'filling soil veins' that are described by Katasonov and Solovyev (1969, 15–16) as being confined to the active layer do not necessarily counter this argument, since cracks below the veins continue into permafrost (cf. section on Frost cracking in preceding chapter).

Crack size and spacing are other possible criteria for distinguishing between permafrost-crack and other polygons. The factors controlling depth and spacing are complex but have been quantitatively considered by Lachenbruch (1961; 1962; 1966) and Romanovskiy (1973*b*, 251–6) and, for desiccation polygons developed in the laboratory, by Corte and Higashi (1964). According to Lachenbruch (1966, 68)

The use of temperatures measured in Alaskan permafrost, a 'power-law model' to calculate thermal stress, and an estimated value of G, [critical rate of strain energy release necessary for fast fracture] to calculate crack depth, lead to computed stress relief compatible with observed polygonal dimensions.

Diameters '. . . vary from a few meters to more than 100 meters (Lachenbruch, 1966, 63). According to Péwe (1966*a*, 77), ice-wedge polygons are 2 (or 3) to 30 m or more in diameter. Although Black (1952, 130) indicated that ice-wedge polygons may be less than 1 m in diameter, he referred to polygons 1–3 m in diameter as resulting from subdivision of larger ice-wedge polygons, and in general it would seem that ice-wedge polygons have diameters exceeding 1 m. Seasonal frost-crack polygons tend to be larger than ice-wedge polygons occurring in the same area according to Hopkins *et al.* (1955, 139) but more

---

[12] A similar distinction between frost-crack and permafrost-crack polygons and related forms is made in the USSR. For instance, Romanovskiy (1973*b*, 256–69) classified 'frost-fissure' polygons (FP) into 4 types: (1) FP with secondary ground [soil] filling, formed in the active layer [frost-crack polygons]; (2) FP with secondary filling of ice, formed in permafrost [ice-wedge polygons]; (3) FP pseudomorphoses [ice-wedge cast or fossil ice-wedge polygons]; (4) FP with primary ground [soil] filling [sand-wedge and soil-wedge polygons].

information is needed. As noted above under Desiccation cracking, polygons with diameters measuring less than 1 m are probably mainly of desiccation origin.

Permafrost-crack sorted polygons can develop by activation of a sorting process along the cracks in the active layer (C. D. Holmes and Colton, 1960, 13–14; Klimaszewski, 1960, 157–8; Péwé, 1964), and perhaps also in fossil forms by melting of ice wedges and concurrent infilling with surface stones (Richmond, 1949, 150–1).[13] Nets initiated by permafrost cracking may exist but are probably less common than polygons because deformation of an originally polygonal pattern would tend to be inhibited by permafrost. Wedge structures of sand, occurring in till and more or less paralleling the gradient, have been described by Goździk (1976), who interpreted them as fossil nonsorted stripes (or possibly greatly elongated polygons) of frost-crack, probably permafrost-crack, origin. Given such occurrences, sorted forms should also occur.

g Frost action along bedrock joints   Frost action in the present context comprises frost wedging and frost heaving. Where sufficiently concentrated along bedrock joint patterns, it can initiate some forms of patterned ground but they are uncommon, judging from the few references to them in the literature (cf. W. E. Davies, 1961a; Fleisher and Sales, 1971; Kunsky and Louček, 1956, 345–7; Washburn, 1956b, 846–7; 1969a, 185). Another illustration, observed by the writer on Manastash Ridge on the Columbia Plateau in Washington, comprised small sorted polygons whose stony borders of angular basalt debris could be traced directly down through a mosaic of fitted fragments into the jointed *in situ* basalt. Control of the patterns by bedrock joints is the essential criterion for all these forms. Sorted polygons are the best-known example; sorted stripes are another (Washburn, 1969a, 185). Sorted circles in bedrock have also been observed (Washburn, 1950, 49; 52, Figures 1–2; M. Y. Williams, 1936). If the latter are controlled by joint intersections as suggested, they could be termed joint-crack sorted circles. The existence of joint-crack nonsorted polygons and nonsorted stripes is strongly suggested by Svensson's (1967) description of a patterned block field in northern Norway in which the rubble, lacking fines, is characterized by tetragonally arranged, shallow furrows whose directions match the joint pattern in the sedimentary bedrock. The furrows contained blocks, some on edge, whose size and shape corresponded to those of the surrounding surface. Svensson (1967, 18) believed the pattern was due in part to frost cracking of the bedrock but that 'A tendency toward the formation of the actual pattern ... existed in the main fracture systems of the sedimentary bedrock.' Given the initial joint pattern, there seems to be no reason to postulate frost cracking rather than just frost action along the joints, including the possible growth of ice wedges, to account for the pattern. If the bedrock fissuring were due to frost cracking, however, the forms would be frost-crack or permafrost-crack polygons. The fact that joint-crack patterns are confined to bedrock joints minimizes the possibility of net patterns developing by deformation of polygons. Joint-crack polygons that have no periglacial significance are also known (Troll, 1944, 593–4, Figures 22–3; 1958, 38, Figures 22–3).

10   Origin – cracking non-essential

a General   A number of hypotheses of patterned-ground origin involve processes that do not depend on initial cracking, and the more probable of these are grouped in Table 5.1. The selection is incomplete and, in view of the uncertainties involved, some important initiating processes may have been omitted. Furthermore, the question exists as to whether or not non-cracking processes can develop strictly polygonal, as opposed to circular, net, or striped patterns. As noted above in discussing cracking processes, cracking can initiate some sorted polygons. On the other hand, observations by Schunke (1975a, 123) convinced him that it is by no means an essential process in this respect, a conclusion also reached by Bout (1953, 76) among others. Polygons have been developed by the mutual impingement of circles on a small scale in the laboratory (Dżulyński, 1963, 145–7), and theroetically this could happen on a natural scale in the field and simulate polygonal patterns of crack origin, especially if secondary bordering cracks developed; also transitional patterns from circles to nets suggestive of true polygons have been frequently observed in the field (for instance, Cegła, 1973, 240). Nevertheless, proof appears to be lacking that impingement develops typical, truly straight-sided polygons in the field. Therefore, polygonal forms in the 'cracking non-essential' section of Table 5.1 carry a question mark. By definition, steps are derived

---

[13] Although Richmond used the term nets, his description specifies polygons (Richmond, 1949, 145).

from other patterned ground by mass-wasting; their origin is therefore keyed to that of other forms even though terracettes similar to steps are initiated by mass-wasting.

b Primary frost sorting Primary frost sorting includes upfreezing and the ejection of stones from fines by repeated freezing and thawing, the sorting effects of needle ice, and the gradual particle-by-particle sorting that Corte termed vertical and lateral sorting (cf. section on Sorting by frost action in preceding chapter). A number of frost-related sorting processes probably modify established patterns to produce sorted forms. The evidence that primary frost sorting can also initiate basic patterns is less clear, and the criteria are correspondingly vague, but the process seems more likely than many to be also a first cause. Recognizing that very few hypotheses fit all reported conditions, R. P. Goldthwait (1976a, 33–4) concluded that this process fits best. In listing primary frost-sorted forms in Table 5.1 the usual 'S' designation for sorting is omitted as redundant. Nonsorted forms of patterned ground are excluded by definition. Primary frost sorting would develop sorted circles or, if closely spaced, sorted nets. The development of clearly defined polygons by impingement is questioned for reasons discussed above (10a). It is also uncertain whether primary frost sorting would produce debris islands, since normally the coarse rubble in which they occur has a scarcity of fines near the surface and the fines of many debris islands appear to be derived from depth, in some places demonstrably so (Figure 5.12).

c Mass displacement Mass displacement as applied to patterned ground is the *en-masse* local transfer of unfrozen mineral soil from one place to another within the soil as the result of frost action, as discussed under Mass displacement in the preceding chapter. Suggested processes include artesian pressure, changes in density and intergranular pressure, cryostatic pressure, differential volume changes in frozen soil, the Mackay effect, and upward freezing from the permafrost table. The relative importance of any one of these processes as applied to the genesis of patterned ground remains to be demonstrated. The processes visualized by Crampton (1977a), Mackay (1979), and Nicholson (1976, 333–9) are inapplicable as a primary process because of being based on the prior establishment of a basic pattern. Although criteria for recognizing mass displacement of a specific kind in patterned ground are uncertain, the demonstrable diapir-like, upward penetration of fines in some forms (Figures 5.6–5.7, 5.12) including steps is proof of the general process (cf. Bibus, 1975,

104–5; Crampton, 1977a, 642–3; Semmel, 1969, 7–11, 39–40; Shilts, 1978, 1060–2, Figure 7; Washburn, 1950, 47–50, Plate 15; 1969a, 108, Figure 108). In some permafrost areas the upward penetration as 'mud-boils' is so intense and widespread that the active layer tends to be overturned (Shilts, 1977, 204–6). In other places, differential frost heaving may well be an adequate explanation for similar but less complex displacements as noted by Schunke, who found no support for the importance of density displacements in the origin of regular sorted patterns he studied in Iceland (Schunke, 1975a, 73, 122–3), although he supported mass displacement, especially cryostatic pressure, as a critical process in the origin of earth hummocks (Schunke, 1977a, 16–18, 38, 43–5). The importance of cryostatic pressure in earth hummocks was also favoured by Tarnocai and Zoltai (1978, 592) but rejected by Mackay (1979) in favour of thawing pressure.

Mass displacement can originate both nonsorted and sorted patterns. Thus movement of a mass of fines through other fines, or through slightly coarser but not obviously stony material, would produce nonsorted varieties, whereas sorted varieties would depend upon movement of fines into distinctly stony material. To the extent this occurs in layered materials, whether by mass displacement or frost heaving, it negates R. P. Goldthwait's (1976a, 30) conclusion that 'The parent material [of sorted patterns] must be a diamicton. Strongly layered regolith is characterized by ice/soil wedge patterns alone.' However, in many places upfreezing of stones from a diamicton probably produces a stony surface overlying relatively stone-free material, thereby forming a layered system. Sorted patterns would then result from mass displacement of fines into the coarse layer. With mass displacement most of the normal geometric patterns are expectable but strictly polygonal ones are questioned in Table 5.1 for reasons previously noted.

d Differential frost heaving Local differential heaving, discussed under Frost heaving and frost thrusting in the preceding chapter, is frost heaving that is significantly greater in one spot than in the area around it because of less insulation or other factors, so that growth of ice lenses and needle ice is especially favoured. Differential growth of vegetation in desiccation cracks or frost cracks may be one cause of a difference in insulation. As applied to patterned ground in Table 5.1, 'differential frost heaving' and the adjective 'frost-heave' imply the resulting blister-like expansion of the ground, which may well start many nonsorted and sorted patterns.

The process is especially frequently invoked for non-sorted circles (cf. Fahey, 1975, 155–8; Frödin, 1918, 21–7). Frost sorting and the fact that sorting can of itself lead to volume increase of soil (Corte, 1966*b*, 213–16) mean that differential frost heaving can be a self-reinforcing process. Criteria for recognizing patterned ground of this origin are largely negative. For instance, forms consisting of non-heaving mineral soil are excluded. Also strictly polygonal forms are more consistent with other origins (unless the development of well-developed polygons by impingement is accepted) as are forms showing mass displacement of material from depth as opposed to surficial or very shallow forms like the nonsorted circles (stony earth circles) described by P. J. Williams (1959*a*; 1961, 345). Earth hummocks with a core stone whose upfreezing caused the hummocks would be of frost-heave origin. Yet, although hummocks are probably of diverse origins (Beschel, 1966) and core stones are locally common and may be responsible for some hummocks (Dionne, 1966*a*), many hummocks lack them and it is difficult to prove that upfreezing of the stones is really the primary factor even where core stones are present.

e Salt heaving . Salt heaving is volume expansion due to growth of salts. The process probably causes several varieties of patterned ground in warm-arid deserts (cf. Hunt and Washburn, 1966). It has many analogies with local differential frost heaving and is probably capable of initiating the same general types of pattern. In combination with desiccation cracking it forms many patterns that simulate periglacial types (cf. Kaiser, 1970, 166–75). Reportedly a distinguishing characteristic of sorted varieties is that bordering stones in warm–arid forms are very rarely on edge whereas the opposite is true of periglacial forms (Kaiser, 1970, 173–4).

f Differential thawing and eluviation   Differential thawing and eluviation in the present context constitute a sorting process with respect to fines and stones, which stems from (1) greater thawing and freezing rates in coarse than fine material as a result of differing thermal properties, other conditions being equal (cf. Ballard, 1973; Journaux, 1973; 1976*a*[14]), and (2) preferential removal of fines by thaw water in stony accumulations. Such sorting probably explains some sorted nets and stripes but criteria for recognizing patterns of this origin are unsatisfactory except for shallow forms developed in thawing ice (Corte, 1959; Washburn, 1956*a*; 1956*b*,

855–6). Nonsorted forms are excluded. Of sorted forms, polygons are again questioned; sorted stripes seem possible but are not known to occur, and other sorted forms seem unlikely.

g Differential mass-wasting   Differential mass-wasting is the faster downslope mass movement of some materials than others. Little is known about the process relative to the initiation of patterned ground but it may account for some sorted steps and stripes (R. B. King, 1971, 384–5; Pissart, 1972*a*, 264; Washburn, 1969*a*, 186–8), and perhaps for some nonsorted ones as well. As discussed under Stripes, a differential downslope movement between the coarse and fine stripes (the coarser moving the slower in the case investigated) was demonstrated by Mackay and Mathews (1974*b*), who showed that stripe movement was caused mainly by downslope collapse of needle ice.

h Slopewash   Rillwork can start small channels and it may thereby also initiate some striped forms of patterned ground. In this context the process is usually limited to small-sorted stripes (cf. French, 1976*a*, 188; Washburn, 1956*b*, 857–8). Although most sorted stripes normally carry drainage, in many places this is probably a result rather than a cause of the pattern. The possibility that some nonsorted stripes are due to rillwork is not excluded; its primary role in some stripes on slopes of less than 10° was cited by Journaux (1976*a*, 271) but he also noted the presence of similarly oriented desiccation cracks in places. French (1976*a*, 266), too, regarded slopewash as an important striping process in periglacial environments, and it has also been held to be such in some warm semiarid environments (Spönemann, 1974), although semiarid forms are larger than most periglacial stripes. Thawing snow often shows a downslope-oriented striped pattern that led Flohr (1935) and Lliboutry (1955; 1961, 215–17) to conclude that small sorted stripes started beneath such snow. The possibility should be investigated that thread-like throughflow of water down a slope, rather than overland flow, may be an important stripe-initiating process.

i Differential weathering   Although not represented in Table 5.1, differential weathering may create local pockets of fines amid coarser debris. This first cause might then be reinforced by sorting processes and heaving. Meinardus (1912*a*, 254–5; 1912*b*, 12–13, 32) suggested that some debris islands result from the weathering of especially susceptible stones in a boulder field. More generally, Nansen (1922*a*, 111–

---

[14] In Journaux (1973; 1976*a*) read frost table for the permafrost.

20) thought that a wide variety of sorted circles and sorted polygons (formed from circles by impingement) were started by differential weathering as the result of moisture variations caused by inequalities of the ground surface. More recently the hypothesis has been applied to large sorted polygons on the basis that large stones weather more slowly than smaller stones, which have a larger surface area to volume ratio and are therefore more subject to frost wedging (R. B. King, 1971, 383). In a slightly different form the hypothesis has also been extended to nonsorted polygons and their development from sorted polygons by comminution of stone borders (Elton, 1927, 173; Huxley and Odell, 1924, 219–23, 228). Without denying the importance of differential weathering in producing fines and some sorting effects in patterned ground, the process seems unlikely to be a basic patterning process in the same sense as the others listed in Table 5.1. Only if it can be demonstrated that differential weathering *in situ* has led to the concentrations of fines within each mesh of a sorted pattern would there be a parallel.

j Stream turbulence Also not represented in Table 5.1 is the hypothesis that some small sorted nets and stone pits on sand bars are the result of stream turbulence during flood stages (Kostyayev, 1973). If developed in this way, such special forms would constitute an illustration of convergent phenomena.

k Wave action Another omission in Table 5.1 is the concept that stone pits can be the result of wave action (Dionne and Laverdière, 1967; Dionne, 1971*b*). If so, this would be still another illustration of convergence, since stone pits have been observed where wave action is lacking, as in places in the Brooks Range, Alaska (T. D. Hamilton, personal communication, 1974) and elsewhere (Stingl, 1974, 257–9).

l Erosion Also omitted from in Table 5.1 is erosion, which has been repeatedly suggested as a cause of silt mounds and stripes, which are widespread on the Columbia Plateau and in some other regions of the United States, including places in California, Oregon, and Oklahoma where, unlike the Columbia Plateau, intense frost action, present or past, is very unlikely. The mounds can be up to about 25 m in diameter and 2 m high. A common characteristic of the mounds and stripes is that they lie above a discontinuity separating them from less erodible material below. Illustrations include silt above gravel, bedrock, or a pedological horizon such as hardpan. According to the hypothesis this dis-

continuity results in a uniform silt cover being dissected into mounds and stripes, depending on gradient, as the cover is stripped by runoff. The hypothesis was advanced by LeConte (1873, 219–20; 1874, 365–7) and supported by Melton (1954), A. C. Waters and Flagler (1929), and recently by M. D. Wilson (1977; 1978). LeConte suggested vegetation might have a role in preserving the mounds, and conceivably in places this possibility might help to counter the frequently voiced objection that erosion alone was unlikely to develop more or less circular, evenly spaced mounds, an objection that was questioned by Melton (1954, 106). A. C. Waters and Flagler (1929, 215–20) stressed that (1) the pattern was not regular in detail, (2) they observed no mounds where there was no gradient, (3) where the gradient was less than 1° the intermound areas were anastomosing, (4) where steeper, the mounds were elongate in the direction of slope, and (5) the mounds were confined to soil less than 2 m (7 ft) thick. Clearly there is room for confusing cause and effect.

Nevertheless the present writer supports the importance of erosion for the Columbia Plateau mounds on the basis of air photographs and field observation, especially the relation of the mounds to the eroding feather edge of the silt,[15] and he accepts the possibility that erosion is equally important as applied to the famous Mima Mounds of western Washington. In no exposures of the latter, or elsewhere in this region, has he ever observed ice-wedge casts that would support the ice-wedge polygon hypothesis (Péwé, 1948), nor despite the strikingly similar surface appearance of the mounds to thawed ice-wedge polygons is there any independent evidence of former permafrost. Rather, there is clear evidence to the contrary. The estimated mean annual temperature reduction in the Cascade Range at the maximum of the last glaciation, based on the glaciation threshold, is 4.2° ± 1° (Porter, 1977, 155). The present mean annual temperature at Olympia, near Mima Prairie, is 10.4° (50.8°F) (National Oceanic and Atmospheric Administration, 1974, 951), which accords very well with the projected temperature of 10°–11° for lowland areas, based on the present glaciation threshold (Porter, 1977, 109, Figure 4). Thus the 4.2° ± 1° reduction for the Cascades also appears to be a reasonable estimate for the lowlands, and is supported by an estimated August sea-surface temperature of 12°–13° for the nearby Pacific Ocean 18 000 years ago

[15] The present writer believes that sorted polygons in the same area may be caused by frost wedging in the joints of the columnar basalt; the coincidence in pattern in small polygons can be demonstrated in places.

(CLIMAP, 1976, 1132, Figure 1), a temperature only 3°–4° less than today's 16.2° (61.2°F) (J. H. Johnson, 1961, 14).[16] If these estimated reductions even approach reality, ice-wedge polygons in the Olympia area would have been impossible, since permafrost would have required a temperature reduction of at least 10°, and ice-wedge polygons even more. This conclusion leaves open the possibility that seasonal frost and/or desiccation cracking helped to determine the erosional pattern. As for the gopher hypothesis (Dalquest and Scheffer, 1942), the present writer does not question that gophers inhabited mounds, only that they built them. Thus he has observed mounds at the bottom of a Mima Prairie kettle hole capable of retaining water – an unlikely gopher habitat – where the mounds give the appearance of having collapsed with the kettle and therefore of having formed while buried ice was still present.

In the above context the erosion hypothesis has been cited as a possible alternative to a periglacial hypothesis for the origin of certain puzzling patterned-ground occurrences. In a periglacial context, the possibility should also be considered that runoff over a frost table or permafrost table may be a more important initial patterning process than has been recognized by its application to rillwork in explanation of stripes. For some sorted forms of patterned ground in periglacial environments, differential thawing and eluviation would be another way of looking at essentially the same process.

III   Periglacial involutions

Periglacial involutions (Denny, 1936, 338) were defined by Sharp (1942b, 115) as 'aimless deformation, distribution, and interpenetration of beds produced by frost action' (Figures 5.43–5.44). Other names include Brodelboden, cryoturbations (Kryoturbate Böden – Edelman, Florschütz, and Jeswiet, 1936, 332), Taschenboden (pocket soil), and Tropfenboden (drop soil), sometimes with different shades of meaning. Actually Sharp did not specifically use the term periglacial involution in his definition but he did elsewhere, and although some investigators restrict the term involutions to periglacial forms, others do not and it is best to be specific (Embleton and King, 1975, 45–7; French, 1976a, 42; Pissart, 1970a, 34).

According to French (1976a, 42, 229), periglacial involutions are more regular than cryoturbations and imply polygons but this is a dangerous assumption even if some periglacial involutions do reflect such a relationship. In many places it is difficult or impossible to prove the periglacial nature of involutions to say nothing of their exact origin. Even if they are correctly identified as periglacial features, it is commonly difficult or impossible to determine the exact type of frost action responsible. Cryostatic pressure (including expulsion of porewater to unfrozen sediments during freezing), and/or changes in intergranular pressure during thawing may be among the most important. Differential volume changes in frozen soil may be a factor. All these processes are discussed under Mass displacement in the chapter on General frost-action processes. Freezing of capillary water at subzero temperatures within frozen skeletons of free soil water may be a critical process according to Pissart (1973a, 108–12) – whose experiments show that significant pressure effects can be created in locally closed systems during freezing simulating that in non-permafrost environments (cf. Pissart, 1970a, 40–4; 1972b) – although he noted that further data are required.

Drop (or pocket) involutions with narrow neck and bulbous rounded bottoms are often interpreted as caused by density differences whereby overlying heavier soil sinks into less dense soil as previously discussed in considering Mass displacement. 'Drops' of coarser material penetrating downward into finer is consistent with the hypothesis but the inverse, involving sand penetrating gravel as described by Gullentops and Paulissen (1978), is more difficult to reconcile with it; yet the evidence for the direction of drop movement was most convincing. These investigators, accepting what they called 'periglacial load-casting', felt compelled to call on '... pressures developed by the *refreezing of the saturated active layer* ... after the summer thaw period' (Gullentops and Paulissen, 1978, 112). Details concerning these (cryostatic?) pressures are lacking. Such downward penetrations of finer material into gravel are very puzzling.

Regardless of their exact origin, the fact that the 'drops' terminated at a constant depth and their bottoms were flat rather than rounded indicates a

5.43 Periglacial involutions in sand and peat, right bank Lena River, 90 km north of Yakutsk, Yakutia, USSR. Scale given by 2-m tape

[16] The 13° July–August terrestrial temperature increase since 18 000 BP indicated in the model presented by Gates (1976, 1142, Figure 7) is misleading for the lowlands considered, since the increase is based on a quadrat that encompasses the State of Washington both east and west of the Cascades.

5.44 Periglacial involution, beach between Peris and Clydan, Wales, Great Britain (*cf. Watson, 1976, 92–6*)

depth control that the authors interpreted as a former permafrost table – a traditional inference (cf. Poser, 1948*a*, 54, 58). Similarly, Watson (1977, 191–2) argued that an involution layer of nearly constant thickness, occurring on a slope of less than 2° and truncating dipping beds, can be reasonably interpreted as having formed above a permafrost table; nevertheless he concluded that 'The case for permafrost remains hypothetical. . . .' Among other possibilities it also seems reasonable that under some conditions a constant depth of seasonal frost action might produce a similar layer. Moreover, contortions appearing very similar if not identical to periglacial involutions can be formed by other processes than frost action, including non-periglacial load casting,

slumping (and perhaps other types of mass-wasting), and volume changes whose results can be strikingly like periglacial involutions (cf. Anketell, Cegła, and Dżułyński, 1970; Butrym *et al.*, 1964; Danilov, 1973*b*; Dionne, 1971*a*; Dżułyński, 1963; 1966; Kaye and Power, 1954; McArthur and Onesti, 1970; McCallien, Ruxton, and Walton, 1964). In some cases the relation of associated structures such as ice-wedge casts may help to identify periglacial involutions (Johnsson, 1962, 386–92). Thus R. B. G. Williams (1975, 102–9), stressing involutions that were associated with ice-wedge casts and a tendency for involutions to bottom out at consistent depths, accepted the occurrences as useful evidence of thaw depth and hence of summer temperature. The need for such supplementary evidence emphasizes the ambiguity of involutions and makes their interpretation problematic.

Although various investigators have regarded periglacial involutions and similar contortions as evidence of permafrost (cf. Gravis and Lisun, 1974, 78–80; Jahn, 1977*b*, 30; Poser, 1948*a*, 54, 58), and some involutions can indeed be suggestive of it, involutions alone should not be accepted as proof of permafrost as emphasized by several investigators (Black, 1969*b*, 233–4; Fries, Wright, and Rubin, 1961, 691–2; Wright, 1961, 943). As stressed by Corte (1969*a*, 141), a critical study is needed of the origin of involutions and similar features.

IV Stone pavements

1 General

Stone pavements are accumulations of rock fragments, especially cobbles and boulders, in which the surface stones lie with a flat side up and are fitted together like a mosaic. Other names that have been applied to periglacial accumulations of this kind include alpine subnival boulder pavements, boulder pavements, dallages de pierres, Pflasterboden, and Steinpflaster.

Well-developed periglacial stone pavements usually occur in valleys in association with lingering snowdrifts or icings (Figure 3.21) but some are in snow-filled depressions such as nivation niches. Less well-developed stone pavements occur on some beaches subject to ice action (Mansikkaniemi, 1976); and pavements, some well developed and striated subparallel to the channel, can occur along rivers subject to ice shove during break-up (Mackay and MacKay, 1977, 2214–20).

Stone pavements caused by deflation are mentioned in the chapter on Wind action.

2 Constitution

Cobbles and boulders are the predominant constituents of periglacial stone pavements but pebbles and smaller grain-size fractions occur locally in the cracks between the larger stones of the mosaic. Any stones beneath the surface tend to decrease in size with depth. The stones may be derived from underlying bedrock or be transported from elsewhere.

3 Origin

The origin of periglacial stone pavements is not well understood but a variety of processes are probably involved. S. E. White (1972, 195) has summarised a widely held view by indicating that they may be the combined result of upfreezing of stones, ground saturation and the removal of fines by meltwater, the rotation and sifting of the stones in the saturated ground under their own weight and the weight of overlying snow, with possibly gelifluction (Hamelin *in* S. E. White, 1972, 199; Schunke, 1975*a*, 139–40, 149; Sørensen, 1935, 60) and snow creep being additional causes. The role of snow weight has been questioned by Bout and Godard (1973, 62–8) and Tricart (1967, 239; 1969, 100) but supported by Fritz (1976, 261–2), Stingl (1969, 45), S. E. White (1972, 199), and others. The weight of icings should be added to this list, since well-developed stone pavements beneath icings have been observed by Porter (1966, 82–3) in the Anaktuvuk Pass area, Alaska (Figure 3.21), and by the writer in the Mesters Vig district, Northeast Greenland. For weight of snow or ice to be a factor, the ground beneath must be thawed (Bout *et al.*, 1955, 491–2), which is less common in permafrost regions than in other periglacial environments.

Stone pavements on beaches have been attributed to the 'compressing and smoothing influence' of an ice foot (Mansikkaniemi, 1976, 1), and those along rivers to the pressure of ice shove including '... the grinding action of rock shod river ice and the push of loose boulders' (Mackay and MacKay, 1977, 2218).

4 Environment

Periglacial stone pavements are widely distributed, occurring in the Arctic (Porter, 1966, 82–4; Poser, 1931, 230–1; 1932, 54–5; Sørensen, 1935, 60–2), Subarctic (northern Scandinavia and Iceland – Bout and Godard, 1973, 62–8), in the middle-latitude highlands (Troll, 1944, 549, 654; 1958, 3, 82; S. E. White, 1972), and in the Antarctic where they have been observed at Marble Point, McMurdo Sound, by the writer. However, their use as evidence of a periglacial environment is handicapped by the fact that they have not been extensively studied and their origin and quasi-equilibrium range is uncertain. Furthermore, various processes in warm-arid regions give rise to desert pavements, which are somewhat similar, although generally characterized by smaller stones. Bout and Godard (1973, 67–8, 77) concluded that although stone pavements are related to seasonal snow cover, they lack zonal significance, at least compared with palsas.

5 Inactive and fossil forms

Fossil stone pavements have been frequently interpreted as of periglacial origin, based in part on the long axis of stones being in the direction of the gradient, suggestive of gelifluction, and on stratigraphic position consistent with cold phases (J. Hagedorn, 1964, 166–72). Nevertheless the value of inactive or fossil stone pavements alone as a criterion of former periglacial conditions is uncertain pending further study. S. E. White (1972, 199) was appropriately cautious in drawing paleoclimatic conclusions from features he studied in the Colorado Front Range, some of which may still be active.

V String bogs

1 General

String bogs (Aapamoore, tourbières réticulées, Strangmoore) are areas of peatland[17] characterized by ridges of peat and vegetation, interspersed with depressions that often contain shallow ponds (Figure 5.45). The ridges are up to about 2 m high and may be tens of metres long, and some are surrounded by palsas (Hamelin, 1957, 91–2). Ridges transverse to the gradient are commonly asymmetrical with the steepest side facing down the regional slope and the gentler, upslope side damming a pond. The regional slope is very gentle, and many string bogs occur on gradients of less than 2°.

Three types of ridge patterns are commonly recognized (cf. Hamelin and Cook, 1967, 164; Tricart, 1967, 229; 1969, 98): (1) Linear transverse ridges, commonly convex downslope, (2) irregular ring-like anastamosing ridges, and (3) ridges forming a net-like pattern. Either vegetation or water may be the major element in a given bog sector. The patterns tend to be much more irregular than patterned ground, and string bogs are usually separately classified.

2 Constitution

The ridges consist of peat containing ice lenses for at least part of the year. Ice lenses as thick as 40 cm have been reported (Dionne, 1968a, 3). The ridges tend to be colonized by xerophytic plants and the depressions by hydrophytic varieties.

3 Origin

The origin of string bogs is still not established and a number of hypotheses exist. These include (1) Solifluction and wrinkling of the bog surface (G. Andersson and Hesselman, 1907, 79; cf. Auer, 1920, 29–31). (2) Raising of ridges by frost thrusting from intervening ponds, combined with ruptures of frozen bog surfaces by water under hydrostatic pressure (Helaakoski, 1912, iii–iv, 69–70, 77–9; cited by Tanttu, 1915, 14). (3) Differential frost heaving combined with detachment and drifting of vegetation into windrows that develop into the ridges or strings (Sohju theory) (Tanttu, 1915, 16–23). This hypothesis was accepted by Auer (1920, 37–40, 118–26) for some occurrences but not as a general explanation, and was favoured by Drury (1956, 66, 72–5) if combined with a tendency for meltwater flowing over a surface to form initial riffles and ridges. This role of meltwater is supported by observations of B. G. Thom (1972), who suggested that this patterning effect progresses upslope so that the oldest ridges or strings are downslope. (4) A variety of botanical and physical processes with stress on gelifluction and changes in moisture regimen (Auer, 1920, 135–43). (5) Disruption of a bog by growth of ice lenses during a cold period and subsequent gelifluction during climatic warming (Hamelin, 1957, 102–5). (6) A 'combination of biological and hydrological factors' with emphasis on the habit of *Scirpus caespitosus* to grow in rows (Jan Lundqvist, 1962, 83). (7) Differential thawing of permafrost and local collapse within bogs (Schenk, 1966, 157–8). (8) Principally differential frost heaving and frost thrusting (Tricart, 1967, 230; 1969, 98).

The above hypotheses encompass three kinds of influences or hypotheses (Moore and Bellamy, 1975, 39–41): (1) Biotic, (2) ice and frost action, and (3) gravity. The growth habit of some plants such as sphagnum to grow in hummocks, thus forming an

---

5.45 String bogs northwest of Moosonee, Ontario, terrain altitude 30–45 m (*Canada Energy, Mines & Resources Photo A 15090–65*)

[17] Muskeg is an equivalent Canadian term (R. J. E. Brown, 1968, 175).

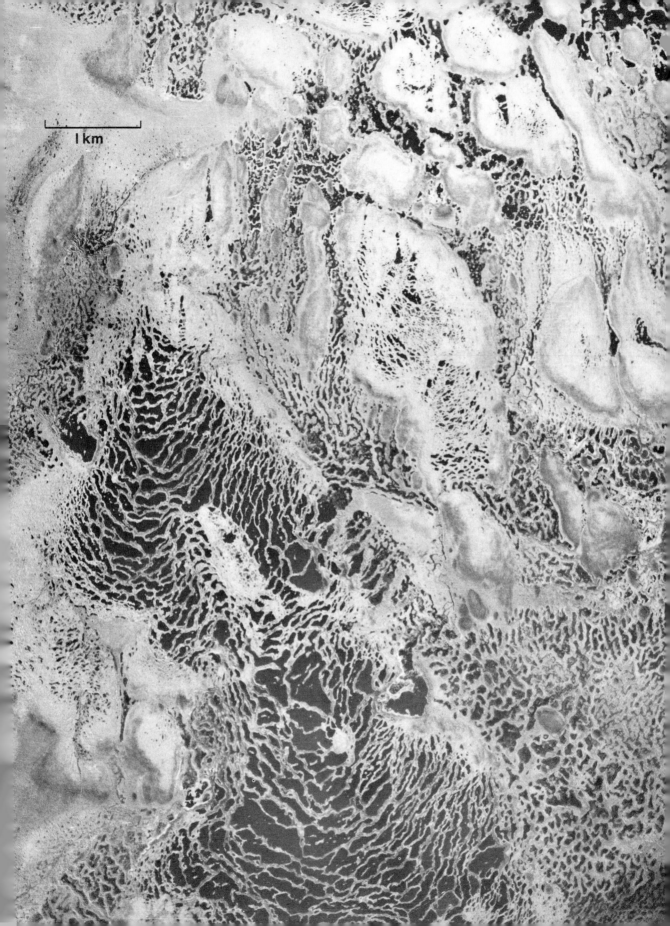

irregular surface with alternating drier and wetter places, is presumably a basic factor. Frost heaving can increase the hummocky nature of such a surface, and gravity can introduce slope-controlled patterning effects. There is a vagueness about most of the hypotheses, and the origin (or origins) of string bogs still remains to be demonstrated as witness the many unanswered questions and problems raised by the various hypotheses and by some commentaries and reports (cf. Henoch, 1960; Mackay, 1958a; P. J. Williams, 1959b). 'There is probably no single explanation and all the mechanisms considered above probably play some part in the process on some sites' (Moore and Bellamy, 1975, 41).

4 Environment

In general, string bogs appear to have a distribution somewhat similar to palsas but to occur slightly farther south in the Northern Hemisphere (Troll, 1944, 639–44; 1958, 72–4), and perhaps generally at a slightly lower altitude range, as in Iceland (Schunke, 1975a, 117, footnote 28). There appears to be a climatic zonation of string bogs and similar forms in Europe, as reflected in the distribution of mires (peat-producing ecosystems). West of the Baltic Sea there are strong east-west components but in Finland and adjacent parts of the USSR there is a more or less regular south to north sequence of concentric domed mires, excentric domed mires, aapamires (string bogs and patterned fens), and palsamires, the last characterized by less strict patterning and the presence of peat plateaus with palsas (Moore and Bellamy, 1975, 12–29; Ruuhijärvi, 1960, 253–94). String bogs are commonly regarded as characteristic of the subarctic taiga. Drury (1956, 63) described them as closely associated with tree line. They are widespread in western Siberia and in the Hudson Bay region of Canada. In the Quebec-Labrador area they may cover 10 per cent of the terrain (Hamelin, 1957, 95). Although string bogs or closely similar features have also been observed far north of tree line and well within the zone of continuous permafrost (Henoch, 1960, 335; Mackay, 1958a), they, together with peatland types, are most common in the zone of discontinuous permafrost as noted by R. J. E. Brown (1968, 175, Figure 1) and Thibodeau and Cailleux (1973, 136–7) among others, and most investigators agree they are not necessarily indicative of permafrost. However, Schenk (1966, 157–8) regarded them as necessarily associated with thawing permafrost and concluded they can be delicate indicators of climatic change. Hamelin's conclusion that net-like string bogs are caused by bog disruption in a cold period and gelifluction in a warmer period also requires that such forms be a record of climatic change but Hamelin (1957, 105–6) wisely emphasized the hypothetical nature of his conclusion.

5 Inactive and fossil forms

The low relief and probably rapid modification of string bogs by environmental changes suggest that inactive and fossil forms would be difficult to recognize. A frequent lateral alternation between xerophytic and hydrophytic plant remnants in stratigraphic section would be suggestive but no such deposits are known to the writer.

VI Palsas

1 General

Palsas, also known as bugors and bulgunniakhs (bulgunyakhs) in the USSR,[18] are mounds or more irregular forms that generally have peat as an important constituent, contain perennial ice lenses, and occur in bogs (Åhman, 1977; Jan Lundqvist, 1969; Seppälä, 1972a; Wramner, 1972, 49; 1973, 123). Although transitions occur between palsas and forms without peat, the latter have not usually been classed as palsas. A contrary view was expressed by Åhman (1976, 29–31; 1977, 129–32, 144–5), who argued that there is no genetic distinction, only that forms without peat would occur in colder environments than would those with it, although he did not actually describe contemporary peat-free forms. The presence of peat beneath and in palsas was emphasized by Seppälä (1977). The present writer agrees with Svensson (1976b, 39) that palsas and pingos (described next) are related through transitional forms but where best developed occur in different environments. The characteristic presence of peat in palsas and its general absence in pingos would seem to be a useful, if arbitrary, distinction.

---

[18] There is no special term for palsas in Russian, and bugor and bulgunniakh, the latter of Yakutian origin, are also used for pingos (A. I. Popov, personal communication, 1976).

Palsa forms include mounds, moderately straight ridges, and winding ridges. Palsas in Iceland have been described as hump shaped, dike shaped, plateau shaped, ring shaped, and shield shaped (Schunke, 1973, 69, 98), those in Norway as palsa plateaus, esker palsas, string palsas, conical or dome-shaped palsas, and palsa complexes (Åhman, 1976, 27–9; 1977, 38–42, 138). Widths are commonly 10–30 m, and lengths 15–150 m. However, lengths up to 500 m have been reported for esker-like palsa ridges running parallel to the gradient of a bog (Åhman, 1975, 223, 227; 1977, 40, 138). Heights range from less than 1 m to 7 m (Åhman, 1977, 144; Forsgren, 1968, 117; Seppälä, 1977), or about 10 m at a maximum (Jan Lundqvist, 1969, 213). Large forms tend to be considerably less conical than small ones (Sten Rudberg and Per Wramner, personal communication, 1971). In places palsas combine to form complexes several hundred metres in extent. Palsa surfaces are frequently traversed by open cracks, caused by doming (dilation cracking), frost cracking, or desiccation. Doming was emphasized by Wramner (1973, 123), frost cracking by Åhman (1977, 43–7, 138–9). When a palsa decays it is often by 'block erosion' along the crack pattern as noted by Åhman.

2   Constitution

The characteristic constituent of palsas is peat containing a permafrost core with ice lenses no thicker than 2–3 cm (Jan Lundqvist, 1969, 208), although locally lenses 10–15 cm thick are common (Forsgren, 1968, 113) and lenses up to almost 40 cm thick have been described (Schunke, 1973, 71–2; Abb. 4). By contrast, pingos have comparatively little peat, if any, and their cores are generally composed of massive clear ice but here again transitions can occur in that some pingos have prominent ice lenses (Mackay and Stager, 1966, 363–7). There are two types of palsas, those with a peat core and those with a core of mineral soil, usually silt. The peat-core type (Figure 5.46) is the most common, the other being regarded as exceptional by some (Jan Lundqvist, 1969, 208–9) but more common than formerly thought (Åhman, 1976, 29–31; 1977, 129–32, 144–5; Sten Rudberg and Per Wramner, personal communication, 1971; Wramner, 1973, 120). The vegetation of a palsa may comprise low shrubs and lichen in addition to the sedges characterizing the peat.

3   Origin

According to Jan Lundqvist (1969, 209), the growth of palsas is caused entirely by the development of the ice lenses but the mode of water transfer to the lenses is not clear. He indicated that hydrostatic pressure is involved in some cases but that the transfer process accompanying the growth of segregated ice is probably the more common. Hydrostatic pressure as an originating process was specifically eliminated by Wramner (1972, 49; 1973, 121) in distinguishing palsas from pingos. In his view, also Åhman's (1976, 29–31; 1977, 129–32) and Seppälä's (1976, 9–10), palsas probably originate by differential frost heaving (for instance, where snow is especially thin), which becomes self-reinforcing as a palsa grows; '... the higher an embryonic palsa is, the deeper the frost penetrates' (Wramner, 1973, 121). The thermal conductivity of peat is an important genetic factor here. For unfrozen peat, the conductivity can range (in cal cm^{-1} s^{-1} °C^{-1}) from about 0.0002 when dry to about 0.0011 when saturated, but for frozen peat it is markedly greater and can approach 0.0056, the value for ice. Consequently freezing tends to be more effective than thawing. Combined with lower conductivity of peat than of mineral soil where present, this favours the development and maintenance of a permafrost core to the extent that palsas lie above the water table (R. J. E. Brown, 1966a, 21; 1968, 176; 1974a, 49–51). As a result the only permafrost in some areas may be associated with palsas (Seppälä, 1972a). It is this factor that explains the common association of palsas with peat as emphasized by Åhman.

According to R. J. E. Brown (1968, 178), palsas and peat plateaus grade into each other in the Hudson Bay Lowland of Canada, apparently as differing forms of the same process. There is a possibility that palsas and turf hummocks (thúfur) are closely related, since both can occur in the higher parts of a bog, the palsas being in the somewhat damper sites (Friedman et al., 1971, 141–4).

Mackay (1965) described mounds domed by gas pressure in the Mackenzie Delta area of northern Canada, and he subsequently suggested that such doming may initiate some palsas (J. R. Mackay, personal communication, 1972).

Not all mounds in bogs owe their form to frost heaving. Mounds can also form by thermokarst processes, such as melting of ice wedges in ice-wedge polygons leaving rounded interior areas standing in relief, or as the result of thawing and collapse

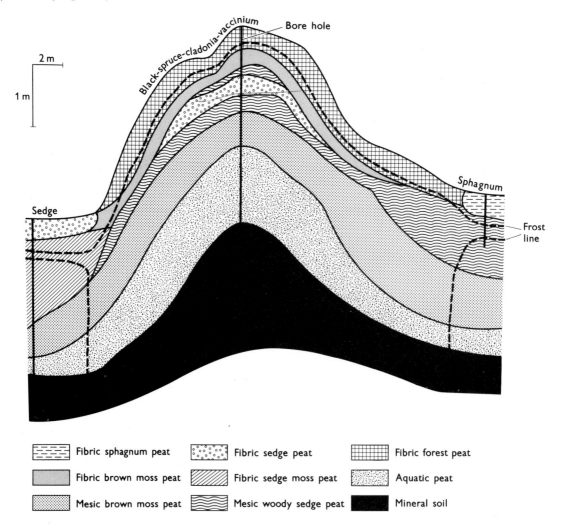

5.46 Cross-section of a peat palsa, lat. 67°06′N, long. 134°17′W, northern Canada (*after Zoltai and Tarnocai, 1975, 35, Figure 5*)

Legend:
- Fibric sphagnum peat
- Fibric sedge peat
- Fibric forest peat
- Fibric brown moss peat
- Fibric sedge moss peat
- Aquatic peat
- Mesic brown moss peat
- Mesic woody sedge peat
- Mineral soil

leaving a more irregular surface with a relief of as much as 10 m (P'yavchenko, 1963, 13, 15–16, 19; 1969, 17, 21–3, 27–8). In the highlands of central Iceland there are isolated mounds, 1–1.2 m high, that are similar in all respects to typical palsas except for having a core characterized by horizontal bedding and ice lensing rather than the usual convex-up attitude corresponding to the mound shape (Schunke, 1973, 71–2; Abb. 4; 1975a, 29, 136–8, 228, 233). These Schunke regarded as a convergent palsa type formed by thermokarst degradation of a peat plateau. He suggested they be designated 'cryokarst mounds' but this term also

describes other thermokarst features, and if these mounds are accepted as a type of palsa as presented by Schunke, the term thermokarst (or cryokarst) palsa would be preferable. The recognition of thermokarst palsas led Jahn (1976b) and Priesnitz and Schunke (1978) to classify palsas into aggradational (frost-heave) and degradational (thermokarst or erosional) types, and Jahn regarded the former as primarily, but not exclusively, an inner (continuous) permafrost type and the latter an outer or peripheral (discontinuous) permafrost type – a distinction stimulated by Grigoriew (1925, 347–51, 358).

4 Environment

Palsas are characteristic of the Subarctic and commonly occur in areas of discontinuous permafrost, and in this zone have been regarded as the only reliable surface evidence of permafrost (R. J. E. Brown, 1974b, 6). It would not be surprising if palsas also occurred in the continuous zone but that the similar forms reported to date (cf. R. J. E. Brown, 1973a, 29; Salvigsen, 1977) are true palsas is not firmly established.

In some places palsas extend into underlying permafrost; in others they rest on an unfrozen substratum (Friedman *et al.*, 1971, 138; Jan Lundqvist, 1969, 208–9). Palsas are almost exclusively associated with bogs, and commonly occur in areas where the winters are long and the snow cover tends to be thin. As noted in the preceding discussion of string bogs, palsas tend to occur slightly farther north than string bogs in Finland and in adjacent parts of the USSR. The Finnish palsas occur in northwestern Lapland in the southern part of the discontinuous permafrost zone where the mean annual temperature and precipitation are below $-1°$ and 400 mm, respectively (Salmi, 1968, 182). In Norway

... the palsa area ... is bounded [on the north] by a mean annual air temperature between $0°$ and $-1°C$, a winter temperature of $-10°C$ for 120 days, a mean annual precipitation below 400 mm/year and less than 100 mm during the Winter (Dec.–March). (Åhman, 1977, 143.)

The upper temperature boundary being on the north reflects the increasing maritime influence toward the Arctic coast. In Sweden 200–210 days with temperatures below $0°$ delimits the general distribution of palsas, and their southern limit very roughly correlates with the $-2°$ to $-3°$ mean annual isotherm (Jan Lundqvist, 1962, 93; cf. Rapp and Annersten, 1969, 67–9). In Iceland the southern limit of palsas (rústs) is probably close to the $0°$ mean annual isotherm (Thorarinsson, 1951, 154–5). Many palsas here formed after 1965 in response to permafrost aggradation (Priesnitz and Schunke, 1978). In Canada palsas – regarded as probably formed within the last 200 years under a climate similar to the present – are reportedly confined to the discontinuous permafrost zone and distributed where the mean annual minimum and maximum temperatures are $-40°$ and $27°$ to $32°$, respectively, and the snowfall (except on the Labra-dor coast) is less than 120 cm (Railton and Sparling, 1973, 1038, 1042–3).[19]

5 Inactive and fossil forms

Palsas appear to go through a developmental cycle that eventually leads to thawing and collapse as discussed, for instance, by Railton and Sparling (1973). The open cracks that commonly accompany palsa growth and the water that tends to accumulate around palsas, possibly as a result of their weight depressing the adjacent bog surface, are important factors here (Wramner, 1973, 122–3), the water promoting thawing because of its high heat capacity (Friedman *et al.*, 1971, 141). The fact that palsas in various stages of growth and decay occur together shows that their collapse is not necessarily indicative of climatic change (Friedman *et al.*, 1971, 141–4; Jan Lundqvist, 1969, 209).

Possibly some degrading forms are characterized by a circular rampart, although this is denied by French (1976a, 246) on the basis that in the absence of a marked ice core degradation would merely restore the former surface. In any event, collapse of a small palsa is more likely to leave such a rampart than in the case of a large palsa, which would tend to have a less conical shape. In possible palsa occurrences illustrated by Svensson (1969), rampart height rarely exceeded 1.5 m, and the ramparts enclosed lakes up to about 35 m across. However, although inclined to favour a palsa origin, he admitted that some of the ring-like forms might be collapsed pingos. Since no active forms were present, he concluded that the collapse was probably climatically determined. These ring-like forms should not be confused with the peat rings of Hopkins and Sigafoos (1951, 76–84), which are typically smaller in diameter and of quite different origin.

The recognition of older fossil palsas without ramparts is problematical. As pointed out by Jan Lundqvist (1969, 209–10) and noted above, the structure of a palsa is normally such that collapse would merely tend to restore the former surface of a bog, and little topographic or stratigraphic evidence of the palsa would remain.

[19] As described by Dionne (1978b, 205, 209) in a publication received while this volume was in press, the main development of treeless palsas on the east side of James Bay in Quebec corresponds to a mean annual air temperature of $-4°$ to $-4.5°$; their southern limit is about $-3°$, whereas the southern limit of palsas with trees is about $-1°$.

VII   Pingos

1   General

Pingos, also called hydrolaccoliths, bugors, and bulgunniakhs (bulgunyakhs), [20] are large, perennial, ice-cored mounds. The term pingo, suggested by Porsild (1938), is an Eskimo name for hill. Pingos range in height from about 3 m (Mackay, 1962, 26) to 70 m (Leffingwell, 1919, 150–1), and in diameter from about 30 to 600 m (Mackay, 1962, 26).[21] Most pingos tend to be more or less circular but elongate forms are also known. On Banks Island in the Canadian Arctic, a pingo with a height of 5 m was 220 m long and 32 m wide; another with a height of 14 m was 190 m long and 83 m wide (Pissart and French, 1976, 939, Table 1). Reports from elsewhere in Arctic Canada describe esker-like forms with lengths of 610 m (2000 ft) (Mackay, 1966a, 72) or even 1300 m (Pissart, 1967, 204) but with heights rarely exceeding 7.6 m (25 ft) for the former and no greater than 8.75 m for the latter. In the absence of a cross section, the distinction between small pingos and some palsas can be nebulous. Gaping dilation cracks radiate from the apex of many pingos as the result of growth of the ice core, and in many the summit area is collapsed because the ice core has thawed, producing thermokarst (Figure 5.47). An excellent, up-to-date review of contemporary pingos has been provided by Mackay (1978b).

2   Constitution

In contrast to palsas, typical pingos are larger and many contain a massive ice core (Figure 5.48) rather than a series of ice lenses (Jan Lundqvist, 1969, 206–10). In fact the ice core of a small pingo described by Rampton and Mackay (1971, 6–11) at Tuktoyaktuk in northern Canada was large enough to develop into a cold-storage facility for the local inhabitants (J. R. Mackay, personal communication, 1976). Drilling shows that some pingos have ice-lensed cores, rather than discrete planoconvex cores of clear ice, to depths greater than the pingo heights (Mackay and Stager, 1966, 363–7). Pingo ice is described briefly in the chapter on Frozen ground.

The material enveloping the core can range from well-sorted clay to gravel, or be a diamicton or even bedrock. Bedrock occurrences in northern Canada include dolomite (Craig, 1969) and well-indurated shale (Balkwill *et al.*, 1974, 1323), and occurrences in sandstone are known from Northeast Greenland (Fritz Müller, 1959, 13–55; 1963, 9–34) and Spitsbergen (Åhman, 1973, 194–6). Deformation of the ice and surrounding material can be prominent, with originally horizontal layers becoming vertical (Pissart and French, 1976, 941–4, Figures 6–7, 9, 11–12).

3   Origin

Pingos are thought to grow in two main ways – by cryostatic pressure and by artesian pressure. There are two varieties of the cryostatic hypothesis: (1) A lake becomes filled with vegetation or sediment, and being in a permafrost environment this leads to the water becoming entrapped by progressive freezing from top, bottom, and sides. Eventual freezing and expansion of the entrapped water body results in heaving and doming of the ground. The role of vegetation is believed to be minor because lakes in the continuous permafrost zone are rarely filled with plant debris (J. R. Mackay, personal communication, 1978). Shallowing of a lake by sedimentation that raises the bottom to within the zone of annual freezing (Porsild, 1938, 55) is a more probable event, whose effect would eventually be comparable to lake drainage. (2) Draining or diversion of a water body that had insulated the underlying high-moisture sediments from freezing, leads to progressive all-sided freezing of the sediments. This causes a massive ice body to form, either by (a) expelling unfrozen water to where it freezes as injection ice, or (b) providing a ready supply of water for development of segregation ice, or both (Figure 5.49).

---

[20] As noted above in discussing palsas, the Russian terms bugor and bulgunniakh are equally applicable to a pingo or palsa. Different investigators have used these terms and hydrolaccolith in different ways as reviewed by Jahn (1975, 90–9), and the present writer agrees with him that it would be well to adopt pingo as the prevailing term for large, perennial, ice-cored mounds. The term hydrolaccolith has been applied both to pingos (cf. Fritz Müller, 1959, 104–5, 111; 1963, 64, 70 – especially the open-system type – Maarleveld, 1965, 15, Table II), and to annual mounds (Tolstikhin and Tolstikhin, 1974, 222) known as frost blisters, whose origin and terminology have been discussed by van Everdingen (1978).

[21] The upper pingo height (*c.* 200 m) and diameter (several thousand metres) cited by Ferrians and Hobson (1973, 483) are erroneous (O. J. Ferrians, Jr., personal communication, 1978).

Small ice-cored mounds less than 50 cm high and ranging in diameter from 2 to 5 m occur within ice-wedge polygons on Banks Island in the Canadian Arctic. Although thought to have originated in essentially the same way as some pingos (closed-system type), these pingo-like forms are not regarded as true pingos because of their small size (French, 1971a).

5.47 Pingo, Wollaston Peninsula, Victoria Island, Northwest Territories, Canada

The pingos in the Mackenzie Delta area of northern Canada, where 98 per cent of the 1380 pingos that have been mapped here (Stager, 1956)[22] are clearly located in (or marginal to) present or former lake basins, support the cryostatic concept. Doming of some of the pingos was accompanied by faulting (Mackay and Stager, 1966, 363–7). The origin of these pingos, known as the closed-system or Mackenzie type, was discussed by Mackay (1962; 1963b; 1966a; 1972a, 17–19; 1978b), Fritz Müller (1959, 97–103; 1963, 59–64), and Porsild (1938)

among others. In addition to (1) filling of lakes and (2) draining of lakes, possible causes of closed-system pingos include (3) changing river channels (for instance, some pingos in the Mackenzie Delta – Mackay, 1963a, 93 – and pingo ridges on Banks Island in northern Canada – Pissart and French, 1977, 149–55 – and ox-bow lake pingos in Mongolia – Rotnicki, 1977), (4) retreat of the sea and, more speculatively, (5) localization along faults (Pissart, 1977b; Pissart and French, 1977, 154–9).

According to the artesian concept, ground water

[22] Some 1450 pingos are now known to occur along the Western Arctic Coast of Canada (Mackay, 1973b, 979). They are also locally common farther west in the Arctic Coastal Plain of Alaska, 732 pingos and pingolike features having been mapped in the National Petroleum Reserve – Alaska (J. P. Galloway and Carter, 1978); 38 pingos were counted from a single vantage point at the head of the Colville Delta (Walker, 1973, 56).

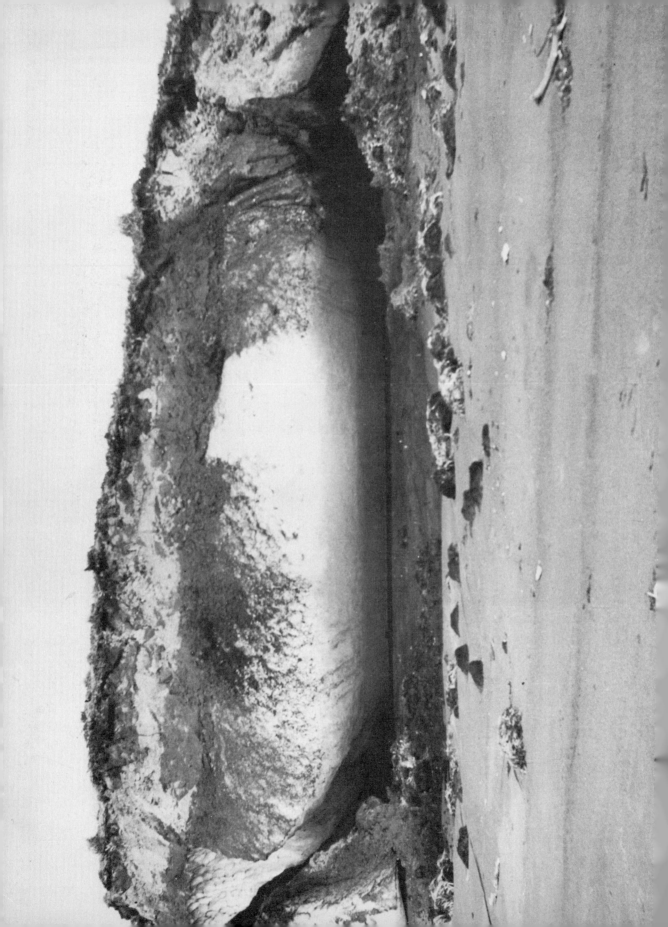

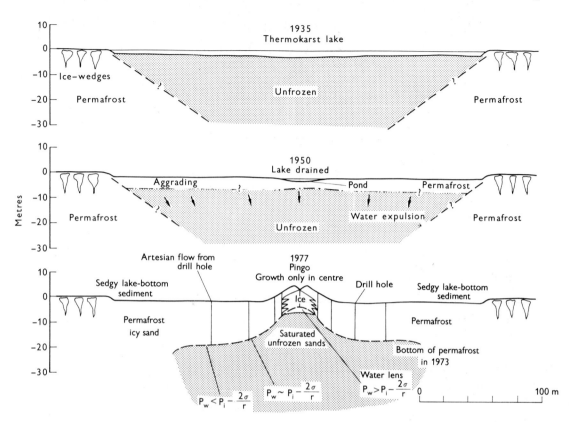

5.49 Growth of closed-system pingo in Mackenzie Delta area, Arctic Canada. Schematic cross-section. p_t = total resisting pressure to uplift (= overburden pressure, p_i, + bending resistance, M), P_w = porewater pressure, σ = surface tension ice-water, and r = radius of largest continuous pore openings. (Cf. discussion of Freezing process and Effect of grain size in preceding chapter, and Figure 5.51.) (*J. R. Mackay, personal communication, 1977; cf. Mackay, 1972a, 19, Figure 22; 1978d, 1222, Figure 3; 1223, Figure 5*)

flowing under artesian pressure below thin permafrost or in taliks within permafrost forces its way to near the surface where it freezes as injection ice, again forming an ice core that heaves the surface (Figure 5.50). Such pingos, known as the open-system or East Greenland type, can develop in either soil or bedrock. Their origin has been discussed in detail by Fritz Müller (1959, 60–72; 1963, 37–47). Other East Greenland occurrences have been described by Allen, O'Brien, and Sheppard (1976), Cruickshank and Colhoun (1965), Flint (1948, 203–6), Lasca (1969, 24–5), O'Brien (1971), O'Brien, Allen, and Dodson (1968), Vischer (1943, 22–3), and Washburn (1969a, 100–5). Isotopic analysis of pingo waters from this region shows that recent

5.48 Wave-cut, cross-section of pingo showing ice core, McKinley Bay, *c.* 100 km northeast of Tuktoyaktuk, Mackenzie Delta area, Arctic Canada. Pingo was about 90 m long and 7–10 m high. Photo by J. R. Mackay (*cf. Mackay, 1973b, 980, Figure 1*)

meteoric water of local origin is the dominant constituent and that pingo groups are not necessarily supplied by a common aquifer (Allen, O'Brien, and Sheppard, 1976). Pingo occurrences in West Greenland were reported by Weidick (1975, 108–9), citing Laursen (1950, 13) and Rosenkrantz (1940; 1942, 41–2) who described 'mud volcanos'. An objection to a purely artesian explanation of pingos is that calculated pressures required to dome a pingo are extreme compared to most measured artesian pressures; therefore additional pressure effects are probably involved, such as pressure magnification as in a hydraulic press (cf. Fritz Müller, 1959, 63; 67, Figure 27; 1963, 39; 98, Figure 27) or as pressures resulting from freezing (G. E. Holmes, Hopkins, and Foster, 1968, H31–2). The hypothesis as presented by Müller has been questioned by Scheidegger (1970, 375) on the basis that water in an artesian tube in permafrost would freeze, and that drilling has failed

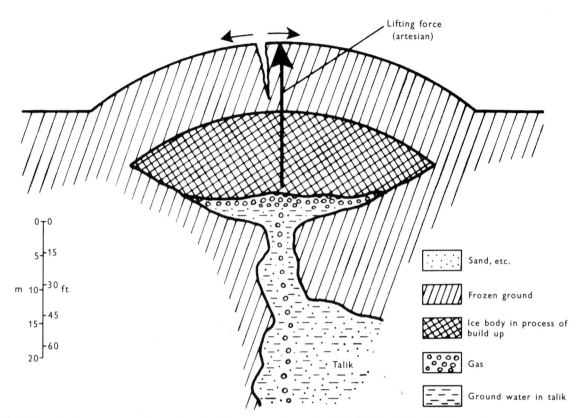

5.50 Growth of open-system pingo. Hypothetical cross-section (*after Fritz Müller, 1968, 846, Figure 2*)

to reveal water pockets beneath pingo ice. In any event, the exact way in which pingos form under artesian open-system conditions is poorly understood (G. E. Holmes, Hopkins, and Foster, 1968, H32; J. R. Williams and van Everdingen, 1973, 440).

The cryostatic, closed-system type and the artesian, open-system type of pingo can differ in continuity of water supply and thereby in pingo habit. Whereas the closed-system pingos tend to be isolated and located on near-horizontal surfaces characterized by lakes, the open-system type is probably most common in drainages where the artesian water pressure is sometimes continuous enough to cause a series of pingos. G. E. Holmes, Hopkins, and Foster (1968, H15) observed that the open-system pingos they studied in Alaska tended to cluster and mutually interfere with each other. According to Black (1976b, 83), open-system pingos tend to be smaller than the closed-system type. Watson (1971, 381–8) reported fossil forms that he believed reflected an upslope migration of injection centres, and Mückenhausen (1960) and Pissart (personal communication, 1971) explained the elongate shape

of some fossil pingos on a slope of 2°–4.5° as reflecting headward growth.

Despite such differences between some closed-system and open-system pingos, these types probably converge. Pissart (1970a, 31–3) has suggested that perhaps the open-system type is really formed under temporary closed-system conditions resulting from winter freezing of open taliks. Mackay (1971b; 1972a, 15–19; 1973a; 1973b) and Mackay and Black (1973, 187) have advanced the intriguing idea that not only are most pingos along the Canadian Arctic Coast the result of ice segregation in a closed system but that there are all gradations between pingos and massive, horizontally layered ice masses such as occur in the same region. Moreover, Mackay (1973b, 999–1001, Figure 4) suggested that some pingos may go through growth stages from build up of pore ice to segregated ice to injection ice (Figure 5.51). Mackay's (1977d) observation that some growing pingos pulsate in height as the result of porewater expulsion and pressure build up in aggrading permafrost and of pressure release by springs supports this view. Water flowing from drill

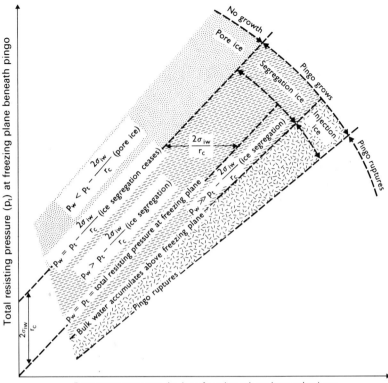

5.51 Relationship between total resisting pressure at freezing front (p_t, kg cm^{-2}) (= overburden pressure, p_i, + bending resistance, M, which varies with thickness of overburden and height of pingo), porewater pressure at freezing front (p_w, kg cm^{-2}), interfacial energy (surface tension) ice-water (σ_{iw}, 30.5 dyn cm^{-1}), and type of ice growth at freezing plane beneath a pingo (cf. Figure 5.49) (*J. R. Mackay, personal communication, 1977; cf. Mackay, 1973b, 1001, Figure 23; 1978b, 139, Figure 2*)

holes in a pingo had a temperature of $-0.1°$ and the flow amounted to an estimated 30 000–40 000 m^3 before the last holes froze from the sides (J. R. Mackay, personal communication, 1977). The resulting loss of pingo height was 60 cm, representative of at least 7 years' pingo growth. Another pingo had a water lens up to 2.2 m thick and at least 30 m in diameter that was due to porewater expulsion, and some pingos can probably contain a water lens for years (Mackay, 1978d). It has been proposed that ice cores having '... irregularly oriented shear zones and fractures may indicate injection ...' (French, 1975a, 459), and the structure of some pingos investigated by Pissart and French (1976) led them to suspect that the pingo ice comprised both injection ice and segregated ice in accordance with Mackay's concept. Ice fabrics can constitute another criterion (cf. Gell, 1975). For instance, unless subsequently modified by recrystallization, elongation of crystals and bubbles in injection ice that is bounded by two

(or more) freezing fronts should be normal to each, whereas in segregation ice there should be only one such front and fabric (Alan Gell, personal communication, 1977; cf. Gell, 1978b).

In addition to cryostatic and artesian pressures, several other, much more questionable causes of pingo growth have been suggested. One hypothesis is that density differences between thick ice masses and overlying heavier material causes upward intrusion of ice diapirs (Gussow, 1954), but the close association of pingos with lakes and former lake basins in the Mackenzie Delta and Anderson River areas, as well as other considerations, cast serious doubt on the hypothesis as applied to these areas (Mackay, 1958b, 55; 1962, 32; 1973b, 987).

Another suggestion is that tectonic subsidence depresses permafrost to sub-permafrost levels at which thawing and resulting compaction of saturated sediments lead to hydrostatic pressures that force water upward through favourable spots in the permafrost to generate pingos (Bostrom, 1967). The hypothesis was proposed for pingos in the Mackenzie Delta area but there are compelling reasons for rejecting it for this area (Mackay, 1968); furthermore some pingos in this area have been observed forming following lake draining and consequent ground freezing in accordance with the cryostatic concept (Mackay, 1973*b*, 987; 1977*b*).

Still another hypothesis postulates that ice segregates during initial freezing in an open system and, as a pingo grows, that more ice is added to the base of the pingo ice and to the permafrost to replenish ice that is lost by thawing at the top (Bobov, 1970; Scheidegger, 1970, 376). Details of the process and the evidence for it appear to be largely lacking in the English literature. A variant calling for pingo growth by ice accumulation at the base of the active layer (Ryckborst, 1975) was effectively countered by Mackay (1976*f*), the reply (Ryckborst, 1976) notwithstanding.

Pingos can be under sufficient pressure to rupture explosively. Bogomolov and Sklyarevskaya (1969; 1973) reported large chunks of ice up to 0.7 × 1.5 × 2 m in size being thrown for distances of 2–8 m, with smaller pieces being cast as far as 22 m. A jet of water up to 1.5 m high accompanied the rupture and spouted for 30 minutes.[23] Small amounts of water and incombustible gas issue from some pingos (Mackay and Stager, 1966, 363–6).

Pingos from two localities some 150 km apart on Banks Island in the Canadian Arctic are probably between 4500 and 7000 ^{14}C years old, thus supporting the view the pingos grew during a cooling trend following the Hypsithermal (postglacial climatic optimum) (French, 1977; cf. Pissart and French, 1976, 943–4). Dating of sediments in pingos in Spitsbergen (Bibus, 1975, 115; Svensson, 1970, 174), the Mackenzie Delta area (Fritz Müller, 1962), and Alaska (G. E. Holmes, Hopkins, and Foster, 1968, H27, H36) provide maximum ages only,[24] almost all Holocene. Except for the youngest pingos, data on minimum ages are largely lacking but actual ages up to a few thousand years are considered probable. The highest annual growth rate observed by Mackay was 21.2 cm yr⁻¹; he considered 1.5 m yr⁻¹ during the first one or two years of growth as possible for the pingos he studied but concluded that

the rate decreases inversely as the square root of time (Mackay, 1973*b*, 994, Table 3; 1003). Ages of several pingos in Siberia have been reported to range from 106 to 162 years (Schostakowitsch, 1927, 418, following Sukachëv, 1912, 82–9). In the Mackenzie Delta area, pingos up to about 6 m high are known to have formed after 1935, following coastal recession that led to draining of lakes. Other pingos have formed since 1950. The vertical growth rate of one pingo, whose crest was characterized by an open dilation crack 0.5–1.5 cm wide and 15–18 m deep, was about 15 cm yr⁻¹ from 1969–71 (Mackay, 1972*b*, 146–7; personal communication, 1972). In general the vertical growth rate of pingos ranges from a minute amount to more than 0.5 m yr⁻¹ (Shumskiy, 1964*b*, 226). According to Jaromír Demek (personal communication, 1969), a 15-m high pingo he studied had an average growth rate of 1.5 m yr⁻¹.

4   Environment

Pingos are necessarily associated with permafrost. Like ice-wedge polygons and sand-wedge polygons, they are key indicators of polar or subpolar environments. However, open-system pingos, believed to be in balance with the present climate, occur in the discontinuous permafrost zone of central Alaska where the mean annual air temperature ranges from −2.2° to −5.6° (22°–28°F) (G. E. Holmes, Hopkins, and Foster, 1968, H7), or −1° to −2° in places (Péwé, 1969, 3). In western Siberia, open-system pingos occur where the mean annual air temperature is below −3° to −4° (Baulin, Dubikov, and Uvarkin, 1973, 15; 1978, 66). Thus open-system pingos do not necessarily imply temperatures as low as that under which closed-system pingos or ice wedges originate in Alaska (−6° to −8° according to Péwé, 1966*a*, 78; 1969, 4; 1975, 52, 58–9) and elsewhere in the Arctic. Thus pingos *per se* should not be taken as indicative of mean annual air temperatures below −7° as is sometimes done (cf. Mitchell, 1977, 205). Because of their association with continuous permafrost, Mackay (1978*b*, 145) cited a mean annual ground temperature of −5° or lower for closed-system pingos. In addition to the

[23] The explosive rupture reported by Plaschev (1956; cf. Swinzow, 1969, 187), which has been cited as another illustration, refers to an icing mound, not a pingo, as pointed out by W. Barr and Syroteuk (1973, 8).

[24] Bibus (1975, 115; also Bibus, Nagel, and Semmel, 1976, 39) cited a ^{14}C age of 1875 ± 470 years BP, which he interpreted as a minimum age although the evidence cited does not appear to support this. He also cited Svensson (1970, 174) for a ^{14}C age of 2400 years BP, whereas the age given by Svensson was a maximum age of 2650 ± 55 years BP. In the case of the Ibyuk Pingo, which Fritz Müller (1962, 284) regarded as probably no older than 7000–10 000 years, surveys show it is still growing, and calculations derived from the 1973–1975 growth rate suggest an age of about 1000 years (Mackay, 1976*e*).

Alaska, Greenland, and Mackenzie Delta occurrences, pingos are reported from the Yukon (O. L. Hughes, 1969), Canadian Arctic Islands (French, 1975*a*; Fyles, 1963; Pissart, 1967; 1970*b*; Pissart and French, 1976; Washburn, 1950, 41, 43), Spitsbergen Åhman, 1973; Liestøl, 1977; Svensson, 1970; and others), and USSR (cf. Frenzel, 1959, 1021, Figure 13; 1967, 29, Figure 17; Grave, 1956). Their distribution in Alaska, Canada, and Eurasia is shown in Figures 5.52–5.54. As illustrated in Figure 5.52, it is striking that most open-system pingos in northwest North America occur in areas that remained unglaciated during the Pleistocene. A few occur in areas that were glaciated prior to 25 000 years ago but open-system pingos are reported to be extremely rare in more recently glaciated parts of the region (R. J. E. Brown and Péwé, 1973, 80–1; Péwé, 1975, 58–9). The reason for this distribution is not established. On the other hand, open-system pingos in the Mesters Vig district of Northeast Greenland are quite common in places that remained glaciated until shortly before 8500–9000 years BP (Lasca, 1969, 24–5, 52; Washburn, 1965, 31; 1969*a*, 100–5).

Numerous features averaging 30 m high and 400 m in diameter and regarded as possible pingos occur at depths of some 30–70 m on the floor of the Beaufort Sea in the vicinity of the Mackenzie Delta (Shearer *et al.*, 1971). If they are pingos, they probably formed beneath sea level as favoured by Shearer *et al.* Any subaerial pingos undergoing submergence would presumably have been quickly destroyed by coastal erosion as pointed out by J. R. Mackay (personal communication, 1977). In any event, the observed sea-bottom temperatures, which range from below −1° to −1.8° (−1.8° being the lowest possible without freezing of the salt water here), are indicative of permafrost. Freezing of fresh water in the sediments would be expected, and in fact fresh-water ice was found in a sea-bottom core taken from a depth of about 35 m (20 fathoms) (Yorath, Shearer, and Harvard, 1971, 244). However, as of 1973, the presence or absence of an ice core in the presumed pingos had not been established by drilling (Mackay and Black, 1973, 186), nor was any confirmation available at a meeting on offshore permafrost that the present writer attended early in 1977.

5 Inactive and fossil forms

Like palsas, inactive pingos, i.e. partially collapsed pingos in permafrost environments, are not necessarily indicators of environmental change. This is because pingo growth can lead to its own destruction by the dilation cracking of the material capping the ice core and exposure of the core to thawing. Fossil pingos, also known as pingo scars – a term that includes disintegrating pingos (Flemal, 1976, 38), in contemporary non-permafrost environments are proof of former permafrost. Fossil pingos are recognized in places by their pattern, rampart-like rim (rampart), and internal collapse structure but where characterized by a thermokarst depression only, they may be very difficult to distinguish from other thermokarst depressions (Black, 1969*b*, 232–3; Flemal, 1976; Kozarski and Szupryczyński, 1973, 306, Figure 52; 307). Flemal considered the distribution of probable and more questionable fossil pingos and possible alternative explanations for such forms, especially glacial and thermokarst processes unrelated to pingos. Among the glacial processes are (1) upward squeezing of saturated material into basal openings by the weight of overlying ice (Stalker, 1960, 16–22), (2) differential deposition in association with disintegrating ice, which may cause ice-contact rings (Parizek, 1969), and (3) formation of kettle holes as noted above. Thermokarst processes – which also include (3) if the glacier ice is buried – can develop depressions resembling pingo scars by (1) melting of widespread ground ice, often with development of thermokarst lakes, and (2) melting of ground ice localized in palsas. Except for palsas being commonly smaller than pingos, the resulting depressions could be very difficult to distinguish from fossil pingos, since ramparts would be characteristic of both. Features similar to pingo scars can be produced by a variety of processes and no single criterion suffices; however, a thermokarst as opposed to glacial origin would be proved if the feature lay beyond the glacial limit. In summary,

Differentiation of pingo scars from other varieties of thermokarst features is most difficult. Form and size are not definitive, although a pingo scar origin is favored for regular, ramparted forms in the dimension range of a few tens of meters to a few hundreds of meters. Perhaps the best criteria are upwarping of strata beneath a rampart or fabrics otherwise indicative of an initial upwarping, and overlapping and superpositioning of individuals indicative of multiple generation of forms. Location on valley sides [for open-system pingos] is also suggestive, as is disposition of the substrate-sediment contact below the level of the adjacent terrain.

Possible modes of origin [for pingo-like scars] other than glacial or periglacial must also be considered. (Flemal, 1976, 48–9.)

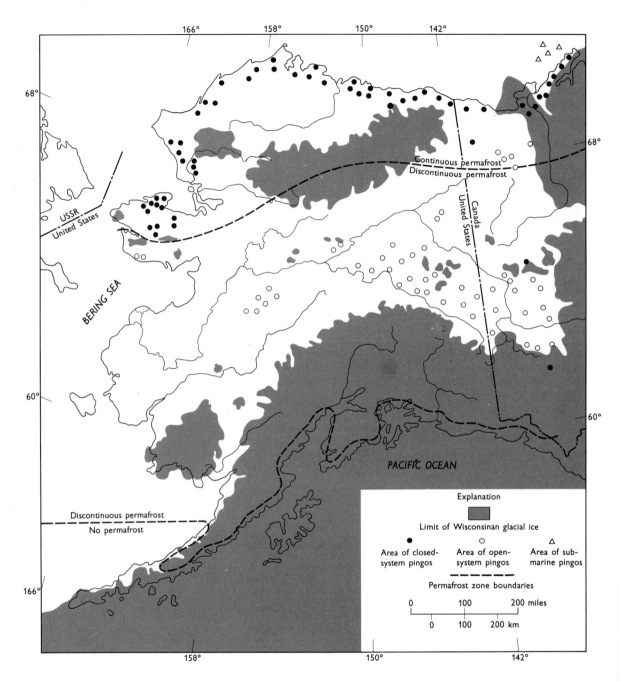

5.52 Pingos in northwest North America. Data from various sources, including personal communications from O. J. Ferrians, Jr., H. Foster, D. M. Hopkins, O. L. Hughes, J. R. Mackay, I. Tailleur, and W. Yeend (*after Péwé, 1975, 58, Figure 31; cf. R. J. E. Brown and Péwe, 1973, 82, Figure 4*). The 'Area of submarine pingos' comprises pingo-like forms whose origin remains to be established (*J. R. Mackay, personal communication, 1977*)

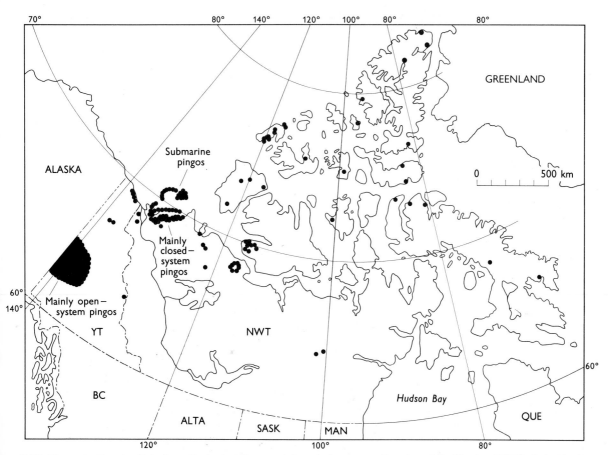

5.53 Pingos in Canada. Data from Geological Survey of Canada base map, French and Dutkiewicz (1976), O. L. Hughes (1969), Pissart (1967), Pissart and French (1976), Shearer (1972), and various publications by Mackay (*after R. J. E. Brown and Péwé, 1973, 81, Figure 3*). The 'submarine pingos' comprise pingo-like forms whose origin remains to be established (*J. R. Mackay, personal communication, 1977*)

Alternative origins include solution depressions in areas of soluble bedrock and deflation hollows where wind action has been prominent. In such places the nature of the associated material would provide critical evidence for or against such possibilities.

According to French (1976a, 246), fossil pingos of the open-system type tend to occur in clusters reflecting repeated generation of pingos where groundwater conditions were favourable. As a result, ramparts tend to intersect and be irregular in plan. Some of the more regular rampart-like features of what are believed to have been open-system pingos are elongate in the direction of slope, and some U-shaped ramparts open out upslope. Typically, open-system pingos are restricted to valley bottoms or to the base of slopes. These features conform to a number of descriptions from western Europe, and French concluded that this region was characterized

by thin or discontinuous permafrost in late-glacial time. However, identification of pingo type must be approached cautiously in view of the difficulties surrounding the recognition of many fossil pingos as such, irrespective of origin.

Fossil pingos associated with glacial deposits are reported from the Ordovician of the Sahara (Beuf *et al.*, 1971, 294-7) but otherwise all known fossil pingos date from the last glaciation or later.

Fossil pingos are becoming increasingly widely reported from the British Isles (Figure 5.55), Europe, and Asia (Cailleux, 1976; Frenzel, 1959, 1021, Figure 13; 1967, 29, Figure 17; 1973a, 32, Figure 17; Maarleveld, 1965; Mitchell, 1971; 1973; Mullenders and Gullentops, 1969; Pissart, 1956; 1963a; 1965; Seppälä, 1972b; Svensson, 1969; Watson, 1971; 1972; 1976, 79–87; Watson and Watson, 1972; 1974; Wiegand, 1965). Thus in France north of lat. 44°30′N,

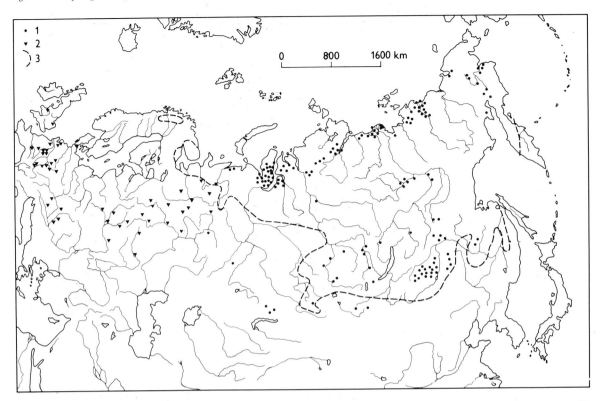

5.54 Distribution of pingos at present and during last cold period in northern Eurasia (*after Frenzel, 1967, 29, Figure 17; 1973a, 32, Figure 17*)

1 Present Pingos
2 Pingos dating from last cold period
3 Present southern limit of permafrost

Cailleux (1976, 374) reported as many as 35 fossil pingos per km² occurring locally, with about one per km² being quite general. Soluble bedrock is present in places, and the origin of some of the forms was regarded as probable rather than proved. Cailleux (1976, 378) associated the time of pingo formation with Dryas III (Younger Dryas, *c.* 10 150–10 950 yr BP).

The widely known fossil pingos described by Pissart from the Hautes Fagnes plateau in Belgium are now believed to have originated by growth of segregated ice rather than injection ice and to be a record of discontinuous permafrost, since their mode of occurrence differs somewhat from typical contemporary pingos (Pissart, 1974*b*; 1976*b*, 134). Occurrences elsewhere may also be subject to critical review as to their implications. The Hautes Fagnes forms postdate the Arcy-Stillfried B Interstade (*c.* 28 000 ¹⁴C years BP) and may well predate the Oldest Dryas (13 300–15 000[?] ¹⁴C years BP) (Pissart, 1976*b*, 134–5; Woillard, 1975, 25).

Possible fossil pingos have been reported from North America (Bik, 1967; 1969; Flemal, 1972; 1976, 44–50; Flemal, Hinkley, and Hesler, 1973; Wayne, 1967, 402), but the evidence is incomplete despite the probability that pingos here were quite common where there was permafrost.

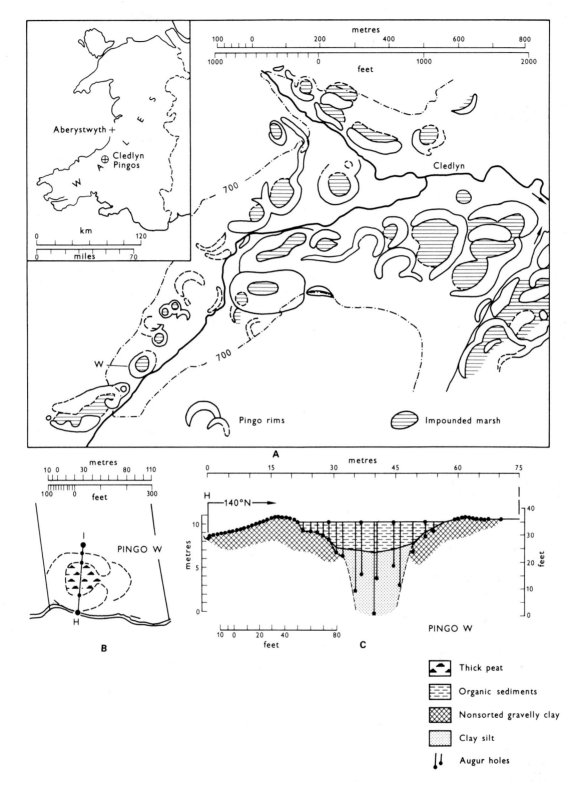

5.55 Fossil pingos, Cledlyn basin, central Cardiganshire, Great Britain. **A**. Map of pingos. **B**. Pingo W. **C**. Cross-section of pingo W (*after Watson and Watson, 1972, 213–14, Figure 1, and Figure 2 – in part*)

6 Mass-wasting processes and forms

I Introduction

Mass-wasting is 'The movement of regolith down-slope by gravity without the aid of a stream, a glacier, or wind' (Flint and Skinner, 1977, G.6 cf. 122). The term was first used by Longwell, Knopf, and Flint (1939, 41).

Although many types of mass-wasting occur in periglacial environments, some types are particularly widespread, including avalanching, slumping, frost creep, and gelifluction. Slushflow is common locally. The role of snow creep in moving rock debris is cited in the Introduction to the chapter on Nivation.

In many places periglacial mass-wasting produces smooth slopes. Different opinions exist where such slopes are bare rock (Glatthänge) (Hagedorn, 1970; Karrasch, 1970, 222–6; 1974), and such slopes can be convergent phenomena formed also under non-periglacial conditions (Poser, 1978). Where slopes are debris mantled and show evidence of gelifluction, the smoothness is commonly attributed to that process and is therefore regarded as typically periglacial.

As with many other types of periglacial phenomena, fossil mass-wasting deposits can provide evidence of climatic change. Features of fossil deposits can also be important in relation to engineering, an illustration being potential slope-stability problems because of shear planes that formed under the former periglacial conditions (Weeks, 1969).

II Avalanching

1 General

Avalanching as the term is used here is the sudden and very rapid movement of snow and/or rock debris down a slope. The term has been applied to free fall, sliding, and flow but some investigators restrict it to movements where flow is dominant (Longwell, Flint, and Sanders, 1969, 166–7). Regardless of the exact mechanism, rapidity of movement is a key characteristic, with calculated rates for rock avalanching ranging from 1 to 100 m s^{-1} (Longwell, Flint, and Sanders, 1969, 166, Figure 8–3). Avalanching is sometimes defined in terms of snow and ice only but avalanches of rock debris (including rockfall and debris avalanches) are also commonly recognized (Varnes, 1958, Plate 1 opposite 40), so that three broad categories can be identified – snow avalanches, rock avalanches, and mixed or snow-rock avalanches.

On a small scale, avalanching can occur on high banks and elsewhere in lowlands but it is essentially a highland phenomenon, characteristic of steep slopes, which if of bedrock can be steeper than the angle of repose. Frost creep and gelifluction on the other hand are confined to debris slopes whose gradient usually does not exceed the angle of repose. The minimum gradient on which avalanches can start depends on conditions and varies widely. Some avalanches attain sufficient momentum to carry them several hundred metres up opposing slopes.

Studies by Shreve (1968*a*; 1968*b*) and the speed of avalanches make it probable that many avalanches are air lubricated in that they ride on a cushion of entrapped air.

Although rock avalanches as well as snow and mixed avalanches are common in periglacial highlands (cf. Rapp, 1960*a*, 97–137), rock avalanches are also common in other highlands and need have no special periglacial significance.

2 Snow avalanches and mixed avalanches

Snow avalanches are of many kinds, ranging from dry powder-snow varieties to wet-snow avalanches. The kind is dependent on climate and particularly on day-to-day weather conditions before, during, and after a snowfall. Because snow-avalanche danger is of critical concern in mountaineering, skiing, and highway maintenance through alpine passes, the prediction and control of avalanching has been intensively investigated and a vast literature has accumulated (Luckman, 1977; Perla and Martinelli, 1976; World Data Center A for Glaciology, 1977).

Many avalanches that start as snow avalanches pick up rock debris *en route* and become mixed avalanches. Both kinds can drastically alter the landscape by ploughing through a forest and leaving a swath of broken and overturned trees or smashed buildings in their wake, but only the mixed avalanche and the rock or debris avalanche can leave a deposit or landform that may become a more enduring record. A characteristic feature of such a deposit is its nonsorted nature and hummocky topography. In places such a deposit is difficult to distinguish from a moraine, and usually there would be still greater difficulty in establishing that an ancient deposit of this kind had been made by a mixed avalanche rather than by a rock or debris avalanche having no periglacial implication.

A type of mixed avalanche deposit known as an avalanche boulder tongue (Rapp, 1959), consisting of angular debris, ranges in the Canadian Rocky Mountains from scattered blocks[1] to tongue-like accumulations up to about 700 m long, 200 m wide, and 25 m thick. Based on Luckman (1977, 37–41; 1978, 270–2), diagnostic aspects include

(1) Position below avalanche tracks or chutes, the upper part of the slope system feeding into avalanche boulder tongues

(2) Loose chaotic surface fabric of angular debris characterized by perched boulders as result of deposition from ablating snow

(3) Smooth surface in middle and upper sections caused by avalanche erosion and consisting of fines and small chips with a few scattered boulders

(4) Sorting gradient at lower end with coarsest debris tending to extend farthest from cliff

(5) Avalanche debris tails (Rapp, 1959, 39–40) consisting of linear wedges of finer debris tapering downslope from blocks

(6) Concave, smooth long profile where tongues are sufficiently thick

(7) Mean slope intermediate between that of alluvial fans and talus (for 21 tongues the mean angle was 20°)

(8) Asymmetric cross profile, usually with steep front facing up valley

Not all these features are unique to avalanche boulder tongues. The fabric, for example, is also typical of slushflow deposits.

In distribution and frequency, snow and mixed avalanches are probably most common in temperate alpine regions and progressively less so in subpolar and polar alpine environments, since temperate alpine highlands would generally have more frequent snow storms and thaw periods than polar highlands. As described by Luckman, avalanche boulder tongues develop best above tree line.

III Slushflow

Slushflow has been defined as the 'mudflowlike flowage of water-saturated snow along stream courses' (Washburn and Goldthwait, 1958). However, the results of essentially the same process have been described as slushers (Koerner, 1961, 1068; Ward and Orvig, 1953, 161–2) and as slush avalanches (Caine, 1969; Jahn, 1967, 220–4; Nobles, 1966; Rapp, 1960*a*, 138–47). They may occur on glaciers (cf. 'slushers' above; also Nobles, 1961, 758–9) where a relationship to stream courses is less well defined or lacking as compared with glacier-free areas, although natural levees of snow are formed in both places. A more comprehensive definition would therefore be simply 'slushflow is the predominantly linear flow of water-saturated snow'.

---

[1] Because of the angularity of the debris, avalanche block tongue would be more descriptive than avalanche boulder tongue in the same way that a distinction can be made between block fields and boulder fields as noted later in this chapter.

6.1 Beginning of slushflow, Mesters Vig, Northeast Greenland. Note saturated snow (dark)

In the present writer's opinion, slushflow differs from most wet-snow avalanching in being more channelled and thus transitional between wet-snow avalanching and fluvial action; similarly, if mixed with rock debris it can be transitional to debris avalanching, or, to use Rapp's (1975, 90, 92) preferred term, debris flow. Looked at from this viewpoint, slushflow and debris flow would be more closely related than in Rapp's view.

Slushflow can be a discontinuous slow process as well as very rapid, and is characteristic of periglacial environments in areas where intense snow thaw in the spring produces more meltwater than can drain through the snow (Figures 6.1–6.2). It is common on Banks Island in the Canadian Arctic (Day and Gale, 1976, 182), and in places like the Mesters Vig district of Northeast Greenland it is the common method of stream break-up and can be an important agent of erosion and deposition (Raup, 1971*b*, 50–68) (Figures 6.3–6.4). Slushflow is also common in Spitsbergen, and the Longyearbyen disaster there in 1953, when several lives were lost and buildings destroyed, has been attributed to it (Jahn, 1967, 220–4). A characteristic depositional feature of

6.2 Slushflow, Mesters Vig, Northeast Greenland

slushflow in mountainous terrain is a whale-backed fan of predominantly nonsorted debris at the mouths of gullies. The fan is formed when a slushflow spews forth from the gully and suddenly drops its debris load. The resulting diamicton can be remarkably till-like, the particle sizes ranging from fine sand to boulders (Figure 6.4). Slushflows are probably more common in arctic and subpolar highlands than in the Antarctic with its infrequent thaw periods, and as noted by Luckman (1977, 43) are not to be expected where slopes are too steep to permit snow to accumulate enough water.

IV Thaw slumping

Slumping as defined by Sharpe (1938, 65) (who used the term slump) is 'the downward slipping of a mass of rock or unconsolidated material of any size, moving as a unit or as several subsidiary units, usually with backward rotation on a more or less horizontal axis parallel to the cliff or slope from which it descends'. In periglacial environments where thaw water from ground ice is such an important factor, the slumping is often mudflow-like and backward rotation is absent. For this reason and

6.3 Lejrelv slushflow, Mesters Vig, Northeast Greenland, 1958. Stones were scattered 20 m above the stream bed and a diamicton and fragmented tundra vegetation were spread over a third of the Lejrelv fan, which was 3/4 km wide at its toe.

6.4 Slushflow fan deposit, Mesters Vig, Northeast Greenland. Scale given by notebook (*c.* 15 × 20 cm) at centre

because the term ground-ice slumps (French, 1976*a*, 119–22) is ambiguous (since it implies that the ice itself partakes in the movement) the terms thaw slumping and thaw slumps are used here.

Thaw slumping tends to characterize steep slopes of unconsolidated material wherever permafrost with an appreciable ice content is subject to thawing. Coastal cliffs and river banks are typical sites. Mackay (1966*b*, 68–80), working in northern Canada, stressed that the pattern of slumping in periglacial environments is very different from that elsewhere. Scarp retreat in the Mackenzie Delta cliff sites he studied ranged from about 1.5 to 4.6 m yr^{-1} (5–15 ft yr^{-1}) where ground ice was especially prominent. In some cases the ice-soil ratio was so high that scarp retreat resulted in very little export of material beyond the toe of the scarps. Several cycles of slumping, separated by scarps averaging 4.6–6.0 m (15–20 ft) in height, could be recognized in places, and he reported that slumps can remain

active for decades. Along 2.5 km of shoreline, from 1964–1971, the mean rate of cliff recession was 2.3 m yr^{-1} (Kerfoot and Mackay, 1972, 119). Surface temperatures, as measured by radiometer, of an ice-rich permafrost face undergoing slumping can be surprisingly high; temperature of a thin water film over clear ice with scattered mud pellets was as high as 5.5°, and was up to 10° for a thin film of mud over ice (Mackay, 1978*a*). Slumping associated with ground ice is reviewed in some detail by French (1976*a*, 119–22). Some of the resulting features are also discussed later in the chapter on Thermokarst.

Slumping (not necessarily thaw slumping) can involve sliding of permafrost blocks whose basal shear surface probably lies in unfrozen or incompletely frozen material. Such block slides, and multiple retrogressive (MR) slides consisting of tiers of blocks, have been described from along river banks 45–60 m (115–200 ft) high in the Mackenzie

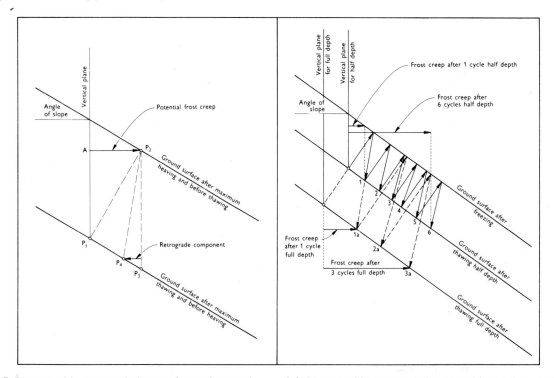

6.5 Diagram of frost creep during one freeze-thaw cycle   **6.6** Diagram of frost creep during several freeze-thaw cycles

Valley where it traverses the discontinuous perma-frost zone (McRoberts and Morgenstern, 1974*b*). The blocks consist of gravels, sands, and silts, whereas the shear surfaces lie in clays, some of which would contain appreciable amounts of unfrozen water at negative temperatures near 0°.

V   Frost creep

1   General

Frost creep 'is the ratchetlike downslope movement of particles as the result of frost heaving of the ground and subsequent settling upon thawing, the heaving being predominantly normal to the slope and the settling more nearly vertical' (Washburn, 1967, 10). The following discussion emphasizes observations in arctic environments and is based largely on Washburn (1967). Detailed studies by Benedict (1970*b*) in an alpine environment provide additional information and support the main points.

Although frost creep is commonly associated with gelifluction and is usually included with it in rate measurements because of the difficulty of distinguishing between their contributions to the total movement, frost creep is regarded as a separate process by most investigators, including Soviet (Iveronova, 1964, 210; Kaplina, 1965, 54; cf. Jahn, 1975, 145). By definition the distinction is based on the difference between creep (in the geologic rather than engineering sense) and viscous flow. However, some researchers emphasize sliding rather than flow as noted later, and others including Jahn (1975, 147) have considered frost creep as a component of solifluction, thus weakening the emphasis on flow,[2] although subsequently Jahn (1978) treated them as separate processes. M. A. Carson and Kirkby (1972, 276–7) emphasized solifluction in periglacial environments but considered it a 'rather rapid form of seasonal creep', yet they also regarded the rate as '... more characteristic of flow or creep in *water films* than of creep in *soils*' (M. A. Carson and Kirkby, 1972, 293) and later, in a thoughtful review of mass-wasting as related to slope development and climate, M. A. Carson (1976, 108) stressed the lack

---

[2] French (1976*a*, 135, 162, 235) also regarded frost creep as a component of what he called solifluction but his definition of the latter is somewhat ambiguous. However, he accepted frost creep as a process distinct from gelifluction.

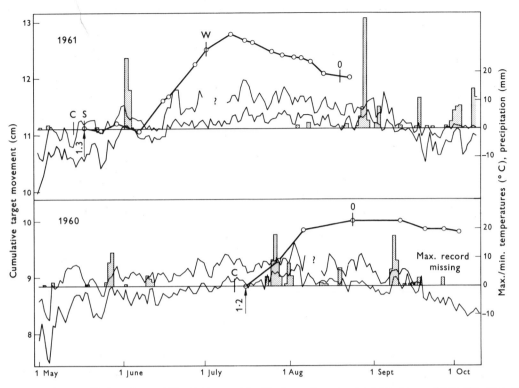

6.7 Movement of target 36, Experimental Site 7, Mesters Vig, Northeast Greenland (*cf. Washburn, 1967, 193, Plate C2–31*)

of information concerning slow earth flow and the need for much more quantitative research on the topic. More information on vertical displacement profiles would be illuminating.

Fleming and Johnson (1975, A26) pointed out that the frost-creep process, originally formulated by Davison (1889) and described below, has been commonly extended to also explain soil creep associated with volume changes caused solely by wetting and drying, but they noted that this extension is purely deductive. They claimed that the general theory has not been successfully tested and is inconsistent with some of their findings. Whether or not creep by wetting and drying or further causes are predominate in other environments, such creep is believed to be quite subordinate to frost creep and gelifluction in most places in periglacial environments. On taluses (scree slopes) lacking fines, creep resulting from temperature changes alone is probably important (Davison, 1888a, 1888b). In the following, frost creep and gelifluction are considered separate processes.

2 Characteristics

Upon freezing, the ground tends to heave at right angles to the slope, since this is the predominant cooling surface. As shown by Taber's (1929, 447–50; 1930a, 308; 1943, 1456) classic experiments, the direction of ice-crystal growth and heaving is normal to the cooling surface. Figures 6.5–6.6 illustrate the results. In Figure 6.5 a particle at P_1 tends to move to P_2; as measured with respect to the vertical plane at P_1, the movement would be the distance A-P_2, which represents the potential frost creep – i.e. the maximum possible movement by frost creep if there were no further freeze-thaw cycles. The potential frost creep would be the true frost creep if the particle at P_2 dropped vertically to P_3 upon thawing of the ground. However, there is a tendency for the movement during thawing to be towards some point P_4, usually between P_1 and P_3, as the result of a retrograde component that is '. . . probably due to cohesion and to interference of particles with each other' (Ahnert, 1971, 36), or may be related to capillary pressures (Washburn, 1967, 109–15). In swelling and shrinking soils, if stresses parallel to a slope are not strong enough to shear '... a significant number of bonds, the soil structure on shrinking will retrace its swelling path' (Kojan, 1967, 237), and thus the slope tends to settle back against itself, reducing the amount of creep that would otherwise be present.

The retrograde component was observed in laboratory work by Davison (1889), who ascribed it to cohesion in a classic paper that first described the mechanism of frost creep. The observation was overlooked for some 70 years until confirmed by field observations like those illustrated in Figure 6.7, where the retrograde component is prominent in the graph of the 1961 movement, starting in early July following gelifluction in June. The arrow with the number 1.3 at the beginning of the graph indicates that the horizontal component of potential frost creep was 1.3 cm when the first observation was made in 1961; the adjusted slope-parallel measurement was 1.6 cm. In view of the absence of later freeze–thaw cycles in the ground, the subsequent downslope movement was gelifluction (as discussed in the next section) rather than frost creep. The retrograde movement starting in July amounted to 0.9 cm (adjusted), so the true frost creep for that year was 1.6 cm less 0.9 cm, or 0.7 cm (cf. Washburn, 1967, 46–50). A recent laboratory experiment, involving a tilting slab at the University of Washington's Quaternary Research Center, not only reproduced retrograde movement during thawing but provided evidence that this movement can exceed that attributable to soil collapse normal to the slope (R. G. Rein, personal communication, 1977). Since there were no desiccation cracks, the probable explanation is that greater desiccation towards the top of the slope than the bottom caused upslope contraction. A similar situation could prevail on many natural slopes.

Retrograde movement has also been discussed by Benedict (1970b), K. R. Everett (1963a, 108, 111, 117–18, 195–6; 1963b, 38, 42), S. A. Harris (1973), Jahn (1975, 145), Kirkby (1967, 360–1; cf. 364, Figure 3), and Anthony Young (1972, 50–1). Jahn (1975, 145) allowed for retrograde movement by introducing a coefficient of cohesion into the equation of movement

$$l = k\,h\,\tan\alpha$$

where l = movement parallel to slope, k = coefficient of cohesion, h = frost heave, and α = gradient.

Figure 6.5 depicts the effect of only a single freeze–thaw cycle. The effect of several cycles is shown in Figure 6.6, which illustrates that the amount of frost creep varies directly with their depth and number of cycles. In general the rate of movement decreases rapidly with depth in both frost creep and gelifluction. There are very few measurements of the rate of frost creep alone. In the special case of surficial frost creep of rock fragments by growth and collapse of needle ice, a movement of 14–18 cm within 36 days was reported by L. A. Zhigarev (cf. Czudek and Demek, 1972, 9; Kaplina, 1965, 68). Rates must be highly variable.

Other factors remaining constant, frost creep should be greater in a subpolar or even temperate climate than in a high polar environment where there tend to be fewer freeze–thaw cycles and the depth of thaw is shallow. Measurements of near-surface movement on a silty diamicton slope of 10°–14° at Mesters Vig, Northeast Greenland, indicate that frost creep tended to exceed other forms of mass-wasting (mainly gelifluction)

... but by not more, and probably less, than 3 : 1 over a period of years, and either process can predominate in a given year.... Absolute values of mass-wasting due to frost creep and gelifluction ... ranged from a mean of 0.9 cm/yr in sectors subject to desiccation during summer, to a mean of 3.7 cm/yr in sectors that remained saturated (Washburn, 1967, 118.)

On a slope with an average gradient of 25° and a stony, sandy soil, measurements by Black and Hamilton (1972, 121, 123, 125, 131) showed that frost creep and creep by wetting and drying could contribute about equally to the surface movement, which was 2–20 mm yr^{-1}. Such measurements are rare. Conditions vary widely, and many more measurements in various regions are needed to assess the significance of frost creep as opposed to other types but it is clearly one of the most important.

VI   Gelifluction

1   General

J. G. Andersson (1906, 95–6) defined solifluction as follows: 'This process, the slow flowing from higher to lower ground of masses of waste saturated with water (this may come from snow-melting or rain), I propose to name *solifluction* (derived from *solum*, "soil", and *fluere*, "to flow").' Gelifluction was defined by Baulig (1956, 50–1; cf. also Baulig, 1957, 926) as solifluction associated with frozen ground, including seasonally frozen ground as well as permafrost, whereas Dylik's (1951, 6) term con-

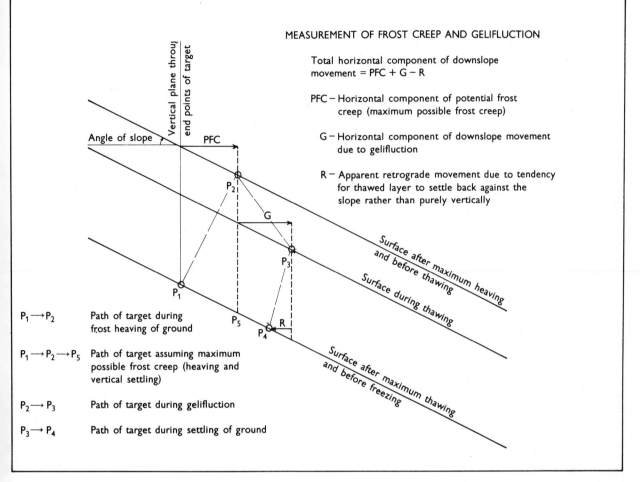

6.8 Relation of gelifluction to frost creep and retrograde movement with reference to a vertical plane through the end points of a line at right angles to movement (*cf. Washburn, 1967, 20, Figure 5*)

gelifluction is necessarily associated with permafrost by definition (cf. Washburn, 1967, 11–13). Thus gelifluction is one kind of solifluction. This terminological distinction is necessary to avoid ambiguity because solifluction, by definition and observation, is not restricted to cold climates, yet its prominence there has led many writers to imply such an association. As a result the sense in which the term solifluction is being used, whether broad or restricted, is not always clear, whereas gelifluction is unequivocally periglacial. It should be noted that solifluction has also been used to encompass both 'macrosolifluction', or solifluction as the term was defined by Andersson, and 'microsolifluction', or small-scale radial movements associated with some forms of patterned ground, including forms on horizontal surfaces (Sørensen, 1935, 7–9; Troll, 1944, 565, 674; 1958, 18–19, 95–6). This usage is unfortunate, especially since it has led some authors to describe patterned-ground features on horizontal surfaces as solifluction forms (Graf, 1973), which can only lead to confusion. Terminological questions are reviewed by Dylik (1967) and Washburn (1967, 10–14).

The relation of gelifluction to frost creep is illustrated by Figure 6.8. Detailed discussions of gelifluction include those of Benedict (1970*b*) and Zhigarev (1967).

6.9 Target line 7, Mesters Vig, Northeast Greenland. View east along target line; target 40 is circled (*cf. Washburn, 1967, 38, Figure 13*)

2 Characteristics

Gelifluction like frost creep must decrease with depth but data are lacking as to whether there is any significant difference between them in this respect. Where both are involved and undifferentiated (the usual case), the vertical displacement profile tends to be concave downslope (P. J. Williams, 1966, 196–201; cf. M. A. Carson, 1976, 107–8).

The predominant characteristic of gelifluction, like solifluction, is its dependency on moisture. The prominence of gelifluction in cold climates is due to (1) the role of the frost table or, especially, permafrost table in preventing downward movement of moisture and thereby promoting saturation of the soil, and (2) the role of thawing snow and ice (including frozen ground) in providing moisture (cf. Schunke, 1974a, 277, 280). Under these conditions, including the possible importance of melting ice lenses in providing glide planes (Beskow, 1930, 624–

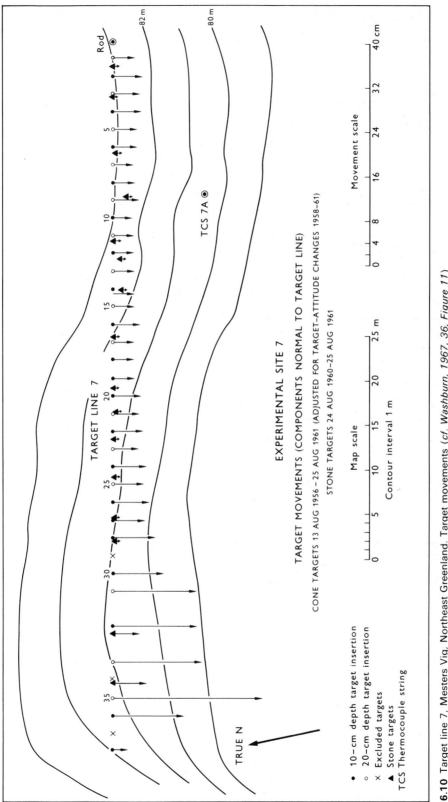

6.10 Target line 7, Mesters Vig, Northeast Greenland. Target movements (*cf. Washburn, 1967, 36, Figure 11*)

5; M. A. Carson and Kirkby, 1972, 277), gelifluction is possible on gradients as low as 1° (St-Onge,
1965, 40) or 1°–2° (Schunke, 1975a, 140–1).

The role of moisture is clearly demonstrated by
observations in many places, including Signy Island
in the South Orkney Islands (Chambers, 1970, 93),
and Mesters Vig, Northeast Greenland. At Mesters
Vig, averaged over a five-year period, rates of
near-surface movement (including both frost creep
and gelifluction) on a 10°–14° gradient ranged from
a minimum of 0.6 cm yr^{-1} in relatively dry spots
to a maximum of 6.0 cm yr^{-1} in wet spots
(Washburn, 1967, Appendix C, Table CII).

The importance of moisture is also demonstrated
by the fact that within limits its influence is more
significant than the binding effect of vegetation or
the effect of gradient. Thus at Mesters Vig the
highest rates of movement at one of the experimental sites was in the best vegetated sector where
the gradient was somewhat less than in the drier
part of the slope (Figures 6.9–6.10). Furthermore,
observations here and at other sites indicate that
significant gelifluction probably occurs only at
moisture values approximating or exceeding the
Atterberg liquid limit – i.e. values at which soils have
little if any shear strength (Washburn, 1967, 104–8).
However, Fitze (1971) on the basis of observations
in Spitsbergen found no consistent correlation in
this respect. Although movement was restricted to
moisture contents above the plastic limit, movement
took place below the liquid limit in some places and
not until well above it in others, depending on the
grain size of the soil. Fitze also regarded gradient
as exerting less influence than grain size.

Gradient is important, since in flow the component of gravity acting parallel to a slope increases
as the sine of the gradient. Observations suggest
a straight-line relation between the sine (or tangent
where low gradients are involved) and the rate of
gelifluction and frost creep in continuously wet
sectors (Table 6.1, Figure 6.11).[3]

Gelifluction is strongly influenced by grain size
(cf. Washburn, 1967, 101–3). The high porosity and
permeability of pure gravel and coarse sand promote
good drainage and do not favour saturated flow
except where porewater pressures reduce intergranular pressures. On the other hand, fines tend
to remain wet longer than coarse grain sizes, and
silt is particularly subject to flow because it lacks
the cohesion of clays and slakes readily. Moreover,
the Atterberg liquid limit is lower in silt than clay
so that less moisture is required for flow. The fact
that silty diamictons are especially prone to flow
has been recognized by many investigators (Johansson, 1914, 91–2; Schmid, 1955, 52–3; Sigafoos and
Hopkins, 1952, 181; Sørensen, 1935, 21–8; and
others). In cold climates mechanical weathering
tends to be more important than chemical weathering (Blanck, Rieser, and Mortensen, 1928, 689–98;
Meinardus, 1930, 45–72; Washburn, 1969a, 43–4),
so that silt tends to predominate over clay and clay-
size particles. This may be one of the reasons for the
prevalence of gelifluction (and frost creep) in such
climates.

Slopewash, discussed in a later chapter, can
eluviate fines from the active layer as shown by
downslope silt accumulations and by the reduction
of fines as compared to the underlying permafrost
(Semmel, 1969, 48–9; cf. Büdel, 1961, 352). Subsurface drainage is frequently audible in coarse
debris, and in many situations it seems obvious that
the debris is a lag concentrate. Where an active
layer is thus depleted of fines, gelifluction will cease.
The concurrent accumulation of fines farther down-

Table 6.1. Mean annual target movements, experimental sites, Mesters Vig, Northeast Greenland (*cf. Washburn, 1967, 94*) (Targets were wood cones on pegs inserted to depths of 10 and 20 cm)

ES		Mean gradient (degrees)		Movement (cm/yr)	
		'Dry'	'Wet'	'Dry'	'Wet'
6	1956–1961	—	2.5	—	1.0
7	1956–1961	12.5	10.5	0.9	3.4
8	1956–1961	12.5	11.5	2.9	3.7
15	1957–1960	—	3.5	—	1.1 (transverse)
		—	3.0	—	3.1 (axial)
17	1957–1959	—	12.0	—	7.6 (transverse)
		—	12.0	—	12.4 (axial)

[3] Frost creep increases with the tangent rather than sine of the gradient (cf. section on Frost creep). However, on gentle slopes the difference is negligible, and where gelifluction is also present the sine function is here arbitrarily applied to the total movement.

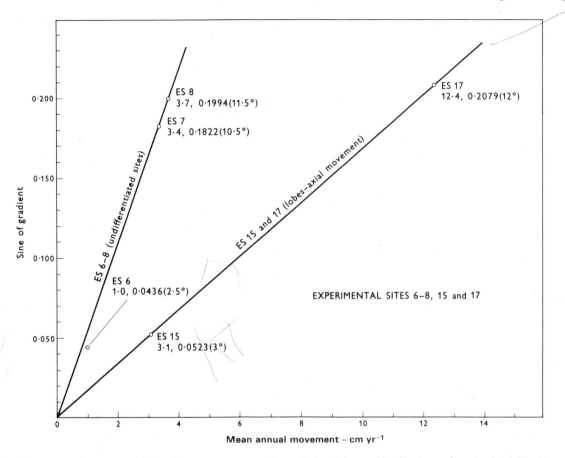

6.11 Mean annual target movements in wet areas at experimental sites, Mesters Vig, Northeast Greenland (*cf. Washburn, 1967, 95, Figure 43*)

slope, for instance where there is a decrease in gradient, can promote gelifluction and patterned ground (Herz, 1964, 46–8; Herz and Andreas, 1966, 191–2; Jahn, 1975, 165), and it seems probable that in many places slopewash has led to an upslope decrease and downslope intensification of gelifluction and patterned-ground processes. A slope lacking permafrost would result in deeper infiltration of water and presumably less effective downslope eluviation of fines.

Considering the above influences, gelifluction should be a function of depth, sine of the gradient (force parallel to the slope), amount of liquid moisture, and grain size of the soil (cf. S. A. Harris, 1973). The function is similar to that for frost creep except for substituting the sine for the tangent of the gradient, liquid moisture for ice, and omitting number and depth of freeze-thaw cycles as factors.

The exact nature of the gelifluction mechanism, whether by flow as in a viscous liquid or by sliding along discrete shear planes or both, remains to be determined. Hutchinson (1974, 439) suggested that gelifluction of clayey soils occurs by sliding, the conditions of limiting equilibrium being given by

$$z = \frac{2c_u}{\gamma} \operatorname{cosec} 2\beta$$

where z = depth, c_u = undrained shear strength at the slip surface, γ = average bulk unit weight of sliding material, and β = inclination of (infinite) slope on which the sliding occurs. To the extent sliding as opposed to flow is important in gelifluction, it seems probable to the present writer that it is primarily on a microscale, along thawing ice lenses (Beskow, 1930, 624–5) and by soil settlement into voids left by thawing lenses (Charles Harris, 1977, 41–2), and that the overall effect is flow, although there are presumably transitions between gelifluc-

6.12 Gelifluction sheet encroaching on emerged strand lines, Mount Pelly, Victoria Island, Northwest Territories, Canada (*cf. Washburn, 1947, Figure 2, Plate 33*)

tion and other processes such as slumping where sliding is usually obvious. An unpublished experiment by J. R. Mackay, Chester Burrous, and Robert Rein, Jr., at the University of Washington's Quaternary Research Center revealed that upfreezing from a permafrost table may favour development of block-like movements. A concentration of segregated ice had formed near the base of the soil (silt) by upward freezing. Subsequent tilting to a 5° gradient and thawing led to a vertical displacement profile characterized by little or no differential movement above a sharp transition to a basal zone of distributed movement. The coincidence of this basal zone with the concentration of segregated ice was apparent from a comparison of frozen cores with cross-sections exposing sand pillars that had been

inserted prior to movement and had been deformed by it.

As discussed below under Frost-creep and gelifluction deposits, gelifluction tends to orient the long axes of stones up and down the slope and to produce several characteristic landforms, including (1) gelifluction sheets (Figure 6.12), (2) gelifluction benches (Figure 6.13), (3) gelifluction lobes (Figures 6.14–6.15), and (4) gelifluction streams (Figure 6.16) (Washburn, 1947, 88–96). They are designated as gelifluction features because of their apparent flow-like nature and because the slopes on which they occur are known as gelifluction slopes, although frost creep, regarded here as a separate process, is an important contributor to movement. For instance, observations at Mesters Vig, Northeast

6.13 Gelifluction bench, Mount Pelly, Victoria Island, Northwest Territories, Canada (*cf. Washburn, 1947, Figure 1, Plate 25*)

Greenland, show that 'Either frost creep or gelifluction can predominate in different places on the same slope, depending on variations in local conditions' (Washburn, 1967, 118).

Data regarding the relative importance of gelifluction and frost creep are very limited. Examples of Mesters Vig observations are illustrated in Figures 6.7 and 6.17A–6.17B. In Figure 6.7, the true frost creep for target 36 in 1961 was 0.7 cm. The gelifluction indicated by the rising graph in June and early July amounted to 2.0 cm so that gelifluction in this year was some three times greater than frost creep. In the preceding year gelifluction and frost creep were nearly equal (1.4 cm and 1.2 cm, respectively). At target 35 (Figure 6.17), 2 m away, they were also nearly equal in 1959–60 (2.3 cm and 2.2 cm, respectively) (Figure 6.17A), but in 1960–61 there was about three times as much gelifluction as frost creep (*c.* 5 cm and 1.8 cm, respectively) (Figure 6.17B). Thus the ratio of gelifluction to frost creep differed in successive years for both targets but each target showed a similar ratio in any one year. The greater amount of gelifluction (and frost creep) at target 35 was due to wetter conditions there. The grain size of deposits can also provide a clue as to the relative importance of frost creep and gelifluction. Thus in places where movement was associated with sandy, relatively non-heaving soil, Schunke (1975a, 146) concluded that gelifluction was the dominant process. McRoberts and Morgenstern (1974a, 459) discounted frost creep because numerous reported rates of movement exceed the maximum expectable from frost creep on a given gradient (although multiple freeze-thaw cycles in a given year were not considered). They believed that although heaving and settling of the soil were very important for movement, this is because extensive thaw water from ice lenses would favour a high thaw-consolidation ratio and therefore gelifluction (cf. discussion of thaw-consolidation ratio in section on Mass displacement in chapter on General frost-action processes).

6.14 Gelifluction lobe, Mount Pelly, Victoria Island, Northwest Territories, Canada. Stakes were used to measure movement (*cf. Washburn, 1947, Figure 1, Plate 26*)

Gelifluction rates quoted in the literature (Figure 6.18; Table 6.2) generally include frost creep and are of the same order of magnitude as mean annual movements for frost creep and gelifluction at Mesters Vig. Considering the variety of factors involved, the consistent order of magnitude is remarkable. Normally movement decreases rapidly with depth. In Iceland, judging from undisturbed tephra layers, there is generally no movement below 20–40 cm (Schunke, 1975a, 144). In Swedish Lapland, based on test-pillar observations at two localities, move-

ment at a depth of 50–70 cm ranged from generally negligible to 2–3 cm (Rudberg, 1958, 116; 1962, 318–19; 1964, 200). In Northeast Greenland, movement on one slope, judging from stake observations, appeared to be negligible below a depth of about 140 cm (Washburn, 1967, 108–9). Where lobes override one another, and presumably in other situations as at the base of slopes, considerably thicker accumulations can form than are actually subject to movement at a given time. Rates on gelifluction lobes rather than on uniform slopes are much more

6.15 Gelifluction lobe, Hesteskoen, Mesters Vig, Northeast Greenland (*cf. Washburn, 1967, 90, Figure 40*)

6.16 Gelifluction streams, Mount Pelly, Victoria Island, Northwest Territories, Canada (*cf. Washburn, 1947, Figure 1, Plate 27*)

variable and are strongly dependent on where they are measured, since the maximum movement occurs along the lobe axis and decreases uniformly towards the margins as illustrated by Figure 6.19. Movement on the lobe tread is commonly greater than that of the front (riser) (Benedict, 1976, 72–3), or of the area immediately downslope where a five- to eight-fold difference has been observed (L. W. Price, 1973, 243). Price also noted that of the lobes he studied the fastest moving were the best developed.

Vegetation has been regarded as impeding gelifluction sufficiently to warrant distinguishing between 'free' (vegetation-free) and 'bound' (vegetation-bound) movement, and recognizing the domain of the former as a frost-debris zone and of the latter as a milder-climate tundra zone. Free gelifluction was thought to be associated with sorted stripes, and bound movement with benches (on gradients of 2° to 15°–20°) and with lobes (on gradients of 15°–20° to 30°) (Büdel, 1948, 30–3, 41–2; 1950, 12–13). According to J. W. Wilson (1952, 249), 'the development of more than a meagre

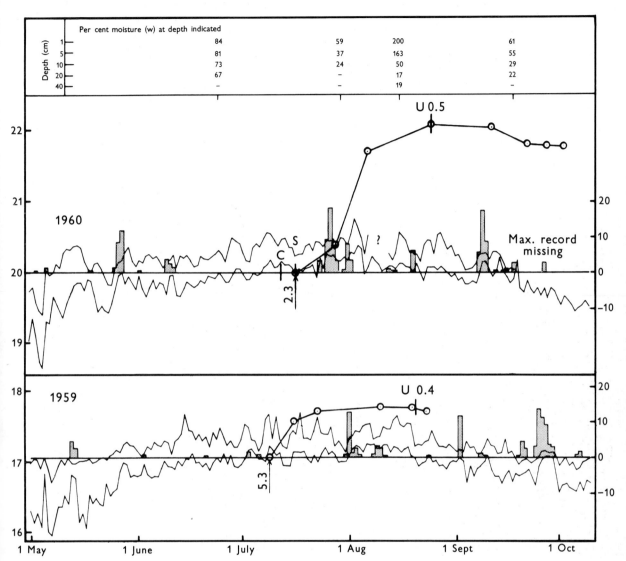

6.17A Movement of target 35, Experimental Site 7, Mesters Vig, Northeast Greenland (*cf. Washburn, 1967, 192, Plate C2–30*)

vegetation, however, generally prevents soil movement (and so patterning) by the stabilizing effect of the roots'. On the other hand, although Bertil Högbom (1910, 46; 1914, 331–2, 360–3) recognized an impeding effect of vegetation, he regarded vegetation as being more likely to be affected by gelifluction than *vice versa*.

There is an intimate interaction between gelifluction and vegetation, with some types of growth being more compatible with gelifluction than others (Frödin, 1918, 1–14; Raup, 1969, 1–21, 26–84, 93–143, 146–159, 199–202; Seidenfaden, 1931;

Sørensen, 1935, 56). Sørensen (1935, 63–4) argued that, even so, plants were generally so strongly influenced by gelifluction that they were a function of it. Others have stated that vegetation, by retaining water and retarding run-off, actually favours slope movements (Sigafoos and Hopkins, 1952, 182). Furthermore, as already noted, measurements at Mesters Vig on vegetation-covered and vegetation-free sectors of the same slope show that the most rapid movement was in the best vegetated sector and that control by moisture was more important there than any binding effect of vegetation (Washburn,

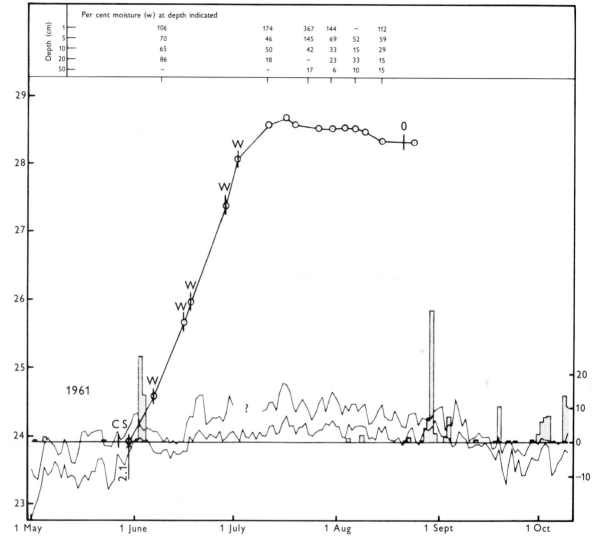

Per cent moisture (w) at depth indicated						
Depth (cm)						
1	106	174	367	144	–	112
5	70	46	145	69	52	59
10	65	50	42	33	15	29
20	86	18	–	23	33	15
50	–	–	17	6	10	15

6.17B Movement of target 35. Experimental Site 7, Mesters Vig, Northeast Greenland (*cf. Washburn, 1967, 192, Plate C2–30*)

1967, 43–4, 104–5). In some places instrumental observations strongly suggest that movement immediately beneath a well-developed vegetation cover in a permafrost environment can be greater than at the surface (L. W. Price, 1973, 237–9, 242–3). The paramount role of bound gelifluction in determining the development of benches and lobes is also questionable. The present writer has seen prominent benches and lobes in vegetation-free areas in both the Arctic and Antarctic. In addition, the location of lobes is controlled not so much by gradient as by concentrations of moisture (Charles Harris, 1972;

1973; Jahn, 1967, 216–18; 1975, 155; Raup, 1969, 137–9; Rudberg, 1962, 316; Washburn, 1967, 96–8). That vegetation can restrict movement and that bound gelifluction tends to be more common than free gelifluction on the lower slopes of some mountains is well known and is not at issue. Nevertheless, it should not be assumed that there is necessarily a zonation of these forms from lower to higher altitude. To the extent vegetation is controlling, the factors influencing its nature and distribution may have an irregular relationship to altitude, regionally and even on the same mountain (cf. J. Hagedorn, 1977,

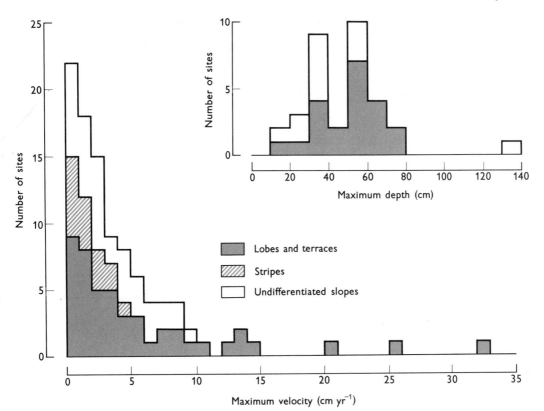

6.18 Maximum rates and depths of frost creep and gelifluction measured in polar, subpolar, and alpine environments. Data from Åckerman (1973), D. J. Barr and Swanston (1970), Benedict (1970*b*), Büdel (1961), Caine (1968*a*), Dutkiewicz (1967), French (1974*a*), Furrer (1972), Charles Harris (1972), Jahn (1961*a*), Kerfoot and Mackay (1972), Pissart (1964*b*; 1973*b*), L. W. Price (1973), Rapp (1960*a*), Rudberg (1962; 1964), Sandberg (1938), Jeremy Smith (1960), Stocker (1973), Washburn (1967), P. J. Williams (1966), and Zhigarev (1960) (*after Benedict, 1976, 57, Figure 2*)

230; Höllermann, 1977, 246–8; Höllermann and Poser, 1977, 334–7).

Although gelifluction is a highly important transportation process, there is doubt, reviewed by Embleton and King (1975, 16–17), that in combination with frost wedging it suffices to explain the erosion of some small valley-like forms along lines of structural weakness (*ravins de gélivation*) as suggested by Boyé (1950, 96–7) among others. That it can contribute to the erosion of such valleys is clear (cf. Washburn, 1947, 93).

Future progress in the study of frost creep and gelifluction will probably make a quantum jump if emphasis is put on detailed laboratory investigations (cf. Higashi and Corte, 1971) and on automatic recording of data in the field as well as in the laboratory. The same is true of patterned-ground research. Traditional baseline studies and periodic comparisons have great value and will still be essential but are unlikely to bring the insights that modern technology now makes possible.

VII Frost-creep and gelifluction deposits

1 General

Gelifluction deposits are widespread periglacial features whose broad dependence on climate has led to many efforts to determine the altitudinal zonation of active and fossil forms. Zone boundaries can be fuzzy and precise definitions are difficult (cf. Höllermann, 1977, 239; Kelletat, 1977*b*, 216–17). Numerous environmental factors complicate the problem but the studies show that active forms on many

Table 6.2. Some near-surface rates of frost creep and/or gelifluction

Location	Gradient (degrees)	Rate (cm yr^{-1})	Reference
Canada			
Ruby Range	14–22	0.9–2.7 (mean; mostly lobes)	L. W. Price (1973, 244, Table III)
Sachs Harbour,	2.5–4 (mean 3)	0.4–1.9 (depth 3 cm)	French (1976a, 290, Table 62.1;
Banks Island		0.4–1.9 (depth 8 cm)	cf. French, 1974a, 76, Table 2)
		0.1–1.7 (depth 15 cm)	
		0.2–1.6 (depth 30 cm)	
Greenland			
Mesters Vig	10–14	0.9–3.7 (mean)	Washburn (1967, 118)
Spitsbergen			
Barentsøya	11	1.5–3 (mean)	Büdel (1961, 365; cf. 1963, 277)
N. side, Hornsund	3– 4	1–3	Jahn (1960, 56; 1961a, 12–13)
N. side, Hornsund	7–15	5–12	Jahn (1960, 56)
Lapland			
Kärkevagge	15 (mean)	2 (mean to 25-cm depth)	Rapp (1960a, 182)
Norra Storfjäll area	5	0.9–3.8 (mean)	Rudberg (1964, 199, Table 2, item 1)
Tärna area	5	0.9–1.8 (mean)	Rudberg (1962, 317, Table II)
Norway			
Okstindan	4–17[1]	0.0–6.0[1]	Charles Harris (1972, 169, Table V; 1973, 24, 26; 1977, 41–2)
Poland			
Karkonosze Range	8–39	0.3–2.1 (mean, depth 0–5 cm) 0.0–1.4 (mean, depth 20 cm) 0–0 (mean, depth 40 cm)	Jahn and Cielińska (1974, esp. 89, Figures 3–4)
South Georgia			
Grytviken	21	3 –5 (mean to 25-cm depth, excluding surface stones)	Jeremy Smith (1960, 75, Table I)

[1] Lower limit is from Charles Harris (1973).

mountains generally lie between the snowline and treeline and trend in the same direction but not necessarily closely parallel to them. Recent profiles illustrating this include those by Karrasch (1977, 161, Figures 3–4) and Höllermann (1977, 241, Figure 1) (cf. also under General (introduction) in section on Patterned ground in preceding chapter). Especially in continental-to-subcontinental climates, the lower limit of gelifluction lies below the treeline in places, but whether this is determined primarily by climate or by variations in treeline caused by human activity requires further research (Höllermann and Poser, 1977, 340).

2   Sheets, benches, lobes, and streams

Frost creep and gelifluction are commonly associated, and the resulting deposit is therefore of joint origin. However, it is common practice to assume that gelifluction is the dominant process or includes frost creep, and that the deposit is therefore primarily a gelifluction deposit. In the following, it is assumed that both processes have contributed to such deposits regardless of their designation. As already noted, volume changes due solely to wetting and drying or to temperature variations may also contribute to movement but in periglacial environments are normally regarded as much less effective than frost creep or gelifluction. However, Black (1973b, 17–18) has suggested that some Antarctic permafrost deposits, previously attributed to gelifluction in a former warmer climate (cf. Calkin and Nichols, 1972, 640) may really be creep phenomena consistent with present conditions.

Frost creep and gelifluction deposits can occur on a gradient of as little as 1°–2° (Louis, 1930, 98; Schunke, 1975a, 140–1; St-Onge, 1965, 40). According to topographic form they can be (1) Gelifluction sheets, characterized by a smooth surface and in places by a bench-like or lobate lower margin (Figure 6.12). (2) Gelifluction benches, characterized by their pronounced terrace form (Figure 6.13). The longest dimension of the benches tends to be parallel

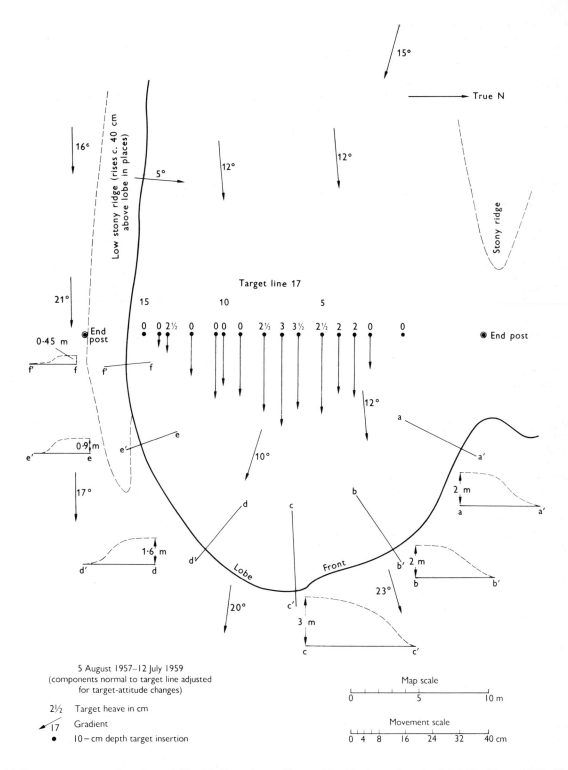

6.19 Target movements, Experimental Site 17, Hesteskoen, Mesters Vig, Northeast Greenland (*cf. Washburn, 1967, 88, Figure 38*)

6.20 Crude stratification in diamicton of gelifluction slope, Schuchert Dal, Northeast Greenland. Scale given by 60-cm rule

to the contour but in places is at angles up to $45°$ (Frödin, 1918, 29, Tafel IV – also cited by Büdel, 1948, 42; and others). A number of factors may be responsible for such cross-contour development, including bedrock structure, exposure to wind (cf. Stocker, 1973), and to sun (cf. Schunke, 1975a, 143), all of which can affect distribution of snow moisture and vegetation and their influence on movement. Risers of some benches, described as fossil gelifluction benches, are up to 15 m high (Pissart, 1963b, 119, 123, 132). (3) Gelifluction lobes, characterized by their tongue-like appearance (Figures 6.14–6.15). Their longest dimension tends to be at right angles to the contour. Risers of some fossil forms are up to 4.6 m (15 ft) high (R. W. Galloway, 1961b, 79). (4) Gelifluction streams, characterized (as opposed to lobes) by *pronounced* linear form at right angles to the contour (Figure 6.16).

Gelifluction benches and lobes commonly have steep fronts, nearly vertical in places in the case of some vegetation-banked forms and up to the angle of repose in stone-banked occurrences. The fronts are subject to erosion, even while benches and lobes are still active, and breaching of fronts can lead to washing out of fines and their accumulation farther down the slope, thus promoting development of gelifluction forms downslope at the expense of forms upslope as previously described. It has been suggested that increased snow accumulation in the lee of lobes whose fronts are growing higher contributes to a developmental cycle by furthering frontal erosion so that such '... lobes carry within them the seeds of their own destruction' (L. W. Price, 1974, 438). Observations by Jahn (1975, 155–6) and the present writer support the view that the breaching of fronts is not necessarily evidence of the continued activity of gelifluction forms, a view contrary to Schunke's (1975a, 147–8).

A relation between gradient, grain size, and form of gelifluction was observed by Schunke (1975a, 145–7), who reported that in Iceland there is a tendency for lobes to occur on the steeper slopes

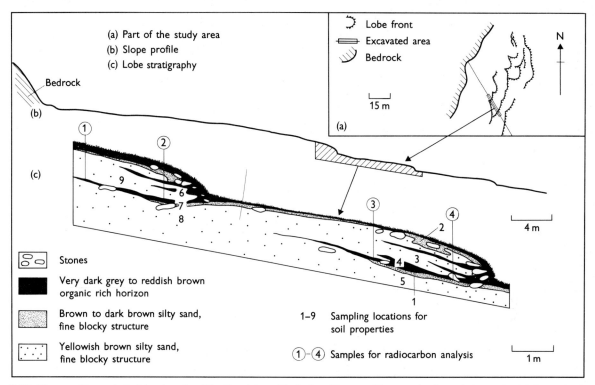

(a) Part of the study area
(b) Slope profile
(c) Lobe stratigraphy

Bedrock

Lobe front
Excavated area
Bedrock

15 m

N

Stones

Very dark grey to reddish brown
organic rich horizon

Brown to dark brown silty sand,
fine blocky structure

Yellowish brown silty sand,
fine blocky structure

1–9   Sampling locations for
       soil properties

①–④   Samples for radiocarbon analysis

4 m

1 m

6.21 Map, profile, and stratigraphy of gelifluction lobes, Okstindan, Norway. The samples from the organic-rich layers had ^{14}C dates ranging from 1420 ± 150 to 2550 ± 80 years BP, the youngest ages being closest to the lobe front (*after Worsley and Harris, 1974, 133, Figure 3*)

$(10°–20°)$ and bench-like forms on the gentler slopes $(5°–15°)$; also that bench-like forms appear to be the more common (and smaller) in sandy than in silty soils. The critical importance of moisture as a variable was recognized by Schunke as by many before him.

Gelifluction deposits tend to be diamictons but some deposits show crude stratification parallel to the slope (Figure 6.20). In some gelifluction benches and lobes the crude stratification is emphasized by organic-rich layers resulting from progressive overriding and burial of the tundra vegetation (Figure 6.21) (cf. Benedict, 1966, 27, Figure 4; P. J. Williams, 1957, 42–4; Worsley and Harris, 1974, 131–3). The material is commonly angular and its arrangement or fabric is characterized by a tendency for the long axis of stones to be oriented in the direction of movement except where movement is restricted (Figure 6.22) (cf. Benedict, 1976, 63–4; Brochu, 1978; French, 1971b, 725–6; Furrer and Bachman, 1968, 5–8; R. W. Galloway, 1961a, 349–50; G. Lundqvist, 1949; K. Richter, 1951, cf. Büdel, 1959, 298; Rudberg, 1958; Schunke, 1975a,

64, 141, 143; Semmel, 1969, 13, 15, 29; Jeremy Smith, 1960, 75–6, Figure 2). The median per cent of downslope-oriented stones in active gelifluction deposits in Europe and North America is 79.3 as compiled by Brochu (1978, 44–6, Tables 1–2). A tendency for stones to lie parallel to the surface has been attributed to frost creep (Embleton and King, 1975, 99; P. J. Williams, 1957, 49) but would seem to apply equally to gelifluction. In association with other evidence, the long-axis orientation has been used to support the gelifluction origin of some deposits (Watson, 1969; 1970, 142–3). In the absence of supporting evidence, including evidence as to topographic form, gelifluction deposits can be easily confused with similar diamictons of other origin (Flint, 1961, 148–9, Table 1; 152; 1971, 152–3; Rapp *in* Watson, 1969, 113). Moreover, as stressed by Black (1969b, 230), in the absence of independent evidence gelifluction deposits may be especially difficult to distinguish from solifluction deposits made in nonfrost climates. However, the former would tend to have constituents that were more angular and less chemically weathered. The

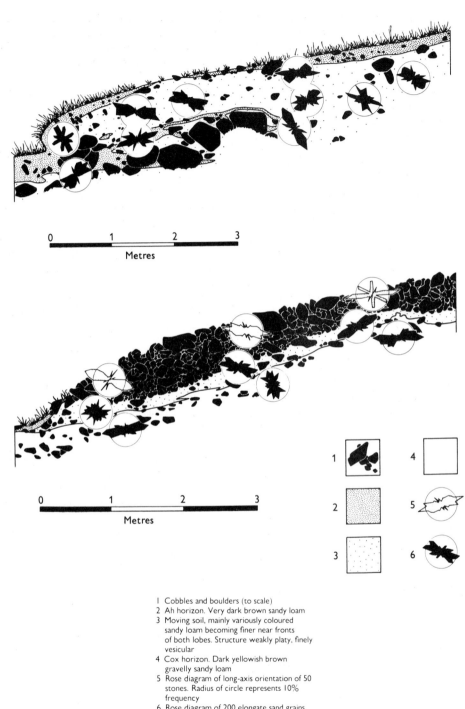

0   1   2   3

Metres

0   1   2   3

Metres

1   Cobbles and boulders (to scale)
2   Ah horizon. Very dark brown sandy loam
3   Moving soil, mainly variously coloured
    sandy loam becoming finer near fronts
    of both lobes. Structure weakly platy, finely
    vesicular
4   Cox horizon. Dark yellowish brown
    gravelly sandy loam
5   Rose diagram of long-axis orientation of 50
    stones. Radius of circle represents 10%
    frequency
6   Rose diagram of 200 elongate sand grains
    measured in thin section. Radius of
    circle represents 10% frequency

6.22 Fabric diagrams for turf-banked and stone-banked lobes, Niwot Ridge, Front Range, Colorado (*after Benedict, 1976, 68, Figure 11; cf. Benedict, 1970b, 183, Figure 21; 210, Figure 51*)

problem of identifying deposits and forms indicative of permafrost is clearly even more difficult. In this connection Benedict (1976, 72) stressed that even though some large-scale frost creep and gelifluction forms are presently active in non-permafrost environments, this does not necessarily prove that they originated independently of permafrost, since their ages suggest they formed when permafrost was more widespread. He therefore argued that 'The traditional viewpoint that lobes and terraces are *not* indicative of former permafrost may be correct, but should be re-examined.'

The suggestion has been made that it is possible to distinguish between periodic (annual) and episodic (occasional) gelifluction deposits (Büdel, 1959, 301–10). According to Büdel, the periodic type is characterized by deposits commonly less than 1 m thick (maximum 2 m) and by laminar structure; the episodic type is characterized by deposits commonly 2 m or so thick (maximum 4 m) and by fold structure including downslope-tipped ice-wedge casts. Büdel associated the periodic type with gradients as low as 1.7° in the present polar environment and with gradients of 4°–6° to 17°–27° in middle-latitude environments during the Pleistocene, the difference being based on an assumed quicker snowmelt and therefore greater desiccation in middle latitudes. However, snowmelt in arctic environments, as at Mesters Vig, Northeast Greenland, can be so rapid that it leads to slushflows and earthflows. Büdel associated Pleistocen episodic gelifluction deposits with gradients of less than 4°–6° in middle latitudes but he found no parallel in present polar environments. The postulated distinction between periodic and episodic deposits, which is not so much one of process as frequency of process, would be a valuable climatic criterion if valid and practical. However, it is largely theoretical and the postulated association is based more on assumed gradient relationships than on the study of deposits forming today.

3   Block fields, block slopes, and block streams

a General   Block fields are considerable areas, broad and usually level or of only gentle gradient (here taken to be 10° or less),[4] covered with moderate-sized or large angular blocks of rock (after Sharpe, 1938, 40). In practice, more than 50 per cent of such a surface could be covered with blocks to qualify as a block field (Rudberg, 1977, 94). Block slopes are similar areas on slopes (Washburn, 1969a, 36) (Figure 4.7). Block streams, made famous by the Falkland Island occurrences described by J. G. Andersson (1906, 97–104), Clapperton (1975), and others, are extensive accumulations confined to valleys or forming narrow linear deposits extending down the steepest available slope. Commonly none of these deposits have a bedrock cliff or free face at their head as occurs with a talus slope. As used by Richmond (1962, 19), the term rubble sheet covers both block fields and slopes, and the term rubble stream is synonymous with block stream. The term block sheet would be synonymous with rubble sheet. The German terms Blockmeer and Felsenmeer include accumulations of rounded boulders as well as of blocks, and the term stone fields, stone slopes, etc., might be used in this sense in English, thereby reserving the terms block for angular, and boulder for rounded, material.

Such accumulations of blocks and boulders have diverse origins but many are so closely related to frost-creep and gelifluction deposits that they are included in this section.

b Constitution   Various lithologies are possible but the distribution of block sheets in a given region can depend on rock type. Thus in Iceland block sheets are essentially confined to basalt and lacking where hyaloclastite and rhyolite occur (Schunke, 1975a, 115). Caine (1968b, 33–42) found a three-layered structure in the non-vegetated block fields and block streams of Tasmania: (1) Blocks with open-work texture at the surface and to depths of over 3 m in places; (2) an intermediate layer, 10–30 cm thick, characterized by interstitial humic mud between the blocks; and (3) a basal layer lying on bedrock and consisting either of blocks with an interstitial filling of silty sand or, in a few places, of silty sand without blocks. Vegetation-covered block fields had the interstitial filling extending to the surface. For

[4] Piirola (1969, 29) applied the term block fields to occurrences on gradients as steep as 30° in Finnish Lapland. Caine (1968a, 6–7) indicated that a gradient of 15°, or 10° (Caine *in* S. E. White, 1976b, 89), was a reasonable upper limit for block fields he studied in Tasmania. The present writer originally suggested 5° as an appropriate boundary between block fields and block slopes (Washburn, 1973, 191) but he now accepts Caine's 10° as more practical. In his excellent study Caine recognized block streams and block glacis (aprons) as types of block fields with block glacis having their greatest extent parallel rather than transverse to the contour. In view of the proposed distinction between block fields and block slopes, it seems best to follow Caine in using the terms block glacis and block streams in the pattern sense but not to define them as types of block fields, since otherwise terminological problems arise for similar features on block slopes.

block fields in more or less level summit areas in northern Scandinavia, Strömquist (1973, 73–5, 145) found a somewhat different three-layered structure showing from top to bottom: (1) Blocks imbedded in finer material; (2) fine-grained material (mainly sand and silt) containing isolated blocks; (3) weathered debris transitional to bedrock and comprising gravelly debris or *in-situ* blocks. The sequence was similar on slopes except that the surface block layer had open-work texture. On the other hand, Strömquist (1973, 128–31, 147–8) found that lowland block fields were characterized by a lack of fine-grained material beneath the surface blocks. The mean thickness of low-gradient deposits in crest areas was 0.7 m; there was greater variability in thickness on slopes, depending on the nature of the slope processes (Strömquist, 1973, 65–72, 145). In both situations grain size of material less than 20 mm in diameter tended to increase towards the ground surface; similar sorting was obtained in experiments after as few as 12–20 freeze-thaw cycles (Strömquist, 1973, 85–90, 145). In some areas the closer a block sheet is to the snowline, the less fine material there is among the blocks, a situation that has been ascribed to the increased effectiveness of slopewash near the snowline (Schunke, 1975a, 116).

In general the texture of block slopes and block streams is characterized by a fabric in which the long axis of individual blocks tends to be aligned in the same direction as the local gradient but less steeply inclined (Cailleux, 1947; Caine, 1968b, 96; 1972b; Klatka, 1961b, 9–13; 1962, 69–83, 120). At the Łysa Gora locality in Poland, the classic site of Łoziński's pioneering periglacial investigations, the fabric is increasingly well developed and the stones are smaller downslope; on a talus-like block slope nearer the source of the blocks, the fabric is less pronounced and the imbrication tends to be directed downslope rather than upslope (Klatka, 1961b, 6, 9–13; 1962, 69–83, 120). A change from an oriented fabric on a slope to an irregular fabric on a lower gradient where mass-wasting was less prominent was reported by Ragnar Dahl (1966a, 65–8).

Caine's (1968b, 43–96, 106; 1972b) observations in Tasmania confirm these textural trends for the upper layer of the block areas he studied. He also noted that at the downslope toe of those areas the orientation of the long axes changed and became transverse to the gradient. However, Strömquist (1973, 99–104, 146) observed that the transverse orientation in the block slopes of northern Scandinavia was most prominent on the steepest slopes, the transition from parallel to transverse orientation

occurring on gradients exceeding 20°. He attributed the change to a decrease in solifluction (i.e. gelifluction) as a result of removal of fines by other slope processes.

c Origin   Weathering *in situ* of massively jointed bedrock can produce both rounded boulders and blocks but only the latter are generally indicative of a predominance of mechanical weathering, usually frost wedging, over chemical weathering, abrasion, and other processes (cf. C. A. M. King and Hirst, 1964). As discussed under Frost wedging in the chapter on General frost-action processes, production of large angular blocks is believed by the present writer to be primarily the result of frost wedging. However, as stressed by Caine (1967, 427; 1968b, 101–3), angularity in dolerite can be associated with subsurface decomposition in places, and on the other hand rounding does not disprove periglacial weathering if the rounding occurred later. Both angular and rounded stones can probably also form as a lag concentrate from erosion of a stony deposit such as till. Eluviation of fines by piping (H. T. U. Smith, 1968) and other processes may help to explain the blocky appearance of a deposit, but time is also a factor in that an accumulation of blocks may reflect insufficient time for frost wedging to comminute material to smaller sizes as noted by Guillien and Lautridou (1970, 45). They also reported that the maximum size of frost-wedged fragments in their experiments was not related to intensity of freezing (Guillien and Lautridou, 1970, 36, 38). However, the production of large blocks would seem to be favoured by deep penetration of freezing into jointed bedrock on a scale difficult to reproduce in the laboratory. Thus Schunke (1977b, 48–9) noted that the location of widespread angular rock debris in periglacial environments appears to be especially dependent on low temperatures rather than just frequency of freezing and thawing.

Block fields and slopes of the La Sal Mountains, Utah, were studied by Richmond (1962, 62–5, 80–1), who recognized four genetic types: (1) Frost-wedged, formed *in situ*; (2) frost-creep, formed with slow downslope movement over bedrock; (3) frost-sorted, formed *in situ* from coarse and fine rubble; and (4) frost-lag, formed with slow downslope movement from coarse and fine rubble. Caine (1968b, 97–105, 116) concluded that the material of the block areas in Tasmania might have originated by weathering under both periglacial and non-periglacial conditions. Whether block fields and slopes can form by frost sorting as indicated by Corte (1966b, 234),

Table 6.3. Effect of gradient on block fields and block slopes (*after Strömquist, 1973, 148*)

Gradient interval	Vertical profile	Block orientation	Form element
0°–10°	Slightly influenced by transport processes.	An increasing number of blocks oriented in the slope direction.	The highest frequency of polygons (0°) and sorted stripes (10°).
Interpretation:	Frost-sorting processes are dominant at low gradients. Slope transport occurs by gelifluction combined with frost heaving.		
10°–20°	An equal influence of slope transport and frost sorting.	Optimal block orientation in the slope direction. Increasing block imbrication.	The highest frequency of sorted steps (18°).
Interpretation:	At this interval a more dominating influence of slope transport can be observed. Most of the transport takes place by gelifluction.		
20°–35°	Strong influence of slope transport. A reduced amount of fine-grained material.	A decreasing number of blocks oriented in the slope direction. Increasing imbrication.	Patterned ground formed by transport through gelifluction is reduced due to washing out of the fine-grained material.
Interpretation:	An increased influence of mudflows and washing transports the fine-grained material downslope. The block orientation is then determined by the force of gravity and by the relationships with other blocks.		
35°	In most cases the vertical profile consists of blocks only.	A block orientation perpendicular to the slope orientation.	No patterned ground.
Interpretation:	The block orientation shows a direct influence of the force of gravity. All the fine-grained material is transported downslope and the block transport is in the form of block creep.		

Richmond (1962, 63), and others remains to be proved, but the demonstrable upfreezing of large stones supports the possibility.

The hypothesis that some block fields are till resulting from glacial plucking and deposition just prior to deglaciation has been advanced by E. H. Muller (1973) for occurrences in Labrador and northern Quebec. Regarding frost weathering (i.e. frost wedging) and frost sorting as primary processes in the origin of block fields and slopes, Strömquist (1973, 148) concluded that changes in gradient explained variations in profile character, block orientation, and patterned ground (Table 6.3). The block streams (stone runs) of the Falkland Islands have been described as being similar in origin to sorted stripes but on a larger scale due to local factors (Clapperton, 1975). Testing of the hypothesis will require, among other things, more data on the origin of large sorted stripes. Corte (1976, 190, 193) thought these block streams might be fossil rock glaciers.[5]

In part following Richmond (1962, 62–5, 80–1) and S. E. White (1976b, 93), the present writer suggests that probably most block slopes and block streams have a developmental history comprising: (1) Production of blocky material by weathering; (2) some frost-sorting effects; (3) movement by mass-wasting; and (4) subsequent removal of some interstitial matrix.

Although stones can undergo significant transport in various ways, it is commonly assumed that any movement of the whole accumulation is due to mass-wasting with emphasis on solifluction, usually gelifluction. However, the role of creep (frost creep and other forms) is not established and may be significant as suggested by Richmond (1962, 63). The build-up of ice in block accumulations can occur by infiltration of snow and water and by condensation of water vapour, and movement can be facilitated by gliding of rock fragments on thawing ice as noted by Czudek and Demek (1972, 9). Caine (1968b, 106–11) interpreted the fabric of the deposits he studied as showing that movement occurred while the blocks were in a matrix of finer material similar to the basal layer he described (cf. above under Constitution). Lack of mixing of these layers and evidence provided by lichens and weathering features indicate no movement at present; the former movement probably occurred by gelifluction and frost creep under periglacial conditions. He concluded that the matrix had been subsequently eluviated from the upper openwork layer and that some blocks had been

[5] Corte's report used the Argentinian name, Malvina Islands, for the Falklands.

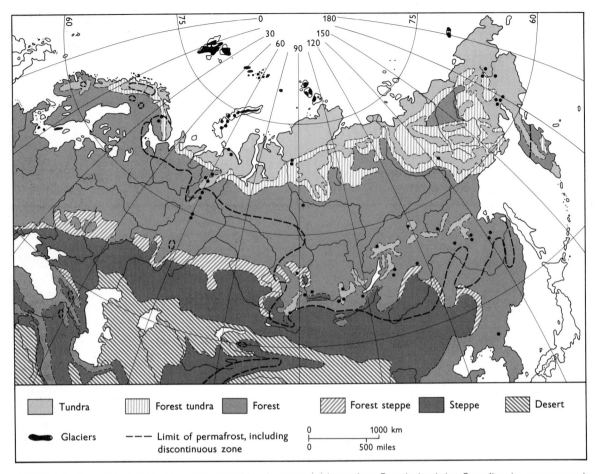

6.23 Distribution of block fields, block slopes, and block streams (●) in northern Eurasia (omitting Scandinavian occurrences) (*after Frenzel, 1959, 1020, Figure 12*)

sorted out of the basal layer by upfreezing as shown by its locally block-free aspect. The lack of thermokarst features and general dissimilarity to rock glaciers were taken as evidence that interstitial ice was not an essential factor in movement.

The time required for development of a block field or block slope is poorly known and obviously depends on many factors. The probability that some block fields can form within as short a time as 6000 years is indicated by a block field in the Åland Islands, Finland. The history of postglacial emergence shows that the site of the block field was submerged until about 6000 years ago, and the angularity of most of the blocks makes it seem unlikely that they were of the same glacial and marine origin as a higher accumulation of rounded boulders (Embleton and King, 1975, 169–71; C. A. M. King and Hirst, 1964, 31–40).

d Environment Active block fields, slopes, and streams are widely distributed in the polar zone, especially but not exclusively in highlands. An inventory of block fields in Scandinavia in relation to altitude was presented by Rudberg (1977). Antarctic occurrences are summarized by Nichols (1966, 26), and Eurasian localities are shown in Figure 6.23.

Activity is commonly indicated by lack of lichens on the stones as opposed to their presence on stones in more stable locations in the same region. Active block accumulations probably also occur in subpolar and middle-latitude highlands but activity or stability is more difficult to prove. To the extent that stability can be achieved by removal of fines or other processes without involving climatic change, stability need not be indicative of such a change (Caine, 1972b, 48); however, such block accumulations would still be a record of periglacial conditions.

More evidence is needed on causes of stability, including evidence from contemporary periglacial environments.

e Inactive and fossil forms   Block accumulations in middle-latitude highlands have been widely interpreted as recording former periglacial conditions but this conclusion requires proof of present inactivity and elimination of alternative possibilities. Schott (1931, 58–71) doubted the periglacial significance of block accumulations, although granting the influence of increased frost wedging during glacial ages. Büdel (1937) regarded many block accumulations below a minimum gradient as stable today, based on various types of evidence but not on instrumental observations, whereas Schmid (1955, 106–8) questioned the fossil nature of many such block accumulations in the German Mittelgebirge. Except under unusual conditions, purely thermal changes seem unimportant as a weathering process responsible for production of large angular blocks, and hydration shattering (except as related to mineralogical changes on a granular scale) requires further evidence as to its effectiveness. It is therefore concluded that widespread accumulations of truly angular blocks are certainly reasonable evidence of former frost wedging if located in an environment where such blocks are not accumulating today, but further interpretations are fraught with difficulty in the absence of additional evidence.

In Alaska evidence provided by periods of dissection permitted Wahrhaftig (1949) to conclude there had been at least five times when rubble sheets, some with features similar to rock glaciers, were active on Jumbo Dome on the north side of the Alaska Range, and he correlated these with times of glaciation in the range. Some of the problems in interpreting block fields and slopes are illustrated by the Åland Islands deposits noted above, a discussion between Ragnar Dahl (1966a; 1966b) and J. D. Ives (1966), and the probability that cold-based ice (i.e. basal ice at a temperature below the pressure-temperature melting point) would tend to minimize glacial erosion and therefore promote the preservation of both tors and block fields beneath ice caps (Sugden and Watts, 1977, 2821–2). French (1976a, 231) reported as a problem that both actively forming and relic block fields are believed to exist in the Appalachian Mountains in the United

States; however, the rubbly accumulations referred to may be more nearly related to talus accumulations than to the types of block fields and slopes discussed above.[6]

In general the present writer believes that the geographic distribution of most demonstrably active block fields, block slopes, and block streams at high altitudes or higher-than-temperate latitudes puts the burden of proof for a non-periglacial origin on those who doubt it (cf. S. E. White, 1976b, 92, 94). In the absence of proof to the contrary, it is concluded that such deposits of angular material are strongly suggestive of frost action even though not necessarily diagnostic of it.

4   Ploughing blocks and braking blocks

a Ploughing blocks   Ploughing blocks (or boulders), known as Wanderblöcke in German, are a special kind of frost-creep and/or gelifluction deposit consisting of isolated, commonly boulder-size stones that leave a linear depression upslope and form a low ridge downslope (Figures 6.24–6.25). The depression and ridge are formed as the stone moves downslope, and their size tends to vary directly with the size of the stone. In general the long axis of the stone lies in the direction of gradient. Particularly studied by Tufnell (1972; 1976), ploughing blocks have been described from many alpine environments (Ball and Goodier, 1970, 212–13; Kuhle, 1974, 473–4; Louis, 1930, 97–8; Marnezy, 1977b, 371–2; Mohaupt, 1932, 10–11, 34; Poser, 1954, 150–2; Schmid, 1958, 258–60; and others). They have also been observed in polar regions by the present writer.

Poser considered ploughing blocks to be gelifluction (flow) phenomena, whereas Schmid explained their movement by frost creep. Mohaupt concluded that sudden fast movement was involved in places but did not exclude slow movement. An observation in Northeast Greenland by the present writer also suggests sudden fast movement in places, since the upslope depression was equally fresh and smooth throughout its length (Figure 6.24) as if it had been formed by a single slip, possibly the same spring. Rates of movement on slopes of 16°–39° in the Karkonosze Mountains of Poland have been measured by Jahn and Cielińska (1974, 92–3, 99),

[6] French cited Hack and Goodlett (1960, 31–2) and Rapp (1967), perhaps reflecting a statement by Judson (1965, 133), quoted by Rapp, that according to Hack and Goodlett (1960) '...these features can form under present climatic conditions'. Although Hack and Goodlett (1960, 31–2) did not discuss this problem, Hack (1965, 32–44) did conclude that talus (scree) is presently forming in Virginia, that frost action could be an important agent, and that movement is occurring today, especially under flood conditions. Rapp, himself, found no presently active block fields or slopes in the region discussed.

6.24 Ploughing block, Mesters Vig, Northeast Greenland

who reported that blocks moved some 2.5–9 times farther than the soil on which they lay, the maximum observed block movement over a 10-year period being 250 cm on the 39° slope. In England, movements of 5 blocks over a 10-year period averaged 2.3 cm yr^{-1} per block, the maximum movement for any block being slightly less than 8 cm yr^{-1}. Gradients were up to 30°, and most of the movement occurred during the winter (Tufnell, 1976).

The altitudinal distribution of ploughing blocks (and turf-banked terraces – Girlanden) has been used in mapping the lower limit of gelifluction in the Swiss Alps; the limit thus determined lies at an altitude of about 2200 m, corresponding to a mean annual temperature of 0° (Furrer and Dorigo, 1972). The method requires a good deal of statistical manipulation. That ploughing blocks may be recognized long after movement has ceased is demonstrated by occurrences in the Harvard Forest at Petersham, Massachusetts, which show post-

movement soil profiles (Lyford, Goodlett, and Coates, 1963, 28–30).

b Braking blocks By contrast with ploughing blocks, blocks (or boulders) that impede gelifluction can be termed braking blocks (Bremsblöcke in German, also known as Staublöcke – dam blocks). If their movement is slow enough relative to the soil in which they are imbedded, soil and vegetation can ride up their upslope side; such braking blocks may be difficult to distinguish from upfreezing blocks, which create a similar effect on steep slopes. Like ploughing blocks, braking blocks are widespread in alpine environments (Graf, 1976, 435–7; Höllermann, 1964, 76–7), and are reported to be more common than ploughing blocks in Iceland (Schunke, 1975a, 143).

6.25 Ploughing block, Cym Tinwen, Wales, Great Britain. Scale given by 5-cm diameter lens cap. Movement probably originated under more rigorous climate than the present

VIII Rock glaciers

1 General

A rock glacier, so referred to in the field by Cross, Howe, and Ransome (1905, 25) and more formally named by Capps (1910, 360), is

... a tongue-like or lobate body usually of angular boulders that resembles a small glacier, generally occurs in high mountainous terrain, and usually has ridges, furrows, and sometimes lobes on its surface, and has a steep front at the angle of repose. (Potter, 1972, 3027; cf. Potter, 1969, 1.)

The front of active rock glaciers can be as steep as 40° and have a height up to 60 m (200 ft) (Vernon and Hughes, 1966, 17), or even 122 m (400 ft) (Wahrhaftig and Cox, 1959, 387). The ridges and furrows are often arcuate and convex downslope but also longitudinal (Figure 6.26). Tongue-shaped rock glaciers have a length-to-ridge

6.26 Rock glacier, northeastern Selwyn Mountains, Yukon Territory, Canada. The bedrock is sandstone and shale. Photo by Clyde Wahrhaftig

ratio > 1.0, and lobate forms a ratio < 1.0 (Dingwall, 1973; Wahrhaftig and Cox, 1959, 389). Some rock glaciers are composite as if one were overriding another (Figure 6.27).

2 Constitution

The surface of rock glaciers tends to be characterized by blocks and smaller stones but all sizes and shapes are possible. The surface appearance is misleading, however, in that the interior of rock glaciers, where known, usually consists of a diamicton in which fines may be plentiful. Probably all active rock glaciers contain ice except possibly some reactivated ones as noted below, and if the necessary information as to its amount and distribution is available, active rock glaciers can be divided into ice-cemented and ice-cored types (Potter, 1969, 13; 1972, 3027). By ice-cored, Potter (1969, 13) meant a core of glacier ice but rock glaciers may contain sizeable ice masses of other origin as discussed later. It may be best to use ice cored without genetic implication as is done in the following where ice cored will imply a volume of perennial ice at least 25 per cent in excess of the amount filling the pore spaces of a rock glacier. Accordingly, ice cemented will be used as implying a lesser volume of excess ice, or interstitial ice (pore ice) alone; this differs from Potter's (1969, 13) usage in which ice cemented is equated with interstitial ice without reference to any excess ice, but excess ice must be common in many rock glaciers lacking an ice core, glacial or otherwise.

The following criteria, mainly thaw features, have been suggested for identifying the ice-cored type (assumed to be derived from a debris-covered

6.27 Composite rock glacier, Sheep Mountain, Kluane Range, Yukon Territory, Canada. Photo by J. P. Johnson (*cf. J. P. Johnson, 1973, 86, Figure 1*)

glacier) (Vernon and Hughes, 1966, 17–22; S. E. White, 1976*b*, 79; cf. Corte, 1976*a*): (1) A spoon- or saucer-shaped depression, elongated downstream, in ice between the base of a cirque headwall and the rock glacier; (2) longitudinal furrows along both sides of the rock glacier upslope from the front; (3) central meandering furrows; (4) conical or coalescing steep-sided collapse pits floored with ice and/or filled with water; (5) absence of prominent transverse ridges and furrows such as occur on rock glaciers without an ice core (although faint transverse ridges at the front may be present); (6) direct observation of an ice body underlying the debris cover (a classical and apparently rare case was described by W. H. Brown, 1925). The possibility that a rock glacier may still flow following partial disintegration of a glacial ice core and thereby destroy evidence of the rock glacier's origin is a complicating factor (cf. S. E. White, 1976*b*, 80).

In some places rock glaciers appear to be transitional to debris-covered glaciers or ice-cored moraines, as illustrated by the exchange between Barsch (1971) and Østrem (1971). Subsequently, Barsch (1977*a*, 234) suggested that large end moraines in a permafrost environment may flow plastically to become a kind of rock glacier and thus account for the outermost ridges of such a complex being the youngest. Benedict (1973, 522) held that where rock glaciers and ice glaciers are transitional and developing concurrently, the environment must favour both glaciation and extensive rockfall, and that the distinction between the resulting forms and deposits is largely artificial. Barsch (1977*a*, 235) discounted the value of differentiating ice-cored from ice-cemented rock glaciers, in part because of the difficulty of distinguishing rock glaciers that may originate in debris-covered glaciers from other types including those containing segregated ice. However, the distinction may be useful, genetically and rheologically. In the southern part of Jasper National Park in Canada, 65 ice-cored (glacial) and 54 ice-cemented rock glaciers have been reported (Luckman and Crockett, 1978). It should be recognized that non-glacial ice cores may have a high enough ice content to simulate, upon thawing, a former core of glacier ice (cf. Barsch and Updike, 1971, 106).

The constituents of rock glaciers can have a variety of origins (Figure 6.28), as a result of which various rock-glacier types can be recognized, including avalanche, moraine, and talus-derived rock glaciers (cf. P. G. Johnson, 1978).

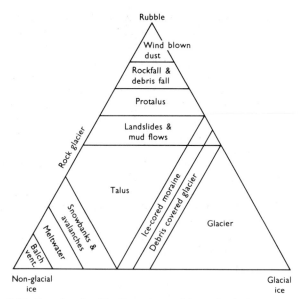

6.28 Genetic classification of possible rock-glacier constituents (*after J. P. Johnson, 1973, 88, Figure 2*)

3   Origin

Rock glaciers can originate in several ways. Hypotheses include (1) avalanching or landsliding (Ernest Howe, 1909, 52; Outcalt and Benedict, 1965, 856), (2) glaciation (several varieties of hypotheses – Capps, 1910, 364; Cross, Howe, and Ransome, 1905, 25; Kesseli, 1941; Richmond, 1952), (3) flow of interstitial ice (several varieties of hypotheses depending on the source of the rock debris – Blagbrough and Farkas, 1968, 821–3; A. C. Spencer, 1900; Wahrhaftig and Cox, 1959, 432–3; and others), (4) basal shearing (Wahrhaftig and Cox, 1959, 427–8), and (5) a combination of (2) and (3) (Flint and Denny, 1958, 131–3). Landsliding in Howe's sense can contribute material to rock glaciers but except for providing debris does not account for their movement, since active rock glaciers move at slow rates only, commonly less than 1 m yr^{-1}, but as discussed by P. G. Johnson (1975, 132–5) and Outcalt and Benedict (1965, 865), ice derived from snow can be important to post-avalanche movement. Basal shearing was suggested to account for re-activation of some rock glaciers whose evidence of movement was concentrated near their base, and Wahrhaftig and Cox thought of this movement as probably not requiring ice but rather being a kind of landsliding that affected only a few of the rock glaciers they examined. All the other hypotheses postulate ice as necessary for movement. Although it has been suggested that 'rock-glacier creep' (Sharpe, 1938, 42–8) is mainly the result of frost creep and gelifluction (Chaix, 1923, 32; Kesseli, 1941, 226–7), movement appears to be dependent on the presence of perennial ice of one kind or another. The basal shear stress calculated from the thickness and surface slope of active rock glaciers considerably exceeds that required for frost creep and gelifluction alone (Wahrhaftig and Cox, 1959, 403–4).

It has been argued that glacier ice can be excluded in the origin of rock glaciers in areas such as Kendrick Peak, northern Arizona, and Capitan Mountains and San Mateo Mountains, New Mexico, which are believed to have remained unglaciated during Wisconsin time because of lying below the snowline (Barsch and Updike, 1971, 99, 103–6, 111–12; Blagbrough, 1976; Blagbrough and Farkas, 1968, 821–2). Proof of such instances was questioned by Whalley (1974, 6–8, 20–8), largely because he believed that evidence provided by flow properties of rock-ice mixtures, even though inconclusive, favours the view that a core of glacier ice is necessary for rock-glacier movement. However, a high ice content can be of other than glacial origin (cf. Barsch, 1977a, 235; Fisch, Fisch, and Haeberli, 1977, 245), and it has been argued, at least for the Swiss Alps, that ice of non-glacial origin is the '*conditio sine qua non*' for rock-glacier movement and that any glacier ice present is near surface and incidental to movement (Fisch, Fisch, and Haeberli, 1977, 255; Haeberli, personal communication, 1979). This conclusion requires more evidence before it can be accepted.

As noted earlier it has been widely held that interstitial ice alone is sufficient for movement – a still popular view (P. G. White, 1976; S. E. White, 1976b, 89; and others). Yet the grain-to-grain contact of rock particles in the case of interstitial ice (pore ice) provides a much stronger rock-ice mixture than where ice keeps rock particles from touching each other, as recognized by S. E. White (personal communication, 1976). Clearly the exact nature of movement – whether by (1) plastic flow of a glacial ice core or ice core of other origin, (2) flow involving even less excess ice, or (3) all these modes as seems reasonable, or (4) flow related to truly interstitial ice only as seems less probable – requires further study. In addition to more information on the flow properties of rock-ice mixtures, further evidence must be sought regarding the occurrence of rock glaciers in unglaciated areas, or in glaciated areas in places where any glacial influence can be excluded.

Tongue-shaped rock glaciers (length-to-width ratio > 1.0) occupy the axis of a valley like a glacier and are much more likely to be of glacial origin than lobate, or valley-wall, rock glaciers (length-to-width ratio < 1.0). The latter commonly originate beneath taluses on valley sides and extend onto valley floors at right angles to the tongue-shaped forms. According to S. E. White (1976b, 89), active lobate rock glaciers in the Colorado Front Range are of the ice-cemented type and move by creep in response to the presence of the 'interstitial ice' and the weight of the rock glacier itself and of the talus on its upper edges.

Calculations based on rate of movement suggest that rock glaciers are a more important mass-wasting process than is usually recognized, a discharge of about $158 \text{ m}^3 \text{ yr}^{-1}$, based on reasonable assumptions, having been calculated for a rock glacier in Pawnee Cirque in the Colorado Front Range (S. E. White, 1976b, 87, 89). On average, each of 994 active rock glaciers in the Swiss Alps has been reported as transporting $1.2–1.6 \times 10^6$ m^3 of sliderock (talus) and ice at a rate of 5–10 cm yr^{-1}, the ice content being about 50–60 per cent of the volume (Barsch, 1978). It is estimated that rock glaciers account for 15–20 per cent of all the mass-wasting in the central and southern Swiss Alps (Barsch, 1977c, 148, 159). Given comparable climatic and topographic conditions, bedrock structure and lithology are important factors in determining the amount of material carried by rock glaciers.

4 Environment

Active rock glaciers are reported from polar, sub-polar, and middle-latitude highlands. A review and list of occurrences has been provided by Glazovskiy (1978). Polar occurrences appear to be relatively rare. An active Antarctic rock glacier has been reported but the origin of the feature and its activity are suggestive rather than firmly established as noted in the report (H. J. H. Harris and Cartwright, 1977). Rock glaciers are widespread in middle-latitude alpine regions. As noted above, there are about 1000 active forms in the Swiss Alps; rock glaciers are especially common in the high central Andes of Argentina where they may be an important water resource (Corte, 1978). The activity of some forms in middle-latitude highlands is shown by movement observations such as most of those in Table 6.4 (cf. also R. L. Ives, 1940, 1277–8). Table 6.4 also shows

Table 6.4. Some rates of rock-glacier movements

Location	Rate (cm yr^{-1})	Observation period (yr)	Reference
Europe			
Gurgler Tal, Ötztal Alps, Austria			Vietoris (1972, 180, 181, Table 3)
Profile 1 (2540 m alt.) (mean max.)	239	5	
Profile 1 (New) (2540 m alt.)			
(mean max.)	500	3	
Profile 2 (2630 m alt.)	42–148	19	
Profile 3 (2680 m alt.) (mean max.)	78	16	
Macun, Lower Engadin, Swiss Alps	25–30	2	Barsch (1969a, 11, 27)
Murtèl-Corvatsch (Murtèl I), Upper			
Engadin, Swiss Alps	7	23 (1932–1955)	Barsch and Hell (1975, 111, 126, Table 8, 139)
	3	16 (1955–1971)	
	4	2 (1971–1973)	Barsch and Hell (1975, 111, 123, Table 7a, 139)
Val Muragle (Muragle I), Upper			
Engadin, Swiss Alps	21	1 (1972–1973)	Barsch and Hell (1975, 123, Table 7b)
Val Sassa and Val 'dell Acqua, Swiss			
Alps	136–158	21	Chaix (1943, 122)
North America			
Banff National Park, Alberta, Canada	0.3–0.6	27	Osborn (1975)
	0.7–0.8	72	
Clear Creek, Alaska Range	36–69	8	Wahrhaftig and Cox (1959, 395, Figure 6)
	(1.2–2.3 ft)		
Colorado Front Range	2–15	5	S. E. White (1971, 55–7, Tables 2–4)
Galena Creek, Wyoming			Potter (1969, 52, Table 7; 130,
Distance from west edge of rock			Figure 21; 1972, 3046, Table 5
glacier (m) 32	52		[Movement line B])
35	55		
53	60		
82	72		
97	76		
106	83		
127	79		
149	78		
154	58		
171	47		
180	40		
191	25		
196	14		
209	6		
217	4		

that rates of the same rock glacier can vary appreciably, not only from place to place (cf. Galena Creek rock glacier) but also from observation period to observation period. Thus the 1955–1973 rate of rock glacier Murtèl I (containing ice lenses but no glacier ice – Barsch, 1977b) was about half its 1932–1955 rate, a change whose cause is probably climatic, although data regarding thickening and thinning of Murtèl I appear to be out of phase.

As discussed above, movement is apparently dependent on the presence of perennial ice, and active rock glaciers are therefore generally accepted as indicative of permafrost (cf. Barsch, 1977a, 231;

1978; Barsch and Treter, 1976, 91; Fisch, Fisch, and Haeberli, 1977; Haeberli, 1975, 96; S. E. White, 1976b, 78).

Because of the great variability in relief in the Swiss Alps, patches of permafrost can exist below timberline at a mean annual air temperature ranging from $-2°$ or $-1°$ to $0°$ or more, far outside their normal distribution. Retaining the term sporadic permafrost for such aclimatic occurrences, Barsch (1978) reported that active rock glaciers in Switzerland are associated with a mean annual temperature less than $-2°$ and that their lower limit is generally coincident with the lower limit of discontinuous permafrost; on north slopes the limit may correspond to $-1°$ (Barsch, 1977*d*). A mean annual surface temperature of about $-1°$ to $-2°$ for Swiss rock glaciers containing permafrost has been suggested on the basis of electrical resistivity data and assumptions regarding the geothermal gradient (Fisch, Fisch, and Haeberli, 1977, 239, 256).

5   Inactive and fossil forms

In general inactive or fossil rock glaciers – the former lacking movement but still containing some ice, the latter without ice (Barsch, 1978) – are geographically associated with valley or cirque glaciers and share their paleoclimatic implication of former lower temperature or greater precipitation or both. However, as noted above, rock glaciers normally indicate at least discontinuous permafrost; therefore – except for aclimatic forms, including any derived from glaciers (cf. Luckman and Crockett, 1978, 548) – rock glaciers should perhaps indicate somewhat lower temperatures than would terminal moraines of most temperate glaciers. As stressed by Potter (1969, 76) correlating the time significance of rock glaciers with that of moraines left by glaciers is dangerous because rock glaciers may remain active much longer than glaciers that wasted rapidly because of being relatively free of debris. Although fossil rock glaciers would normally lack the prominent outwash trains expectable from the rapid melting of such comparatively 'clean' glaciers (Potter, 1969, 77), this contrast might not hold in some high-latitude arid environments where runoff from glaciers is restricted.

According to Corte (1976, 192) inactive rock glaciers in Argentina are consistent with at least four Holocene climatic fluctuations. Marnezy (1977*a*, 157–65) reported two ages of rock glaciers in Switzerland – one late glacial or younger, the other probably *Petit âge glaciaire alpin* (*c.* 1590–1850 AD – cf. Ladurie, 1971, 220–3) Luckman and Crockett (1978, 547–9) recognized at least two age phases in the southern part of Jasper National Park, Canada, the oldest in their opinion probably correlative with early Holocene (pre-6600 years BP) rock glaciers in nearby Banff and Yoho National Parks.

IX   Taluses

Taluses, also known as scree slopes, are apronlike accumulations of sliderock (scree) that can be many metres thick, about 30 m having been reported by Brunner and Scheidegger (1974, 3). Accumulations can also be no more than a veneer over bedrock despite a massive appearance, as the present writer once observed in Greenland. The debris is typically angular (Figure 6.29).

Taluses are due to mass-wasting by definition. The breaking away of rock from the cliff face may result from various wedging effects, such as hydration and root growth, but frost wedging is commonly regarded as the most important, as discussed under Frost wedging in the chapter on General frost-action processes. Rates of cliff recession above a talus slope are not well known but bedrock structure and lithology can be critical factors. For a cliff of dolomitic limestone in the Canadian Arctic, Souchez (1969) estimated a long-term rate of 0.5–0.8 mm yr^{-1} except where he believed frost wedging was

Table 6.5. Some rates of talus movement

Location	Gradient (degrees)	Rate (cm yr^{-1})	Observation period (yr)	Reference
Canada				
British and northern Richardson Mountains, Yukon Territory	⩽25	2.5 (mean) 5.8 (max)	1	Rampton and Dugal (1974)
Lake Louise district, Alberta	25 (mean) 34 (max)	6.6–111.0	4–7	Gardner (1973, 99, Table 1)
Europe				
Chambeyron, French Alps		1–450	2	Michaud (1950, 188)
Kärkevagge, Lapland	14–38.5	0–22 25–500 (rolling or sliding)	2–6	Rapp (1960*a*, 172–5, incl. Tables 30–1)

6.29 Talus near Holman, Victoria Island, Northwest Territories, Canada

intensified because of moisture from snowdrifts, the rate there being up to 1.3 mm yr^{-1}.

Rates of talus movement are difficult to measure (cf. Barnett, 1966, 379–81). The movement is discontinuous and is commonly regarded as occurring by (1) subsidence, (2) creep, (3) rolling or gliding of individual stones, (4) small talus slides, or (5) a combination of the foregoing (cf. Rapp, 1960a, 171–6). Field experiments discussed by Brunner and Scheidegger (1974) support the view that in many places the movement is concentrated within half a metre of the surface, with deep-seated creep being negligible or absent. Some rates of movement are shown in Table 6.5.

Where there is a sufficiently high cliff face, size sorting of an active talus is commonly from finer sliderock near the top of a talus to coarser near the base because of the greater kinetic energy of large fragments. This fall sorting is widespread in periglacial environments, perhaps more so than where chemical weathering contributes more importantly to breakdown of sliderock at the base of a slope.

6.30 Protalus rampart, Mountain Lakes Wild Area, Oregon. A remnant snow patch lies immediately upslope on concave inner side of the protalus rampart. Photo by G. A. Carver

Nevertheless, in places this fall sorting is accompanied or replaced by reversed fall sorting in which the largest fragments accumulate near the top. This reverse sorting appears to be increasingly prominent the lower the cliff face (Rapp, 1960a, 102). Zonal (downslope) sorting of both types can be strongly modified by various processes such as debris flow, snow avalanching, and slushflow (slush avalanches).

Avalanche talus, compared with rockfall talus, has been reported to have lower average gradient, greater profile concavity, greater range of particle size, poorer sorting, and less stability, including perched blocks (Luckman, 1971, 106). Despite modification by various processes, zonal sorting is apparently far more common than lateral (across-slope) sorting (Bones, 1973).

Stratification of talus tends to be poor to absent but is not as well known as would be desirable.

Similarly, talus fabric is not well established. In places elongate clasts tend to parallel the gradient direction but dip upslope, the degree of development being a function of surface roughness, decreasing with increasing size of sliderock (McSaveney, 1972, 181).

Although taluses are most common in periglacial environments they are not limited to them. Taluses have been reported as forming in Virginia today (Hack, 1965), and as having no special climatic significance (Wilhelmy, 1958, 17–18). However, stabilized taluses can indicate former greater frost wedging as suggested for the Baraboo, Wisconsin, area in the United States (H. T. U. Smith, 1949b, 199–203). Two ages of talus have been reported in the White Mountains of Arizona (Merrill and Péwé, 1977, 40–1), and also in northern Cyrenaica, Libya, where two sets of stabilized taluses, the older cemented, provide evidence of two Weichsel cold phases (Hey, 1963). Partially cemented talus slopes in the Jura Mountains of France probably record Würm frost wedging (Judson, 1949). Layers of bouldery debris alternating with uncemented finer debris on hillsides in North Staffordshire, England, have been interpreted as Pleistocene talus deposits recording former intense frost action alternating with milder conditions (Prentice and Morris, 1959). Described as éboulis ordonnés, the North Staffordshire deposits may be related to grèzes litées, which are thought to be associated with meltwater from snow as described in the chapter on Slopewash. Probably transitions occur between normal talus deposits and éboulis ordonnés.

X   Protalus ramparts

A protalus rampart (named by Bryan, 1934, 656) is 'A ridge of rubble or debris that has accumulated piecemeal by rock-fall or debris-fall across a perennial snowbank, commonly at the foot of a talus' (Richmond, 1962, 20) (Figure 6.30). Sliding is usually also involved (Richmond, 1962, 61). Protalus ramparts occur in many cirques. The material is dominantly angular, coarse, and non-oriented, and much of it is produced by frost wedging. Some sorting may be present in view of the greater kinetic energy of larger blocks. Fines tend to be absent because they lack the momentum to bounce, slide, or roll across the snow as pointed out by S. E. White (*in* Blagbrough and Breed, 1967, 768). Protalus ramparts up to 9 m (30 ft) high have been repeatedly reported (Behre, 1933, 630; Daly, 1912, 593), and heights up to 24 m (80 ft) have been observed by Blagbrough and Breed (1967, 766), who reported basal widths up to 14 m (45 ft) and average lengths of 91 m (300 ft). Sharp (1942c, 496) observed widths up to 15 m (50 ft) and lengths up to 122 m (400 ft), and he noted that the ramparts were within 15–90 m (50–300 ft) of the base of the slope from which their constituents came. Possibly the largest feature that has been described as a protalus rampart complex has a height of 55 m and a length, along its outer edge, of 1 km (Sissons, 1976, 184). Evidence indicates that the deposit, located in northeast Scotland, postdates the last glaciation of the locality. However, two closed depressions led Sissons (1976, 189–90) to infer there had been under

lying ice, and he suggested that this part of the deposit may have been analogous to a rock glacier. That some protalus ramparts evolve into rock glaciers is known (cf. Corte, 1976, 176). Probably most protalus ramparts are arcuate and concave upslope but some form linear benches; convex-upslope forms have also been reported (Watson, 1966, 82; Watson and Watson, 1977, 24).[7]

Some unusual conical mounds, 0.6–7.6 m high and consisting of rubble from the adjacent cliff, have been described from the Brooks Range, Alaska, by Yeend (1972), who concluded they were somewhat similar to ramparts in origin and referred to them as winter protalus mounds. It seems possible that, unlike ramparts, they originated in a single episode of sliding and evolved their form as the result of unequal ablation of underlying snow or ice.

In addition to the influence of climate, including its indirect effects such as vegetation cover, the location of protalus ramparts is strongly influenced by topography and by the susceptibility of the bedrock or unconsolidated material on the slope above to disintegration and mass-wasting. In location, material, and form, protalus ramparts are probably similar to many small end moraines, especially in cirques. Except in some cirques, a distinguishing characteristic would be the orientation of the ridge, which in a protalus rampart would be oriented parallel rather than transverse to a valley. Confusion might arise in the case of a lateral moraine fragment but the lithology of ridge stones as between local material and that from up the valley might be diagnostic. Protalus ramparts can be readily confused with the toes of some taluses that have moved by creep or solifluction but some ramparts can be distinguished by their comparative lack of fines (Blagbrough and Breed, 1967, 769).

If located where they are no longer forming, protalus ramparts indicate a former cooler climate (H. T. U. Smith, 1949a, 1503), greater snowfall, or both. They do not necessarily prove former greater frost action, since frost wedging activity might remain unchanged while locus of deposition varied with change in size and/or position of the snowbank. Protalus ramparts indicate a snow environment and the high probability of frost wedging, and they may be a guide to a former approximate orographic snowline, since they are clearly related to it. Multiple protalus ramparts in southern Utah appear to record two Wisconsin glaciations (Blag

---

[7] Watson used upslope convexity as evidence against a morainal origin but the argument in the case described is more suggestive than conclusive in the opinion of the present writer, who is indebted to Dr Watson for the privilege of visiting the features with him.

brough and Breed, 1967, 771–2). Ramparts of two different ages also occur on San Francisco Mountain, Arizona (Péwé and Updike, 1976, 31–2; Sharp, 1942c, 496–8); they are common on cirque slopes facing north to northeast and probably formed within the last 4000 years (Neoglacial time) according to Péwé and Updike. In the Argentine Andes, four levels of protalus ramparts recording four Holocene climatic fluctuations are reported by Corte (1976, 192).

Protalus ramparts have been observed in the Antarctic by the present writer, and are probably reasonably common in polar, subpolar, and middle-latitude highlands (cf. Blagbrough and Breed, 1967; Bryan, 1934; Bryan and Ray, 1940, 35; Richmond, 1964, D28, D30).

7   Nivation

I Introduction: II Nivation benches and hollows: III Cryoplanation terraces and cryopediments

I   Introduction

Nivation is the localized erosion of a hillside by frost action, mass-wasting, and the sheetflow or rill-work of meltwater at the edges of, and beneath, lingering snowdrifts. The term was introduced by Matthes (1900, 183).

As discussed by Embleton and King (1975, 130–44) in an excellent summary of nivation, the effectiveness of nivation is strongly influenced by the thickness of the snowdrift and the presence or absence of permafrost beneath it. To the extent that the ground surface remains frozen beneath the snow, as is usual in permafrost regions except at the feather edge of a snowdrift, frost wedging and gelifluction are inhibited and nivation is limited to the feather edge and the immediately adjacent snow-free area. As a snowdrift becomes thinner and contracts, the site of maximum effectiveness of nivation follows and eventually traverses the area freed from snow. Where permafrost is lacking and extensive thaw areas occur beneath the snow, nivation, including gelifluction (Boch, 1946, 216–17), can affect the snow-free and snow-covered areas simultaneously. There is some difference of opinion as to the role of snow creep but that it can displace mobile particles is evident (Costin et al., 1964; Haefeli, 1953, 248; 1954, 62; Jennings, 1978; Jennings and Costin, 1978; Mackay and Mathews, 1967; Mathews and Mackay, 1963; 1975; Thorn, 1974a, 119–27, 277–85; P. J. Williams, 1962, 358). If sufficiently deep, snow can exert creep forces strong enough to break steel rods having breaking strengths of 1.6 to over 15.9 bars as observed for snow depths of 6 m to over 30 m (Costin et al., 1973).

Quantitative studies of nivation are conspicuously few. On the Wasatch Plateau in Utah, a record snow accumulation in 1952 resulted in persistence through the summer of large snowbanks over weak materials. This led to undermining of sod and the development that summer of 'incipient cirques' 15–500 m in diameter and up to almost 1 m deep (Flint, 1957, 100). Detailed studies of sites lacking permafrost have been carried out by Thorn in the Colorado Front Range. He reported that compared with nearby snowfree sites, nivation (1) increases physical erosion (mainly by sheetflow and rillwork) by an order of magnitude; (2) increases chemical weathering (not necessarily erosion) two to four times; (3) contrary to most views, does not increase the number of freeze-thaw cycles but presumably enhances their effectiveness by supplying moisture; (4) does not initiate but only modifies hollows; (5) is probably most effective during deglaciation; and (6) does not appear to be a continuum process leading to development of cirques (Thorn, 1974a, 155–223; 1974b; 1976b; 1976c). To what extent these conclusions are generally applicable remains to be determined. Increased chemical weathering was also reported by Boch (1946, 213) on the basis of rocks having thicker weathering rinds inside than outside the areas of contracting snowdrifts.

The effect of nivation is to countersink lingering snowdrifts into hillsides and thereby produce nivation hollows. Thus the distribution of such features is strongly influenced by topography and is probably closely related to the position of the orographic snowline. Nivation has been reported from many periglacial environments and has a wide quasi-equilibrium range. Nivation is not confined to highlands but can occur at low altitudes in polar environments, for instance, along coastal cliffs characterized by seasonal snowdrifts (Grigor'yev, 1966, 74–5; 1976, 77–8). Boch's (1948, 609) view that it is active only in a subarctic climate, not in a continental arctic climate, is certainly too restrictive, although nivation is probably less important in arid polar climates than in other periglacial environments. Nevertheless, it has been suggested that nivation in arid Southern Victoria Land, Antarctica, is important in producing rectilinear slopes (Shaw and Healy, 1977, 50–4). Reports concerning nivation include those of Ballantyne (1978), Berger (1967), Boch (1946; 1948), Ekblaw (1918), Gardner (1969), E. P. Henderson (1959*b*, 70–81), W. V. Lewis (1939), Matthes (1900, 179–85), McCabe (1939), Nichols (1966, 29–30), Ohlson (1964, 85–95), Schunke (1974*a*; 1975*a*, 25–6), St-Onge (1969), Thorn (1974*a*), and many others.

As noted in the chapter on Slopewash, it is widely believed that frost wedging and flow of meltwater from lingering snowdrifts are the key processes in the production and deposition of grèzes litées. They have therefore been ascribed to nivation (Guillien and Lautridou, 1970, 42). Some grèzes litées are probably of this origin but it remains to be demonstrated that all deposits with the characteristics of grèzes litées are necessarily nivation phenomena.

II Nivation benches and hollows

From observations in the Canadian Arctic, St-Onge (1969) concluded, contrary to Boch (1946, 220), that nivation is strongly influenced by lithology, as reflected in various nivation landforms. Thus nivation terraces, the largest features, were in gabbro that produced coarse debris; nivation benches were in sandstone that disintegrated into somewhat finer material; and other forms such as semicircles, hollows, and ledges were in shale that broke down into fine sand and silt. Although the choice of terms is open to criticism, since one man's terrace may be another man's bench, and hollow is often used for a variety of landforms, the general concept that lithology plays a critical role seems sound. St-Onge (1969, 5–6) summarized the evolution of a nivation terrace in gabbro or bedrock of similar hardness as follows

Snow accumulates in an area oriented to the wind or in an initial irregularity in the slope. Conditions favourable to frost shattering [frost wedging] are created. This process

yields large quantities of coarse fragments and few fines. The fines are removed by water percolating through the boulders. The backwall recedes leaving an apron of coarse fragments. These boulders act as the base level of the terrace since they can no longer be saturated. As the terrace widens, the effects of percolating water and sheet wash are greatly reduced. Away from the outer edge, the open-work boulder field is gradually clogged with a matrix of fine material. The moisture retention capacity increases with a corresponding increase in frost shattering. Eventually the percentage of fine material is sufficiently high to render geliturbation possible.[1]

'Fossil' nivation depressions may be related to former colder climate, greater snowfall, or both. Their identification is complicated by similar-appearing features such as earthflow scars. Nivation depressions constitute a useful periglacial criterion if their significance relative to present and former snowlines can be established for a region.

III Cryoplanation terraces and cryopediments

Cryoplanation terraces (Demek, 1968*a*, 92; Pécsi, 1964, 61–3; cf. 1963, 178) are hillside or summit benches that are cut in bedrock, predominantly transect lithology and structure, and are confined to cold climates (Figure 7.1).[2] They are also known as altiplanation terraces (Eakin, 1916, 77–82), less commonly as equiplanation terraces (cf. Cairnes, 1912, 344–8), and in Europe and Asia as goletz (or golec) terraces (Jorré, 1933).

Cryopediments are much like cryoplanation terraces; however, rather than being located on the middle to upper sections of slopes and on summit areas they are confined to the lower sections of slopes (Czudek and Demek, 1970*a*; 1973*a*; French, 1973; 1976*a*, 155–7). Here fluvially related processes can become a more important factor than with cryoplanation terraces. Cryopediments are regarded as a periglacial analogue of warm-desert pediments (cf. Demek, 1969*a*, 66), although the origins are quite disparate. Allowing for the difference in position of cryopediments, most of the following discussion of cryoplanation terraces applies to both.

---

[1] Geliturbation = '*Congeliturbation* = frost-action including frost-heaving and differential and mass movements' (Bryan, 1946, 640).
[2] Cryoplanation as a term (Bryan, 1946, 639–40) has come into increasingly wide use, and cryoplanation terrace and other derivatives, such as Kryoplanationsterrasse, are supplanting altiplanation terrace in the literature (Embleton and King, 1975, 159; French, 1976*a*, 157; Jahn, 1975, 164; Karrasch, 1972, 159; Reger and Péwé, 1976). Although not generally specified, the qualification that cryoplanation terraces predominantly transect lithology and structure (although to some extent influenced by them) is critical to the definition, otherwise cryoplanation terraces could be merely lithologic or structural benches in periglacial environments. The distinction is clearly made by Reger and Péwé (1976, 101).

7.1 Cryoplanation terrace and tor near Eagle Summit, Alaska. Scale given by figure. Photo by Troy L. Péwé

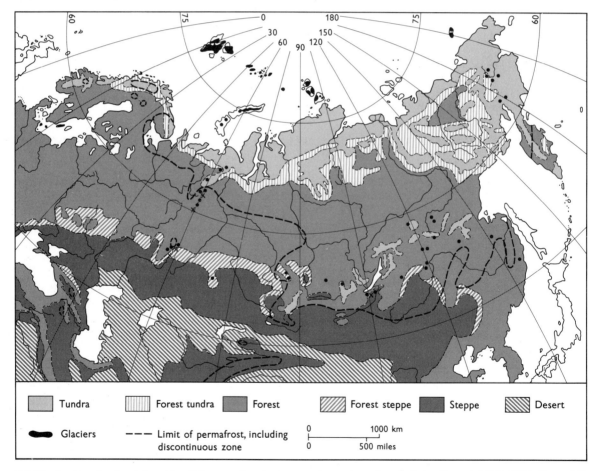

	Tundra		Forest tundra		Forest		Forest steppe		Steppe		Desert

Glaciers   – – – Limit of permafrost, including discontinuous zone

0 ___ 1000 km
0 ___ 500 miles

7.2 Distribution of active (●) and fossil (x) cryoplanation terraces in northern Eurasia (*after Frenzel, 1959, 1019, Figure 11*)

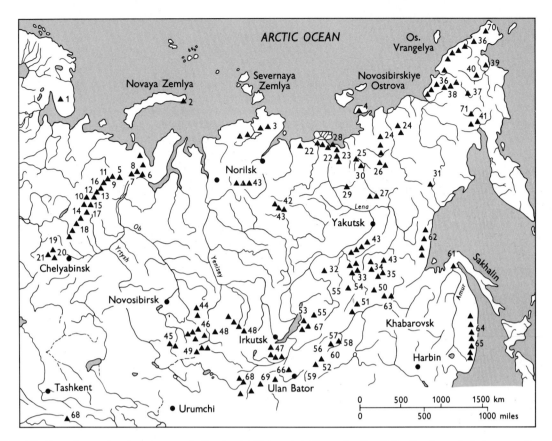

7.3 Distribution of cryoplanation terraces in the Soviet Union and Mongolia. No distinction is made between active and fossil forms (*after Demek, 1969a, 11, Figure 3*)

USSR – 1 Koliskiy Peninsula (*Perov, 1959; Kozlov, 1966*); 2 Novaya Zemlya (*Miloradovich, 1936*); 3 Taymyr Peninsula-Gory Byrranga (*Kaplina, 1965*); 4 Novosibirskye Islands-Bolshoy Lyakhovskiy Islands (*Byelorusova, 1963*); 5 Polyarniy Ural Mts (*Baklund, 1911; Panov, 1937*); 6 Polyarniy Ural Mts.–Ray-Iz Massif (*Zavaritskiy, 1932*); 7 Polyarniy Ural Mts. (*Sofronov, 1945*); 8 Polyarniy Ural Mts.–western slope of the Pechora River Basin (*Kaletzkaya and Miklukho–Maklay, 1958*); 9 Pripolyarniy Ural Mts. (*Aleshkov, 1936; Chernov and Chernov, 1940; Dobrolyubova and Sochknai, 1935; Dolgushin, 1961; Lyubimova, 1955*); 10 Severniy Ural Mts.–Upper Vishera River Basin (*Duparc and Pearce, 1905; Duparc, Pearce, and Tikanowich, 1909*); 11 Severniy Ural Mts. (*Aleshkov, 1936; Edelshteyn, 1936; Panov, 1937*); 12 Severniy Ural Mts. (*Varsanofeva, 1932*); 13 Severniy Ural Mts. (*Boch and Krasnov, 1943, 1946; Tarnogradskiy, 1963*); 14 Severniy Ural Mts.–Visherskiy Ural (*Suzdalskiy, 1952*); 15 Severniy Ural Mts.–Khrebet Chistop (*Gorchakovskiy, 1954*); 16 Severniy Ural Kamen Mt. (*Strigin, 1960*); 18 Sredniy Ural Mts. (*Aledsandrov, 1948*); 19 Yuzhniy Ural Mts. (*Bashenina, 1948*); 20 Yuzhniy Ural Mts.–Iremel Mt. (*Tsvetayev, 1960; L. O. Tyulina, 1931*); 21 Yuzhniy Ural Mts.–Yaman Tau (*Gorchakevskiy, 1954*); 22 Yakutia–Kryazh Chekanovskogo (*Gakkelya and Korotkevich, 1962; Sochava in S. V. Ohruchev, 1937*); Kryazh Prontchistcheva (*Rusanov et al., 1967*); 23 Yakutia–Gory Kharaulakh (*Sochava and Gusev in S. V. Obruchev, 1937*); 24 Yakutia–Alazeyskoye Ploskogorye, Polousniy Khrebet (*Gakkelya and Korotkevich, 1962; Rusanov et al., 1967*); 25 Yakutia–Khrebet Kular (*Demerk, 1967; Rusanov et al., 1967*); 26 Yakutia–Khrebet Cherskogo (*Kropachev and Kropacheva, 1956*); 27 Yakutia Verkhoyanskiy Khrebet (*Korofeyev, 1939; Lazarev, 1961; Yegorova, 1962*); 28 Yakutia–Tuora-Sis (*Lazarev, 1961*); 29 Yakutia–Verkhoyanskiy Khrebet (*Yegorova, 1962*); 30 Yakutia – surroundings of the town of Batagay (*Demek, 1967*); 31 Nyerskoye Ploskogorye Mts. (*Tskhurbayev, 1966*); 32 Yakutia–Olekma–Vitim region (*S. V. Obruchev, 1937*); 33 Yakutia–Olekma River Basin (*Timofeyev, 1965*); 34 Yakutia–Aldanskoye Nagorye Mts. (*Dolgushin, 1961; Rabotnov, 1937*); 35 Yakutia–Aldanskoye Nagorye Mts. (*Kornilov, 1962*); 36 Arctic Ocean Coast (*S. V. Obruchev, 1937*); 37 Upper Anadyr River Basin (*L. N. Tyulina, 1936*); 38 Severniy Anyuykiy Khrebet (*S. V. Obruchev, 1937*); 39 Zolotiy Khrebet (*S. V. Obruchev, 1937*); 40 Ust-Bielskiye Gory (*Kaplina, 1965*); 41 Kolymskiy Khrebet (*Sochava in S. V. Obruchev, 1937*); 42 Anabarskiy Massif (*Yermolov, 1953*); 43 Gory Putorana, Anabarskiy Massif, Aldanskoye Ploskorgorye Mts., Prilenskoye Plateau (*Krasnov and Kozlovskaya, 1966*); 44 Kuznetskiy Alatau (*Il'in, 1934; Tolmachev, 1903*); 45 Altai–Terektinskiy Khrebet (*Keller, 1910*); 46 Altai (*Il'in, 1934; Makerov, 1913*); 47 Sayanskoye–Dzidinskoye Nargorye Mts. (*Lamakini and Lamakini, 1930*); 48 Vostochniy Sayan-Zapadnyi Sayan (*Kushev, 1957*); 49 Zapadniya Tuva (*Kozlov, 1966*); 50 Olekma region (*Kozmin, 1890; Sukachev, 1910*); 51 Olekminskiy Stanovik, Cheromnagovyiy Khrebet, Mogochinskiy Khrebet, Tungirskiy Khrebet (*Korzhuyev, 1959; Makerov, 1913*); 52 Patomskoye Nagorye (*Bashenina, 1948*); 53 Barguzinskiy Khrebet–Severo–Muyskiy Khrebet–Khrebet Verchniye–Angarskiy (*Dumitrashko, 1938, 1948; L. N. Tyulina, 1948*); 54 Stanovoye Nagorye-Khrebet Udokan (*Preobrazhenskiy, 1959, 1962*); 55 Khrebet Udokan, Khrebet Ulan-Burgasy, Itatskiy Khrebet, Tsipinskiye Gory (*Gravis, 1964*); 56 Yablonoviy Khrebet (*Dengin, 1930*); 57 Borschchovochniy Khrebet (*Ryzov, 1961*); 58 Upper Amur River Basin (*Kaplina, 1965*); 59 Khentey, Yablonoviy Khrebet, Borschchovochniy Khrebet (*Bashenina, 1948; Nikol'skaya, Timofeyev and Chichagov, 1964*); 61 Lower Amur River Basin-Gora Praul (*Ganeshin, 1949*); 62 Khrebet Kzugdhur (*Kaplina, 1965*); 63 Tukuringra–Dzagdy (*Nikol'skaya and Shcherbakov, 1956*); 64 Sikhote Alin (*Solov'yev, 1961*); 65 Sikhote Alin (*Pryalukhina, 1958*); 66 Daurskiy Khrebet (*Zamoruyev, 1967*); 67 Juzhno-Muyskiy Khrebet (*Rudavin, 1967*); 68 Pamir (*Sekyra, 1964*). Mongolia–69 Khangai (*H. Richter, Haase, and Barthel, 1963*); 70 USSR – Chukotka (*Zhigarev and Kaplina, 1960*); 71 USSR Penzhina River Basin (*Sochava, 1930*)

Cryoplanation terraces have a veneer of solifliction debris, which may be imprinted with patterned ground. Lithology, and structure such as jointing, may help to influence location and degree of development of some terraces. In Jahn's (1975, 164) experience jointing is a critical factor in that it permits break up of the bedrock and development of cliffs, which then recede leaving the terrace. Although the importance of blocky break up of rock was also emphasized by Karrasch (1972, 162), it was not assigned a major role by Reger and Péwé (1976, 101). Cryoplanation terraces range in size from about 10 m across to widths of 2–3 km and lengths exceeding 10 km. The gradient of the terrace tread is $1°–10°$ (Reger and Péwé, 1976, 101) or up to $12°$ (Demek, 1969a, 42, 44, 64). The terrace riser has a height ranging from 3 to 76 m, and a gradient of $9°–32°$ where debris covered but nearly vertical where bedrock appears[3]; the thickness of debris on a terrace ranges from <1 m to 3 m (Reger and Péwé, 1976, 99, 101).

Nivation, combined with gelifluction, is believed to be an essential process in the origin of cryoplanation terraces in Alaska (Péwé, 1970, 360; 1975, 61; Reger and Péwé, 1976, 105), Siberia, and elsewhere (Demek, 1968a; 1969a; 1969b). Their distribution in Eurasia, Europe, and North America is shown in Figures 7.2–7.5. Cryoplanation terraces have been cited as being the most widespread periglacial erosional landform in Alaska (Péwé, 1975, 61), but this disregards gelifluction slopes as erosional forms. Although cryoplanation terraces can occur on various kinds of bedrock, they are best developed on closely jointed resistant rock such as andesite and basalt; they are thought to form perhaps slightly below the snowline, except for being able to form above it on isolated peaks and ridges that are too small to carry large glaciers. Altitudes of well-formed terraces in Alaska rise from west to east on a line parallel to and below past and present snowlines (Péwé, 1975, 161).

Three stages of terrace development were described by Demek (1969b): (1) The first stage is characterized by nivation, which produces a nivation hollow or bench. Elongate snow patches are particularly favourable sites for terrace development. The developmental sequence of a nivation terrace as described above by St-Onge (1969, 5–6) probably corresponds to this stage. (2) The second stage is characterized by nivation operating on a frost-wedged cliff and by a complex of processes that transport the resulting debris across the tread at the base of the cliff. These processes vary, depending on lithology and other factors, but include gelifluction, slopewash, eluviation, and piping. The tread commonly has a gradient of about $7°$. If the cliff rises above the snow patch, frost-wedged debris falls on the snow patch and tends to accumulate at its base, forming protalus ramparts. Eventually the cliff is worn back. Demek (1969a, 64) split this stage into the initial-terrace and the mature-terrace stages. (3) The third stage is characterized by a summit flat whose gradient may be less than $2°$. Most processes become less active because of the low gradient; however, those responsible for patterned ground continue unabated, and deflation may be locally significant.

It has been argued that the presence of sorted forms of patterned ground (except sorted stripes) indicates a lack of significant material transport across terrace treads by gelifluction and surface drainage, and therefore that material must have been removed by deflation and subsurface eluviation (Schunke, 1975a, 66, 176–8). However, the argument is only valid on demonstrably truly horizontal surfaces, since gelifluction can occur on gradients as low as $1°–2°$ (Schunke, 1975a, 140–1, 179) and the transition gradient from polygons and nets to sorted stripes ranges from about $2°$ to $7°$ (cf. R. P. Goldthwait, 1976a, 33; Hussey, 1962; Washburn, 1956b, 836–7).

The rate of terrace development is not well known but is probably on the order of tens of thousands of years for large well-developed terraces (cf. Demek, 1969a, 66–7). In Alaska Péwé (1973a) reported some subdued forms to be as old as early to middle Quaternary, based on evidence of overlying Illinoian loess. According to Reger and Péwé (1976, 107), well-developed terraces are cut on glaciated surfaces of Illinoian age and are younger than 50 000 years, and on Seward Peninsula there is evidence that some terraces postdate early Wisconsin glaciation and are 10 000–35 000 years old. The question of age is important. Büdel (1977, 57–9) has argued that bedrock erosion surfaces attributed to periglacial processes are really much older and have been only 'overprinted', not formed by such processes. His evidence is not specific enough to be compelling but the danger of misinterpretations because of overprinting is obvious and pinpoints the need for age determinations and other detailed studies.

---

[3] The height of 3 to 76 m supersedes Péwé's (1975, 61) estimate of 1 to 30 m based on earlier work.

7.4 Distribution of cryoplanation terraces in Europe. No distinction is made between active and fossil forms (*after Demek, 1969a, 10, Figure 2*)

1 Northern shore of southwestern England (*Guilcher, 1950*); 2 Southwest England – Dartmoor (*Demek, 1965; Te Punga, 1956a; R. S. Waters, 1962*); 3 South-west England – Mendip Hills (*R. S. Waters, 1962*); 4 England – Northeast Yorkshire (*Gregory, 1966*); 5 France – Brittany (*Guilcher, 1950*); 6 France – Vosges; 7 West Germany – Harz (*Hövermann, 1953*); 8 West Germany – Niedersächsisches Bergland (Lower Saxonian Highland) (*Spönemann, 1966*); 9 West Germany – Odenwald; 10 Belgium – Ardennes (*Gullentops et al., 1966*); 11 Poland – Karkonosze Mts. (*Jahn, 1965*); Czechoslovakia – Krkonoše (*Sekyra, 1964*); 12 Czechoslovakia – Rychlebské hory Mts. (*Ivan, 1965; Panoš, 1960*); 13 Czechoslovakia – Hrubý Jeseník Mts. (*Czudek, 1964;*

Czudek and Demek, 1961; Demek, 1964b, 1964c; Hrádek, 1967); 14 Czechoslovakia – Českomoravská Vrchovina (Bohemian – Moravian Highlands) (*Demek, 1964a, 1964b*); 15 Czechoslovakia – Novochradské hory Mts. (*Demek, 1964d*); 16 Czechoslovakia – Český Les (Bohemian Forest) (*Balatka in Demek, 1965*); 17 Czechoslovakia – Slavkovsky Les, Tepelská Vrchovina (Slavkov Forest, Tepelská Vrchovina Highlands (*Czudek in Demek, 1965*); 18 German Democratic Republic – Erzgebirge, Czechoslovakia – Krušné hory (*Czudek in Demek, 1965; Kral, 1968; H. Richter, 1965*); 19 Czechoslovakia – Nízký Jeseník (Low Jeseník Mts.) (*Czudek in Demek, 1965*); 20 Czechoslovakia – Bobravská Vrchovina (Bobrava Highlands) (*Demek, 1965*); 21

Czechoslovakia – Chřiby Hills (*Czudek, Demek, and Stehlík, 1961*); 22 Czechoslovakia – Hostýnske Vrchy (Hostýn Hills) (*Czudek, Demek, and Stehlík, 1961*); 23 Czechoslovakia – Moravskoslezské Beskydy (Moravian-Silesian Beskydy Mts.) *Stehlík, 1960*); 24 Hungary (*Pecsi, 1963, 1964, 1965*); 25 Bulgaria–Vitosha (*Demek and Šmarda, 1964; Maruszak, 1961*); 26 Bulgaria – Rila (*Demek and Šmarda, 1964*); 27 Faeroe Islands (*C. A. Lewis, 1966*); 28 Sweden – surroundings of Kiruna (*Bashenina, 1967*); 29 German Democratic Republic – western vicinity of Dresden (*H. Richter and G. Haase, personal communication*)

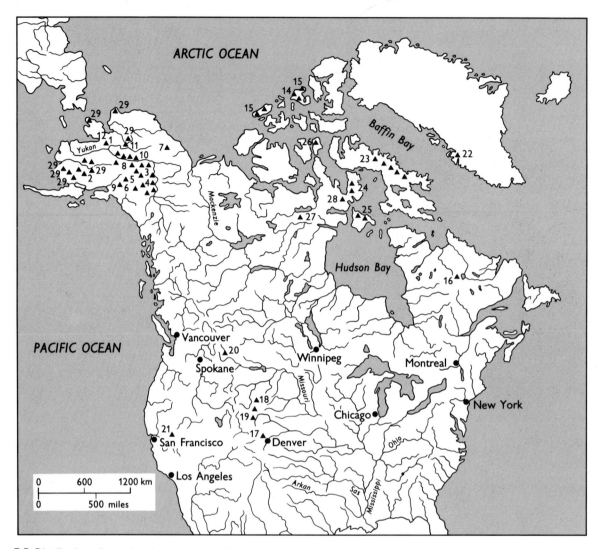

7.5 Distribution of cryoplanation terraces in North America. No distinction is made between active and fossil forms (*after Demek, 1969a, 13, Figure 4; and Reger and Péwé, 1976, 102, Figure 2*)

1 Alaska–Yukon–Koyukuk region (*Eakin, 1916*); 2 Alaska–Central Kuskokwim region (*Cady et al., 1955*); 3 Alaska–Yukon–Tanana region (*Mertie, 1937; Prindle, 1905*); 4 Alaska–Yukon–Tanana region (*Jahn, 1961b, 1966*); 5 Alaska – northeast of Fairbanks (*Hanson, 1950, Taber, 1943*); 6 Alaska – Alaska Range (*Jahn, 1961b*); 7 Alaska – Brooks Range (*Jahn, 1961b*); 8 Alaska – Cosna Nowitna region (*Eakin, 1918*); 9 Alaska – Talkeetna Mts. (*Wahrhaftig, 1965*); 10 Alaska – Indian River Upland (*Wahrhaftig, 1965*); 11 Alaska – Lockwood Hills, Zane Hills (*Wahrhaftig, 1965*); 12 Alaska – Selawik Hills (*Wahrhaftig, 1965*); 13 Alaska – Gulkana Upland (*Wahrhaftig, 1965*); 14 Canada – Ellef Ringnes Island (*St-Onge, 1965*); 15 Canada – Meighen Island, Ellef Ringnes, Prince Patrick Islands (*Heywood, 1957; Pissart, 1966c; Robitaille, 1960; Thorsteinsson, 1961*); 16 Canada – Labrador (*Derruau, 1956*); 17 USA – Colorado (*Russell, 1933*); 18 USA – Bighorn Mts. (*Mackin, 1947*); 19 USA – Colorado Front Range (*Bradley, 1965; B. W. Scott, 1965*); 20 USA – Glacier National Park, Montana; 21 USA – Monitor Pass, California; 22 Greenland (*Ekblaw, 1918*); 23 Canada – Baffin Island (*Bird, 1967*); 24 Canada – Melville Peninsula (*Bird, 1967*); 25 Canada – Southampton Island (*Bird, 1967*); 26 Canada – Somerset Island (*Bird, 1967*); 27 Canada – Marjorie Hills (*Bird, 1967*); 28 Canada – Rae Isthmus (*Bird, 1967*); 29 Alaska (*Reger and Péwé, 1976, 102, Figure 2*)

Assuming the reality of cryoplanation surfaces, as the present writer does, it is logical that fossil forms should be widespread, yet their distribution as distinct from active forms (cf. Figure 7.2) remains to be determined in most regions. Presumably many of those reported in Figures 7.3–7.5 are relic.

Although the development of cryoplanation surfaces is confined to cold climates and is facilitated by the presence of permafrost, terraces also form without permafrost according to Demek (1969a, 56–7). On the other hand more recent work by Reger and Péwé (1976, 107–8) in Alaska convinced them that

The presence of stable cryoplanation terraces is undoubtedly an indication of former shallow permafrost [i.e. shallow permafrost table].... [and] it appears that the terraces indicate a more rigorous climate than that required for the formation of ice-wedge polygons or open-system pingos.

The evidence cited is that

Terraces observed in photographs and examined in the field in the subarctic ... appear to be inactive. Our work in subarctic Alaska indicates there is generally an absence of fresh surfaces on the scarps of cryoplanation terraces, suggesting that nivation is no longer causing scarp retreat; even the sharpest terraces are now inactive.

Photographs taken of similar terraces at an interval of 50 yr indicate no detectable changes in the positions and forms of ascending scarps.

They estimated

... that when cryoplanation terraces were active on the uppermost slopes of Indian Mountain, the mean summer temperature was probably between +2 and +6°C (Reger, 1975).... Although terrace formation is probably not [directly] related to mean annual air temperature, lowland locations with mean summer temperatures between +2 to +6°C have mean annual temperatures of about −12°C. (Reger and Péwé, 1976, 107–8.)

In view of (1) Demek's report that cryoplanation terraces can form in the absence of permafrost, (2) Péwé's (1975, 62) earlier view that some terraces might still be active in Alaska, and (3) Büdel's doubts regarding origin, it would be best to withold judgment as to whether cryoplanation terraces are proof of present or former permafrost – and even if generally indicative of permafrost, whether they imply a more rigorous climate than that required for ice-wedge polygons or open-system pingos. Much remains to be learned about the origin and distribution of cryoplanation terraces but clearly they constitute an important criterion of periglacial conditions.

8 Slopewash

I Introduction: II Grèzes litées

I Introduction

Slopewash, here taken as including sheetflow and rillwork, is less likely than streams to contribute to typical periglacial features, but is probably a significant process in places and has certain periglacial aspects worthy of note. Dylik (1972, 171–2) and French (1976a, 142) emphasized the melting of ice in the active layer as the most important source of moisture for slopewash. Probably an even more important source locally is the melting of long-lasting snowdrifts, but in either case the moisture supply and the presence of an impervious frost or permafrost table is conducive to erosion of vegetation-free slopes. Using sediment traps on such slopes in Spitsbergen, Jahn (1961a, 15–19) measured a sediment removal rate of up to 16.3 g m^{-2} yr^{-1}, not including the majority of grains <0.2 mm in diameter, which bypassed the traps. Most of the traps collected sediments from clayey rubble but the highest rate in one year came from an 11°–23° bedrock slope of sandy shales. On Banks Island in the Canadian Arctic, Lewkowicz, Day, and French (1978) measured an erosion rate of up to 10.04 g m^{-2} yr^{-1} on a 5° slope of silty colluvium that had a 60–80 per cent vegetation cover.

French (1976a, 142) noted that some Soviet investigators regard slopewash in conjunction with thawing as a major factor in flattening of relief but he did not believe this view had been adequately demonstrated. Little of the pertinent literature has been translated. The importance of slopewash in eluviating fines in the active layer and thereby influencing gelifluction is discussed in connection with that process in the chapter on Mass-wasting processes and forms. In addition, slopewash and frost wedging are often cited as responsible for grèzes litées as discussed below.

II Grèzes litées

Grèzes litées are bedded slope deposits of angular, usually pebble-size rock chips and interstitial finer material, in which the bedding is manifested by more or less regularly repeated alternation of grain-size characteristics. This may include a change in grain-size limits or, more commonly, in proportion of finer to coarser material (Dylik, 1960; Guillien, 1951; 1964). Grèzes litées (Figure 8.1) are described by Guillien (1951) as occurring on gradients of 7°–45° and consisting of fragments up to 2.5 cm in diameter, which he considered as the upper limit of such deposits (Guillien, 1954, 2250; 1964, 103). Grèzes litées occur on gradients as low as 2°–3° in the Crau area near the mouth of the Rhône in France as demonstrated to the present writer by E. Bonifay in 1978. Malaurie (*in* Malaurie and Guillien, 1953, 712) reported grèzes litées as having an alternation of granules and larger fragments 3–5 cm in diameter. Fragments of cobble size have also been reported.

S. A. Harris (1975) described alternating layers of slide-rock and loess, which he included with stratified scree, or éboulis ordonnés (Tricàrt, 1956b), terms that have been regarded as equivalent to grèzes litées (Brosche, 1978, 89; French, 1976a, 234; Tricart, 1967, 239). The opportunity for confusion is rife inasmuch as the term éboulis ordonnés commonly denotes sliderock (cf. Guillien, 1951, 155), whereas some grèzes litées may have little if any direct relation to such deposits (cf. Dylik, 1960, 37). Watson (1965, 16) intimated it would be desirable to restrict the term grèzes litées to deposits characterized by fragments 2.5 cm or less in diameter as described by Guillien, and to use éboulis ordonnés for deposits with larger clasts. Clearly a sharpening of terminology is in order, especially since deposits of disparate origin may be involved. Pending further

8.1 Grèzes litées near Cemmaes, Wales, Great Britain. Scale given by 16-cm rule

investigations, it would seem best to restrict the term éboulis ordonnés to deposits directly related to talus.

The origin of grèzes litées has been ascribed to frost wedging, sheetflow, and rillwork, in association with perennial snowdrifts (perhaps beneath them) on vegetation-free slopes (Guillien, 1951, 159–61). An association with niveo-eolian deposits (discussed in the chapter on Wind action) was emphasized by Jahn (1972b, 98). Additionally it has been suggested that an alternation of processes may be involved – creep, sheetflow, rillwork, and gelifluction (after Bout, 1953), and/or differential creep (after Baulig)

(Malaurie *in* Malaurie and Guillien, 1953, 712–13).

A primary role in transportation and deposition has also been assigned to a combination of gelifluction and of meltwater derived mainly from thawing of frozen ground rather than snow (Journaux, 1976b). In general the importance of sheetwash is widely accepted (cf. Boardman, 1978, 32–3). Contemporary deposits have been recorded from southern Banks Island, Arctic Canada, by French (1976a, 235); from West Greenland by Malaurie, and from Spitsbergen by Dutkiewicz, (1967, 67–8) and Jahn (1961a, 19–22). Grèzes litées have been described as constituting a subarctic cryonival phenomenon by Guillien, and laboratory experiments by Guillien and Lautridou (1970; 1974) indicate that frost wedging of certain limestones can produce the same kind of material that constitutes the Pleistocene grèzes litées of the Charente region of France. These investigators reaffirmed the belief that the sorting and beading show that meltwater from snowdrifts was the essential agent in depositing the coarser material and eluviating the finer, and they regarded grèzes litées as indicative of nivation. Accordingly grèzes litées (along with such uncertain evidence as 'nivation relief' and asymmetrical valleys) have been cited as evidence that snow rather than frost action was the main Pleistocene periglacial influence in western Europe (Gullentops, 1977). This is an interesting possibility but the exact nature and origin of grèzes litées do not yet appear to be firmly established, and more observations are needed of such deposits in course of formation before far-reaching conclusions can be safely inferred from them. There is doubt as to whether daily or seasonal events or both are responsible for the bedding, and whether or not high-latitude occurrences are really exceptional as suggested by French (1976a, 235). Even though grèzes litées are common in periglacial environments and may provide a useful characteristic of them (cf. Tricart, 1967, 242–3), grèzes litées are still too poorly known to be considered diagnostic.

9 Fluvial action

I Introduction: II Icings: III Break-up phenomena: IV Flat-floored valleys: V Asymmetric valleys: *1 General; 2 Origin.* **VI Dry valleys and dells:** *1 General; 2 Origin*

I Introduction

There is a tendency to minimize the role of fluvial action in periglacial environments but in many places it is highly significant, particularly during snowmelt (cf. M. Church, 1972; French, 1976a, 167–83; Jenness, 1952, 238–42; McCann and Cogley, 1973; McCann, Howarth, and Cogley, 1972; Wilkinson and Bunting, 1975). The study by Church is particularly detailed. Compared with its effects in other environments, fluvial action differs mainly in the degree to which it is influenced by frost action. Icings and thick river ice in winter, break-up phenomena in the spring, frost wedging and gelifluction on adjacent slopes, and permafrost – all contribute to fluvially related features typical of a periglacial environment. It should be noted that permafrost *per se* does not necessarily advance or retard stream erosion compared with nonpermafrost regions; much depends on more local factors (K. M. Scott, 1978).

II Icings

Icings were briefly discussed in the chapter on Frozen ground. A wide-ranging series of articles on icings in the USSR has been authored by Alekseyev (1978), Alekseyev *et al.* (1969; 1973), and Anisimova *et al.* (1973a; 1973b; 1978); an annotated bibliography and a comprehensive discussion of icings in general have been presented by Carey (1970; 1973).

Icings are usually formed by (1) groundwater coming to the surface in natural seepages or springs (Figure 3.19) or in artificial cuts that intersect the groundwater table (Carey, 1973; Eager and Pryor, 1945), or by (2) a shallow river breaking through

its ice cover during freeze-up because of hydrostatic pressure developed by continued flow of water beneath the ice as freezing approaches the river bed (Figures 3.20–3.21). Icings are favoured by underlying permafrost, and where freezing is intense icings formed by seepages are most common at the toe of south-facing slopes (Kachurin, 1959, 396; 1964, 27).

Except for perennial springs, groundwater icings are relatively small, usually less than 0.5 km² in area and 1 m thick (French, 1976a, 68), and river icings are seldom more than several hectares ('a few acres') in area but are reported to have areas as large as 21 850 hectares (5400 acres) in Siberia (Ferrians, Kachadoorian, and Greene, 1969, 34–6). The largest icing in Yakutia, in the Moma River basin, has been variously reported as having an area of 76 to 112 km², with the total area of 82 large icings in the same basin being 428 km² (Nekrasov and Gordeyev, 1973, 38). In Alaska, floodplain icings can be up to 5 km wide and as much as 50 km long (Mayo, 1970, 310), and thicknesses up to 10 m are cited in the above accounts. The volume of icings in the Sagavanirktok River basin has been conservatively estimated at 1.23×10^8 m³ (J. R. Williams and van Everdingen, 1973, 439). Shumskiy (1964b, 194) cited accumulations up to 27 km long, 10 m thick, and containing up to 5×10^8 m³ of ice.

Icing mounds developed by hydrostatic pressure in association with river icings occur in many regions. They form not only isolated mounds, some with sediment, but also winding ridges indicative of continued water flow beneath the ice (Froehlich and Słupik, 1978). Icing mounds have been described as having vertically oriented 'candles' up to 50 cm long, and as capable of picking up considerable debris (including large cobbles) which becomes subject to ice rafting (M. Church, 1972, 101, 103–4). According to Czudek and Demek (1972, 18–19),

9.1 Fluvial ice rafting. Mass-wasting of rock debris on to river ice, Mesters Vig, Northeast Greenland

icings can erode directly by exerting pressure against valley sides and undermining them, as well as indirectly by forcing a stream against the valley side with similar results. Icings often lead to flooding as well as erosion and '... may cause problems ... when they form on or near the [trans-Alaska] pipeline, roads, and other pipeline facilities' (Sloan, Zenone, and Mayo, 1976, 1).

III Break-up phenomena

Ice break-up along rivers in the spring is usually accompanied by floods, especially as the result of downstream ice jams in rivers flowing from warmer towards colder environments. However, the downstream progress of break-up, although common, does not always pertain because of the influence of tributaries and other factors (Mackay, 1963c). This downstream flow of ice under flood conditions can cause both erosion and deposition along banks at levels well above the normal summer stage. The strip affected, called a Bečevnik in Siberia – a term adopted by Hamelin (1969) – tends to be bench-like and may show a number of micro features that are formed under such conditions. These features include ice-rafted stones, striated stones, ice-shove indentations and ridges, stone pavements, and disrupted vegetation, among others. A detailed list, based on observations in Siberia, has been presented by Hamelin (1969), who coined the term glaciel to refer collectively to these and other phenomena associated with drift ice (cf. Introduction in chapter on Lacustrine and marine action). Striking boulder pavements occur in places along the Mackenzie River in northern Canada, as do fluvial ice-shove ridges transverse to the channel (Mackay and MacKay, 1977). Channel-parallel ice-shove ridges as high as 5–6 m above normal river level have been reported from Grande Rivière in subarctic Quebec Province, Canada (Dionne, 1976a, 137–9, 145, Figures 22–25).

Ice rafting is the transport of rock debris or other material by floating ice, whether in streams, lakes, or the sea. Ice in streams can pick up bottom material (cf. Washburn, 1947, 83–4), or carry material slumped on to its surface from the banks (Figure 9.1). Following break-up of the ice, the debris may be transported far downstream before being deposited. If such erratic debris can be established as ice-rafted, it is clear evidence of a climate sufficiently cold to form ice farther upstream.

Rafting can also occur in other ways, as in tree roots (Figure 10.5), and even if ice rafting is established, river ice occurs so far south in temperate regions, and the distance involved in fluvial ice rafting may be so great, that the significance of the evidence is minimized. Moreover, many of the features listed by Hamelin (1969), such as ice-shove indentations and ridges, are subject to destruction by any subsequent ice-free floods and are unlikely to be preserved as a record of former periglacial conditions.

Many of the features are subject to alternative explanations. Striated stones, for instance, might be derived from a local deposit of glacial or landslide origin, or might have been rafted far downstream from another environment. It must be concluded that however striking the results of the break-up of rivers, little diagnostic evidence is likely to remain long enough to be of value in environmental reconstructions. The whale-backed slushflow fans of mountain streams, discussed in the chapter on Mass-wasting processes and forms, may be an exception.

influences valley development as in Iceland (Schunke, 1975a, 156, 229, 234). The runoff can be voluminous, then diminish to a braided pattern and finally cease altogether. Snow distribution by drifting and otherwise can not only determine the volume of runoff but also soil saturation and thus places where gelifluction can modify the valley sides and contribute debris to the valley floor (Schunke, 1975a, 156–7). In places the predominant bed material is derived from valley sides rather than from upstream and commonly it is very stony. As noted below, gelifluction can strongly influence valley asymmetry.

Flat-floored valleys of the kind described are often regarded as characteristically periglacial but Mortensen (1930, 153–6) and Poser (1936, 72, 83) pointed out that they are not universally so, an important consideration being that concentrated periods of runoff are not uniquely periglacial. According to French (1976a, 173–5), a dominance of bedload transport and therefore braiding is an important explanatory factor, since he regarded such transport as relatively minor in nonperiglacial environments. Certainly permafrost favours development of flat-floored valleys by providing a near-surface limit to downward percolation of water and thereby promoting runoff. Yet, however typical of permafrost environments, such valleys are not necessarily diagnostic of them. Moreover, in Schunke's (1975a, 157–8, 188) opinion, the shallow depth of postglacial weathering in places and the postglacial age of some flat-floored valleys cast doubt on Büdel's (1969) view, noted previously under Frost wedging in the chapter on General frost-action processes, that permafrost-related break up of the underlying bedrock is an important factor in the origin of such valleys.

IV Flat-floored valleys

Distinctive broad, flat-floored valleys (Sohlentäler in German) are common in periglacial environments. Unlike many floodplains, the valley floor tends to lack a pronounced channel, whether the valley is eroded into bedrock or unconsolidated debris. Valley size is highly variable. In Iceland, depths generally range from 5 to 20 m and widths from 5 to 100 m or more; the average gradient ranges from about 2.3° to 6.3° (4–11%) (Schunke, 1975a, 154).

Runoff is usually meltwater from snow rather than glaciers and is typically episodic, which critically

V Asymmetric valleys

1 General

Asymmetric valleys are valleys that have one side steeper than the other (Figure 9.2). In places the asymmetry can be ascribed to the influence of the periglacial environment on fluvial action.

Asymmetric valleys in the Northern Hemisphere have been variously described as having the steepest slopes facing north or east, or west or southwest (Poser, 1947a, 12–13), north or northeast (Malaurie, 1952; Malaurie and Guillien, 1953), south or south-

9.2 Asymmetric dry valley of dell type, Grocholice, Poland. Note steeper slopes at left than at right of older valley, incised by younger valley

west (Troll, 1947, 171), and northwest, among other directions (Büdel, 1944, 503; 1953, 255). According to Poser and Müller (1951, 26–8), climatically determined asymmetric valleys are exceptional in polar latitudes because temperature differences due to exposure are less marked and less heat is received than farther south, but French (1976a, 178–9) accepted steep north-facing slopes as normal for high northern latitudes up to about lat. 70°N. There is a considerable literature on asymmetric valleys, particularly in Europe; Karrasch's (1970) monographic study is especially comprehensive. Kennedy (1976, 180–201) has briefly reviewed the 'vexed question' of valley asymmetry with emphasis on North American examples.

2   Origin

Periglacial explanations of asymmetry take several forms. For places where the steepest slopes face south or southwest in a northern environment, it has been argued that maximum insolation and thawing are concentrated on these slopes, thereby allowing maximum stream erosion at the slope base (Poser, 1947a, 13; cf. H. T. U. Smith, 1949a, 1503). Greater gelifluction on one slope may force a stream over to the opposite slope, which is then undercut (Büdel, 1953, 255). This last explanation involving greater gelifluction on south-facing slopes has been invoked to explain steep north-facing slopes in Alaska (Ogotoruk Creek – Currey, 1964, 95),[1] Northeast Greenland (Poser, 1936, 87 – who stressed the asymmetrical effect of local snowdrifts), central United States (Wayne, 1967, 401), and Siberia (Gravis, 1969, 240–1), among others, and has been cited as the most probable explanation of such asymmetry (French, 1976a, 178–9). Greater gelifluction on one slope and consequent shifting of a stream and undercutting of the opposite slope has also been invoked to explain steep westerly facing slopes in western Czechoslovakia (Czudek, 1973a) and steep southwest-facing slopes on Banks Island in the Canadian Arctic (French, 1971b). In another area of Banks Island, steep north-facing slopes predominate (Day and Gale, 1976, 179). For valleys lacking an eroding stream and having the steepest slopes facing north or east, it has been suggested that greater gelifluction on the opposite slopes has reduced their gradient (Malaurie, 1952; Poser,

1947a, 13). Reduction of gradient by periglacial processes regardless of fluvial undercutting has been advanced as the main explanation for gentle slopes facing east, northeast, southwest, and north, compared with steeper westerly facing slopes in 70 per cent of the cases of valley asymmetry in the northern part of the Moravská brána (Moravian Gate) in Czechoslovakia (Czudek, 1973b).

Currey (1964, 92–4) discussed eight ways in which periglacial processes could produce asymmetric valleys, and there are probably others. A number of variables exist, some working in opposition to each other. For instance, it can not always be assumed that the depth of thawing on north-facing slopes is less than on south-facing slopes receiving more insolation, since the vegetation on the latter may produce an overriding insulating effect (L. W. Price, 1971, 645–7). With respect to gelifluction, the distribution of snowdrifts and hence wind direction and topography are important variables influencing its location and effectiveness (Büdel, 1944, 503–4; 1953, 255; French, 1971b, 727–9; 1972a; Semmel, 1969, 56; Troll, 1947, 171; Washburn, 1967, 93–8). In places gelifluction may be of minor significance. Study of opposite valley asymmetries in a small area in the vicinity of Inuvik in the Canadian Arctic convinced Kennedy and Melton (1972) that various contemporary processes can interact in different ways to produce valley asymmetry and that there is no simple periglacial style of asymmetry. Moreover, in addition to periglacial processes there are various other causes of asymmetry, especially lithology and structure, that must be eliminated before a periglacial origin can be accepted. Considering the number of variables and different explanations, and the possibility of complex origin, the interpretation of asymmetric valleys invites extreme caution.

The difficulty is compounded with fossil forms because of the various complicating effects of climatic change, depending on slope orientation (French, 1976a, 219). Yet, in western Europe there appear to be a surprising number of asymmetric slopes with the steepest slope facing west or southwest, even though the explanations vary (Table 9.1). In discussing this, French (1976a, 253–6) indicated that in western Europe the steepest slope is the warmer, whereas at high latitudes it is the colder regardless of orientation. He presumed that Pleistocene freeze-thaw processes had been most active on

[1] K. R. Everett (1966, 212–13), referring to steep southeast-facing slopes in the same area, suggested that structure and lithology were responsible.

Table 9.1. Examples of slope asymmetry attributed to Pleistocene periglacial conditions in Europe (*after French, 1976a, 254-5, Table 12.1*)

Location	Reference	Orientation of steeper slope	Processes involved[1]					Mechanism involved[1]	
			(1) Differential insolation and freeze-thaw	(2) Differential solifluction	(3) Wind and snow	(4) Wind and loess	(5) Lateral stream erosion	Decline of N and E facing slope	Steepening of S and W facing slope
United Kingdom Chiltern Hills	Ollier and Thomasson (1957)	W/SW	X	x			x		x
Hertfordshire	Thomasson (1961)	W	X	x			x		x
Chalk, southern England	French (1972b; 1973)	W/SW	X	x	x		x		x
France Gascony	Taillefer (1944)	W		x	X		x	x	x
Gascony	Faucher (1931)	W				X	x		x
North France	Gloriod and Tricart (1952)	W	x	X	x		x	x	x
Netherlands Haspengouw	Geukens (1947)	W	X	x	x		x	x	x
Veluwe	Edelman and Maarleveld (1949)	W/SW		x	X		x	x	x
Belgium Hesbaye	Grimberbieux (1955)	W	X		x		x		x
Ardennes	Alexandre (1958)	W		X			x	x	x
Germany Erzegebirge	Lösche (1930)	W	x				X		x
South Germany	Büdel (1944; 1953)	W/SW		x	x	X	x	x	x
Muschelkalk	Helbig (1965)	W		X			x		x
Czechoslovakia Bohemia	Czudek (1964)			X				x	
Poland Łódź plateau	Klatkowa (1965)	W	X	x	x		x	x	x

[1] X – dominant process; x – secondary process.

the 'warmer' middle-latitude slopes, thereby causing their steepness, but that predominance of gelifluction on the 'warmer' high-latitude slopes had displaced drainage to the opposite slopes, which became undercut. Additional data are required to confirm these conclusions.

Granted that valley asymmetry is probably favoured by a periglacial environment, it is by no means clear that climatically-determined asymmetry is necessarily restricted to it. Kennedy's (1976, 180–201) review suggests a wider distribution.

In any event, as also emphasized by others (Black, 1969b, 231; M. A. Carson and Kirkby, 1972, 327; French, 1976a, 256), the usefulness of asymmetric valleys as indicators of former periglacial conditions (to say nothing of permafrost) is very limited at present. However intriguing, far-reaching inferences based on such evidence (cf. Crampton, 1977b) invite extreme caution.

VI Dry valleys and dells

1 General

Dry valleys as the name indicates are valleys now devoid of streams (Figure 9.2). Dells[2] are small shallow dry valleys commonly having a gentle, broadly U-shaped cross profile but steep-sided valleys have also been included (French, 1976a, 266–7).

Valleys believed to be zonal prototypes have been discussed by Poser (1936) among others. Büdel (1948, 39, 43–4) described them as being particularly common in the 'tundra zone', and contemporary examples of a dell type, associated with thermokarst development and initiated by thawing ice wedges and mass-wasting, have been described from Siberia (Czudek and Demek, 1973b, 58–60). The alluvium in dry valleys of periglacial origin tends to be predominantly angular rubble (Büdel, 1953, 255).

Most descriptions refer to valleys in nonperiglacial regions today, and these valleys have been interpreted as disequilibrium features in Britain and Europe (Bull, 1940; Geike, 1894, 395–6; Kessler, 1925, 186–91; Clement Reid, 1887). Among the best evidence that some dry valleys are really fossil and formed in a former environment is the presence of dunes and other eolian deposits that are datable and eliminate any possibility that contemporary floods, however rare, are responsible for the valleys rather than merely for spotty erosional effects within them (J. Hagedorn, 1964, 104, 119–20).

2 Origin

Dry valleys have been regarded as due to linear erosion by streams and/or mass-wasting, (accompanied by an abundant supply of debris from the mass-wasting) in areas where subsurface drainage was inhibited by permafrost or seasonally frozen ground (Büdel, 1953, 255; Clement Reid, 1887, 369–71). The present lack of streams is usually attributed to thawing of the frozen ground and consequent infiltration of the surface water to depth. Dry valleys, including dells, have been frequently cited as evidence of former periglacial conditions, and many discussions of periglacial valley asymmetry refer to dry valleys (cf. Klatkowa, 1965).

The dangers inherent in interpreting dry valleys as indicative of a single climatic zone have been very perceptively evaluated by Mortensen (1930, 153–6). As in the case of asymmetric valleys, alternative possibilities must be eliminated before dry valleys can be accepted as indicators of a former periglacial environment. Other possible explanations include solution (in limestone areas), beheading of streams, and, especially, former greater precipitation, which Dury (1965, C15–C40) regarded as the primary cause. Some dry valleys occur where the underlying beds are highly impermeable today and would promote surface stream flow given adequate precipitation, an illustration being the Warwickshire Itchen in England (Dury, 1970, 266).

---

[2] The widely used German term Dellen (sing. Delle), introduced by Schmitthenner (1923, 37) lacks an exact counterpart in English (Fischer and Elliott, 1950, 10). Although use of dell in this context has been criticized with some justification as being misleading (Black, 1974b, 669), it is internationally recognized (Demek, 1978, 148; Embleton and King, 1975, 16; French, 1976a, 266–72; Jahn, 1975, 172; Katasonova, 1963; Klatkowa, 1965, 9, 11, 127–8), and is continued here.

10 Lacustrine and marine action

I Introduction

Periglacial lacustrine and marine action are similar in some ways, whether associated with lakes or the sea. Ice rafting and ice shove,[1] for instance, are similar in both environments.

Most periglacial aspects of lacustrine and marine action result from the effects of drift ice. The term glaciel, defined as '*Terme générique s'appliquant à tout ce que se rapporte aux glaces flottantes en hydrologie, morphologie, géographie humaine ...*' (Hamelin, 1959, 42; 1961, 201), has been applied to the assemblage of phenomena, whether fluvial, lacustrine, or marine (Hamelin, 1976). The study of these phenomena has been recently emphasized by Dionne (1968*b*; 1970*a*; 1972; 1973*a*; 1976*b*; 1978*a*) and Hamelin (1969), including preparation of an annotated bibliography (Dionne, 1974*b*), and the proceedings of the First International Symposium on the Geological Action of Drift Ice (Dionne, 1976*c*).

II Lacustrine action

1 General

There are only a few geomorphic processes and features that characterize some lakes as periglacial. They include (1) development of thaw lakes as described in the chapter on Thermokarst, (2) formation of ice-shove ridges, and (3) deposition of varved sediments. Each of these characteristics may be difficult to distinguish from similar features of non-periglacial origin.

2 Varves

a General 'A varve is a pair of laminae deposited during the cycle of the year' (Longwell, Flint, and Sanders, 1969, 380, 661).

There are various kinds of varves. The cold-climate variety, known as glacial varves, consists of a sandy to silty lamina overlain by a finer-grained, silty to clayey lamina, the pair forming a single varve; the transition to the next varve is very abrupt compared to the usual graded bedding within a varve. Because of the difference in grain size, the fine-grained lamina tends to be much darker than the coarser lamina (Figure 10.1).

b Origin Cold-climate varves commonly reflect the annual freeze-thaw cycle of lakes receiving abundant sediment from glacial streams. After the ice of streams and lakes breaks up, sediment is poured into the lakes. The coarser material settles first to form the coarse 'summer' lamina, but the finest is kept in suspension until freezing stops the influx of sediment and creates the quiet-water conditions that permit the suspended material to settle out and form the fine-grained 'winter' lamina.

Unfortunately the recognition of such varves is not simple, especially with respect to proof of annual character. The term varve has been loosely and improperly used to designate laminated silts for which the annual origin is not proved. Even for laminae pairs whose glacial origin is demonstrable, some pairs may record variations in meltwater flow and episodes of storm and quiet during the summer rather than reflect the full annual cycle (Sigurd Hansen, 1940, 418, 422, 468–75).

---

[1] The term ice shove rather than ice push is preferred here because the latter is commonly applied to glacial features.

10.1 Cold-climate varves near Uppsala, Sweden. Scale given by 15-cm pencil. The varves (at least 16 visible) were deposited in the Baltic Ice Lake. Photo by Richard F. Flint

c Significance Because varves vary considerably in thickness from year to year and form distinctive sequences, varves lend themselves to correlation between a series of former lake basins in a given region. Such correlations are the basis of a Late-Glacial and post-Glacial chronology in Sweden. Established by DeGeer (1912a; 1940) and subsequent workers, and checked with radiocarbon dates in places, this varve chronology as proposed spans almost 17 000 years dating back from AD 1900, but is commonly accepted for only about the last 12 000 years (Flint, 1971, 403–5). A varve chronology has also been attempted for parts of North America (New England and eastern Canada) but with less success because of gaps in the record (Antevs, 1925).

Regardless of their chronologic significance, varves of the kind described are cold-climate features. With the exception of varves deposited in supraglacial lakes, they are deposited in ice-free sites and are therefore, strictly speaking, periglacial phenomena. Moreover, some varves in cold climates

may well develop without glaciers being in the immediate vicinity.

Combined with their chronologic implications, cold-climate varves provide information as to how long these climatic conditions persisted in a given locality, and by correlation with deposits in adjacent basins, the spatial as well as temporal march of the conditions may be ascertained. Unfortunately, aside from proving appreciable periods of freezing and thawing, cold-climate varves do not in themselves provide much quantitative information on temperature ranges or on precipitation.

3 Ice rafting

Ice rafting is most common in lakes that abut against glaciers. Icebergs calving from a glacier carry rock debris of various sizes to places where otherwise only fine sediments are deposited. Such ice-rafted material, often consisting of isolated stones as well

10.2 Lacustrine ice-shove ridge at base of Mount Pelly, Victoria Island, Northwest Territories, Canada (*cf. Washburn, 1947, Figure 2, Plate 20*)

as pockets of nonsorted debris, is a useful criterion in identifying varves or laminated silts of glacial origin. Rafting by break-up of icefoot complexes is another type of ice rafting (Evenson, 1973). If ice rafting is clearly established as opposed to other processes, the resulting deposit is in itself good evidence of a periglacial environment.

4   Ice shove and ice-shove ridges

Ice-shove ridges are embankments of rock debris pushed up on the shore of a lake or sea by ice. Such ridges are commonly concentrated on points and form distinctive features (Figure 10.2) that differ from massive beach ridges in being more irregular in plan and profile. Ice-shove ridges 4–5 m high have been reported from Selawik Lake on the Seward Peninsula, Alaska (D. M. Hopkins, personal communication, 1962, *in* J. D. Hume and Schalk, 1976, 110), and up to 7 m high from lakes

in the Somes Bay area of subarctic Canada, although the prevailing heights here were 1–2 m (Dionne, 1978*a*, 20). Where associated with beach ridges, they consist of the same material but, as a result of its rearrangement, are usually less well sorted. Some ice-shove ridges show imbricate structure (Figure 10.3). A detailed description of lacustrine ice-shove ridges and related features has been presented by Dionne and Laverdière (1972).

There are two modes of formation of ice-shove ridges: (1) Ice-shove can result from the grinding action of ice floes driven against the shore by wind drag and wind-generated waves and currents (cf. Tyrrell, 1910; Ward, 1959; Washburn, 1947, 76–7). In places lacustrine ice-shove ridges can be several metres high and extend several metres above water level. (2) Ice-shove ridges can develop by thermal expansion of the ice cover with rise of temperature, following any cracking of the ice and freezing of water in the cracks (cf. I. D. Scott, 1927, 107–13). The ice cover must be rigid enough

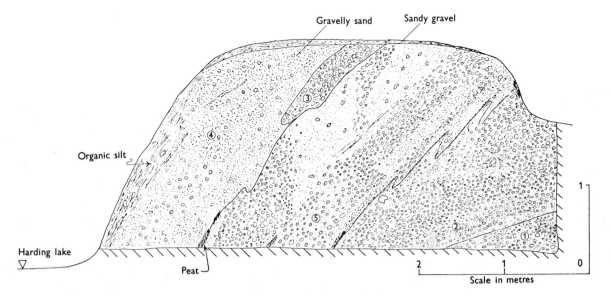

Gravelly sand   Sandy gravel

Organic silt

Harding lake

Peat

Scale in metres

10.3 Cross section of ice-shove ridge showing imbricate structure, Harding Lake, central Alaska (*after Blackwell, 1965, 83, Figure 27*)

to act as a strut in transmitting stress, and the extent of thrusting in a given year is cumulative and depends on the number of times the ice cover cracks, heals, and expands (W. H. Hobbs, 1911, 158). According to Zumberge and Wilson (1953, 374, 381), a rise in air temperature of about $0.5°$ ($1°F$) h^{-1} for twelve hours and ice at least 20 cm (8 in) thick are sufficient to cause ice thrusting. Pessl (1969) reported that a rise in ice temperature of about $1° h^{-1}$ for six hours was sufficient and that the shoreward movement of ice during a thirty-day period amounted to 1 m at Gardner Lake, Connecticut, in the United States.

Views differ as to whether wind or thermal expansion of ice is the most effective agent. Long ago Gilbert (1890, 71–2; 1908) and (as he noted) others before him argued for thermal expansion; Tyrrell (1910) supported the role of wind, and later workers have continued the discussion. Lawrence Goldthwait (1957) mapped the direction of movement of ice-pushed stones on New Hampshire lake shores and concluded from the direction and from direct observation of the effect of thermal expansion that the latter was the most important. A somewhat similar study at an Ontario lake in Canada led Adams and Mathewson (1976) to favour wind-generated ice shove. Ward (1959) attested to the efficacy of wind in building ridges on a Canadian Arctic lake. These views are not necessarily con-

flicting, since lake depth, location, shape, and size can be controlling factors in places. Tyrrell (1910, 18) made the point that ice can be so firmly frozen to the shore that expansion is inhibited. Any lateral stress created by rise of temperature and expansion of ice formed in thermally induced cracks is taken up by crushing of new ice in the cracks, or by buckling elsewhere, until thawing creates a moat around the shore and wind can become effective in shoreward movement of the ice. Presumably freeze-anchoring of this kind would be most effective where temperatures are low and shores gently shelving. The importance of shape and size was emphasized by I. D. Scott (1927, 113) whose observations led him to conclude that an ice cover exceeding 3 km (2 mi) in shortest diameter is seldom if ever rigid enough to cause significant on-shore expansion, whereas the long fetch of large lakes favours wind-generated ice shove. It would appear that ridges produced by ice expansion are probably most prominent in regions where permafrost is lacking and freeze-anchoring of lake ice would be minimized (cf. Washburn, 1947, 76–7).

As with ice-rafted debris in lakes, lacustrine ice-shove ridges are good evidence of a cold climate but, unlike debris that is rafted by icebergs, ice-shove ridges may have no relation to glaciation. In fact most icebergs calved from glaciers ground too far from shore to produce subaerial ice-shove ridges.

10.4 Marine ice rafting. Iceberg with dirt layers, Wolstenholme Fjord, Northwest Greenland

III Marine action

1 General

Marine action that is characteristic of periglacial environments and is likely to leave an identifiable record includes (1) formation of ice-shove ridges, (2) irregular striation of coastal bedrock, and (3) deposition of glaciomarine drift. Evidence of other characteristic aspects, such as the development of an icefoot (a type of shore-fast ice), is less likely to leave an enduring record, except as it may contribute to marine erosion as discussed below under The strandflat problem.

Drift ice and its thawing in the tidal zone and on the beach can simulate certain patterned-ground forms (Jahn, 1977*a*). Marine transgression by thawing of ground ice and subsidence along shore exposed to marine erosion can give the impression of eustatic rise of sealevel or tectonic subsidence. In the Cape Simpson area of Alaska present transgression rates would be capable of developing a 200-m wide submarine bench in only 7 years (Lewellen, 1973, 133–6).

2 The strandflat problem

Long ago Nansen (1904, 102–30) argued that a coastal platform of predominantly marine erosion, aided by subaerial agencies, is preferentially developed by a combination of frost wedging, shore ice (especially an icefoot), and wave action on a shore subject to glacially controlled, alternating submergence and emergence. This concept was applied primarily to the Norwegian strandflat, a low coastal flat including islands (skerries), but similar strandflats occur along some other coasts in periglacial environments. Ahlmann (1919, 93–105, 143, 196–8), on the other hand, argued that the strandflat is the downwarped margin of a subaerial erosion surface ('the distal base-levelled plain') rather than being developed in the manner advocated by Nansen. In rejecting Ahlmann's explanation of the strandflat

10.5 Rafting by tree roots, Mesters Vig, Northeast Greenland. The stone around which the tree root grew must have been carried from a great distance, possibly Siberia. Scale given by 16-cm rule

and amplifying his own concept, Nansen (1922*b*, 7–54) stressed even more strongly the efficacy of 'shore erosion by frost' while downplaying marine abrasion as such, compared with his earlier views. Others have had still different views, and the subject has generated much discussion (cf. E. Dahl, 1946; H. Holtedahl, 1960; Olaf Holtedahl, 1929, 15–28, 146–48).

That an icefoot can contribute to marine erosion has been observed by Moign (1976). In a subsequent abstract Fairbridge (1977) concluded, on the basis of distribution, age, and deposits, including drift-ice deposits, that the 'so-called wave-cut shore platforms of the middle latitudes' are of periglacial origin and he advocated that 'The classic concept of a wide platform of marine abrasion created without the aid of ice floes should be abandoned.' Detailed evidence will be required to substantiate this view. In the meantime the origin and possible periglacial significance of strandflats must remain problematical.

3   Ice rafting

Ice-rafted sediments may be deposited in the sea by icebergs, floating shelf ice and other drift ice, or even by tree roots (Figures 10.4–10.5). If of glacial origin, such ice-rafted deposits are glaciomarine drift. However, identification of these deposits as glaciomarine drift may be difficult to establish unless they are associated with marine shells *in situ*. Although extensive deposits of glaciomarine drift are periglacial in the sense of being cold-climate phenomena, they are much more closely identified with glacial deposits, and further discussion is omitted. Drift-ice deposits have been comprehensively reviewed by Vanney and Dangeard (1976); they can be highly significant in reconstructing environmental changes (cf. Ruddiman, 1977).

4 Ice shove and ice-shove ridges

Ice-shove ridges on polar coasts are similar to those on lake shores but the marine environment can produce larger ridges. They tend to be located on unprotected low-lying capes exposed to violent currents and drifting sea ice during the majority of the year (Bird, 1967, 217–20) (Figure 10.6). Ridge heights as great as 9 m are reliably reported (Bretz, 1935, 219) but reports of 18 m (60 ft) and more are suspect as noted by Nichols (1953). The presence of any permafrost extending into the intertidal zone tends to limit the amount of beach debris that can be ploughed up to form ice-shove ridges (Owens and McCann, 1970, 412–13). Deep water close to shore favours the shoreward thrust of ice on a gentle beach. Icebergs generally ground before reaching the shore but sea ice may be pushed tens or perhaps even hundreds of metres inland to form massive ridges (cf. Washburn, 1947, 78–80). Leffingwell (1919, 173) reported that ice at Point Barrow, Alaska, reached buildings nearly 60 m (200 ft) inland, and according to James Ross, ice on the north coast of King William Island in Arctic Canada '. . . travelled as much as half a mile beyond the limits of the highest tide-mark' (Ross, 1835, 416). An ice finger on the north coast of Somerset Island reached inland as far as 185 m, leaving a continuous line of gravel ridges, although few were more than 1 m high (R. B. Taylor, 1978, 139, Figure 6; 145, 147). Low 'ice push boulder barricades' formed at or below mean low tide and paralleling the shore for several hundred metres are common on some rock coasts of Melville Island. Frost wedging and wave action during emergence tend to modify these boulder barricades to form gravel beaches (McLaren, 1977, 56, 211–12; 217–23). The ice can striate, gouge, or plane beach deposits, and the ridges or mounds that are formed may be either ice cored or ice free (J. D. Hume and Schalk, 1976; R. B. Taylor and McCann, 1976). The force required to move stones has been investigated by Mansik-kaniemi (1970, 18–23), who concluded that an average force of 80 per cent of a stone's weight was needed for the conditions in the area he studied in Finland.

Unlike lake ice, thermal contraction and expansion of sea ice is probably a minor factor in producing ice-shove ridges. Sea ice is fragmented into floes and is much more mobile and subject to move-ment by winds and currents throughout the year than lake ice that remains solidly frozen during the winter. Nevertheless the general effect of near-shore drift ice is to protect the shore from wave action. The protective effect of shore-fast ice such as an ice-foot or shelf ice is of course absolute in this respect, although an icefoot can contribute to cliff erosion in other ways as discussed under The strandflat problem.

An observation that ice-shove ridges are either more common or less common along a modern beach than along an emerged beach in the same area needs to be interpreted with great care. Possible causes, in addition to climate change, include differences in topography at different levels, exceptional storms that in the absence of ice could wipe out ridges built over a period of many years, and mass-wasting and other processes that tend to lower and level old ridges. However, some ice-shove ridges, especially those made of gravel, can persist for a long time. Well-preserved ice-shove ridges up to 8 m high, constructed of emerged beach deposits and located at an altitude between 110 and 120 m at Skeldal, Northeast Greenland, have a radiocarbon age exceeding 8000 years, judging from the age of near-by shells (Lasca, 1969, 27, 45–8).

5 Other beach effects

Numerous small-scale beach features result from wave action combined with freezing and thawing. During freeze-up, slush is spread over the swash zone where it can freeze and become buried by bedded sand and gravel. Snow accumulations, and ice blocks deposited by storm waves, may be buried in similar fashion. Thawing of the snow and ice masses leads to collapse features, including kettles that can cover 20–50 per cent of a beach and be up to 10 m in diameter and 1 m deep; subsequently the depressions become filled with sediments (Short and Wiseman, 1973). As a result, an irregular, deformed strati-graphy is left that may be preserved as a clue to its periglacial origin.

6 Striated bedrock

Sea ice grinding against a bedrock coast produces striae somewhat similar to glacial striae but commonly much more discontinuous and irregular (Dionne, 1973a). Under certain circumstances they may record the former presence of sea ice in places where it is now absent.

10.6 Ice-shove ridges, Point Barrow, Alaska. The ridges shown were not ice-cored but somewhat similar ridges containing ice cores occurred nearby

11 Wind action

I Introduction: II Loess: III Dunes: IV Niveo-eolian deposits and forms: V Deflation: VI Ventifacts

I Introduction

Wind action is common in periglacial environments and, although equally prominent in many others, it has certain aspects that are of great value in reconstructing past periglacial environments. It is therefore commonly considered along with other more distinctive periglacial processes.

The outstanding aspect of wind action is the transport and deposition of snow, which has important periglacial results. Thus some areas are swept bare of snow, exposing them to intense frost action; others become sites of large drifts that are protective in this respect but provide a source of moisture that can facilitate various processes, especially gelifluction. These and other effects of snow distribution are covered elsewhere, and the following discussion is directed towards wind action on soil and rock.

The prominence of wind action in some periglacial environments is attributed in part to the sparseness of vegetation, especially in broad valleys with braided streams from glaciers, in part to the influence of glaciers in generating katabatic winds. In places erosion and deposition by wind in the vicinity of outwash plains has probably been essentially continuous since deglaciation, as in the Matanuska Valley, Alaska (Trainer, 1961, C22–C32). Descriptions of contemporary periglacial wind action include, among many others, observations in Alaska (Black, 1951; Péwé, 1951; 1955, 720; Trainer, 1961, C4–C16; Tuck, 1938), Arctic Canada (Bird, 1967, 237–41; French, 1972a; Mackay, 1963a, 43–6; Pissart, 1966b; Pissart, Vincent, and Edlund, 1977), Greenland (Belknap, 1941, 235–8; Flint, 1948, 208–10; Fristrup, 1953; W. H. Hobbs, 1931; Nichols, 1969, 80–4; Teichert, 1935; Troelsen, 1952), Iceland (Bout, 1953, 35–52; Bout et al., 1955, 510–15),

Scandinavia (Seppälä, 1974), and the Antarctic (Lindsay, 1973; McCraw, 1967, 399, 404; Nichols, 1971, 314–15; Selby, Rains, and Palmer, 1974). An important early overview was provided by Samuelsson (1927).

There are also numerous places where abrasion and deflation by wind are of minor importance, as on some of the islands of the Canadian Arctic (Pissart, 1966b; St-Onge, 1965, 13; Washburn, 1947, 74–6), locally at least in West Greenland (Boyé, 1950, 112–17) and in the Mesters Vig district of Northeast Greenland. Locally important wind action is generated in a variety of periglacial environments by needle ice disrupting vegetation and exposing the soil to deflation, with the result that the turf is stripped away – a process described as turf exfoliation by Troll (1973). In Iceland periods of intensified wind effects have been attributed, in part, to such disruption of vegetation by increased frost action during cold periods, and, inversely, intensified frost action in places has been caused by wind disrupting vegetation and exposing bare ground (Schunke, 1975a, 208–10).

Applying the term periglacial in the restricted sense to ice-marginal phenomena, H. T. U. Smith (1964, 178) listed the following criteria for the periglacial origin of eolian sands

(1) interfingering or conformable relations with glacial deposits; (2) derivation from source areas which can be best attributed to ice-border conditions or to eustatically lowered sea level; (3) derivation from beaches of ice-dammed lakes; (4) paleontological or paleobotanical data; (5) associated indications of periglacial frost action; or (6) dune form and orientation recording wind direction divergent from that of the present and consistent with that which might occur in a periglacial environment.

In discussing periglacial wind action, Cailleux

(1973*b*, 51–2, 55, 57–9) cited three criteria as uniquely indicative of eolian deposits in periglacial environments: (1) Transport of stones (20–60 mm in diameter) by gliding on icy surfaces. (2) Interstratification of sand and snow. Deposits in which the sand has settled as the result of melting of the snow tend to have undulatory bedding. (3) Eolian sand filling frost cracks of the same age.

Comprehensive studies devoted to former periglacial wind action and deposits include Ivar Högbom's (1923) discussion of European dunes and Cailleux's (1942; 1973*a*) studies of wind effects on Quaternary sand grains. In connection with the latter, the reliability of frosted (lustreless) sand grains as proof of wind abrasion has been shown to be problematical because of comparable, chemically induced frosting (Kuenen and Perdok, 1962). However, Cailleux (1973*b*, 54), while recognizing chemical frosting, argued that the scale can distinguish between them in that chemical frosting can affect all grain sizes, including those <0.2 mm in diameter, whereas wind frosting does not. There are also a number of regional surveys (cf. Dylik, 1969, for Poland, and Demek and Kukla, 1969, for Czechoslovakia) and collected papers (Association Française pour l'Étude du Quaternaire, 1969; Galon, 1969; Schultz and Frye, 1968; Velichko, 1969), but much of the literature is scattered in various studies of Quaternary stratigraphy. The status of research on periglacial eolian phenomena in the United States was reviewed by H. T. U. Smith (1964), who emphasized its elementary character, especially with respect to eolian sands and ventifacts.

II Loess

Defining loess has been a problem because of different views as to its origin. Consequently the American Geological Institute glossary (Gary, McAfee, and Wolf, 1972, 416) defined it on the basis of its physical characteristics only, although recognizing the probable importance of wind. Others make deposition by wind part of the definition. Following the latter, loess is here regarded as 'Wind-deposited silt, commonly accompanied by some clay and some fine sand' (Flint and Skinner, 1977, G.5, cf. 214). For example, a typical loess in Alaska has a grain-size range of 0.005–0.5 mm for 80–90 per cent of its particles (Péwé, 1975, 38). Loess tends to be highly cohesive and to form nearly vertical exposures. Stratification is apparently lacking or poorly developed in many exposures but very careful cleaning of an exposure often shows faint, well-developed beds, and stratification is probably more common than is generally supposed. Particles are angular, and the porosity of loess usually exceeds 50 per cent. Much loess is calcareous but there are large areas as in New Zealand where even fresh loess lacks $CaCO_3$ (Fink, 1976, 221). When completely fresh it is grey but it weathers easily and most exposures are characterized by a buff colour. Loess deposits commonly blanket the landscape, covering both hills and valleys. In Alaska, hilltop loess at Gold Hill near Fairbanks is 61 m thick, and valley-bottom loess in the area is up to 95 m thick but much of it was redeposited from adjacent slopes (Péwé, 1975, 40).

Loess is believed to originate in two principal environments – warm deserts and cold deserts. In both places the absence of vegetation is the main factor allowing the wind to pick up and export the silt, and in both the silt tends to be but little weathered chemically. However, associated fossils and the geographic location and distribution of loess deposits generally serve to distinguish in which type of climate the loess originated; also periglacial loess is described as better sorted (being predominantly in the 0.01–0.05 mm range) than loess from warm deserts (Grahmann, 1932, 10, 12–13, 21–3; cf. Embleton and King, 1975, 182; Flint, 1971, 254–5).

Most loess may be periglacial. According to Troll (1944, 573–4, 674–5; 1958, 23, 96), all loess is derived from frost climates, some of it from glaciofluvial sediments in which glacial rockflour has been concentrated, and possibly some (in the grain-size range of 0.01–0.1 mm) from frost wedging. Smalley (1972, 535), too, advocated a cold-climate origin but assumed glacial grinding as opposed to frost wedging was '... the only effective way of producing loess material in sufficient quantities to form the great deposits of China, North America, and Europe'. Subsequently Smalley and Krinsley (1978) modified this view to include cold-climate weathering as an important source of loessial silt, and then accepted the possibility that some loess is of warm-desert origin.

Laboratory experiments support the view that the lower limit to which wind action can abrade quartz grains is about 0.04–0.05 mm (Kuenen, 1960, 446), whereas much loess is smaller, which would be consistent with a glacial origin. Nevertheless a warm-desert origin is not excluded to the extent pre-existing finer-grained deposits can be reworked by the wind. The fact remains that loess has been observed being derived from alluvial fan deposits in the Sinai Desert (Yaalon, 1969). According to Flint (1971, 265–6) most of the Chinese loess is probably non-glacial, but an increase in grain size of sediments towards areas affected by glaciation and by periglacial influences is leading Chinese investigators to the view that much of the loess is the cold-desert type (Bowler, 1978, 158–9). Clearly the origin of loess can be complex, yet the main issue is not so much the existence of warm-desert versus cold-desert loess but rather the distribution and exact climatic significance of each.

Cold-induced deserts can be either widespread where climate directly inhibits vegetation, or more local where climate favours broad, vegetation-free floodplains, especially the braided outwash plains of glacial streams from which much periglacial loess is commonly derived as has been long recognized (Bryan, 1945; V. A. Obruchev, 1945, 57–8). It is because of this derivation from previously sorted sediments that periglacial loess is believed to be generally the better sorted. Much loess is blown far beyond its source and can be deposited in a quite different environment, usually where there is sufficient grass or tree growth to anchor it. Thus while cold-climate loess is derived from a periglacial environment, it is not necessarily indicative of such an environment in the area of deposition.

III Dunes

Like loess, sand dunes form in various environments characterized by little or no vegetation. They develop along sandy shores and in warm deserts and cold deserts. In the Antarctic dry-valley environment, dunes in Victoria Valley are 61–91 m (200–300 ft) long (Webb and McKelvey, 1959, 127), up to 13 m high, and sufficiently extensive to have been termed a desert erg comprising sand sheets, barchan dunes, transverse dunes, and whaleback dunes or mantles, a characteristic feature being interstratified layers of snow and ice-cemented sand occurring as permafrost (Calkin and Rutford, 1974; Selby, Rains, and Palmer, 1974, 550, 553–5). The size of wind-transported grains in the Antarctic is exceptional. Granules 3 mm in diameter have hit a tent 2 m above its base, and pebbles 19 mm in diameter have been collected from dune-ripple crests in the Victoria Valley. Calculations support H. T. U. Smith's (1966) view that the high density of cold Antarctic winds, compared with warm desert winds, are an important factor in such occurrences (Selby, Rains, and Palmer, 1974, 547–50).

On the Canadian Shield, sand dunes over 24 m (80 ft) high occur in the middle Thelon Basin (Bird, 1951, 27), and dunes up to 30 m (100 ft) or more high cover a large area on the south side of Lake Athabasca (Hermesh, 1972, 67–148; Rowe and Hermesh, 1974). In both areas unusually extensive accumulations of sand, reflecting underlying sandstone and the glacial and postglacial history, are available to strong winds.

Although mobile, dunes are unlike loess in that they come to rest in essentially the same kind of environment in which they originated or in one located very close to it. In this respect they are a more useful environmental indicator than loess. To establish the periglacial nature of dunes requires the elimination of other possibilities, and as in the case of loess this can usually be done by regional considerations. If sand-grain textures typical of glacial abrasion are retained as revealed by scanning electron microscopy, this could be diagnostic evidence (cf. Krinsley and Cavallero, 1970). Periglacial dunes occur in the same kind of areas where periglacial loess originates – broad, vegetation-free floodplains and, especially, glacial outwash plains. They also occur along some cold-climate lake shores and sea coasts but such dunes are probably in a minority.

Dunes are particularly significant because of the information that their morphology, orientation, bedding, and sand-grain characteristics may provide as to periglacial wind strength and direction. However, it should be recognized that the direction of the most effective dune-building winds may differ from that of the prevailing winds. With information from enough regions at hand it may be possible to reconstruct wind conditions marginal to the Pleistocene ice sheets as attempted for Europe (Figure 11.1), and thereby help establish the character of the periglacial climates then prevailing. One trouble with reconstructions such as in Figure 11.1 is that the time period may be so broad as to cover changing conditions. Thus, as kindly pointed out by Gerard Maarleveld (personal communication, 1975), Late-Glacial wind directions during dune building changed from northwest and west to west and southwest from Older Dryas to Younger Dryas time in the Netherlands (Maarleveld, 1960, 55–7) and Poland (Dylikowa, 1969, Table 1, opposite 64). A number of periods of dune building with changing wind directions from the Oldest Dryas into the Holocene are recorded in the Warsaw–Berlin Pradolina – a late glacial meltwater-channel complex (Krajewski, 1977).

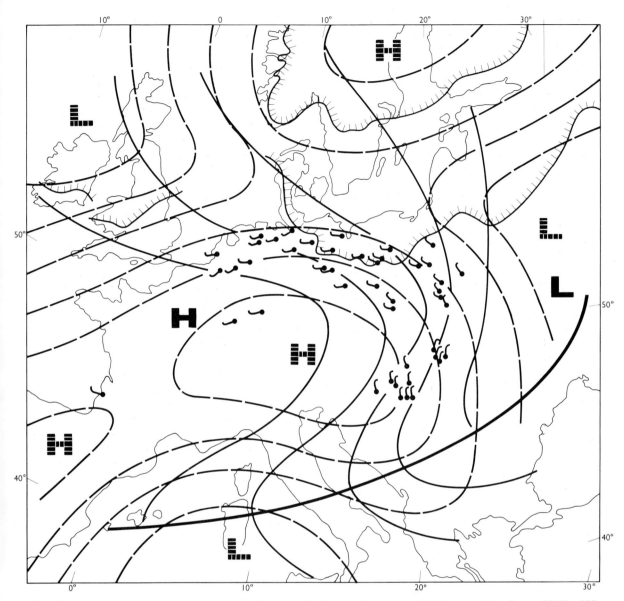

11.1 Summer atmospheric pressure and wind directions in Europe during Late-Glacial time (*after Poser, 1948b, 308, Figure 5; 1950, 81, Figure 1; as reinterpreted by Reiter, 1961, 380, Figure 7.42; 1963, 407, Figure 7.42*). Ice margins at beginning and end of Late-Glacial time, wind directions, and dashed lines after Poser; continuous lines after Reiter

H = Highs
L = Lows

In North America, investigation of sand dunes covering an area of 57 000 km² (22 000 mi²) in Nebraska led H. T. U. Smith (1965) to infer that some of them were built under a former periglacial wind regime, and a similar conclusion was reached by E. P. Henderson, (1959*a*, 45–7) for sand and silt dunes in Alberta, Canada. Peculiar ridges of clay (paha), whose origin has long been debated, occur in Illinois and Iowa. Flemal, Odom, and Vail (1972) concluded that the Illinois paha are loess dunes, formed during deposition of the late-Wisconsin Peoria loess. As such they would be periglacial. Large stabilized dunes occur in Arctic Alaska (Carter and Robinson, 1978).

11.2 Pleistocene ventifact, Warthe moraine, Daszyna, Poland

IV Niveo-eolian deposits and forms

The term niveo-eolian, apparently first suggested by Van Straelen (Rochette and Cailleux, 1971, 36), is defined as designating '... mixed deposits of snow and wind-driven sand or other detritus, and the forms and microforms generated by them' (Cailleux, 1974, 437).

Microforms are the most common and are of considerable variety but are predominantly seasonal features that are rarely if ever preserved from one year to another (Cailleux, 1972; 1974). Larger forms

in polar environments occur as persistent dunes of various kinds, consisting of interstratified sand and snow preserved as permafrost as noted in the discussion of dunes. It is doubtful if the niveo-eolian origin of such dunes could be recognized following disappearance of the permafrost.

Niveo-eolian deposits are common in polar and subpolar environments and are also widespread in temperate highlands as, for instance, in the Sudety Mountains in Poland (Jahn, 1972b). As noted by Cailleux (1972, 377, 406–7), niveo-eolian deposits are not necessarily confined to the Earth and may be represented by Martian polar accumulations of mineral dust, interstratified with snow or ice consisting mainly of H_2O (Briggs *et al.*, 1977, 4144–8; Cutts *et al.*, 1976), with each identifiable stratum probably representing many thousands of years (C. B. Leovy, personal communication, 1978).

V Deflation

In many periglacial environments the principal wind effect is to remove silt and sand. In barren, wind-swept areas this can result in a stone pavement comparable to those of warm deserts (cf. Fristrup, 1953, 57 – photo).

Deflation along stream-cut banks can cause blow-outs and climbing dunes with consequent vegetational changes as along sandy banks of the Meade River in northern Alaska (Peterson and Billings, 1978, 17–20). Where there is frequent deposition of sand and silt on the adjacent tundra surface, the result can be a soil characterized by alternating layers of wind-blown material and organic matter as observed by Ugolini (1966, 14) and the present writer in Northeast Greenland.

Deflation can also be important on tundra surfaces where the vegetation has been disrupted, an important agent here being needle ice (Troll, 1973) as previously noted. In such spots the original surface tends to be undercut and the deflated areas enlarged, leaving the unaffected, vegetation-covered areas standing in relief. Striking examples have been described from Iceland (cf. Schunke, 1975*a*, 148–51).

VI Ventifacts

Ventifacts are stones faceted and polished by wind. Like dunes they can provide much information about periglacial wind conditions. In places like the Antarctic dry valleys, ventifacts present a startling array of sculptured forms rivalling or exceeding those of warm deserts in their degree of development (Lindsay, 1973; Nichols, 1966, 35–6; Selby, 1977). In this connection it may be significant that snow at low temperatures has sufficient hardness to abrade soft rocks (Fristrup, 1953; Teichert, 1939; Troelsen,

1952). Silt-size ice particles with hardness 2–3.5 (Mohs scale) at $-10°$ to $-25°$ have been reported to abrade minerals as hard as 5.5 at only moderate wind velocities (*c.* 6.2–10.3 m s^{-1}; 14–23 mi h^{-1}) (Dietrich, 1977).

The role of wind in the origin of the cavernously weathered rocks (honeycomb forms) in the Antarctic, which are especially prominent in the dry valleys, is not entirely clear. According to Cailleux and Calkin (1963), these forms owe their origin less to wind than to weathering processes, and they are remarkably similar to forms that have been ascribed to weathering elsewhere as in Australia (Bradley, Hutton, and Twidale, 1978). On the other hand, Sekyra (1972, 670–1) concluded that the frequency and size of the forms are directly related to the intensity and duration of wind erosion. In any event deflation is probably an important genetic factor in the Antarctic.

Pleistocene ventifacts of periglacial origin (Figure 11.2), usually associated with deflation horizons, are locally prominent in North America (Mather, Goldthwait, and Thiesmeyer, 1942, 1163–73; Powers, 1936; Sharp, 1949) and elsewhere. They appear to be especially widespread in northern Europe (Johnsson, 1958, 239–40; Paepe and Pissart, 1969; Schönhage, 1969; and many others) but this may reflect, at least in part, the many investigations compared to most other regions. Large boulder ventifacts that have not been subject to movement since deposition are especially valuable indicators of paleo-wind directions. They can show that the present wind regime has probably remained the same for much of Quaternary time in some places (Selby *et al.*, 1973), or has changed markedly in others, and they can be very useful in reconstructing periglacial environments. Small ventifacts are subject to changes of position and are commonly multifaceted. Prominent causes of disturbance can be the wind itself and, in periglacial environments, frost heaving and gelifluction (Cailleux, 1942, 45, 126). If wind disturbance can be discounted, such multifaceted ventifacts can be used as evidence of periglacial conditions, as Sharp (1949, 185, 193) did where his observations showed that the ventifacts were one-to-two thirds buried at the time of cutting and that many had been rotated on a vertical rather than horizontal axis.

12　Thermokarst

I Introduction: II Collapsed pingos: III Thaw slumps: IV Linear and polygonal troughs: V Beaded drainage: VI Thaw lakes: *1 General; 2 Oriented lakes.* **VII Alases**

I　Introduction

As noted in the chapter on Frozen ground, the term thermokarst designates topographic depressions resulting from the thawing of ground ice. There are many kinds of thermokarst, including collapsed pingos, ground-ice mudslumps, linear and polygonal troughs, thaw lakes, and alases. Some thermokarst features are due to climatic change but many are caused by such changes of surface conditions as disturbance of tundra vegetation, shift of stream channels, fire, or other non-climatic equilibrium changes that promote thawing.

A schematic model of the interaction of thermal and denudational balances as they affect a permafrost environment and the presence or absence of thermokarst has been elaborated by Jahn (1972c; 1976a, 126–7, 130–5).

Man-induced changes may cause continued thermokarst development over many years, as at the Sachs Harbour airstrip on Banks Island in Canada's Western Arctic where thawing and subsidence have apparently been active for 10 years or more after the initial disturbance (French, 1975b; 1976a, 130–2). In another area the thickness of the active layer was shown to be 136–393 per cent greater two years after burning than before (Mackay, 1970, 426–8; 428, Table 1), although there were other places in the same area that exhibited less drastic changes (Heginbottom, 1973, 651, Table 1). The opposite effect can apparently also occur where changes in moisture conditions and vegetation lead to a thinner active layer after a fire (Kryuchkov, 1968; 1969; cf. Jerry Brown, Rickard, and Vietor, 1969, 7). Removal of vegetation for placer mining can lower the permafrost table by 4.6 m (15 ft) or more in 1–2 years (O. L. Hughes, 1969, 6). Just the cross-tundra movement of vehicles can cause serious thaw effects (French, 1975a, 463; Kerfoot, 1973; Rickard and Brown, 1974). Nonclimatic thermokarst occurrences are sufficiently common to be a warning against interpreting isolated ice-wedge casts or truncated ice wedges in a present permafrost environment as proof of an earlier warm period, or as proof of the date of a climatic change if they are now in a non-permafrost environment (cf. Danilov, 1972, 35–44; 1974, 36–43).

Given a disturbance that causes removal of a surface insulating layer such as peat, and knowing the original thickness of the active layer and the nature of the terrain and underlying soil, the new depth of thawing in the absence of flowing water can be roughly predicted by use of a diagram (Figure 12.1) developed by Mackay (1970, 430, Figure 13). By comparison with undisturbed adjacent areas, surface disturbance by fire or bulldozing in the Inuvik area of northern Canada appears to cause a marked decrease of ice content in the top 2 m of frozen ground, an increase in the next metre, and little if any change at greater depth (Heginbottom, 1974). Water can greatly accelerate thawing. As pointed out by Czudek and Demek (1973b, 14–16), once thawing of permafrost starts, it tends to be more extensive in wet areas than dry, an important factor being the twofold higher specific heat of water than ice; furthermore for most dry soils the difference is fourfold. Water temperatures averaging 16° in ice-wedge polygon troughs up to 1 m or more deep were measured by Jahn (1972a, 286) in the Mackenzie Delta area during sunny weather when the air temperature was only 10°. In the surface layer of shallower water the temperature was 19°.

Mackay (1970) made a case for differentiating

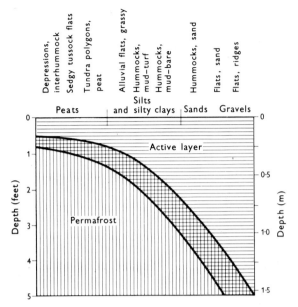

12.1 Approximate range in depth of thaw in common types of periglacial terrain. The diagram can be used to estimate new thaw depth consequent on disturbance of the ground surface (see text) (*after Mackay, 1970, 430, Figure 13*)

between thermokarst subsidence and thermal erosion, the former being without the intervention of flowing water, the latter always with it. Although no such distinction in use of the term thermokarst is made here, it is important to recognize that the evolution of a thaw depression may differ depending on whether or not flowing water is present. The importance of flowing water in thermal erosion along river banks was emphasized by Gill (1972, 128–30) and Walker and Arnborg (1966), the latter stressing the effectiveness of thermal erosion where ice-wedge polygons exist. In both investigations, reported rates of bank retreat were up to about 10 m yr⁻¹. As described by Grigor'yev (1976, 80; cf. 1966, 77) along the arctic coast of Yakutia facing the Laptev Sea

As a result of the abrasional and thermal effect of the sea, at the base of the shore slopes of the remnants of the ancient alluvial plain thermoabrasional niches with an extent of 50–100 m[1] are created. Locally the height of these niches attains 3 m and their penetration into the seashore attains up to 20 m (Fig. 22).

Erosion rates of 5–20 m yr⁻¹ and even 100 m yr⁻¹ (citing Khmyznikov, 1937) were given by Grigor'yev (1966, 79; 1976, 83). Rates up to 10 m yr⁻¹

have been reported in the vicinity of Point Barrow, Alaska (Lewellen, 1970, 2). The role of thawing in retreat of lake and sea cliffs has been briefly reviewed by F. E. Are (1972).

In discussing these and other varieties of thermokarst development, Czudek and Demek (1973*b*) recognized the following forms, most of them characterized by effect of some specific erosion process that accelerates the thawing of ground ice. (1) Thermoabrasion forms related to wave action along lake and sea cliffs; (2) thermoerosion forms involving fluvial action on thawing river banks; (3) thermosuffosion forms developed by subsurface erosion of soil by percolating water (piping) from thawing ground ice; (4) thermoplanation forms consisting of (a) broad benches resulting from continued lateral development of thermoabrasion or thermoerosion forms, often as the result of merging ground-ice slumps (discussed later), or (b) broad depressions initiated by downward rather than lateral thawing (although often extended by the latter as in alases, discussed later); (5) certain dells. A somewhat different scheme is followed below, the emphasis being on some especially characteristic forms.

Much thermokarst research has been carried out in the Soviet Union, and Kachurin's (1961) comprehensive review of the phenomena there is one of the few monographs on thermokarst (cf. Kachurin, 1962, for brief summary). Work is also progressing elsewhere, especially in Canada. According to French (1974*b*, 785), thermokarst activity in Canada's Western Arctic is probably primarily the result of local climatic factors, judging from the general constancy of climate since 4000 ¹⁴C years BP. On the other hand, dating of bottom deposits in basins in the Mackenzie Delta area suggests extensive development of thermokarst due to postglacial warming in the period 8000–11 500 ¹⁴C years BP (Rampton and Bouchard, 1975, 14).

II Collapsed pingos

Collapsed pingos may form shallow, usually round or oval depressions, many of which have a slightly raised rim. However, whether such a rim is an essential characteristic whose absence militates against a pingo origin (Pissart, 1958, 80) is not clear. As described in the section on Pingos in the chapter on Some periglacial forms, partial collapse of a pingo does not necessarily imply environmental

---

[1] 50–100 m is from the original text (Grigor'yev, 1966); the translation (Grigor'yev, 1976) cited 500–100 m.

change, since pingo growth frequently causes dilation cracking of the material above the ice core and, consequently, exposure of the core to thawing.

More or less completely collapsed pingos are widely reported from Pleistocene periglacial regions in Europe and Asia (Frenzel, 1959, 1021, Figure 13; 1967, 29, Figure 17; Maarleveld, 1965; Mullenders and Gullentops, 1969; Pissart, 1956; 1965; 1974*b*; Svensson, 1969; Wiegand, 1965). Among other places they probably also occur in North America but the reports are less definitive (Flemal, Hinkley, and Hesler, 1973; Wayne, 1967, 402).

Depressions caused by collapse of pingos may be very difficult to distinguish from other thermokarst features or from some glacial kettles.

III   Thaw slumps

Thaw slumping[2] can form semicircular hollows opening downslope that are initiated by exposure and thawing of ground ice, and are excavated by mudflow or gelifluction of the resulting saturated debris, commonly leaving a headwall of ground ice. Headwalls observed by French and Egginton (1973, 205-7) were rarely over 2.5 m high and retreated an estimated 7-10 m each summer; others described by Kerfoot and Mackay (1972, 120-3, Figure 3, Table 1) were up to about 8 m high and retreated a mean of 3.5 m yr^{-1} and a mean maximum of 7.4 m yr^{-1}. Such features coalesce, and most probably stabilize with time.

Thaw slumping is widespread in the Arctic, and French regarded 'ground ice slumping' as probably one of the most rapid erosion processes in the high Arctic. Thaw slumps in the Canadian Arctic have been described from Banks Island (French, 1973, 205-7; 1974*b*, 789-91; Washburn, 1947, 85-6); Ellef Ringnes Island (Lamonthe and St-Onge, 1961), Melville Island (Heginbottom, 1978), and the Mackenzie Delta area (Kerfoot and Mackay, 1972, 120-3; Mackay, 1966*b*, 67-78). Since ground ice includes ice wedges, the large-scale slumps associated in places with thawing ice wedges, as in Siberia where the resulting features have been called thermocirques (Czudek and Demek, 1970*b*, 106-8), should also be regarded as thaw slumps.

IV   Linear and polygonal troughs

Where ice wedges thaw they commonly leave troughs, which may carry water that can accelerate thawing as noted earlier. Where ice-wedge polygons are degrading, whatever the reason, pronounced inter-trough mounds can be left standing in relief. Some high-centre ice-wedge polygons are of this origin. In Siberia where ice wedges have melted out along river banks, the inter-trough mounds become accentuated, forming conical hills known as baydjarakhs (Czudek and Demek, 1970*b*, 106-7) whose cross-sections parallel to the river bank exhibit striking triangular facets (Figure 12.2). Impressive examples of degrading ice-wedge polygons and baydjarakhs have been described from Banks Island in the Canadian Arctic (French, 1974*b*, 791-3), and in many cases, as in the clearing of farmers' fields in the Fairbanks area, Alaska (Péwé, 1954), they provide warnings of the dangers inherent in disturbing vegetation where permafrost exists.

V   Beaded drainage

Beaded drainage '... consists of series of small pools connected by short watercourses' (Hopkins *et al.*, 1955, 141). The pools result from the thawing of ice masses, which in many places are ice-wedge polygon intersections, and the connecting drainage is commonly along thawing ice wedges and therefore tends to comprise short straight sections separated by angular bends. The pools are some 1-3 m deep (2-10 ft) and up to 30 m (100 ft) in diameter. They have been described from northern Alaska and elsewhere in permafrost regions. In places artificial disturbance of the tundra vegetation will initiate or enlarge beaded drainage (Lawson and Brown, 1978, 14-24). A somewhat similar drainage pattern occurs in some swampy areas lacking permafrost, and the following criteria are useful in distinguishing beaded drainage from other forms (Hopkins *et al.*, 1955, 141)

(1) Beaded-drainage pools and channels are sharply defined; swamp-drainage courses unrelated to permafrost have indistinct, gradational borders. (2) Beaded-drainage channels generally are straight or consist of series of straight segments separated by angular bends; swamp-drainage courses are straight or smoothly curved. (3) Beaded-drainage courses generally are associated with ice-wedge polygons and locally with pingos and thaw lakes.

---

[2] For terminology, see discussion of Thaw slumping in chapter on Mass-wasting processes and forms.

12.2 Baydjarakhs, left bank Aldan River, 244 km above junction with Lena River, Yakutia, USSR

VI Thaw lakes

1 General

Thaw lakes, defined by Hopkins (1949, 119), are lakes whose basins have been formed or enlarged by thawing of frozen ground (Black, 1969c; Ferrians, Kachadoorian, and Greene, 1969, 12; Hussey and Michelson, 1966; Sellmann *et al.*, 1975). They are thus a kind of thermokarst feature, and are sometimes referred to as thermokarst lakes (cf. Klassen, 1979). Some lake basins are entirely the result of thawing, others are merely modified by it. Since ground-ice masses are especially common in silts, thaw lakes tend to have a similar association and to indicate the presence of such sediments (Wallace, 1948, 174). Disregarding pre-existing topography, the depth of thaw lakes depends on the amount of ground ice and depth of thawing. In the Barrow area of Arctic Alaska, ground ice is particularly common in the upper 3–5 m, and thaw settlement curves based on representative values suggest that lakes due to thaw-ing alone should rarely exceed 3 m in depth even if an extensive thaw bulb lies beneath them, and in fact many of the lakes are very shallow. Also lakes here that are less than 2 m deep commonly freeze to the bottom and have no potential for increasing their depth by thawing (Sellmann *et al.*, 1975, 14–16).

Because of drainage changes and concurrent erosion and deposition, thaw lakes in a permafrost environment are dynamic features that tend to change their configuration and, in places, to migrate slowly over the tundra at estimated rates at one location of 6–19 cm (2.3–7.5 in) yr^{-1} (Wallace, 1948, 178–9), and at another of about 30 cm (1 ft) yr^{-1} (Tedrow, 1969, 339). A thaw-lake cycle and the resulting deposits have been described by Britton (1957, 54–9). Lakes occupying collapsed pingos, some oriented lakes, and alas lakes are varieties of thaw lakes.

In glaciated areas especially, it may be difficult to distinguish between kettle holes formed by melt-ing of buried glacier ice (glacial thermokarst), and thaw depressions caused by melting of other ground

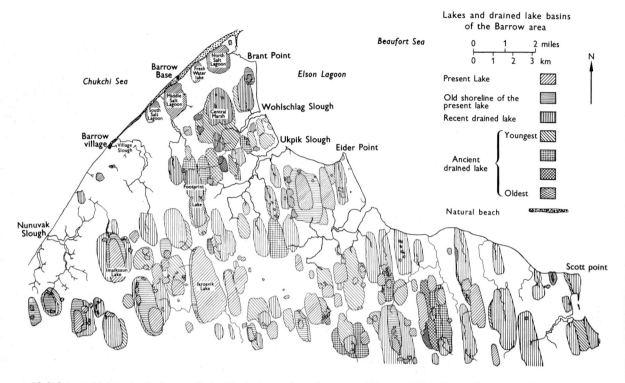

12.3 Oriented lakes, Arctic Coastal Plain, Alaska (*after C. E. Carson and Hussey, 1962, Figure 2 opposite 420*)

ice, including pingo ice. The environmental implications of the modes of origin can be quite different. The following criteria are based on observations by D. M. Hopkins (personal communication, 1976). Glacial thermokarst is confined to ice-contact deposits, commonly till or gravelly to sandy outwash. Other thaw depressions are usually associated with fine-grained sediments. Where lake deposits occupy these depressions, the sediments commonly have (1) a basal layer of plant debris derived from collapse of vegetated, migrating lake shores, and (2) above the plant debris, a layer of micro cross bedding resulting from shore currents. Thaw basins of whatever origin tend to have steep banks and irregular shorelines (oriented lakes and some collapsed pingos are exceptions) so that morphology alone is not always a reliable criterion for distinguishing between types of thaw depressions. In the case of collapsed pingos, ramparts and rampart groupings provide suggestive evidence, as discussed in the section on Pingos in the chapter on Some periglacial forms.

2   Oriented lakes

Oriented lakes are lakes that have a parallel alignment and are commonly elliptical in plan. They are described from various environments and as caused by various processes (W. A. Price, 1968), but some are probably primarily due to differential thawing of permafrost under the influence of predominant winds.

The oriented lakes of the Arctic Coastal Plain of Alaska (Figure 12.3) have been described as an illustration. Black and Barksdale (1949), who first drew attention to the problem, suggested that the lakes had formed parallel to a former predominant wind direction by their gradual extension downwind. A subsequent, widely accepted view is that they lengthened at right angles to the present predominant (and seasonally opposite) winds because of bars forming on the lee shores and protecting them from erosion by waves, currents, and thawing as the lake enlarged, mainly by thawing (C. E. Carson and Hussey, 1962). The general role of insolation in maintaining steep south-facing slopes was emphasized by Jahn (1975, 107).

12.4 Alas, Krest-Khal'dzhay, above right bank Aldan River, 360 km above junction with Lena River, Yakutia, USSR. Note small pingos, left centre

Similar lakes occur on the Old Crow Plain in northern Yukon Territory, Canada (Bostock, 1948, 76–7) and in the Liverpool Bay area east of the Mackenzie Delta (Mackay, 1956; 1963*a*, 46–55). A large group of lakes on the west side of Baffin Island shows a directionally similar but less well-developed orientation (Bird, 1967, 215). Oriented lakes also occur in other parts of the world and a variety of causes have been suggested (W. A. Price, 1968), among which wind has been widely accepted as the predominant influence in the orientation of the Alaskan and Canadian examples, and it has been applied more generally as well (Kaczorowski, 1976; 1977).

Nevertheless problems regarding the Alaskan and Canadian occurrences remain (cf. Black, 1969*d*, 293–4). On the basis of multispectral satellite imagery, it has been suggested that (1) NNW–SSE and ENE–WSW bedrock lineaments may be controls, with the prominent NNW–SSE lakes reflecting the predominant (additional) influence of wind (operating as described by Carson and Hussey) and (2) a previously unrecognized ENE–WSW orientation of small lakes reflecting the structural control alone (Fürbringer and Haydn, 1974). Also, unpublished work by A. D. Horn and M. L. Schuh covering the Prudhoe Bay area suggests that structural control, except for the largest and smallest lakes, is more important than previously recognized (A. D. Horn, personal communication, 1978). Although favouring a cross-wind hypothesis as a general explanation, Mackay (1963*a*, 54–5) concluded that

… the precise mechanism of lake orientation remains un-explained. The amount of littoral transport, the aspect of a two-cell circulation, the possibility of thermal effects in a permafrost region, the preference for vegetation growth under favored microclimatic and topographic conditions, and the effect of lake ice on lake orientation need further study.

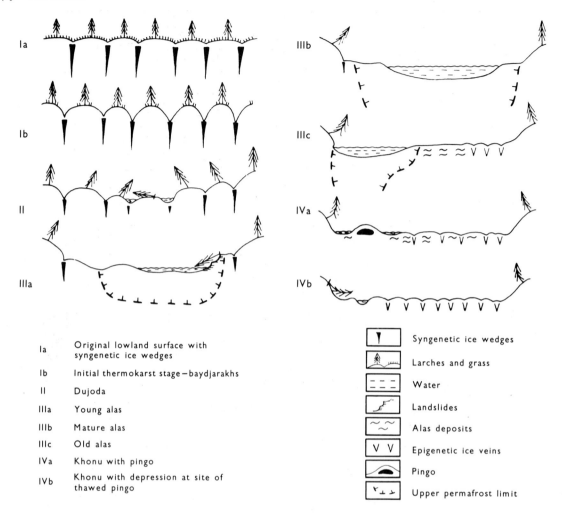

12.5 Sequential alas forms (*after Czudek and Demek, 1970b, 111, Figure 9*)

VII   Alases

An alas is a thermokarst depression with steep sides and a flat grass-covered floor (cf. Czudek and Demek, 1970*b*, 111). The term is of Yakutian origin. Alases (Figure 12.4) are commonly round to oval and many contain shallow lakes. In Central Yakutia, alases range in depth from 3 to 40 m and in diameter from 0.1 to 15 km (P. A. Solov'yev, 1962, 48). Alas valleys develop as individual forms coalesce. Except for lack of major subsurface valleys, alases resemble true karst topography. They are particularly well developed in Siberia, where they have been extensively studied by Czudek and Demek (1970*b*), Demek (1968*b*, 114–17), P. A. Solov'yev (1962; 1973*a*; 1973*b*; 1975), and others.

Alases may start by a forest fire in the taiga disturbing the thermal equilibrium of permafrost and causing a sequential development of ice-wedge troughs, baydjarakhs, broad collapse areas, lakes, filling in of lakes, growth and collapse of pingos, and joining of adjacent alases into vast alas valley systems tens of kilometres long (Figures 12.5–12.6), the Moro Valley alone having an area of 12 000 km² according to Jahn (1975, 108), who regarded the presence of ice wedges as indispensable for alas development. In the central Yakutian lowland, 40 to 50 per cent of the Pleistocene surface has been destroyed by alases,

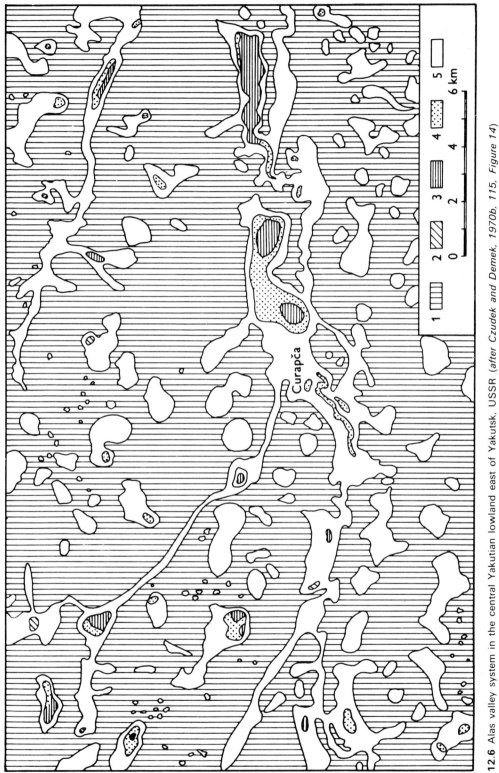

12.6 Alas valley system in the central Yakutian lowland east of Yakutsk, USSR (*after Czudek and Demek, 1970b, 115, Figure 14*)

1 Original lowland level with syngenetic ice wedges; 2 Young lakes; 3 Lakes in alases; 4 Recently desiccated lakes in alases; 5 Old alases

many of which originated during the Hypsithermal interval (*c.* 2500–9000 ^{14}C years BP) but some within a human generation (Czudek and Demek, 1970*b*, 113). Although both fluvial and lacustrine action may contribute to alas development, the origin is always as thermokarst. Alas lakes, especially those sufficiently deep to remain unfrozen below their seasonal ice cover, accelerate thawing and alas growth. On the other hand, drainage will reduce growth, and in place alases have been artificially drained to develop agricultural land (Jahn, 1975, 109).

It may well be that further investigation will show that many European and North American lakes in permafrost environments are akin to alas types, including some of the oriented lakes (as noted by Jahn, 1975, 105), and many of the irregular lakes in the drift-covered portions of the Canadian Shield. Cailleux (1971) suggested that the shallowly submerged margins of many lakes on outwash in northern Quebec may be due to melting out of segregated ground ice, whereas the sharply defined, central deeps may be kettle holes. Once all the ice has melted out of the ground in an alas environment where ground ice had been widespread, the alas relief would tend to collapse and its former presence could be very difficult to identify and reconstruct from the topography, although lacustrine deposits and other stratigraphic evidence could supply a clue.

13 Environmental overview

Table 13.1 is an attempt to summarize the environmental implications of some of the periglacial processes and features discussed in the preceding chapters. It follows the style of an overview by Lee Wilson (1968*b*, 379, Table 2), which in turn was based on a 1964 Yale Seminar review by the present author. The table is tentative and highly subjective.

A generally increasing persistence of snow cover at high altitudes and a decreasing number of freeze-thaw cycles at high latitudes are taken into consideration. Although based on contemporary situations, the table is presumably equally applicable to Pleistocene conditions.

Table 13.1. Quasi-equilibrium range of periglacial processes and features
Predominant ranges are suggested by (1) and lesser ranges in decreasing order by (2), (3), (4). R indicates rare or absent

Processes	Lowlands			Highlands			
	Polar	Subpolar	Middle latitude	Polar	Subpolar	Middle latitude	Low latitude
Frost action							
Frost wedging	2	1	3	2	1	2	3
Frost heaving and frost thrusting	1	2	3	2	1	3	4
Mass displacement	1	2	R	2	1	3	R
Seasonal frost cracking	2	1	3	2	2	3	R
Permafrost cracking	1	2	R	2	3	R	R
Mass-wasting							
Slushflow	1	2	R	2	3	4	R
Frost creep	2	1	3	3	1	2	4
Gelifluction	1	2	3	2	1	3	4
Nivation	1	2	R	2	1	2	3
Fluvial action							
Ice rafting	1	1	3	4	3	3	R
Lacustrine action							
Ice shove	1	2	3	4	3	3	R
Ice rafting	1	2	3	4	3	3	R
Marine action							
Ice shove	1	2	R	—	—	—	—
Ice rafting	1	2	R	—	—	—	—
Wind action	1	2	3	1	2	3	3
Features							
Permafrost	1	2	R	1	2	3	4
Patterned ground							
Nonsorted circles	1	2	R	2	2	3	R
Sorted circles	1	3	R	2	1	R	R
Small nonsorted polygons and nets	1	2	3	2	1	2	2
Large nonsorted (ice-wedge or sand-wedge) polygons and nets	1	2	R	2	3	R	R

Table 13.1. (continued)

Features	Lowlands			Highlands			
	Polar	Subpolar	Middle latitude	Polar	Subpolar	Middle latitude	Low latitude
Small sorted polygons and nets	1	2	4	2	1	3	2
Large sorted polygons and nets	1	3	R	2	3	R	R
Small nonsorted and sorted stripes	1	2	4	2	1	3	2
Large nonsorted and sorted stripes	1	3	R	2	1	4	R
Involutions	1	2	R	2	3	4	R
String bogs	2	1	3	R	2	4	R
Palsas	2	1	R	R	2	3	R
Pingos	1	2	R	R	R	—	—
Slushflow fans	1	2	R	2	3	4	R
Thaw slumps	1	1	R	3	3	R	R
Small gelifluction lobes, sheets, streams	1	2	3	2	1	3	3
Large gelifluction lobes, sheets, streams	1	2	R	2	1	3	4
Block fields, block slopes, and block streams	2	2	R	1	1	3	4
Rock glaciers	R	R	R	2	1	1	4
Taluses	3	3	4	2	1	2	3
Protalus ramparts	R	R	R	2	1	1	3
Nivation benches and hollows	1	2	R	2	1	3	4
Cryoplanation terraces	2	2	R	1	1	R	R
Grèzes litées	2	3	R	2	1	3	4
Flat-floored valleys related to permafrost	1	2	R	1	2	R	R
Asymmetric valleys related to permafrost	1	2	R	1	2	R	R
Dry valleys and dells related to permafrost	1	2	R	1	2	R	R
Cold-climate varves	2	1	3	3	2	2	R
Lacustrine ice-shove ridges	1	2	3	4	3	3	R
Marine ice-shove ridges	1	2	R	—	—	—	—
Beaded drainage and thaw lakes	1	2	R	R	R	R	R
Loess	2	1	3	4	4	4	4
Dunes	1	2	4	R	3	4	R
Ventifacts	1	2	R	1	2	3	4

14 Environmental reconstructions

I Introduction: II Britain: *1 General; 2 Fossil pingos; 3 Frost cracking; 4 Periglacial involutions, depth of thawing.* **III Continental Europe:** *1 General; 2 Fossil pingos; 3 Frost cracking; 4 Involutions, depth of thawing; 5 Gelifluction; 6 Asymmetric valleys; 7 Dunes and loess; 8 Overview.* **IV USSR: V Japan: VI North America: VII Southern Hemisphere: VIII Some estimated temperature changes**

I Introduction

Environmental reconstructions utilizing the evidence of relic or fossil periglacial features are necessarily based on the nature and present distribution of comparable active features. The most critical features are those indicative of permafrost because of the negative mean annual temperature required for permafrost, whereas many other cold-climate features are much less informative as to temperature implications.

With attention thus focused on permafrost, it is natural to seek analogs in present-day permafrost environments. However, most such environments, being in high latitudes, are associated with a quite different solar-radiation regime than accompanied the development of periglacial features in middle latitudes. This difference is stressed because the temperature change between night and day in middle latitudes entails in many places a greater frequency of freeze-thaw cycles than that accompanying the more seasonal periodicity of daylight and darkness in high latitudes, whereas the latter regime favours longer and more severe cycles and deeper ground freezing. In French's (1976a, 218–19) view, if this situation resulted in 50–100 freeze-thaw cycles yr^{-1} in middle latitudes, frost wedging, frost creep, and gelifluction would probably have been '... several times more intense than in high latitudes today....' This view is subject to question because frequency is not necessarily a measure of

intensity but his stress on recognizing important differences between high-latitude environmental conditions and former middle-latitude periglacial conditions is unassailable. These differences also include the probability that fluvial activity in middle latitudes extended over longer periods and that the presence of ice sheets caused differences in circulation patterns and in precipitation and wind effects (French, 1976a, 220–1).

Fossil periglacial indicators of permafrost occur in many parts of the world and permit estimates of temperature change. Among the most important indicators are fossil frost cracks[1] and pingos. Fossil pingos are proof of former permafrost and hence of mean annual ground temperatures below 0°. Although frost cracks can occur at mean annual ground temperatures above 0°, fossil cracks in the form of ice-wedge casts are an acceptable proof of former continuous permafrost. Large, well-developed fossil soil wedges of frost-crack origin can be reasonably assumed to indicate a permafrost environment, including perhaps continuous permafrost in the present writer's opinion, whereas smaller, poorly developed soil wedges are possibly evidence of sufficiently intense freezing to be consistent with discontinuous permafrost. Permafrost cracking responsible for polygons requires mean annual air temperatures that have been variously estimated to range from about − 2° to about − 10°. Different aspects of these and related matters are discussed in the section on Nonsorted polygons (under Pat-

[1] As noted under Frost cracking in the chapter on General frost-action processes, the term frost crack as used in this volume implies origin only, whereas in the European literature on fossil phenomena it is often restricted to very narrow features as distinct from wedge forms.

terned ground) in the chapter on Some periglacial forms. Péwé's (1966a, 78; 1966b, 68; 1975, 52) estimate of $-6°$ to $-8°$ for the origin of ice-wedge polygons, based on Alaskan conditions, is especially widely quoted. However, extrapolation to other regions is dangerous because of the critical importance of the rate of temperature change in the ground and variations in environmental factors.

A more reliable approach is to compare present and fossil occurrences in the same region. Thus near the Sella Mountains in the Italian Alps, Kelletat and Gassert (1974) reported ice-wedge casts as low as 2000 m, an altitude at which the present mean annual air temperature is about $2.7°$ compared with a temperature of about $-2.7°$ at the probable lower limit of present permafrost some 800–900 m higher. They therefore concluded from this and other evidence that the temperature was $5°–6°$ lower than now when the polygons formed, which they believed was during Younger Dryas time [c. 10 150–10 950 yr BP]. Presumably development of the polygons required a still lower temperature than that needed for development of permafrost, so an even greater temperature change than $5°–6°$ would be implied. Black (1976a, 10) has suggested that rather than extrapolating with respect to mean annual air temperatures, it would be preferable to compare degree days of freezing or some other index that would better reflect long cold spells.

Fossil or relic cryoplanation terraces and cryopediments should provide important evidence of former periglacial conditions, but as previously noted their distribution is not well established and opinions differ as to their significance with respect to permafrost (cf. discussion of Cryoplanation terraces and cryopediments in chapter on Nivation). In many places structure appears to exert a predominant control on terrace development, as in southwest England (R. S. Waters, 1962), and although such terraces may carry periglacial deposits they are hardly true cryoplanation forms cutting across structure. The identification and use of true cryoplanation forms in environmental reconstructions is still a largely undeveloped potential.

Environmental reconstructions based on periglacial phenomena date back to Łoziński (1909; 1912), the father of periglacial studies. Bertil Högbom's (1914) classic paper on frost action and its effects and Leffingwell's (1915; 1919) excellent observations on permafrost in Alaska, along with Soviet observations, provided much useful information on processes and their climatic associations and thereby aided later reconstructions. Compre-

hensive reconstructions of this period include those of Kessler (1925) and Soergel (1919). Two decades later the availability of additional field observations and studies, including Troll's (1944; 1958) world-wide survey of periglacial processes and features, permitted improved reconstructions. Among the notable efforts were studies, including maps, by Poser (1947a; 1947b; 1948a; 1950) and Büdel (1951) showing Würm environmental zones in Europe, and a similar effort for Asia by Frenzel and Troll (1952) and Frenzel (1959). The maps, based in part on periglacial phenomena, are very sketchy and subject to considerable modification, but are still among the most noteworthy of their kind.

In the Northern Hemisphere far more information for environmental reconstructions is available from Europe (cf. Wright, 1961) and the USSR than from North America. Although fossil periglacial features are widespread in parts of North America, they have been less investigated and there is consequently better stratigraphic control and dating of the European features, but even for the latter many stratigraphic problems remain (cf. Frenzel, 1973b; Semmel, 1973a). In the following, Würm and Weichsel are regarded as essentially equivalent time terms referring to the last glaciation, the former pertaining to the Alpine Ice Sheet, the latter to the Scandinavian. Yet correlations between deposits of these ice sheets are still problematical, as was emphasized by Frenzel at the symposium on 'Problèmes de Stratigraphie du Quaternaire en France et dans les Pays Limitrophes' in Dijon in 1978. In Britain the last glaciation is known as the Devensian, in the Soviet Union as the Valdai.

Much less information is available from the Southern Hemisphere than from the Northern but periglacial studies are increasing not only with respect to contemporary features as noted in preceding chapters but also with respect to fossil forms and their climatic implications. Remote sensing techniques are proving especially helpful in discovering fossil occurrences such as large nonsorted polygons, which are proving to be widespread in places.

II   Britain

1   General

Most environmental reconstructions of Europe that are based on periglacial criteria apply to the Continent itself rather than Britain. Nevertheless, periglacial features are also widespread in Britain and

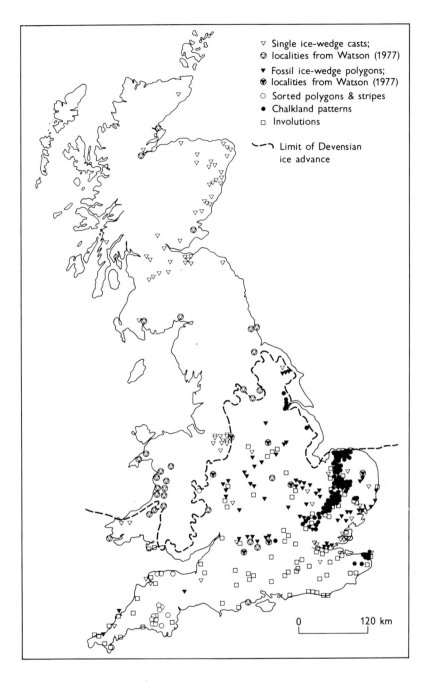

Legend:

▽ Single ice-wedge casts;
⊘ localities from Watson (1977)
▼ Fossil ice-wedge polygons;
◉ localities from Watson (1977)
○ Sorted polygons & stripes
● Chalkland patterns
□ Involutions

⌐ Limit of Devensian ice advance

0   120 km

14.1 Distribution of patterned ground of Devensian age in Britain (*after West, 1968, 275, Figure 12.21, from Sissons, 1965; R. B. G. Williams, 1969; Worsley, 1966. Additional localities from Watson, 1977, 189, Figure 4*)

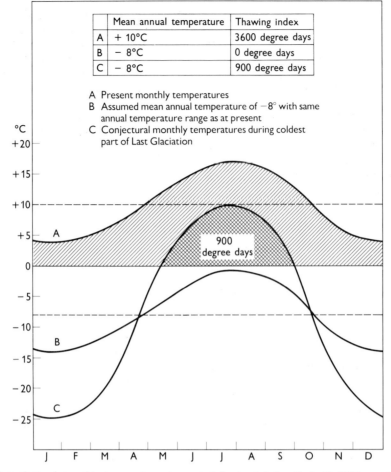

14.2 Present and conjectural monthly temperatures in central England (*after R. B. G. Williams, 1975, 108, Figure 2*)

are being increasingly studied. Regional summaries include those of R. Evans (1976), R. W. Galloway (1961*a*; 1961*b*; 1961*c*), Kelletat (1970*a*; 1970*b*), Sugden (1971), Te Punga (1957), Tufnell (1969), Watson (1977), West (1968), and R. B. G. Williams (1964; 1965; 1968; 1969; 1975), and there is a wealth of local information becoming available.

A map showing distribution of patterned ground of Devensian (Weichsel) age, compiled by West (1968, 275, Figure 12.21) with additions by Watson (1977, 18, Figure 4), is shown in Figure 14.1, and a reconstruction of British climate during the last glaciation, based on periglacial phenomena, was presented by R. B. G. Williams (1975), who recognized that his interpretation of the evidence is rather speculative as noted later.

2   Fossil pingos

According to Watson (1977, 195),

On the evidence of pingos, the mean annual air temperature of southern England and Wales in the Younger Dryas was between −4° and −5°C, representing a fall of 13–14°C compared with the present day.

The basis for this conclusion is that the fossil pingos appear to have been open-system forms whose distribution in Alaska generally ranges from −3° to −6°, although temperatures as high as −1° are not excluded (Watson, 1977, 193, Table 1; 194). In view of this range and the assumption that Alaskan conditions are representative, the temperature implication is rather flexible.

3 Frost cracking

In southeast Suffolk, ice-wedge casts and other peri-glacial features are reported from Middle Pleistocene deposits (Rose and Allen, 1977, 83–4, 91), but in Britain as elsewhere most fossil periglacial features date from the last glaciation.

Based on polygonal patterns with meshes some 8 m (25 ft) to perhaps 30 m (100 ft) in diameter near Evesham, England, and on a requirement of at least −3.5° for Alaska ice-wedge polygons, Shotton (1960) suggested a minimum temperature reduction of 13.5° for this area during the Devensian maximum.

As previously noted (cf. section on Nonsorted polygons in chapter on Some periglacial forms), R. B. G. Williams (1975, 99–100) regarded a mean annual air temperature of −8° to −10° or lower as probably necessary for the formation of ice-wedge polygons. He therefore concluded that during the maximum of the Devensian, −8° to −10° was near the mean annual temperature in the Midlands and East Anglia, where ice-wedge casts are common. Based on a present mean annual air temperature of 10° (Figure 14.2), this implies a temperature rise to the present of at least 18°. Warmer temperatures or thicker snow cover were thought to have prevailed near the south-west coast where ice-wedge casts are rare (R. B. G. Williams, 1975, 100–1). Furthermore, accepting a snow depth of 25 cm as sufficient to inhibit frost cracking in central Alaska, he suggested that when ice wedges were growing in England, the precipitation in the eastern part was less than half, and in the Midlands less than a third, of the present amount (R. B. G. Williams, 1975, 109–11). However, the idea (cf. R. B. G. Williams, 1969, 407–8) that there was an east-west paleoclimatic contrast has been regarded as questionable, at least without better dating of critical sequences (Worsley, 1977, 214–16).

On the basis of −7° air temperature for the southern limit of continuous permafrost in Alaska and on ice-wedge polygons as indicative of continuous permafrost, Watson (1977, 183, 195–6) concluded there was a minimum change of 16°–17°, but he estimated that the change since the Devensian maximum was as much as 25°. He reasoned that former ice wedges of this age on the Continent extended some 500 km south of England and that, by analogy with Alaskan conditions, this implies that mean annual temperatures in England at the time were −12° to −14° in a tundra environment, and at least −16° to −17° in the English Midlands where he believed full polar con-ditions existed. Because of the assumptions the evidence for such a large temperature change does not seem compelling. Again by analogy with Alaska, Watson suggested that the July temperature change since the Devensian maximum was relatively small (10°). Citing evidence from earlier and later Devensian deposits, he noted that ice-wedge casts in Scotland suggest that continuous permafrost existed there as late as the Younger Dryas.

4 Periglacial involutions, depth of thawing

Despite the uncertainty as to how periglacial involutions originate and how they can be distinguished from similar-appearing features unrelated to frost action, R. B. G. Williams (1975, 102–9) argued for their usefulness as evidence of thaw depth. By comparing actual thaw depths in Alaska and Siberia with presumed depths as indicated by features he accepted as involutions in England, Williams judged that thaw depths there during the last glaciation were significantly greater than present depths in northern Alaska and northern Siberia but comparable to those in central Alaska and central Siberia. He therefore concluded that the thawing index must also have been similar to that characterizing the transition from tundra to boreal forest, involving a range of 900 to 1500 degree days (°C). Furthermore, if former ice wedges formed at the same time as the presumed involutions, the association of (1) the assumed mean annual air temperature of −8° required for frost cracking and the growth of ice wedges, (2) 900 degree days of thawing, and (3) a temperature for the warmest summer month of 10° (which Williams accepted as the lowest probable temperature for 900 degree days of thawing) would require an annual temperature range of 30°; i.e. a distinctly continental climate (Figure 14.2, Curve C). An interpretation based on a temperature range like the present (Curve A) would require mean monthly temperatures below freezing (Curve B) – an unrealistic distribution if involutions are thaw-layer phenomena. The many assumptions make the conclusions highly conjectural as Williams recognized, but he noted they are supported by a similar reconstruction based on independent biological evidence (Coope, Morgan, and Osborne, 1971, 98–9; cf. Coope, 1977, 333–4). The methodology should become increasingly valuable as reliable periglacial evidence becomes more readily available.

III   Continental Europe

1   General

Reconstructions pertaining to the Continent were very prominent several decades ago. Since then much additional local information has accumulated regarding periglacial features (for instance, Michel, 1975), including fossil frost cracks at different stratigraphic horizons (Haesaerts and Van Vliet, 1974, 558; Paepe and Pissart, 1969, 323–5), especially in loess (Jersak, 1975; Rohdenburg, 1967). Brief summary discussions of periglacial phenomena as related to the Pleistocene glaciations of central Europe include those of Liedtke (1975) and Marcinek and Nitz (1973, 66–89).

2   Fossil pingos

As noted in the chapter on Some periglacial forms, fossil pingos are clear evidence of former permafrost and are becoming increasingly widely known in Europe, including the British Isles (Cailleux, 1976; Frenzel, 1959, 1021, Figure 13; 1967, 29, Figure 17; Maarleveld, 1965; Mitchell, 1971; 1973; Mullenders and Gullentops, 1969; Pissart, 1956; 1963a; 1965; 1974b; Pissart *et al.*, 1972; Svensson, 1969; Watson, 1971; 1972; Watson and Watson, 1972; Wiegand, 1965). Their distribution supports in a general way Poser's reconstruction of the distribution of permafrost during the Würm maximum, which is reviewed below.

3   Frost cracking

In a series of papers, Poser (1947a; 1947b; 1948a) surveyed the distribution of fossil frost cracks and loess or loam wedges in Europe. The wedges he equated with ice-wedge casts – although perhaps some formed without ice wedges as suggested by Péwé (1959) – and he accepted them and the much narrower simple fossil frost cracks as evidence of former permafrost. He assumed that the latter cracks indicate a warmer permafrost climate than the wedges. In plotting the features (Figure 14.3), Poser (1948a, 63, Figure 4) noted their depth and width, and found that the deepest and widest were in central Europe, especially central Germany, where maximum depths of wedges were 7–8 m and the greatest widths were 3 m. He considered but, because of regional contrasts, discarded Soergel's (1936, 239, 246) view that the width of wedges is an indicator of age as well as climate. From regional trends Poser (1947b, 264; 1948a, 64) concluded that during the Würm maximum middle-central Europe was by far the coldest region, followed by southeast Europe and northwest-central Europe, the last with an appreciably warmer winter climate.

Taken alone this evidence afforded by depth of cracking is open to question because so many variables besides temperature control frost cracking. As previously discussed it can be assumed that most fossil frost cracks indicate former permafrost and require for their development a lower mean annual air temperature than that needed for permafrost alone. How much lower depends on conditions. Therefore to define temperatures more closely on the basis of degree of crack development is problematical unless other factors are known to be constant. In addition to terrain and moisture factors, it is not immediately apparent why regional contrasts eliminate age as a variable in the cases considered. Nor is it always clear that reports of wedge width reflect the true width rather than the apparent as seen in an exposure at an angle to the minimum cross-section.

An occurrence of large, well-developed fossil nonsorted polygons at Crau, south of lat. 44°N in eastern Basse-Provence, France (Cailleux and Rousset, 1968) is of special interest. Their diameters range from 10 to 30 m, and the bordering wedges, consisting of eolian silt, are up to 1.5 m wide at the top and extend to depths of 2–3 m. The fact that the polygons occur in gravel strongly supports a frost-crack origin by eliminating desiccation cracking as a factor. Cailleux and Rousset concluded, on the basis of Péwé's criterion of −6° to −8°, that the mean annual air temperature at time of formation must have been −5° or colder. The present mean annual temperature at Marseille, just east of Crau, is 14° so that a temperature increase of at least 19° is indicated if Cailleux and Rousset's −5° premise is accepted. The premise seems conservative because frost cracking of gravels normally requires a lower temperature than finer material, yet the change seems large. Since they thought Würm deposits in this region show no such features, Cailleux and Rousset concluded that the polygons are of Mindel or Riss age and record a colder climate than the Würm in this region. However, these or nearby polygons were regarded as of Würm age by Frenzel (cf. Figure 14.12), and the Crau features are confidently dated

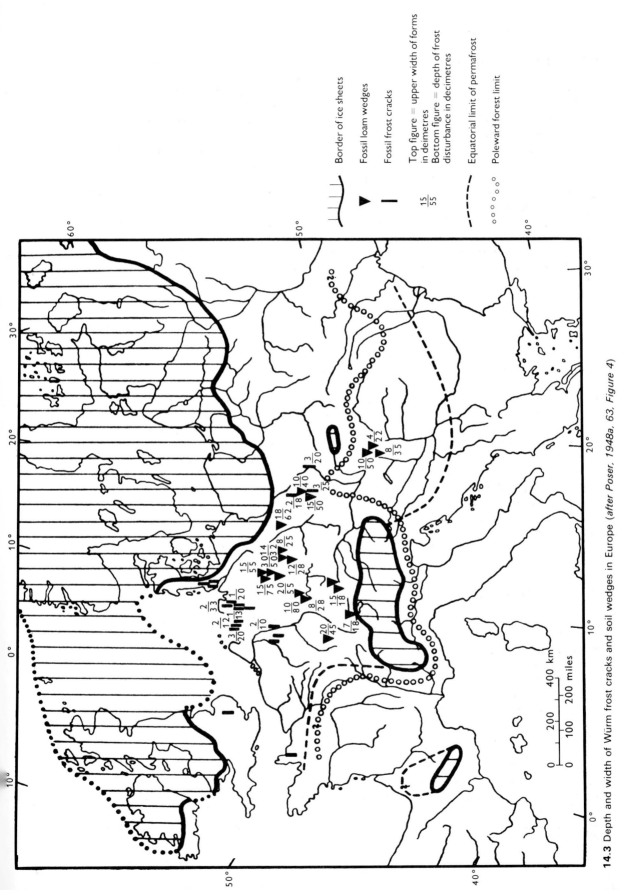

Border of ice sheets

Fossil loam wedges

Fossil frost cracks

Top figure = upper width of forms in deimetres
Bottom figure = depth of frost disturbance in decimetres

$\frac{15}{55}$

Equatorial limit of permafrost

Poleward forest limit

14.3 Depth and width of Würm frost cracks and soil wedges in Europe (*after Poser, 1948a, 63, Figure 4*)

as Würm by E. Bonifay (personal communication, 1978).

In Switzerland, ice-wedge casts and other evidence of fossil patterned ground led Furrer (1966, 495–6) to infer permafrost in the vicinity of Rafzerfeld (near Zürich) during and shortly after the Würm maximum. Assuming an upper limit of −2° as necessary for permafrost and comparing this with the present mean annual air temperature of 8.5°, he concluded there has been a temperature change of at least 10.5°. Using the present temperature lapse rate of 0.5° per 100 m and comparing the altitudes of present patterned ground in mountains of the Vorderrhein Valley with fossil forms of the Rafzerfeld area, Furrer arrived at a temperature increase for this area of 11°–11.5°.

At Gimbre near Münster (Westf.) in the Federal Republic of Germany, H.-M. Müller (1978) reported well-developed ice-wedge casts dating from the Weichsel maximum, underlying a younger set less securely dated as Older Dryas. The present mean annual temperature at Münster is 9.3° (US Department of Commerce, National Oceanic and Atmospheric Administration [NOAA], National Climatic Center, personal communication, 1979), and allowing −5° as a maximum mean annual temperature for the development of ice-wedge casts as discussed later, the temperature increase here since the Weichsel maximum would be at least 13.3°.

Following review of the literature and extensive study of periglacial features in central Poland, Goździk (1973, 92–5, 113–17) accepted −6° as the maximum mean annual air temperature necessary for the development of the frost-crack phenomena. He concluded that mean annual air temperatures in the region were no higher than this during the Würm maximum and that the presence of fossil sand wedges and ventifacts indicates aridity at the time (Goździk, 1973, 93–4, 117). Based on the present mean annual temperature at Warszawa (8.3°) and Wrocław (8.8°), the inferred temperature increase would be about 15°.

As shown by Figure 14.3 the fossil frost cracks and soil wedges generally lie outside the limit of Weichsel (Würm) Glaciation. This is expectable because features formed in front of an advancing glacier are likely to be overrun and destroyed, and normally glacial maxima reflect temperature minima whereas glacial retreat accompanies a warming trend that would be less favourable for frost cracking. Nevertheless, late-glacial frost cracks occur north of the Weichsel limit (cf. Figure 14.11 and Kozarski, 1974). Christensen (1974) reported fossil ice-wedge

polygons of late-glacial age (as well as older) in southwest Jutland, Denmark, where the present mean annual temperature is 8°–8.5°, and he inferred a temperature rise since their origin of 14°–16.5°, based on Péwé's −6° to −8° criterion. According to Svensson (1976c, 46, 53), some large nonsorted polygons on Sweden's west coast are fossil ice-wedge polygons formed during deglaciation, and as such record a temperature increase to the present of 12°–13°. Farther north in the Voss area of western Norway there are fossil wedges of presumed permafrost-crack origin that Mangerud and Skreden (1972, 91–4) regarded as evidence of an 8° temperature increase since their formation. Probably fossil frost-cracks are more common north of glaciation limits in both Europe and North America than is usually recognized (cf. Jahn, 1975, 24–5).

It should be noted that Kostyayev (1966) and Morawetz (1973) have not accepted such features as proof of permafrost, Kostyayev viewing them as convection phenomena. Morawetz, accepting a mean annual temperature of below −3° as necessary for permafrost where the winter snow cover exceeds 10 cm, believed that snowline data indicate Würm snowfall had been sufficiently great to inhibit permafrost in western Europe, even north of the northern limit as drawn by Poser. However important the role of snow cover, more evidence is needed before apparent ice-wedge casts can be accepted as convection phenomena.

The evidence of fossil frost cracks in Europe is reviewed further in the Overview discussion and at the conclusion of this chapter in the section on Some estimated temperature changes.

4   Involutions, depth of thawing

In the same series of papers in which Poser discussed fossil frost cracks, he reviewed the distribution of other fossil periglacial features (Figure 14.4). In discussing involutions, he stated that they do not form under present climatic conditions in middle and eastern Europe, where freezing extends to depths of 1.2–1.6 m. Citing Russian workers, especially Sukachëv (1911), Poser concluded that involutions are cryostatic phenomena caused by forces set up in the active layer as downward freezing from the surface begins to merge with the permafrost table. He therefore argued that the depth to which involutions extend approximates the depth of the active layer – i.e. the depth of summer thawing in a permafrost environment. Poser also regarded Stiche as pro-

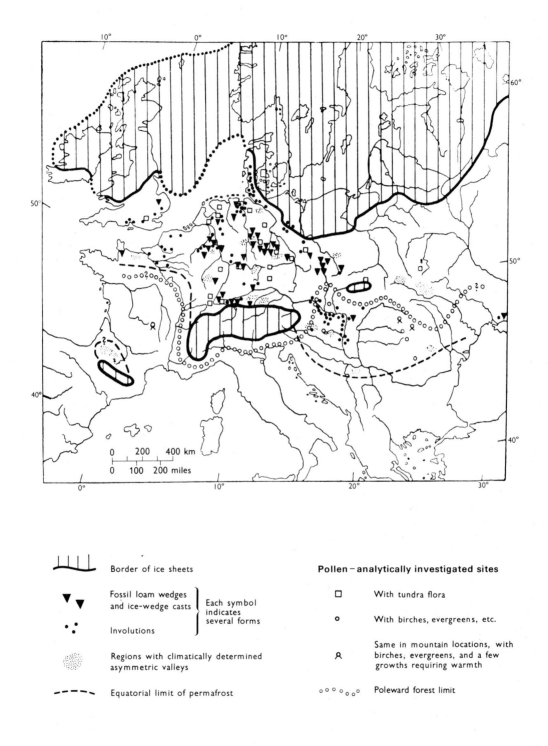

Border of ice sheets	

Pollen−analytically investigated sites

▼ ▼ Fossil loam wedges and ice-wedge casts ⎫ Each symbol indicates several forms

▫ With tundra flora

∴ Involutions ⎭

○ With birches, evergreens, etc.

⠿ Regions with climatically determined asymmetric valleys

Я Same in mountain locations, with birches, evergreens, and a few growths requiring warmth

– – – – Equatorial limit of permafrost

○ ○ ○ ○ ○ ○ Poleward forest limit

14.4 Distribution of Würm periglacial features in Europe (*after Poser, 1948a, 55, Figure 1*)

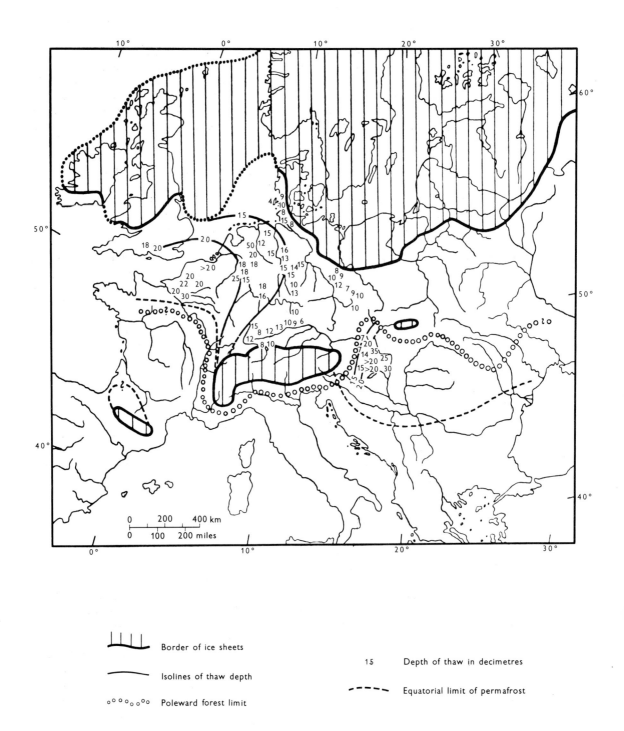

14.5 Depth of Würm summer thaw in Europe, based on involutions (*after Poser, 1948a, 59, Figure 2*)

viding information on depth of thawing. He described them as narrow, chevron-like anticlinal deformations of platy bedrock, which occur along frost cracks or joints and have an abrupt downward termination. Interpreting Stiche as due to frost action along frost cracks or joints, he accepted their downward termination as controlled by, and therefore indicative of, a former permafrost table. On these premises he reconstructed the depth of summer thaw during the Wurm (Figure 14.5).

Although accepting most involutions as evidence of permafrost, Kaiser (1960, 133) argued that they could not be used to infer depth of thaw, since there were too many uncertainties involved such as the level of the original surface, contemporaneity of the features, and local climatic influences. Involutions in some places may indeed be associated with permafrost but, as discussed in the chapter on Some periglacial forms, they probably also form by seasonal frost action in a non-permafrost environment. Moreover, very similar or identical-appearing features can probably form in environments that have no relation to frost action.

Also it is not demonstrated that Stiche are caused by frost cracking or, if they are, that they terminate at the permafrost table. On the other hand, if they are due to frost wedging along cracks or joints, their depth may merely indicate depth of seasonal freezing and not the presence of permafrost. Pending further investigations, the interpretation of Stiche as permafrost features is problematical. Thus the evidence that Poser adduced for depth of thawing is uncertain.

An increased thaw depth was invoked by Büdel (1977, 59–60) to explain, in eastern Svalbard, a widespread organic-rich layer at the top of the permafrost just below the present active layer. Radiocarbon dating gave an age of 3000–3100 years and the thaw was assigned to the close of the postglacial warm period.[2] However, no specific temperature parameters were inferred.

5   Gelifluction

If correctly identified, gelifluction deposits are clearly evidence of cold climates, although not necessarily of permafrost. Stratigraphic sections showing a sequence of gelifluction and other deposits can thus provide useful evidence of environmental conditions and changes.

An interesting illustration is provided by a study of numerous profiles in the Harz and the Thüringer Mountains of the German Democratic Republic. Correlation of the sections with one another led Schilling and Wiefel (1962) to the conclusion that a basal sequence of gelifluction and slopewash deposits record cool moist conditions during Würm I, followed by erosion during Würm I-II and II-III. The succeeding main sequence of gelifluction deposits, correlated with intensely frost-wedged debris at higher altitudes and loess at lower, record a cold dry climate during Würm III, with colder conditions in the Harz Mountains than in the Thüringer as indicated by the nature and distribution of associated involutions. This was followed during the later part of Würm III by erosion and then a cover deposit of fine-grained gelifluction facies or eolian facies, depending on altitude and exposure, recording somewhat warmer and moister conditions. Regardless of whether these conclusions stand, the detailed local and regional correlation of gelifluction sequences presented by Schilling and Wiefel (1962, 456–7, Tables 1–2) illustrates a challenging approach.

In the Iberian Peninsula, comparison of the lower limit of contemporary, blocky solifluction (presumably gelifluction) deposits with the lower limit of similar fossil deposits led Brosche (1978, 83–7, Table 2, Figures 2–4; 94–5) to infer a possible temperature rise of 8°–10° (or less in the northern part) since the last glaciation.

6   Asymmetric valleys

As noted in the chapter on Fluvial action, the distribution of asymmetric valleys has also been cited as evidence of a periglacial environment. Poser (1948a, 54) reported observing two types of asymmetric valleys in the same valley system: (1) Upper parts of valleys where the volume of water was inadequate to carry away all the debris contributed by solifluction from the sunlit slope; consequently the stream was displaced to the opposite side and the gradient of the sunlit slope was preferentially lowered by the mass-wasting. (2) Lower valley sectors where the water volume was sufficient to transport material, and the stream therefore preferentially eroded and steepened the thawed sunlit slope. Poser believed that both types reflect perma-

[2] The postglacial warm period is apparently time transgressive globally but as defined by the time-stratigraphic term Hypsithermal it ended some 2600–2800 years ago (Porter and Denton, 1967, 198–205).

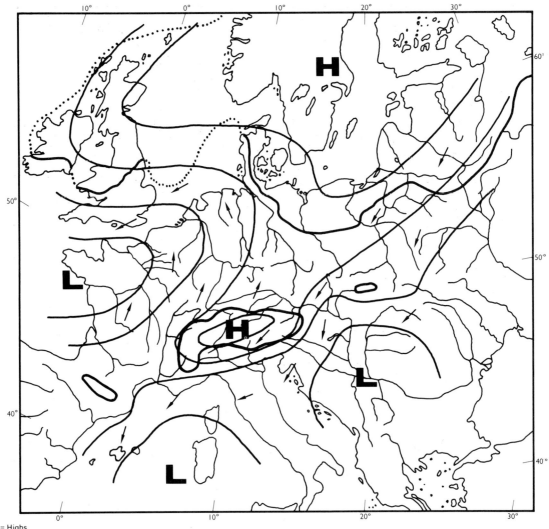

H = Highs
L = Lows

14.6 Summer atmospheric pressure and wind directions in Europe during Würm Glaciation (*after Poser, 1948a, 61, Figure 3*)

frost conditions, and he therefore drew the Pleistocene permafrost boundary in Europe south of them (Figure 14.4). The distribution and significance of asymmetric valleys in southern Germany and eastern Austria have been reviewed by Helbig (1965). As previously discussed, asymmetric valleys are subject to so many interpretations that their value as evidence of permafrost is doubtful; however, this does not deny their climatic significance in places.

7   Dunes and loess

Poser (1947*b*; 1948*a*) attempted a reconstruction of atmospheric pressure and wind conditions for summer and winter during the Würm maximum (Figures 14.6–14.7). Involved were mainly general theoretical considerations, supplemented by conclusions as to temperature derived from his analysis of depths of frost cracking and thawing, and by information from distribution of dunes (Poser, 1947*b*, 237; 1948*a*, 61).

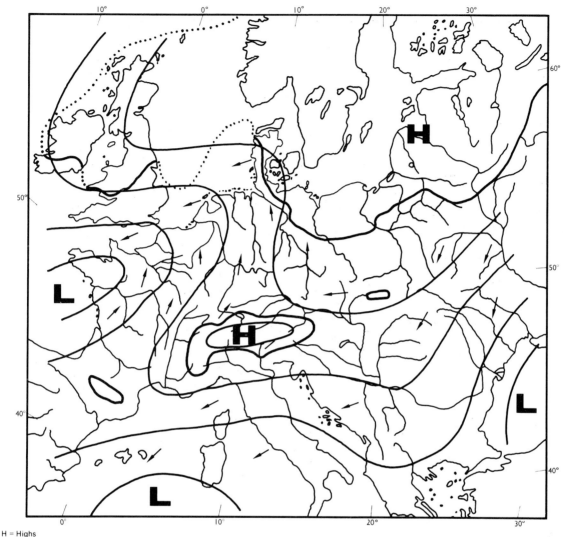

H = Highs
L = Lows

14.7 Winter atmospheric pressure and wind directions in Europe during Würm Glaciation (*after Poser, 1948a, 65, Figure 5*)

The reconstruction of atmospheric pressure and wind conditions during Late-Glacial time (Figure 11.1) was directly based on dunes (Poser, 1948*b*; 1950) and supplementary information from loess studies (Poser, 1951). The difficulty of such reconstruction is indicated by the problems cited by Poser, additional data (Maarleveld, 1960), and varying interpretations (Büdel, 1949*b*, 89–90; Reiter, 1961, 380; 1963, 407).

According to a more recent study (Seppälä, 1973), the distribution and orientation of ancient parabolic sand dunes in northern Fennoscandia, which are now being destroyed, demonstrate that the dunes were developed by west and northwest winds whereas the effective winds today are from the southwest. Seppälä concluded that the dunes formed under periglacial conditions as the result of a high-pressure system over the retreating Scandinavian Ice Sheet; also that the period of eolian activity was brief as indicated by the short distance of dune migration, and that the climate at the time permitted the dunes to be partially vegetated as shown by their parabolic shape.

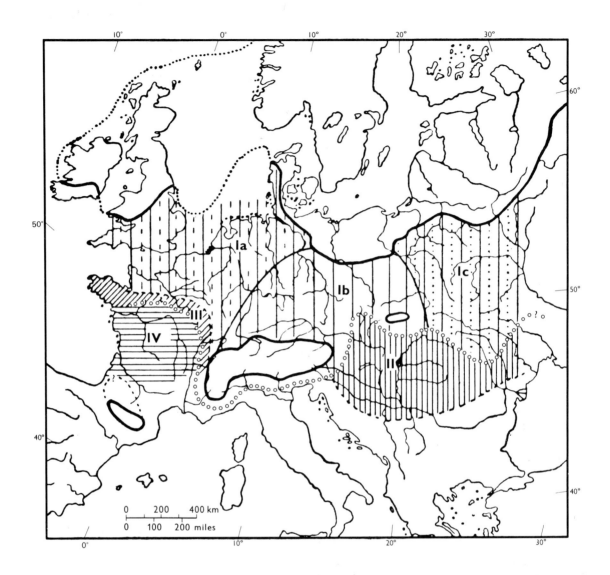

		Ib	Intermediate glacial province
	Border of ice sheets	Ic	Glacial – continental province
	Equatorial limit of permafrost	II	Continental permafrost – forest climate
°°°°°°°	Poleward forest limit	III	Maritime tundra climate without permafrost
I	Permafrost – tundra climate		
Ia	Glacial – maritime province	IV	Maritime forest climate without permafrost

14.8 Climatic regions in Europe during Würm Glaciation (*after Poser, 1948a, 65, Figure 6*)

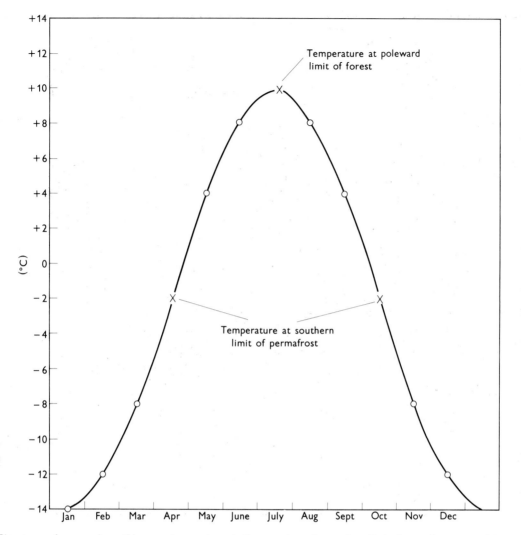

14.9 Sine curve of assumed monthly mean temperatures in Europe where the southern limit of permafrost crossed the northern limit of forest during the Würm maximum (*cf. Poser, 1947a, 16; 1948a, 57*)

8 Overview

Based mainly on conclusions regarding depth of frost cracking and depth of thawing, and on palynological data regarding the nature of the vegetation, Poser (1947*b*; 1948*a*) recognized four climatic provinces in his reconstruction of European environmental conditions for the Würm maximum (Figure 14.8). These were (Poser, 1948*a*, 65–6) (1) Permafrost – tundra climate with subdivisions: (*a*) glacial-maritime province in the west characterized by relatively moist and warm conditions; (*b*) intermediate-glacial province in central Europe charac-

terized by the lowest temperatures and a summer precipitation maximum; (*c*) glacial-continental province farther east characterized by continental conditions the year around but with warmer summers and winters than in central Europe. (2) Continental permafrost – forest province with varied continental conditions. (3) Maritime-tundra province without permafrost, transitional between (1*a*) and (4). (4) Maritime-forest province without permafrost.

Poser was able to draw some conclusions regarding actual temperatures by making the following premises: (1) Permafrost today requires a mean

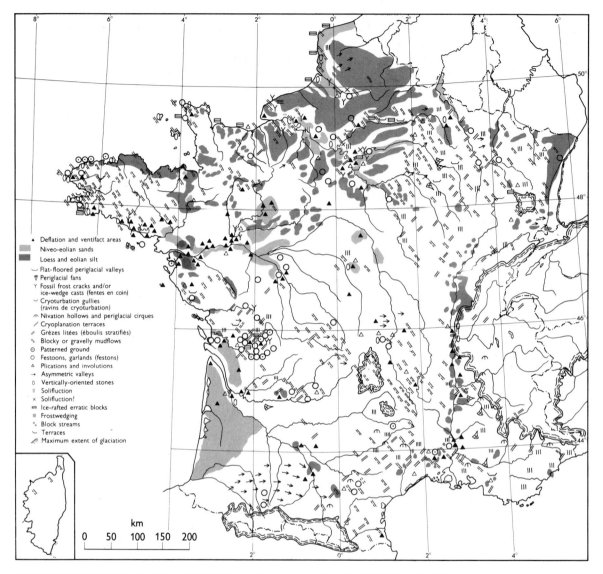

14.10 Periglacial features in France (*after Cailleux, Guilcher, and Tricart, 1956, Plate III, 34*)

annual temperature of $< -2°$. (2) The poleward limit of forest in the Northern Hemisphere approximates the 10° isotherm for July. (3) The distribution of monthly mean temperatures approximates a sine curve. (4) The mean annual temperature approximates the mean temperatures for April and October. (5) The foregoing premises apply to the Würm maximum as well as to the present. (6) The southern limit of permafrost during the Würm maximum is given by the distribution of fossil frost cracks, involutions, Stiche, and asymmetric valleys. (7) Palynological evidence for the position of the forest limit during the Würm maximum is reliable (an unstated premise). All the premises are open to question.[3]

Taking the permafrost limit and the forest limit

[3] Premise 1 suffers from the fact that the mean annual temperature required for permafrost is subject to some regional variation as discussed in the chapter on Frozen ground. Premise 2 does not hold for some regions. The applicability of premises 3–5 to late-Pleistocene conditions was questioned by Butzer (1964, 283, footnote). (Cf. also Frenzel, 1967, 137; 1973a, 141–2.) Premise 6 is based in part on the unreliable evidence of involutions, Stiche, and asymmetric valleys. Premise 7 (with the evidence then available)

as providing two isotherms ($-2°$ and $+10°$), Poser found that they crossed in the eastern foreland of the Alps.[4] Here, then, the mean annual temperature (and therefore April and October temperature) was $-2°$ and the July temperature $10°$. By virtue of the sine-curve premise, this gave the monthly temperatures in Figure 14.9. Thus the mean summer (June–August) and winter (December–February) temperatures were, respectively, $8.7°$ and $-12.7°$. Therefore, the mean July and January temperatures during the Würm maximum would have been, respectively, some $8°–9°$, and $12°$ lower than at present (Poser, 1948a, 57). Comparison of the position of the Würm permafrost limit and forest limit elsewhere shows that the temperature reduction was different in different places, seasonally as well as regionally (Poser, 1947a, 16).

Poser's attempt to put quantitative parameters on his reconstructions illustrates the possibilities as well as the difficulties, and it remains a signal pioneering effort, however uncertain some of the premises and however major the eventual refinements.

Subsequently Büdel (1953, 250) concluded the Pleistocene temperature decrease in central Europe must have been as much $13°–14°$, based on mean annual temperatures of $-4.8°$ to $-8.6°$ at the southern boundary of Eurasian permafrost, and thus that the temperature was at least $-5°$ in central Europe where there are ice-wedge casts in areas that now have a mean annual temperature of $8°–9°$. Büdel's conclusion for central Europe is very close to Poser's if allowance is made for the fact that Poser assumed $-2°$ for the southern limit of permafrost and that his mean annual temperature reduction would be approximately $10°$ (i.e. $\frac{8.7° + 12°}{2}$, so that his comparable annual estimate would be a Würm temperature reduction of at least $13°$, taking $-5°$ for the southern limit of permafrost.

Somewhat later Cailleux, Guilcher, and Tricart (1956) produced a comprehensive map or periglacial features in France, showing ice-wedge casts and/or fossil frost cracks (fentes en coin) of presumed Würm age as far south as lat. $45°$N (Figure 14.10). Fossil cryoplanation terraces are shown between lat. $48°$ and $48°30'$N in Finistière (Dep.) and at lat. $46°$N

in Charente (Dep.). If correctly identified the ice-wedge casts demonstrate that permafrost existed far south of where it was mapped by Poser and more nearly as shown by Kaiser (1960).

Kaiser's map (Figure 14.11) of western and southern Europe shows periglacial features he regarded as indicative of permafrost. These were ice-wedge casts, involutions, sorted patterned ground (Strukturböden), and pingos. The map is notable for taking age differences into account and indicating the relationship of the features to various Pleistocene glacial borders. Kaiser (1960, 137; 1969, 30) inferred a maximum Pleistocene temperature reduction in central and western Europe of $15°–16°$, but he assumed $-2°$ for the southern limit of permafrost as related to periglacial features in areas for which he gave a present mean annual temperature of $13°–14°$. Kaiser also cited ice-wedge casts near Bordeaux, France, and Beograd, Yugoslavia, where the present mean annual temperature is about $12°$ (US Department of Commerce, 1975), although Kaiser indicated about $11°$ for Beograd. If $-7°$ is accepted as necessary for the development of ice wedges, Kaiser's $15°–16°$ temperature reduction for central and western Europe would be revised to $20°–21°$, which seems very large, although Kaiser (1969, 30) suggested that the maximum Pleistocene temperature reduction was considerably greater than $15°–16°$ in more northern regions.

A map by Frenzel (1967; 1973a) (Figure 14.12), based on periglacial features such as those cited by Kaiser but with the addition of cryoplanation terraces and palsas and the omission of involutions and patterned ground – except for 'giant' (ice-wedge) polygons (Frenzel, 1967, 28–30; 1973a, 30–1) – agrees with Kaiser's map in showing former permafrost in southern France. Frenzel's reconstruction is confined to the Würm and distinguishes between forms developed during the different stades.

The distribution of fossil pingos and other permafrost indicators (Figures 5.54, 14.12), and the assumption of $-2°$ for the southern limit of permafrost, led Frenzel (1967, 136, 206; 1973a, 141, 209) to conclude that the mean annual temperature reduction during the coldest part of the Würm (immediately after Stillfried B Interstade) was at least $12°$ in East Anglia and Wales, $15°–16°$ in

---

would not be acceptable to modern palynologists; in fact the northern limit of forest during the Würm maximum is not well established as pointed out by G. C. Maarleveld (personal communication, 1977). For instance, there is little indication of its position in France despite study of numerous pollen profiles (van Campo, 1969).

[4] Poser's evidence for the southern limit of permafrost east of here is based mainly on the questionable evidence of involutions and asymmetric valleys.

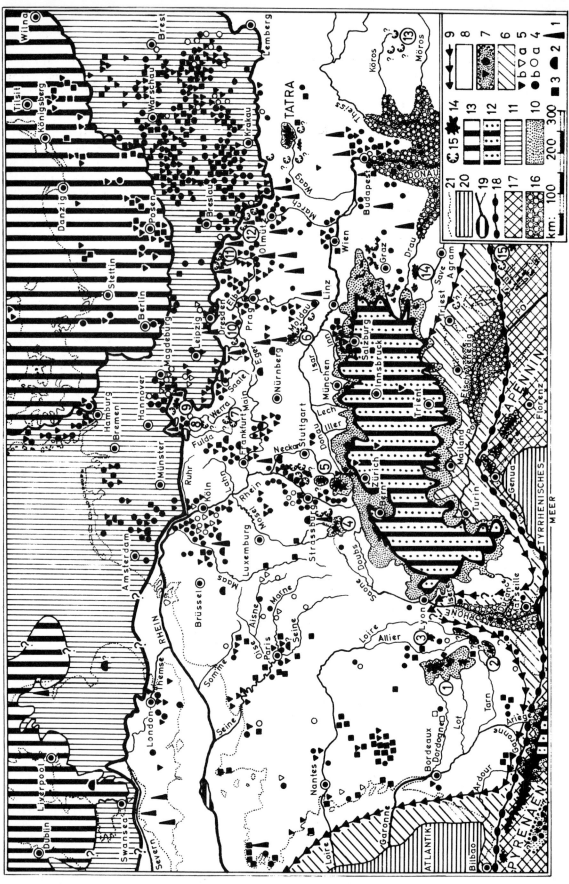

14.11 Borders of Pleistocene glaciations and fossil features indicative of permafrost in western and central Europe (Karlheinz Kaiser, personal communication, 1979; cf. Kaiser, 1960. Plate 1 opposite 136)[5]

southern France, and 13° in central Hungary. Other figures are cited for other stades. Information is lacking as to the exact nature and location of the features on which these conclusions are based.

A map of Würm periglacial provinces in Europe, based on the earlier work of Büdel, Tricart, and Troll, shows regions of widespread polar desert, including southern England and southern France (Tricart, 1967, 35, Figure 5). A generalized map of Würm permafrost in Europe and the Soviet Union was published by Velichko (1972, Figure 3, opposite 68).

A more recent regional map of Würm permafrost (Figure 14.13) was compiled from miscellaneous sources by Maarleveld, who assumed that a maximum mean annual air temperature of $-2°$ was implied by fossil frost cracks and $-6°$ by ice-wedge casts. Thus the southern permafrost boundary implies the Würm $-2°$ isotherm. The map also shows the present-day 10° and 13° isotherms. Figure 14.14 is the latest compilation of Würm permafrost features.[6] It shows ice-wedge casts and pingo scars in the Federal Republic of Germany and is notable for including only carefully evaluated and reasonably well-dated features from among the several hundred others that are reported in the literature.

1 Proved cryoplanation terraces (mainly Würm, after Demek, 1969*a*; cf. Kaiser, 1960); 2 Proved pingos (mainly Würm); 3 Proved sorted forms of patterned ground (mainly Würm); 4 Proved periglacial involutions (pocket soil, etc.) a Pre-Würm b Würm; 6 Regions beyond continuous permafrost (probably mainly tree and shrub tundra); 7 Islands of permafrost beyond continuous permafrost (Ebro Basin); 8 Permafrost zone; 9 Southern limit of continuous (possibly also discontinuous) permafrost; 10 Maximum extent of valley-glacier complexes and piedmont lobes (Günz, Mindel, Riss–Alps, Pyrenees); 11 Maximum extent of Scandinavian Ice Sheet (Elster and Salle Glaciations); 12 Valley-glacier complexes with piedmont lobes during Würm Glaciation (Alps, Pyrenees); 13 Regions covered by Scandinavian Ice Sheet during Weichsel Glaciation; 14 Plateau and valley glaciations; 15 Cirque glaciations 16 Larger areas of trees (along floodplains) north of forest limit; 17 Boreal forest, northerly part of southern Europe; 18 Poleward limit of permafrost; 19 Inland seas (in part as ice-dammed lakes); 20 Ocean; 21 Present coastline

An increase in mean annual air temperature of 14°–16° since the Weichsel maximum is indicated (Johannes Karte and Herbert Liedtke, personal communication, 1979).

The foregoing discussion illustrates that correct identification of permafrost indicators and correct assumptions as to the temperatures they imply are central questions that permit a considerable spread of opinion. Nevertheless the focus is sharpening and it is becoming increasingly feasible to compare conclusions with evidence from other sources, especially in Europe where so much periglacial research has been conducted. A stimulating attempt at such a comparison was made by Maarleveld (1976), who assumed maximum mean annual air[7] temperatures of about 0° for small seasonal frost cracks, 0° to $-6°$ for other seasonal frost cracks, 0° for drop-like involutions, $-2°$ for chaotic involutions and for pingos, $-6°$ for ice wedges in permafrost, and $-2°$ to $-7°$ for certain snow-meltwater deposits he associated with frozen ground. In considering seasonal frost cracking versus permafrost cracking, he applied the Stefan equation (discussed under Seasonally frozen ground in the chapter on Frozen ground) to reported depths of fossil cracks to estimate the former surface freezing index[8] (Maarleveld, 1976, 65). Using these guidelines, Maarleveld prepared a temperature curve for the Netherlands for the last 50 000 years and compared it with climate-related curves from elsewhere that are based on a variety of evidence. Figure 14.15 shows that in some broad aspects the curves match reasonably well but that there are also notable differences.

IV   USSR

Frenzel in reviewing the Quaternary environments of northern Eurasia, plotted the distribution of

[5] Figure 14.11 is an updated, unpublished version of Kaiser's 1960 map, and was substituted for the latter at the last moment while this volume was in proof. The writer is most grateful to Professor Kaiser for his kindness in drawing attention to the map and permitting its publication.

[6] The writer is much indebted to Drs Johannes Karte and Herbert Liedtke who compiled the map, and to Drs Stefan Kozarski (IGU) and Andrei Vilichko (INQUA) under whose auspices it was prepared, for their kindness in permitting advance publication and expediting the map so it could be inserted while this volume was in proof.

[7] G. C. Maarleveld, personal communication, 1978.

[8] The *surface* temperature freezing index was compared directly with the *air* temperature freezing index in drawing the conclusion that seasonal frost cracking prevailed in the central Netherlands (G. C. Maarleveld, personal communication, 1978). The comparison would be conservative in not favouring permafrost cracking, since air temperatures would normally be lower than ground-surface temperatures (cf. section on Geothermal gradient in chapter on Frozen ground).

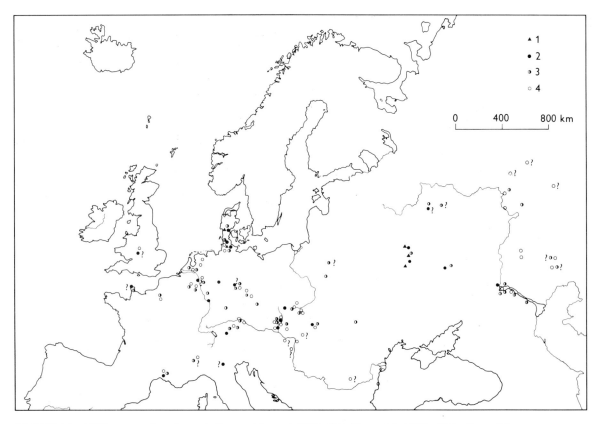

14.12 Weichsel/Würm permafrost forms in Europe (*Frenzel, 1967, 204, Figure 98; 1973a, 208, Figure 98*)

1 Prior to Amersfoort Interstade
2 Between Amersfoort and Brørup Interstades
3 Immediately after Brørup Interstade
4 After Stillfried B Interstade

some fossil periglacial features (Frenzel, 1959, 1005, 1010, 1019, 1020, Figures 6–7, 11, 13; cf. 1960; 1967). He was careful to note that thermokarst phenomena were probably largely caused by human or other non-climatic disturbances rather than by climatic change. Although using periglacial evidence, his reconstructions were based mainly on detailed comparisons between the distribution of former and present vegetation and on the assumption that the differences were climatically controlled (Frenzel, 1968).

A map of Pleistocene periglacial features in the USSR, including the former extent of permafrost, was presented by Markov (1961) and Popov (1961) and is reproduced as Figure 1.2. A more recent map by Kostyayev (1966) shows the southern limit of permafrost much farther north but is based on the dubious assumption that most polygonal systems of

soil wedges (Bodenkeile) in Quaternary deposits are due to convection and have no necessary relationship to frost action.

Permafrost in the USSR at least as old as early Pleistocene is indicated by syngenetic ice-wedge casts in the Olerskaya Formation of the Kolyma Lowland in the Far Northeast (Arkhangelov and Sher, 1973; 1978), and the continuous presence of permafrost in Yakutia since the Lower Pleistocene (Katasonov, 1977), or first half of the Pleistocene (Grigor'yev, 1973; 1978), has been reported.

According to Velichko (1973; 1975; 1978), permafrost in eastern Europe during the Upper Pleistocene reached its maximum limit at the end of the last (Valdai) glaciation, which was much less extensive than the preceding (Dnieper) glaciation. This and a lack of parallelism between the limits of the permafrost and of the ice sheet (Figure 14.16) were

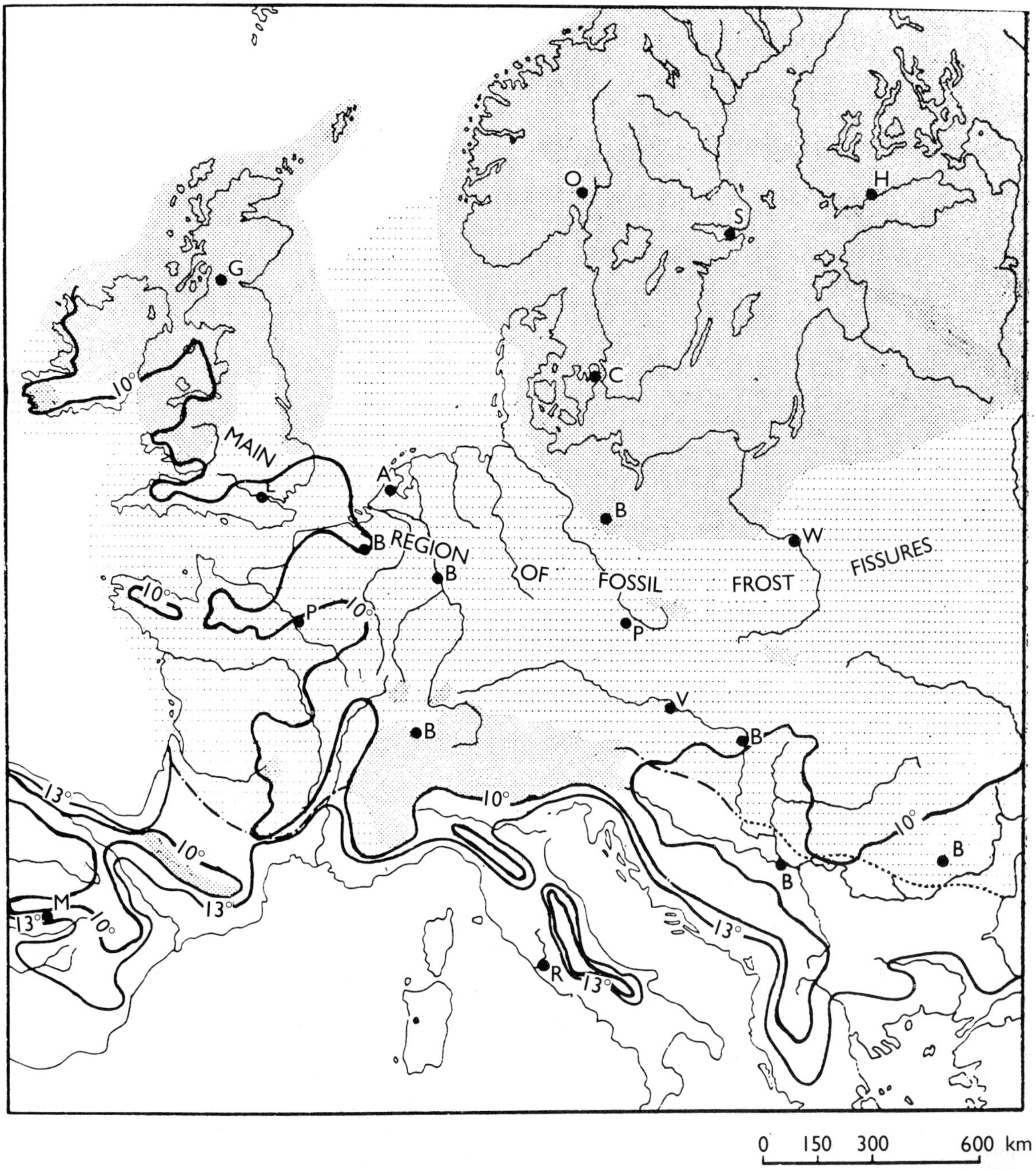

Permafrost area

Ice-covered area

a b Southern limit of region with fossil frost fissures and frost cracks
Indefinite part of the outline is shown by b

—13°— Present mean annual air temperature

14.13 Weichsel/Würm permafrost in Europe (*after Maarleveld, 1976, 72, Figure 9*)

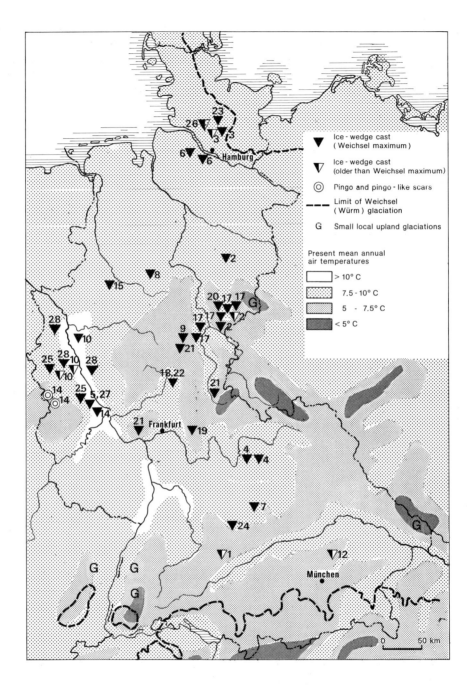

Sources: 1 Bleich and Groschopf (1959); 2 Brüning (1966); 3 Dücker (1954); 4 Emmert (1965); 5 Frechen and Rosauer (1959); 6 Hagedorn (1961); 7 Haunschild (1967); 8 Henke (1964); 9 Horn (1971); 10 Kaiser (1958); 11 Kaiser (1960); 12 Karrasch (1970); 13 Lüttig (1960); 14 Mückenhausen (1960); 15 H.-M. Muller (1978); 16 Picard (1956); 17 Rohdenburg (1967); 18 Sabelberg *et al.* (1976); 19 Schirmer (1967); 20 Selzer (1936); 21 Semmel (1968); 22 Sabelberg, Rohdenburg, and Havelberg (1974); 23 Svensson (1976*a*); 24 Zeese (1971); 25 Löhr and Brunnacker (1974); 26 Picard (1956); 27 Schirmer (1970; cf. 5 – Frechen and Rosauer, 1959); 28 Thoste (1974); 29 Lüttig (1960)

14.14 Features indicative of permafrost in the Federal Republic of Germany during the Weichsel maximum (*courtesy Johannes Karte and Herbert Liedtke, 1978*)

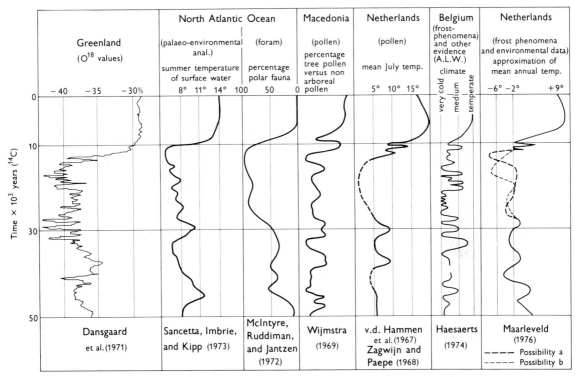

14.15 Some climate-related curves derived from various types of evidence (*after Maarleveld, 1976, 74, Figure 10*)

¹ And other evidence (A.L.W.)

taken to indicate that the permafrost was not controlled by the climatic influence of the ice. From the distribution and dating of various periglacial features, Velichko further concluded that the permafrost increased continuously during Valdai time rather than waxing and waning with the known fluctuations of the ice sheet (Figures 14.17–14.18). He also believed that the distribution of past and present permafrost was associated with, and controlled by, the distribution of sea ice, which contributed to the development of widespread anticyclonic polar air masses.

An interesting point in Velichko's (1975) stimulating discussion is the reported lack of relation between the limits of Valdai ice and of permafrost. This could perhaps be explained by the former being dependent on temperature and precipitation, whereas by definition the latter is directly dependent on ground temperature alone. The conclusion regarding *continuously* increasing permafrost was left somewhat open in Velichko's (1973; 1978) discussion, and more data would be helpful regarding the climatic interpretation of the periglacial features

he cited; it is not immediately apparent why the increased diameter of fossil nonsorted polygons in Loess II as compared with Loess I might not be related to the increased thickness of Loess II (Figure 14.17). Similarly, detailed evidence would be helpful regarding the dependency of permafrost on distribution of sea ice, including the disappearance of sea ice as being a major factor in the sudden degradation of some 10 000 000 km² of permafrost, the majority within 1000–1500 years (Velichko, 1973, 23–4; 1978, 71).

The presence of a nonfrozen layer at a depth of 80–100 m sandwiched between permafrost in western Siberia has been cited as a consequence of the Holocene Hypsithermal (thermal maximum), with the overlying permafrost having developed within the last 3000–5000 years (Kudryavtsev, 1966, 107; cf. Jahn, 1975, 29).

Looking at the USSR as a whole, Popov (1976, 56; 1977, 51–2) reported that fossil periglacial features are much more widespread and reflect much greater shifts of natural zones in the west than in the east during the Pleistocene and Holocene.

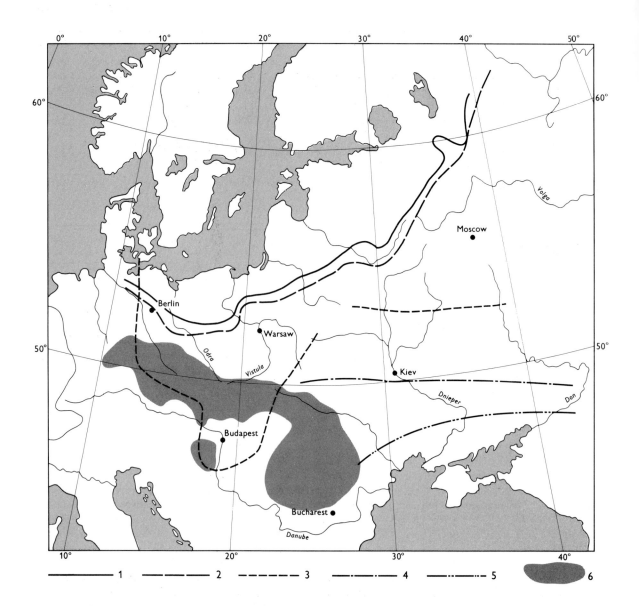

——— 1	— — — 2	--- 3	—·—·— 4	—··—··— 5	⬭ 6

1. Limit of Bologovsk stade of Valdai glaciation;
2. maximum limit of Valdai glaciation;
3. limit of permafrost at start of Valdai glaciation;
4. limit of permafrost in post-Bryansk time;
5. limit of permafrost at end of Valdai glaciation;
6. periglacial environment in mountains

14.16 Main stages in development of permafrost in Eastern Europe during Upper Pleistocene (*after Velichko, 1975, 94, Figure 2, after Velichko and Berdnikov, 1969, 625, Figure 5*)

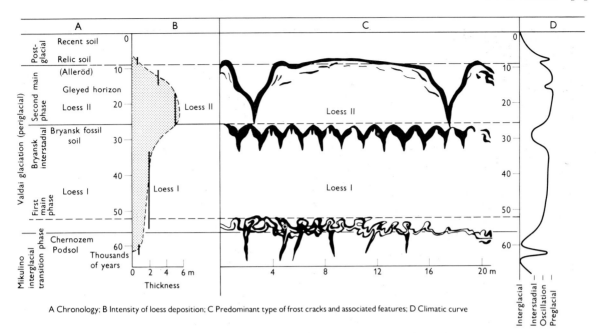

A Chronology; B Intensity of loess deposition; C Predominant type of frost cracks and associated features; D Climatic curve

A Chronology; B Intensity of loess deposition; C Predominant type of frost cracks and associated features; D Climatic curve

14.17 Changes in predominant type of frost cracks and associated features in Eastern Europe during Upper Pleistocene (*after Velichko, 1975, 97, Figure 4*)

V Japan

Periglacial features in Japan are primarily highland occurrences but lowland ice-wedge casts have been reported from Hokkaido. According to the report some are V-shaped, up to 2 m deep and wide, margined by aligned stones and upturned beds, and form polygonal patterns (Koaze, Nogami, and Iwata, 1974*a*, 181–2; 1974*b*). The features contain volcanic sand, radiocarbon dated at 32 200 years BP and interpreted as formerly part of the active layer. Citing a present mean annual temperature of nearly 6° and Péwé's (1966*b*) conclusion that Alaskan forms required a mean annual air temperature of −6° to −8° to develop, the report estimated that the temperature in the lowlands of eastern Hokkaido was 12°–14° colder than at present. Further data are required for confirmation of such occurrences and extrapolations.

VI North America

Fossil periglacial features are reported from many places in North America.[9] However, the origin of some features is suggestive or debatable rather than proved (Black, 1969*b*; 1976*a*, 21–4; Flemal, Hinkley, and Hesler, 1973; R. W. Galloway, 1970; Malde, 1961; 1964; Prokopovich, 1969; Ruhe, 1969, 177–82), and much remains to be learned. Brunnschweiler (1962; 1964) was the first to attempt an environmental reconstruction of North America based on periglacial features. Field work and a review of the literature originally convinced him that in many places periglacial features extend far south of previous estimates. His revised map (Figure 14.19) is more conservative but still speculative in showing the distribution of supposed thermokarst basins on the Atlantic Coastal Plain as far south as Virginia. Even the existence of thaw basins in New Jersey

[9] Antevs (1932, 51–5), Barsch and Updike (1971), Berg (1969), Black (1965*a*; 1969*a*; 1969*b*), Borns (1965), Brookes (1971), Clark (1968), L. Clayton and Bailey (1970), DeGraff (1976), Denny (1951; 1956, 30–42), Dionne (1966*b*; 1966*c*; 1969; 1970*b*), Fleisher and Sales (1971), Flemal (1972), R. W. Galloway (1970), R. P. Goldthwait (1940, 32–9; 1969), Hartshorn and Schafer (1965, 17), Horberg (1951, 11–13), W. H. Johnson (1978), Lee (1957, 2), Mears (1966), Michalek (1969), Morgan (1972), Potter and Moss (1968), Rapp (1967, 233–44), Raup (1951, 113, 114–15, Figure 4–5), Schafer (1965), Schafer and Hartshorn (1965, 124), Sevon (1972), H. T. U. Smith (1962), Walters (1978), Wayne (1967), Wolfe (1953; 1956), and others.

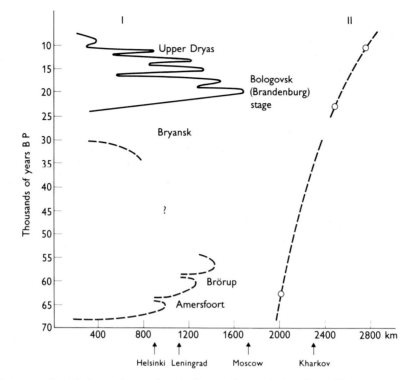

14.18 Changes in extent of glaciation and permafrost in Eastern Europe during Upper Pleistocene (*after Velichko, 1975, 95, Figure 3*)

nearer the Wisconsin glacial border (Wolfe, 1953; 1956) remains to be confirmed (Rasmussen, 1953). The evidence is meagre and far from clear. Their appearance is suggestive and coring of one of the basins (Szabo Pond) on the inner edge of the New Jersey Coastal Plain showed a basal vegetation zone consistent with a spruce woodland with dwarf birch, tentatively dated at about 12 000 years BP (W. A. Watts, personal communication, 1978). Although wedge features in shale about 45 km south of the ice border in central New Jersey have been interpreted as possible fossil ice-wedge polygons (Walters, 1978), this interpretation conflicts with a review that concluded, pending additional evidence, that mean annual air temperatures at the border in this region were perhaps as much as 2° (Sugden, 1977, 28–9, Figure 4). Some other features mapped by Brunnschweiler as probably associated with permafrost are also of debatable significance, including the Mima Mounds of Washington whose origin is far from established (cf. section of Erosion under Patterned ground [Origin – cracking non-essential] in chapter on Some periglacial forms). Consequently the ex-

tensive tundra (ET) and other vegetation zones as reconstructed by Brunnschweiler (1964, 230, Figure 8) are suspect to the extent they are based on such evidence.

More recently R. J. E. Brown and Péwé (1973, 90–4) compiled a map and list of ice-wedge cast occurrences (Figure 14.20, Table 14.1), and discussed some of the paleoenvironmental inferences (cf. Péwé, 1975, 56–7). Here again the origin of some of the features is controversial as noted by the authors. Despite problems of correctly identifying periglacial features, and greater difficulty of dating them than in Europe, periglacial features are beginning to provide scattered evidence regarding former temperature and moisture conditions in North America. According to Péwé (1973*b*, abs., 22; cf. R. J. E. Brown and Péwé, 1973, 89–91), ice-wedge casts from widely separated localities in temperate North America show that permafrost existed along the ice-sheet border during late-Wisconsin time, 10 000 to 20 000 years ago, and that

14.19 Pleistocene periglacial features in United States (*after Brunnschweiler, 1964, 224, Figure 1*)

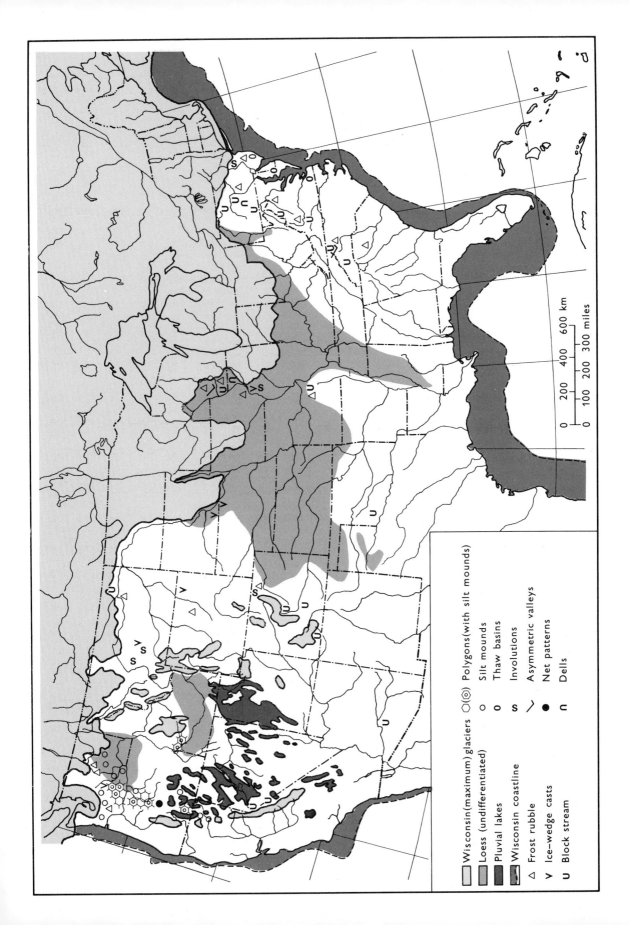

	Wisconsin (maximum) glaciers	◯(◎)	Polygons (with silt mounds)
	Loess (undifferentiated)	o	Silt mounds
	Pluvial lakes	⊙	Thaw basins
	Wisconsin coastline	S	Involutions
	Frost rubble	⟩	Asymmetric valleys
△	Ice-wedge casts	●	Net patterns
⋁	Block stream	∩	Dells

0   100   200   300 miles

0   200   400   600 km

Table 14.1. Location and age of ice-wedge casts in North America (*after R. J. E. Brown and Péwé, 1973, 92, Table 1*)

Map location number	Reference	Locality	Time of ice-wedge formation (yr)	Distance from glacial border (km)
United States – outside Wisconsin border				
1	Péwé (1948)	Thurston Co., Washington[1,2]	Wisconsin ~14 000	At border
1	Newcomb (1952)			
1	Ritchie (1953)			
2	Schafer (1949)	Near Vaugn, Montana	Wisconsin	16
3	Lee Clayton and Bailey (1970)	SW North Dakota	Early Wisconsin (?)	10–100 (?)
4	Ruhe (1969)	NW Iowa	Wisconsin	10–100 (?)
4a	Ruhe (1969)	Tama Co., Iowa	Wisconsin	40 (?)
5	Black (1965b)	SW Wisconsin	~20 000	0.5–100
6	Wolfe (1953)	New Jersey[a]	Wisconsin	<20
United States – inside Wisconsin border				
7	Birman (1952)	Rhode Island[a]	Late Wisconsin	~70
8	Denny (1936)	S. Connecticut[a]	Late Wisconsin	~50
9	Totten (1973)	Richland Co., Ohio[2]	<14 000	5
10	Wayne (1965; 1967)	W. central Indiana[2]	<14 500	75
11	Horberg (1949)	Bureau Co., N. Illinois[2]	~14 000	30
12	Black (1956b)	Outagamie Co., and Columbia, Co., Wisconsin	<12 000	50
Southern Canada				
13	Westgate and Bayrock (1964)	Edmonton, Alberta	>31 000	600
14	Berg (1969)	Edmonton, Alberta		600
15	Morgan (1969)	Calgary, Alberta	Pre- to late-Wisconsin	~300
16	Morgan (1972)	Toronto, Ontario		~300
17	Dionne (1971c)	15 km NW Quebec City, Quebec	11 500–12 000	~600
18	Dionne (1969)	South shore of St Lawrence River	~12 000	~600
19	Lagarec (1973)	SE Quebec	~12 000	~600
20	Borns (1965)	N. Nova Scotia	~12 000	~200
21	Brookes (1971)	Newfoundland		
Alaska				
22	Ager (1972)	Healy Lake	Wisconsin	
23	Péwé, Church, and Andresen (1969)	Big Delta	Wisconsin	
24	Péwé (1965b)	Shaw Creek	Pre-Illinoian	
25	Blackwell (1965)	Tanana River	Wisconsin	
26	Péwé (1965a)	Fairbanks	Pre-Illinoian	
27	Hopkins et al. (1955)	Bristol Bay	Wisconsin	
28	Hopkins[b]	Kvichak Pen.	Wisconsin (?)	
29	Hopkins and Einarsson (1966)	Pribilof I.	Pre-Wisconsin [Illinoian (?)]	
30	Hopkins[b]	Pribilof I.	Pre-Wisconsin	
31	Guthrie and Matthews (1971)	Northern Seward Pen.	Wisconsin	
32	McCullough, Taylor, and Rubin (1965)	Baldwin Pen. NW Alaska	Pre-Illinoian Pre-Wisconsin [Illinoian (?)]	
33	McCullough and Hopkins (1966)	NW Alaska	Pre-Wisconsin [Illinoian (?)]	
34	Hopkins, MacNeil, and Leopold (1960)	Nome	Wisconsin	

[a] Doubtful ice-wedge casts.
[b] Personal communication.

Notes added by A. L. Washburn:
  [1] Presence of former ice wedges based on mound topography but no ice-wedge casts reported to date; origin of mounds uncertain.
  [2] Regarded as unconvincing by Black (1976a, 22–3).

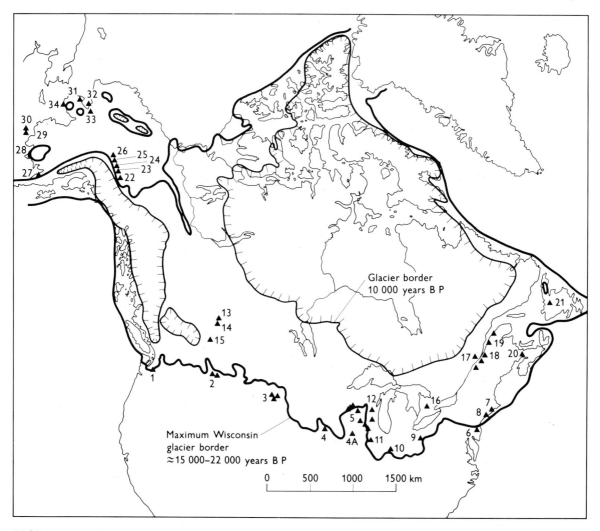

14.20 Location of ice-wedge casts in North America. Glacier borders generalized from Prest (1969) (*after R. J. E. Brown and Péwé, 1973, 90, Figure 8*)

at the Wisconsin maximum the mean annual isotherm of $-6°$ to $-8°$ (or $-7°$) was probably at least 2000 km south of its present position. French (1976a, 241–5) concluded that the distribution of fossil wedge features suggests a former permafrost belt 80–250 km wide along the retreating ice sheet; he believed it was much narrower than the belt in Europe because of the farther south position of the ice margin in North America and the influence of large proglacial water bodies in reducing the permafrost area and in ameliorating the climate.

In the Arctic Coastal Plain of Canada, relic permafrost studied by Mackay, Rampton, and Fyles (1972) had a mean annual temperature of $-7°$ to

$-10°$ at the depth of zero annual amplitude. Since this temperature could not have risen above $0°$ for a prolonged period without thawing the permafrost, they concluded that it had never been more than $7°–10°$ higher than now following development of the permafrost, which they believed occurred much more than 40 000 years ago. On the other hand, inactive (relic) ice wedges below a younger generation of ice wedges in the same area are evidence of a somewhat warmer period than the present (Mackay, 1975a). This warm interval is the Hypsithermal on the basis of numerous ^{14}C dates subsequent to the thaw unconformity and relic wedges; these dates range from 7950 ± 280 years BP

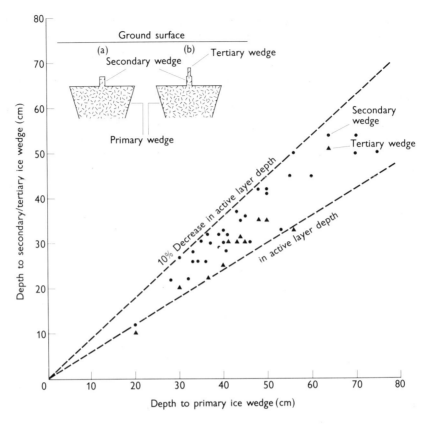

14.21 Depth to top of secondary and tertiary ice wedges as related to decrease in depth of active layer, Mackenzie Delta area, Arctic Canada *(after Mackay, 1976b, 233, Figure 47.1)*

(GSC-2305, Pelly Island) to 10 330 ± 150 years BP (GSC-516, Garry Island) (J. R. Mackay, personal communication, 1978), with intermediate dates from other samples (Garry Island, Hooper Island, Pelly Island, and Tuktoyaktuk – Mackay, 1977c, 461).

A decrease in mean annual air temperature of 10.4° since 9890 ± 130 years BP has been proposed by Delorme, Zoltai, and Kalas (1977, 2031, Table 1; 2041, Table 8; 1978, 463) for their site 3, some 70 km southwest of Tuktoyaktuk, based on paleontologic evidence, a present temperature of −10.8°, and (in the light of Mackay's [1978c] criticism) an inferred, minimum former temperature of −0.4°. The inferred temperature decrease is unexpectedly large and requires further research. Along the coast in the same area, thin secondary and tertiary ice wedges extending upward from contemporary primary wedges indicate a 10–40 per cent decrease in the thickness of the active layer

(Figure 14.21), probably consequent on a post-1950 climatic cooling (Mackay, 1976b).

To the south in central Yukon Territory two generations of fossil sand wedges associated with ventifacts have been reported – one Illinoian or early Wisconsin, the other classical (late).Wisconsin – from which cold dry climates were inferred. Both generations cut paleosols developed on glacial or glacio-fluvial deposits, and since the older wedges were the best developed it was concluded that the climatic cooling they recorded was the more severe of the two (either longer or more intense) (Foscolos, Rutter, and Hughes, 1977, 1, 5, 9, 11).

In the Edmonton, Alberta, area of western Canada, Berg (1969) described fossil sand wedges in Wisconsin sand and gravel beneath till of the last ice advance. The present mean annual precipitation here is 40 cm and the temperature 2° (Berg, 1969, 326). Citing observations that active sand wedges in the Antarctic occur where the annual precipitation is about 16 cm, and Péwé's (1966*b*, 68) conclusion that frost cracking associated with active ice-wedge polygons in Alaska requires at least − 5°, Berg (1969, 331–2) concluded that the Edmonton sand wedges record both a drier and colder climate, with precipitation being about half the present and the temperature lower by 7° or more.[10] Farther south at Calgary, Alberta, Morgan (1969) described ice-wedge casts believed to predate the earliest Laurentide ice advance. Citing Péwé's (1966*b*) report that active ice wedges in Alaska are associated with mean annual air temperatures of − 6° to − 12°, he concluded that the average annual temperature of the area when the ice wedges formed was at least 9° and possibly up to 14° cooler than now.

In eastern Canada, Morgan (1972) reported ice-wedge casts near Kitchener, Ontario, which he originally regarded as having probably originated some 13 000–13 500 ^{14}C years BP. Again citing Péwé's observations, and comparing them with the present mean annual temperature of 7° in the Kitchener area, Morgan concluded that there has been a temperature rise of at least 13°–14° since the Kitchener polygons formed. Subsequently Morgan and Morgan (1977) found many more ice-wedge casts on various older tills in the area but not on younger, and they concluded that the temperature had risen by probably 6° to 8° within about 200 years. The evidence included beetle faunas that are incompatible with permafrost, and later work dated the interval at about 12 750–12 950 years BP (Alan and Anne Morgan, personal communication, 1978).

Farther east, Dionne (1966*b*; 1969; 1970*b*; 1973*b*) reported some 180 ice-wedge casts in Quebec in the vicinity of the Highland Front Moraine south of the St Lawrence River. They occur mainly in fluvio-glacial material. Dionne (1966*b*, 97; 1966*c*, 26) estimated that the former ice wedges required a mean annual temperature on the order of − 4° to − 5°, based on Péwé's (1966*a*, 78; 1966*b*, 68) observation of active ice wedges at − 6° to − 8° in Alaska and preliminary observations by Rapp, Gustafsson, and Jobs (1962) who reported probably inactive ice-wedge (?) polygons at − 3° to − 4° in Swedish Lapland. Subsequent investigations showed that these polygons are characterized by ice-wedge casts and are indeed inactive but that frost cracking can affect the active layer (Rapp and Annersten,

1969; Rapp and Clark, 1971) and extend into permafrost in exceptionally cold winters (A. Rapp, personal communication, 1971). Revising earlier estimates and citing the occurrence of over 300 ice-wedge casts, including some in southern Quebec on the north side of the St Lawrence, Dionne (1975, 73) concluded that a mean annual air temperature lower than − 6° was required for forming ice wedges and, since the present temperature range in the region is 2° to 5.5°, that the temperature was 8°–12° colder when the ice wedges originated. Nevertheless, he believed that the permafrost was probably discontinuous, although mean annual temperatures less than − 6° usually imply continuous permafrost. As previously discussed, application of Péwé's Alaskan observations to other environments, and the difficulty of inferring mean annual temperatures from fossil ice wedges, allow a considerable margin of error. The association of (1) ice-wedge casts with the St Narcisse moraine complex north of the St Lawrence (Dionne, 1971*c*) and with the Highland Front Moraine south of it (moraine complexes dated, respectively, at 11 000–11 500 and 12 500 ^{14}C years BP or slightly older), and (2) the lack of evidence whether the ice wedges formed at both periods or perhaps only at the time of the St Narcisse moraine complex, led Dionne (1975, 71–3) to bracket the occurrence of lower temperature between 11 000 and 13 000 ^{14}C years BP.

Brookes (1971) reported ice-wedge casts in western Newfoundland that he believed to be probably 11 200–11 500 years old. Taking Péwé's − 6° as a base, he suggested they formed when the mean annual air temperature was 12° less than at present, although such a temperature difference seemed somewhat excessive to him. Subsequently Eyles (1977, 2803–5) reported ice-wedge casts in north-central Newfoundland from which he inferred a temperature rise of at least 10.4° – again based on the − 6° criterion – since a time tentatively bracketed at 10 000–12 000 years BP.

In the United States, pre-Illinoian ice-wedge casts and therefore former permafrost of this age have been reported from Cape Deceit and the Fairbanks area in Alaska. The Cape Deceit features (Guthrie and Matthews, 1971, 475–82) were thought to be about 1 ± 0.5 million years old (Hopkins, 1972, 125), and the Fairbanks features, including gelifluction sheets, at least 1 million years old (Figure 14.22) (R. J. E. Brown and Péwé, 1973, 91; Péwé, 1973*b*,

---

[10] Actually Péwé's (1966*b*, 68) − 5° temperature referred to the level of zero annual amplitude where the temperature is somewhat warmer than the mean annual air temperature, which in the situation described by Péwé is − 6° to − 8°, so that the difference if the Alaska comparison is accepted would be at least 8°.

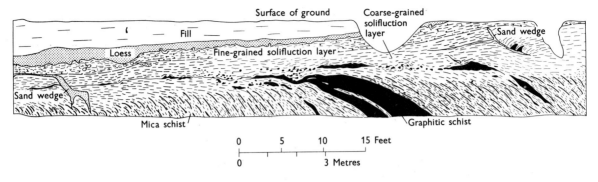

14.22 Early Quaternary gelifluction deposits and ice-wedge casts of sand, University of Alaska campus (*after Péwé, 1975, 63, Figure 33*)

19; 1975, 63). Younger pre-Wisconsinan (Illinoian?) ice-wedge casts have also been reported (Table 14.1).

Buried ice wedges in silt help to reconstruct the Wisconsin environment of Alaska. Although reported from a number of places, they have been particularly studied, along with other stratigraphic features, in a 110-m (360-ft) long permafrost tunnel containing a 14-m (45-ft) vertical ventilation shaft, excavated near Fairbanks by the US Army Cold Regions Research Engineering Laboratory (CRREL). Here a lower zone of large ice wedges, 1–2 m (3–6 ft) across, unconformably underlies a zone, some 6 m (20 ft) thick, of smaller wedges that are up to 30 cm (1 ft) across (Figure 14.23) (Sellmann, 1967; 1972). Radiocarbon dates of material from the gravel beneath the lower zone of large wedges range from 33 750 (± 2000) to >39 900 ^{14}C years BP (Sellmann, 1972, 8, Table I), dates from the lower zone itself range from 30 700 (+ 2100–1600) to 33 700 (+ 2560–1900) ^{14}C years BP, and dates from the upper zone of small wedges range from 8460 (± 250) to 14280 (± 230) ^{14}C years BP (Sellmann, 1967, 17, Table II). Chemical analysis of the cation content of extracted soil water showed a pronounced increase below the unconformity (Figure 14.23), suggesting that the lower zone had been less subject to thawing and leaching than the upper.[11] Another unconformity separates the upper zone of small wedges from some 3 m (10 ft) of overlying silt. Radiocarbon dates above and below this unconformity bracket its age between 6970 (± 135) and 8460 (± 250) years BP. The upper unconformity, which is reported elsewhere in the Fair-

banks area as well (Péwé and Sellmann, 1973), was believed by Sellmann to record the lower limit of thawing during the Hypsithermal interval. The lower unconformity, which also appears to occur elsewhere in the area, may represent the lower limit of thawing during a Wisconsin interstade but confirmation is lacking. A chemical profile from nearby Eva Creek showed no confirming variation between 10 000 and at least 56 900 ^{14}C years BP (Péwé and Sellmann, 1973, 168, Figure 2). The lower-lying large ice wedges suggest epigenetic permafrost (i.e. slow sedimentation rates and a relatively stable land surface), rather than the syngenetic permafrost indicated by the small size and vertical frequency of the overlying wedges.

Ice wedges have also been stratigraphically investigated in other places in central Alaska (Péwé, 1965a; 1965b), in the Barrow area (Jerry Brown, 1969, 122), and elsewhere. According to Messenger and Péwé (1977), mapping of periglacial and other features in Alaska shows that mean annual air temperatures there during Wisconsin time were 4° –10° lower than at present. The surface has barely been scratched but there is ample evidence to indicate the potential significance of periglacial studies for the Quaternary history of Alaska.

Three intersecting generations of ice-wedge casts have been reported by Mears (1966) from the Laramie Basin of Wyoming where the mean annual temperature is now 5.2° (41.4°F). He considered the oldest and largest ice-wedge casts – some 1.2 m (4 ft) wide at the top and 2.4 m (8 ft) deep – and the youngest ice-wedge casts as recording, respectively, pre-Wisconsin and late-Wisconsin ice wedges. How-

---

[11] Other chemical analyses of soil water associated with ice wedges in Alaska are given by Jerry Brown (1966) and Kinosita *et al.* (1978). It should be noted that isotopic fractionation by freezing alone can cause vertical differences in the chemistry of soil water (Mackay and Lavkulich, 1974).

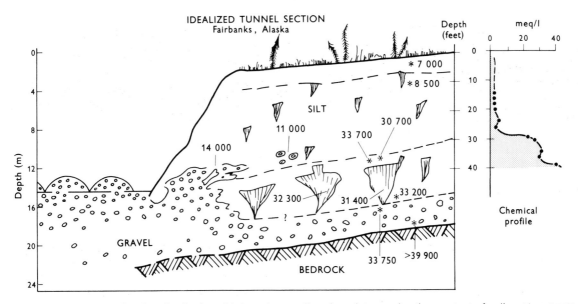

14.23 Idealized section showing distribution of ice wedges, radiocarbon dates, and cation content of soil-water extracts at CRREL permafrost tunnel near Fairbanks, Alaska (*after Sellmann, 1967, 15, Figure 13 with supplementary data from Sellmann, 1972, 9, Figure 9*)

ever, according to Péwé (1975, 56), pre-Wisconsin ice wedges in North America outside of Alaska remain to be proved. Other features that Mears (1973) interpreted as probable ice-wedge casts and periglacial contortions, some in bedrock, occur in the Bighorn Basin where the present mean annual temperature is 7.4°. Mears (1966) accepted a temperature of −6° (21°F) or lower as necessary for the development of ice wedges in Alaska, and he concluded that there had been a temperature rise of at least 14° in Wyoming intermontane basins since Pleistocene glacial maxima (Mears, 1973).[12]

In the mountains of the southwest United States there are slope deposits, believed to be periglacial and to date from the last glaciation, that carry important paleoclimatic implications if correctly interpreted. According to R. W. Galloway (1970), this distribution implies a timberline depression of 1300–1400 m, a mean annual temperature 10°–11° lower than today, and a drier rather than wetter (pluvial) climate. Galloway (1970, 256) also concluded that

... unlike snowlines, which are dependent on both warmer temperatures and winter snowfall, the timber line and the lower limit of periglacial action are fairly closely related

to summer temperature over a wide range of precipitation and so are potentially more reliable guides to past climates.

According to Barsch and Updike (1971, 112–13), the timberline depression on Kendrick Peak in northern Arizona must have been at least 1000 m during late Pleistocene time.

In Wisconsin the distribution of ice-wedge casts, where the mean annual air temperature is now about 5°, led Black (1969b, 234–5) to infer late Wisconsin mean annual temperatures of −5° to −10°. He argued that the growth of the ice wedges required adequate moisture and temperatures 10°–15° lower than now; their destruction was consequent not only on warmer but also on drier conditions, judging from the fact that the ice-wedge casts consist of eolian sand.

In the eastern United States, repeated measurements spanning a third of a century led R. P. Goldthwait (1969; 1976b) to conclude that gelifluction features and large-scale sorted patterned ground on Mt Washington (alt. 1917 m), New Hampshire, are inactive. Except for sporadic patches, permafrost is now confined to the summit area above 1825 m. The large-scale periglacial features he studied extend to much lower altitudes where the

---

[12] According to Mears (1973) '... creation of permafrost probably requires annual average temperatures of at least −6 C.' This is clearly a misstatement; presumably growth of ice wedges was meant (cf. Mears, 1966).

present mean annual temperature is 0°. Since he believed that the features must have developed under conditions of widespread continuous permafrost, Goldthwait (1976*b*, 38) concluded that the mean annual air temperature at that time must have been $-4°$ or colder and he inferred a temperature increase of 4° to 6° since their origin. The $-6°$ would be consistent with the $-6°$ to $-8°$ temperature at the southern limit of continuous permafrost in Alaska as reported by Péwé (1975, 52). Goldthwait suggested that the large-scale patterned ground originated during a Mt Washington nunatak stage 9000–15 000 years ago and that much of the patterned ground was reactivated during Neoglacial cooling some 300–3000 years ago.

As noted earlier in this section, Walters (1978) reported possible ice-wedge casts in central New Jersey about 45 km south of the Wisconsin glacial border. The present mean annual temperature here is about 10°–11°, so following Péwé's $-6°$ to $-8°$ criterion, he suggested a temperature increase since the Wisconsin maximum of some 16°–19° (Walters, 1978, 52). Such evidence is suggestive only but warrants further investigation.

A critical appraisal and reconstruction of North American periglacial environments is clearly becoming increasingly practicable.

VII   Southern Hemisphere

In the Southern Hemisphere, late-Pleistocene periglacial features have been reported from several parts of southern Africa as reviewed by Butzer (1973) but questioned by him except for some occurrences in the Drakensberg highlands where contemporary periglacial phenomena are also present.

In Australia a variety of fossil periglacial features occur on some of the highest areas as at Mount Kosciusko (2230 m; 7316 ft) but former gelifluction is the most widespread. The lower limit of gelifluction today is some 100 m above treeline, and based on the lower limits of former gelifluction there has been a rise in treeline of about 975 m in the Victorian Alps, 1000 m in the Snowy Mountains, and 600 m in Tasmania following the last glaciation (R. W. Galloway, 1965, 605). Here the position of treeline approximates the 10° isotherm for the warmest summer month, and if the annual temperature range is known, changes in mean annual temperature can be inferred from changes in treeline. On this basis, and assuming a former annual tempera-

ture range 3° greater than the present because of other considerations, R. W. Galloway (1965, 606) concluded that there has been a temperature rise of at least 9° in the Snowy Mountains–Canberra region. However, from the presence of fossil block streams in the Snowy Mountains area, and the belief that their movement had required an ice matrix and hence a mean annual temperature below 0° as compared with a present temperature of 5°–5.5°, Caine and Jennings (1968, 100–1) concluded that the temperature rise in this area since the maximum of the last glaciation could have been as little as 6°. Costin and Polach (1971, 229, 234) and Costin (1972, 588), on the basis of now stable, coarse, angular deposits blanketing slopes in the Snowy Mountains in places where the present mean annual temperature is about 7°–10°, inferred a postglacial temperature rise of at least 8°–10°, thus bracketing Galloway's estimate. Similar deposits occur in the Black Mountain area, Canberra, at altitudes requiring a temperature rise of up to 14° if similar temperature criteria are accepted. On the assumption of a former thin snow cover, comparable to that at some 1200 m higher today, the minimum mean annual temperature rise would be reduced to 10° (Costin and Polach, 1972, 22–3). Obviously some uncertainty exists in the basic assumptions, including the exact temperature significance of such slope deposits. A temperature rise of at least 2° (4°F) in the Snowy Mountains within the last 2000–3000 years is suggested by nonsorted steps of this age, although former more intense frost action because of greater snow removal by wind is an alternative possibility (Costin *et al.*, 1967; cf. Costin, 1972, 589).

For Tasmania, R. W. Galloway (1965, 605) concluded that the temperature rise since the last glaciation has been at least 5°. Subsequently, because of abundant evidence of former periglacial conditions, including inactive rock glaciers where the present mean annual temperature is estimated to be about 5.3°, Derbyshire (1973, 145) suggested a temperature rise in Tasmania of at least 6.5°.

New Zealand has been the site of considerable periglacial research. Te Punga (1956*b*) interpreted wedge-like features in weathered bedrock in North Island as of ice-wedge origin but they do not have the characteristics of ice-wedge casts, and the present writer suggests they may be weathering phenomena along joints. It is notable that Soons (1962, 83, 87) reported no ice-wedge casts or other evidence of permafrost in South Island, where a more rigorous periglacial climate would be expectable. Nevertheless, features interpreted as ice-wedge casts were

Table 14.2. Reported ice-wedge casts, mostly within a few hundred metres of sealevel in Argentina (*after Gonzalez and Corte, 1976, 25, Table 1*)

Location	S. Latitude	Reference
San Antonio Oeste (Chubut)	40°	Czajka (1955, 139–40)[1]
Bajada Colorado (Chubut)	42°	A. Figueroa (plane obs.)
Pampa del Castillo (Chubut)	46°	Corte (1967*b*), Czajka (1955, 138–9)[2]
South Río Deseado (Sta Cruz)	48°	Auer (1970, 29)
South Río Deseado (Sta Cruz)	48°	A. Figueroa (personal communication, 1973)
E. Lago Cardiel (Sta Cruz)	49°	A. Figueroa (aerial photos)
Chalia River (Sta Cruz)	50°	Auer (1970, 28)
Pediments (cryopediments) between Río Gallegos and Santa Cruz	51°	A. Corte (plane obs.)
South of Río Gallegos	51.30°	Corte (1967*b*)

[1] Located near sealevel but regarded by Czajka (1955) as less convincing than Pampa del Castillo occurrence.
[2] Located at altitude of 720 m (Czajka, 1955, 138, Figure 7).

reported on Banks Peninsula in South Island by S. A. Harris (1976), who inferred a change of at least 11°. This was based on a present temperature of 12° at sealevel, a 2° decrease to allow for altitude (100 m) and exposure, and a maximum of −1° for the existence of permafrost (S. A. Harris, personal communication, 1978). The report is intriguing but the origin of the features requires confirmation. If true ice-wedge casts are present, a greater temperature change than 11° would seem to be indicated. Also deposits believed to be of gelifluction origin occur in both islands, and in North Island have been tentatively interpreted as correlative with two Wisconsin stades (Cotton and Te Punga, 1955, 1011; Soons, 1962, 76–9). Deposits of grèzes litées (or éboulis ordonnés) have been described from South Island (S. A. Harris, 1975; Soons, 1962, 79–83).

In South America, periglacial research has been largely confined to the Andes. In Argentina, Corte (1976, 192) reported that inactive rock glaciers at different altitudes were consistent with at least four Holocene climatic fluctuations. In addition, Gonzalez and Corte (1976) reported excavating ice-wedge casts in bedrock at an altitude of 200 m at lat. 38°S where the present mean annual temperature is 14.5°, and they listed other reported occurrences of ice-wedge casts, mostly within a few hundred metres of sealevel to just beyond lat. 51°S, many of them being based on observations from the air (Table 14.2). Because of the forms at lat. 38°S and the assumption that at least −5° to −6° is required for ice-wedge polygons, Gonzalez and Corte (1976, 29–32) inferred a temperature rise of about 20° since the last glaciation. However, all the lowland reports require confirmation, including the bedrock occurrence in which the possible influence of jointing was not discussed.

Evidence from the Venezuelan Andes, based on relation of periglacial features to former and present snow lines, led Schubert (1975, 205–10) to conclude that in places the lower limit of the periglacial zone has risen some 1200 m since the maximum of the Late-Pleistocene Mérida Glaciation.

VIII Some estimated temperature changes

Some temperature increases to the present as inferred from fossil periglacial features are listed in Table 14.3 and their distribution is shown in Figures 14.24–14.25. The figures and Table 14.3 help to summarise the discussion only and are in no sense complete, since a number of other examples could be extrapolated from the literature by comparing the occurrence of critical and adequately dated fossil periglacial features with the present climate. Because of the variety of temperatures that have been assumed as an upper limit for development of ice-wedge polygons, the column headed 'Adjusted temperature increase (min)' in Table 14.3 has been added to provide a common −5° maximum-temperature criterion for their development. This temperature adjustment is included in Figures 14.24–14.25. An important consideration in addition to soil conditions is that paleo snow cover and vegetation are commonly major unknowns that can significantly affect ground temperature. As repeatedly noted, mean annual air temperatures must be inferred from incomplete data unless independent information on these factors is available. In the absence of such paleo data, the −5° criterion is believed to be a *conservative upper limit* for the mean annual air temperature under which

Table 14.3. Some suggested temperature changes inferred from periglacial features

No. (cf. Figures 14.24–14.25)	Location	Latitude	Age of features	Mean annual temperature increase to present (min)[1]	Comments	Adjusted temperature increase (min)[2]	Reference
	Britain						
1	Evesham	52°N	Devensian (Weichsel) (max)	13.5°	−3.5° max assumed for ice-wedge polygons	15°	Shotton (1960)
2	Central England	52°–53°N	Devensian (max)	18°	−8° assumed for ice-wedge polygons	17°	R. B. G. Williams (1975, 108, Figure 2)
3	East Anglia	52°–53°N	Devensian (max)	12°	Based on pingos and permafrost forms; −2° max assumed for southern limit of discontinuous permafrost		Frenzel (1967, 136, 206; 1973a, 141, 209)
4	East Anglia	52°–53°N	Devensian (max)	>16°–17°	−7° max assumed for southern limit of continuous permafrost; former continuous permafrost based on ice-wedge casts	15°	Watson (1977, 183, 195–6)
5	Wales	52°–53°N	Devensian (max)	12°	Based on pingos and permafrost forms; −2° max assumed for southern limit of discontinuous permafrost		Frenzel (1967, 136, 206; 1973a, 141, 209)
6	Wales	52°–53°N	Devensian (max)	>16°–17°	−7° max assumed for southern limit of continuous permafrost; former continuous permafrost based on ice-wedge casts	15°	Watson (1977, 183, 195–6)
7	Wales	52°–53°N	Younger Dryas	13°–14°	Based on −4° to −5° mean annual air temperature for open-system pingos in Alaska	11° (Because open-system pingos can occur at −2°)	Watson (1977, 183, 195–6)
	Continental Europe						
8	Iberian Peninsula	37°–41°N?	Last glaciation	8°–10°	Based on lower limit of presumed gelifluction		Brosche (1978, 94–5)
	Spain						
9	Zaragoza	42°N	Riss?	20°?	Based on −5° max for presumed ice-wedge casts; verification of features required (cf. van Zuidam, 1976)	20°?	Johnsson (1960, 76–9)
	France						
10	Crau, Basse Provence	44°N	Würm (max) (see text)	19°	−5° max assumed for ice-wedge polygons	19°	Cailleux and Rousset (1968) (cf. Frenzel, 1967, 206 (1973a, 209)

No.	Location	Latitude	Period			Notes	Reference
11	Bordeaux	45°N	Würm (max) (presumed)		17°	Based on ice-wedge casts and/or fossil frost cracks (fentes en coin)	Cailleux, Guilcher, and Tricart (1956). Present temperature (11.8°), courtesy US Dept. Commerce, Natl. Oceanic and Atmospheric Administration, Natl. Climatic Center, 1979
12	Angoulême area	45°–46°	Würm (max) (presumed)		17°	Based on ice-wedge casts and/or fossil frost cracks (fentes en coin)	Cailleux, Guilcher, and Tricart (1956). Present temperature (12.3°), courtesy US Dept. Commerce, Natl. Oceanic and Atmospheric Administration, Natl. Climatic Center, 1979
13	Angers area	47°–48°	Würm (max) (presumed)		17°	Based on ice-wedge casts and/or fossil frost cracks (fentes en coin)	Cailleux, Guilcher, and Tricart (1956). Present temperature (11.5°), courtesy US Dept. Commerce, Natl. Oceanic and Atmospheric Administration, Natl. Climatic Center, 1979
14	Le Havre area	49°–50°	Würm (max) (presumed)		17°	Based on ice-wedge casts and/or fossil frost cracks (fentes en coin)	Cailleux, Guilcher, and Tricart (1956). Present temperature (11.7°), courtesy US Dept. Commerce, Natl. Oceanic and Atmospheric Administration, Natl. Climatic Center, 1979
15	Yugoslavia Beograd	45°N	Pleistocene (max)	14°³	17°₁³	Based on ice-wedge casts and −2° for southern limit of continuous permafrost	Kaiser (1960, 137, footnote 2)
16	Eastern foreland of Alps	46°–47°N	Würm (max)	10°	13°	−2° assumed for southern limit of continuous permafrost	Poser (1948a, 57)
17	Central Hungary	46°–47°N	Würm (max)	13°	16°	−2° max assumed for southern limit of permafrost	Frenzel (1967, 136, 206; 1973a, 141, 209)
18	Switzerland Ratzerfeld area	48°N	Würm (max)	11°–12°		Based on lapse rate and difference in altitude of fossil and present-day patterned ground	Furrer (1966, 496)
19	Ratzerfeld	48°N	Würm (max and shortly after)	11°	14°	−2° max assumed for permafrost (and for ice-wedge polygons)	Furrer (1966, 496)
20	Central Europe	48°–52°N	Würm (max)	13°–14°	13°	−5° assumed for southern limit of continuous permafrost (and for ice-wedge polygons)	Büdel (1953, 250)

Table 14.3. (continued)

No. (cf. Figures 14.24–14.25)	Location	Latitude	Age of features	Mean annual temperature increase to present (min)[1]	Comments	Adjusted temperature increase (min)[2]	Reference
21	Central Europe	48°–52°N	Pleistocene (max)	15°–16°	−2° assumed for southern limit of continuous permafrost	18°	Kaiser (1960, 137; 1969, 30)
22	Poland Central	51°–53°N	Würm (max)	15°	−6° assumed for fossil ice-wedge polygons in area (−5° for ground temperature at depth of zero annual amplitude)	14°	Goździk (1973, 92–5, 112–17)
23	Federal Republic of Germany Munster (Westf.)	52°N	Würm (max)	See text		14°	H.-M. Müller (1978). Present temperature, courtesy US Dept. Commerce, Natl. Oceanic and Atmospheric Administration, Natl. Climatic Center
24	Netherlands	52°–53°N	Weichsel (max)	15°	−2° to −7° assumed for certain snow-meltwater deposits, −6° max for ice-wedge polygons, −2° max for pingos and chaotic involutions, +9° for present mean annual temperature	14°	Maarleveld (1976, 71–4, Figure 10)
25	Denmark Southwest Jutland	55°–56°N	Weichsel (late glacial)	14°–17°	−6° to −8° max assumed for ice-wedge polygons	13°	Christensen (1974, 172–3)
26	Sweden West coast	56°–57°N	Weichsel deglaciation	12°–13°	−5° to −6° max assumed for ice-wedge polygons	12°	Svensson (1976c, 46, 53)
27	Norway Voss	61°N	Weichsel (Younger Dryas?)	8° (11°–13°?)	−3° to −4° max assumed for ice-wedge polygons, possibly −6° to −8°	10°	Mangerud and Skreden (1972, 91–4)
28	Asia Japan Hokkaido	43°–45°N	>32 000 14C yr BP	12°–14°	−6° to −8° max	11°	Koaze, Nogami, and Iwata (1974a; 1974b)
29	North America Canada Kitchener, Ontario	43°N	13 000–13 500 14C yr BP	13°–14°	−6° to −8° max assumed for ice-wedge polygons	12°	Morgan (1972)

No.	Location	Latitude	Age	Angle	Description	Angle	Reference
30	Southern Quebec	45°–49°N	11 000–13 000 ^{14}C yr BP	8°–12°	−6° max assumed for ice-wedge polygons but nevertheless discontinuous permafrost was considered probable	7°	Dionne (1975, 71–3)
31	Newfoundland West coast	48°–50°N	11 200–15 000 ^{14}C yr BP	12°	−6° max assumed for ice-wedge polygons	11°	Brookes (1971)
32	North-central coast	49°30′N	10000–12000 ^{14}C yr BP	10°	−6° max assumed for ice-wedge polygons	9°	Eyles (1977, 2803–5)
33	Calgary, Alberta	51°N	Pleistocene	9°–14°	−6° to −12° assumed for ice-wedge polygons	8°	Morgan (1969)
34	Edmonton, Alberta	54°N	Wisconsin (prior to last ice advance)	7°	−5° max assumed for ice-wedge polygons	7°	Berg (1969)
35	United States New Mexico Sacramento Mts.	32°–34°N	Last glaciation (max)	10°–11°	Based on slope deposits and timberline depression		R. W. Galloway (1970, 247–8)
36	New Jersey Central	40°30′N	Wisconsin (max)	16°–19°	−6° to −8° max assumed for ice-wedge polygons. Verification of features required	15°	Walters (1978, 52)
37	Wyoming Intermontane basins	41°–45°N	Pre-Wisconsin and late Wisconsin	14°	−6° max assumed for ice-wedge polygons	13°	Mears (1966; 1973)
38	Wisconsin	43°–45°N	Late Wisconsin	10°–15°	−5° to −10° assumed for ice-wedge polygons, depending on constitution	10°	Black (1965b; 1969b, 234–5)
39	New Hampshire Mt. Washington	44°N	Probably 9000–14 000 (or 15 000) ^{14}C yr	4°–6°	Based on large sorted patterns and −4° as max temperature for 'active permafrost'	5°	R. P. Goldthwait (1969; 1976b, 38–9)
40	Southern Hemisphere Argentina Gonzalez Chavez, Buenos Aires Province	38°S	Last glaciation	20°?	−5° to −6° max assumed for permafrost cracking. Verification of features required	20°?	Gonzalez and Corte (1976, 26–9, 32)
41	Rio Gallegos, Santa Cruz Province	51°30′S	Last glaciation[4]	17°	−10° assumed for ice-wedge polygons	12°	Corte (1967b, 11–13)
42	Australia Canberra-Snowy Mountains area	36°–37°S	Last glaciation (max)	9°	Based mainly on differences in altitude of fossil and active solifluction features and their relationship to treeline		R. W. Galloway (1965, 605–6)

Table 14.3 (continued)

No. (cf. Figures 14.24– 14.25)	Location	Latitude	Age of features	Mean annual temperature increase to present (min)[1]	Comments	Adjusted temperature increase (min)[2]	Reference
43	Tasmania	41°–43'S	Last glaciation (max)	5°	Based mainly on differences in altitude of fossil and active soli- fluction features and their relation- ship to treeline		R. W. Galloway (1965, 605–6)
44	Tasmania	41°–43'S	Last glaciation (max)	7°	Based on inactive rock glaciers and other periglacial evidence where present mean annual temperature is estimated to be 5.3°		Derbyshire (1973, 145)
45	New Zealand	44°S	Probably older than last glaciation	11°?	Based on inferred ice-wedge casts and −1° max for permafrost. Verification of features required	15°?	S. A. Harris (1976; personal communication, 1978)

[1] Temperature increase according to assumptions of author cited and present mean annual temperature *to nearest degree (0.5° is rounded to 1°)* as stated by author cited or recorded by representative meteorological station(s) in area (US Department of Commerce, 1975)

[2] Based on a −5° maximum temperature criterion for development of ice-wedge polygons (see text)

[3] Kaiser (1960, 137, footnote 2) cited a mean annual temperature of *c.* 11° for Beograd, but according to later information (US Department of Commerce, 1975) the temperature is 11.8° (i.e. 12°), and Kaiser's estimate has been adjusted accordingly

[4] From Gonzalez and Corte (1976, 32)

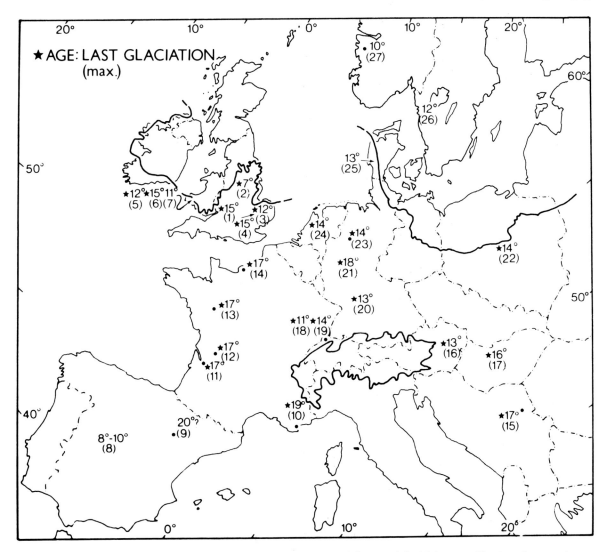

14.24 Some minimum temperature increases in Europe as estimated from periglacial features. Numbers in parentheses are keyed to Table 14.3. Temperatures based on ice-wedge casts are adjusted to a common temperature base of −5°C (max) as explained in the table. Glaciation limits (heavy lines) – Devensian *after Figure 14.1 and Boulton et al., 1977, 238, Figure 17.4; Weichsel and Würm after Figure 14.11*

ice-wedge polygons and permafrost soil-wedge polygons might form. Presumably mean annual air temperatures would be usually lower but very rarely higher, judging from studies to date, including the relation of air temperature to continuous permafrost (Figures 3.4–3.8 and accompanying discussion) and the common association of active ice-wedge polygons with the latter. Reasons for questioning a lower temperature as a generalized upper limit for all regions (including those with little snow and vegetation) are discussed in the section on Nonsorted

polygons (under Patterned ground) in the chapter on Some periglacial forms.

The maps might suggest that minimum temperature increases to the present tend to be greater in temperate regions of Europe than of North America. However, the evidence is quite insufficient; the features are not all of the same age and quality, and the environmental factors are not necessarily comparable.

In some high-latitude regions such as northeastern Siberia permafrost may have existed since

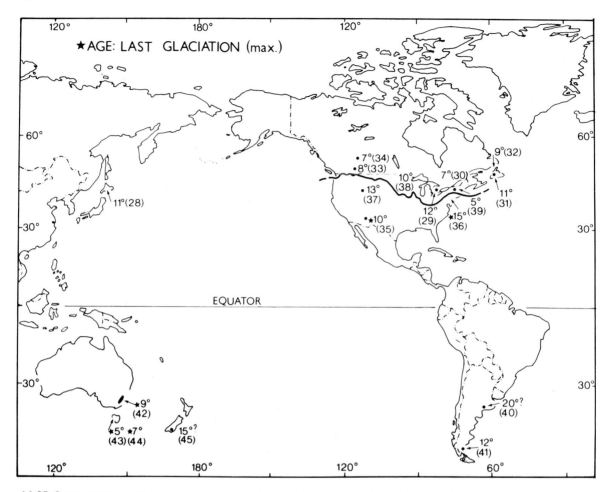

14.25 Some minimum temperature increases outside of Europe as estimated from periglacial features. Numbers in parentheses are keyed to Table 14.3. Temperatures based on ice-wedge casts are adjusted to a common temperature base of −5°C (max.) as explained in the table. Wisconsin Glaciation limit (heavy line) *after Flint, 1971, 490, Figure 18.11*

the Lower Pleistocene as previously discussed, and as is almost certainly true for an even longer time in the Antarctic. Periglacial evidence of climatic change during the Quaternary would be minimal in such regions except for any evidence of changes in thickness of the active layer.

Although periglacial evidence offers the exciting prospect of quantitative comparisons, more research and information are needed for them. Some of the necessary data are already available, mainly in the European, Soviet, and North American literature. Required are (1) critical re-evaluation of reports of contemporary and fossil features of known age in various parts of the world with respect to origin and associated climatic and other environmental parameters, and (2) similarly focussed, new and carefully designed field and laboratory investigations. Given sufficient quantitative data, the rewards could be exciting indeed!

References

All references have been examined and verified in original except (1) those cited for Figures 7.3–7.5 and 14.14 (in part), and (2) a few that were not located but are keyed to a secondary citation in the text.

Transliteration of citations in Russian (except for author's names in translations) follows the system adopted by the U.S. Board on Geographic Names and, with minor changes, by the American Geographical Society.

Please see Addendum at end for additional references.

AARTOLAHTI, TOIVE, 1970, Fossil ice-wedges, tundra polygons and recent frost cracks in southern Finland: *Acad. Scien. Fennicae Annales, Ser. A, III. Geol.-Geog.* **107**. (26 pp.)

1972, Dyynien routahalkeamista ja routahalkeama-polygoneista (Frost cracks and frost-crack polygons on dunes in Finland): *Terra* **84**, 124–31.

ACADEMIA SINICA, 1975, *Permafrost*: Lanchow, Research Institute of Glaciology, Permafrost and Desert Research. (124 pp.) [To be available in English translation by National Research Council of Canada in 1980.]

ADAMS, W. P., and MATHEWSON, S. A., 1976, Approaches to the study of ice-push features, with reference to Gillies Lake, Ontario: *Rev. Géog. Montréal* **30**(1–2) (*Le Glaciel* – Premier colloque international sur l'action géologiques des glaces flottantes, Québec, Canada, 20–24 April 1974), 187–96.

ADLER-VIGNES, M., and DIJKEMA, K. M., 1975, Modèle destiné à la simulation du gel dans milieux poreux – application en cas réel: *Fondation Française d'Études Nordiques, 6ᵉ Congrès International, 'Les Problèmes Posés par la Gélifraction. Recherches Fondamentales et Appliquées'* (Le Havre, 23–25 Apr. 1975), *Rept.* **301**. (7 pp.)

AGER, T. A., 1972, *Surficial geology and Quaternary history of the Healy Lake area, Alaska*: Univ. Alaska, MS Thesis. (127 pp.)

AGUIRRE-PUENTE, J., 1975, Contribution à l'étude du méchanisme physique du gel des roches: *Fondation Française d'Études Nordiques, 6ᵉ Congrès International, 'Les Problèmes Posés par la Gélifraction. Recherches Fondamentales et Appliquées'* (Le Havre, 23–25 Apr. 1975), *Rept.* **302**. (13 pp.)

1978, État actuel des recherches sur le gel des roches et de materiaux de construction: 599–607 in *Third International Conference on Permafrost* (Edmonton, Alta., 10–13 July 1978), Proc. **1**: Ottawa, Canada Natl. Research Council. (947 pp.)

AGUIRRE-PUENTE, J., VIGNES, M., and VIAUD, P. [AGIRPUENT, ZH., VIN', M., and VIO, P.], 1973, Issledovaniye strukturnykh izmeneniy v gruntakh pri promerzanii: 161–75 in *Akademiya Nauk SSSR, Sektsiya Nauk o Zemle, Sibirskoye Otdeleniye, II Mezhdunarodnaya Konferentsiya po Merzlotovedeniyu, Doklady i soobshcheniya* **4** (Fizika, fiziko-khimiya i mekhanika merzlykh gornykh porod i l'da): Yakutsk, Yakutskoye Knizhnoye Izdatel'stvo. (245 pp.)

1978, Study of the structural changes in soils during freezing [Issledovaniye strukturnykh izmeneniy y gruntakh pri promerzanii]: 315–23 in Sanger, F. J., ed., *USSR Contribution Permafrost Second International Conference* (Yakutsk, USSR, 13–28 July 1973): Washington, DC, Natl. Acad. Sci. (866 pp.)

AHLMANN, H. W., 1919, Geomorphological studies in Norway: *Geografiska Annaler* **1**, 1–148, 193–252.

1936, Polygonal markings: 7–19 in Scientific results of the Swedish–Norwegian Arctic Expeditions in the summer of 1931 **2**(12): *Geografiska Annaler* **18**, 1–19.

ÅHMAN, RICHARD, 1973, Pingos i Adventdalen och Reindalen på Spetsbergen: *Svensk Geografisk Årsbok* **49**, 190–7.

1975, Palstrukturer och palsmorfologi i Nordnorge: *Svensk Geografisk Årsbok* **51**, 223–32.

1976, The structure and morphology of minerogenic

palsas in northern Norway: *Biuletyn Peryglacjalny* **26**, 25–31.

1977, Palsar i Nordnorge. En studie av palsars morfologi, utbredning och klimatiska förutsättningar i Finnmarks och Troms fylke: *Lunds Universitets Geografiska Inst. Medd. Avh.* **78**. (165 pp.)

AHNERT, FRANK, 1971, A general and comprehensive theoretical model of slope profile development: *Univ. Maryland Occasional Papers in Geog.* **1**. (95 pp.)

AKADEMIYA NAUK SSSR, 1964, *Fiziko-geograficheskiy atlas mira*: Moskva. (298 pp.)

1973, II Mezhdunarodnaya konferentsiya po merzlotovedeniyu – II International Conference on Permafrost: *Doklady i soobshcheniya* **1–7**: Yakutsk, Yakutskoye Knizhnoye Izdatel'stvo. (103, 154, 102, 246, 127, 111, 272 pp.) (*See* US Army Corps of Engineers, Cold Regions Research and Engineering Laboratory, 1974.)

ÅKERMAN, JONAS, 1973, Preliminära resultat från undersökningar massrörelser vid Kapp Linné, Spetsbergen: *Lunds Universitets Naturgeografiska Inst., Rapporter och Notiser* **18**, 1–13.

ALEKSANDROV, A. I., 1948, Kamennyye morya i reki Urala: *Materialy po geomorfologii Urala, vyp.* **I**, 264–7.

ALEKSEYEV, V. R., 1973, Prichiny i faktory naledeobrazovaniya: *Akad. Nauk SSSR, Sibirskoye Otdeleniye, Institut Geografii Sibiri i Dal'nego Vostoka, Doklady* **39**, 12–23.

1974, Causes and factors of ground-ice formation [Prichiny i faktory naledeobrazovaniya]: *Soviet Geog., Review and Translation* **15**(7), 395–407.

1978, *Naledi i nalednye protsessy (voprosy klassifikatsii i terminologii)*: Novosibirsk, Nauka. (188 pp.)

ALEKSEYEV, V. R., *et al.*, 1969, *Naledi Siberii*: Akad. Nauk SSSR, Sibirskoye Otdeleniye, Inst. Merzlotovedeniye: Moskva, Izdatel'stvo Nauka. (206 pp.)

1973, Siberian naleds [Naledi Siberii]: *US Army Corps of Engineers, Cold Regions Research and Engineering Laboratory Draft Translation* **399**. (300 pp.)

ALESHKOV, A. N., 1936, Über Hochterrassen der Urals: *Zeitschr. Geomorphologie* **9**, 143–9.

ALEXANDRE, JEAN, 1958, Le modelé quaternaire de l'Ardenne Centrale: *Soc. Géol. Belgique Annales* **81**, 213–31.

ALLEN, C. R., O'BRIEN, R. M. G., and SHEPPARD, S. M. F., 1976, The chemical and isotopic characteristics of some East Greenland surface and pingo waters: *Arctic and Alpine Research* **8**, 297–317.

ALTER, A. J., 1969a, Water supply in cold regions: *US Army Cold Regions Research and Engineering Laboratory, Cold Regions Science and Engineering Mon.* **III**–C5a. (85 pp.)

1969b, Sewage and sewage disposal in cold regions: *US Army Corps of Engineers, Cold Regions Research and Engineering Laboratory, Cold Regions Science and Engineering Mon.* **III**–C5b. (107 pp.)

AMERICAN METEOROLOGICAL SOCIETY, 1953, Bibliography on frost and frost forecasting: *Meteorol. Abs. and Bibliography* **4**(3), 273–420.

ANDERSEN, B. G., 1963, Preliminary report on glaciology and glacial geology of the Thiel Mountains, Antarctica: *US Geol. Survey Prof. Paper* **475**-B, B140–B143.

ANDERSLAND, O. B., and ANDERSON, D. M., eds., 1978, *Geotechnical engineering for cold regions*: New York, McGraw-Hill. (566 pp.)

ANDERSON, D. L., 1960, The physical constants of sea ice: *Research* **13**, 310–18.

ANDERSON, D. M., 1968, Undercooling, freezing point depression, and ice nucleation of soil water: *Israel J. Chem.* **6**, 349–55.

1970, Phase boundary water in frozen soils: *US Army Corps of Engineers, Cold Regions Research and Engineering Laboratory Research Rept.* **274**. (19 pp.)

1971, Remote analysis of planetary water: *US Army Corps of Engineers, Cold Regions Research and Engineering Laboratory Spec. Rept.* **154**. (13 pp.)

1977, General aspects of the physical state of water and water movement in frozen soils: 2–16 in *Frost action in soils*: Univ. Luleå, Internat. Symposium (Luleå, Sweden, 16–18 Feb. 1977) Proc. 2. (119 pp.)

ANDERSON, D. M., GATTO, L. W., and UGOLINI, F. C., 1973, An examination of Mariner 6 and 7 imagery for evidence of permafrost terrain on Mars: 499–508 in *North American Contribution, Permafrost Second International Conference* (Yakutsk, USSR, 13–28 July 1973): Washington, DC, Natl. Acad. Sci. (783 pp.)

ANDERSON, D. M., and MORGENSTERN, N. R., 1973, Physics, chemistry, and mechanics of frozen ground: A review: 257–88 in *North American Contribution, Permafrost Second International Conference* (Yakutsk, USSR, 13–28 July 1973): Washington, DC, Natl. Acad. Sci. (783 pp.)

ANDERSON, D. M., and TICE, A. R., 1970, Low-temperature phases of interfacial water in clay-water systems: *US Army Corps of Engineers, Cold Regions Research and Engineering Laboratory Research Rept.* **290**. (17 pp.)

1972, Predicting unfrozen water contents in frozen soils from surface area measurements: 12–18 in Highway Research Board, *Frost action in soils*:

Natl. Acad. Sci. – Natl. Acad. Eng., Highway Research Record **393**. (88 pp.)

1973, The unfrozen interfacial phase in frozen soil water systems: 107–24 in Hadras, A., *et al.*, eds., *Physical aspects of soil water and salts in ecosystems*: Ecological Studies **4**: Berlin – Heidelberg – New York, Springer-Verlag. (460 pp.)

ANDERSSON, GUNNAR, and HESSELMAN, HENRIK, 1907, Vegetation ocy flora i Hamra kronopark: Stockholm, *Statens Skogsförsoksanstalt Medd.* **4**(2), 41–110.

ANDERSSON, J. G., 1906, Solifluction, a component of subaerial denudation: *J. Geol.* **14**, 91–112.

ANDREWS, MARTHA, 1978, Selected bibliography of disturbance and restoration of soils and vegetation in permafrost regions of the USSR (1970–1977): *US Army Corps of Engineers, Cold Regions Research and Engineering Laboratory Spec. Rept.* **78–19**. (175 pp.)

ANISIMOVA, N. P., *et al.*, 1973*a*, Podzemnyye vody kriolitosfery: *Akademiya Nauk SSSR, Sektsiya Nauk o Zemle, Sibirskoye Otdeleniye, II Mezhdunarodnaya Konferentsiya po Merzlotovedeniyu, Doklady i soobshcheniya* **5**: Yakutsk, Yakutskoye Knizhnoye Izdatel'stvo. (128 pp.)

1973*b*, Ground water in the cryolithosphere [Podzemnyye vody kriolitosfery]: *US Army Corps of Engineers, Cold Regions Research and Engineering Laboratory Draft Translation* **437**. (172 pp.)

1978, Groundwater in the cryolithosphere [Podzemnyye vody kriolitosfery]: 363–442 in Sanger, F. J., ed., *USSR Contribution Permafrost Second International Conference* (Yakutsk, USSR, 13–28 July 1973): Washington, DC, Natl. Acad. Sci. (866 pp.)

ANKETELL, J. M., CEGŁA, JERZY, and DŻUŁYŃSKI, STANISŁAW, 1970, On the deformational structures in systems with reversed density gradients: *Rocznik Polskiego Towarzystwa Geologicznego (Soc. géol. Pologne Annales)* **40**(1), 1–30.

ANTEVS, ERNST, 1925, Retreat of the last ice-sheet in eastern Canada: *Canada Geol. Survey, Mem.* **146**. (142 pp.)

1932, *Alpine zone of Mt Washington Range*: Auburn, Maine. (118 pp.)

ARAKAWA, KIYOSHI, 1966, Theoretical studies of ice segregation in soil: *J. Glaciology* **6**(44), 255–60.

ARCTIC INSTITUTE OF NORTH AMERICA, 1953–1975, *Arctic Bibliography* **1–16**.

ARE, A. L., and DEMCHENKO, R. YA., 1976, Some results of long-term studies on soil thawing in Yakutia: *HEAT TRANSFER – Soviet Research* **8**(2), 117–26.

ARE, F. E., 1972, The reworking of shores in the permafrost zone: 78–9 in Adams, W. P., and Helleiner, F. M., eds., *International Geography 1972*

1 (Internat. Geog. Cong., 22d, Montreal): Toronto, Univ. Toronto Press. (694 pp.)

1976, O subakval'noy kriolitozone Severnogo Ledovitogo Okeana: 3–26 in Akad. Nauk SSSR, Sibirskoye Otdeleniye, *Regional'nyye i teplofizicheskiye issledovaniya merzlykh gornykh porod v Sibiri*: Yakutsk, Yakutskoye Knizhnoye Izdatel'stvo. (223 pp.)

1978*a*, Subsea cryolithozone of the Arctic Ocean [O subakval'noy kriolitozone Severnogo Ledovitogo Okeana]: *US Army Corps of Engineers, Cold Regions Research and Engineering Laboratory Draft Translation* **686**. (26 pp.)

1978*b*, Subakval'naya kriolitozona Aziatskoy Arktiki – Offshore permafrost in the Asiatic Arctic: 336–40 in *Third International Conference on Permafrost* (Edmonton, Alta., 10–13 July 1978), Proc. **1**: Ottawa, Canada Natl. Research Council. (947 pp.)

ARKHANGELOV, A. A., and SHER, A. V., 1973, K voprosu o vozraste mnogoletney merzloty na Kraynem Severo-Vostoke SSR: 5–11 in *Akademiya Nauk SSSR, Sektsiya Nauk o Zemle, Sibirskoye Otdeleniye, II Mezhdunarodnaya Konferentsiya po Merzlotovedeniyu, Doklady i soobshcheniya* **3** (Genezis, sostav i stroyeniye merzlykh tolshch i podzemnyye l'dy): Yakutsk, Yakutskoye Knizhnoye Izdatel'stvo. (102 pp.)

1978, The age of the permafrost in the Far Northeast of the USSR [K voprosu o vozraste mnogoletney merzloty na Kraynem Severo-Vostoke SSSR]: 155–9 in Sanger, F. J., ed., *USSR Contribution Permafrost Second International Conference* (Yakutsk, USSR, 13–28 July 1973): Washington, DC, Natl. Acad. Sci. (866 pp.)

ASSOCIATION FRANÇAISE POUR L'ÉTUDE DU QUATERNAIRE, 1969, *La stratigraphie des loess d'Europe*: Supplément au Bulletin de l'Association. (176 pp.)

ASSOCIATION GÉOGRAPHIQUE D'ALSACE, 1978, *Colloque sur le Périglaciaire d'Altitude du Domaine Méditerranéen et Abords* (Strasbourg – Université Louis Pasteur, 12–14 May 1977): Strasbourg, Assoc. Géog. d'Alsace. (366 pp.)

AUER, VÄINÖ, 1920, Über die Enstehung der Stränge auf den Torfmooren: *Acta Forestalia Fennica* **12**, 1–145.

1970, The Pleistocene of Fuego-Patagonia; Part V: Quaternary problems of southern South America: *Acad. Scien. Fennicae Annales, Ser. A, III. Geol.-Geog.* **100**. (194 pp.)

AUGHENBAUGH, N. B., 1958, Preliminary report on the geology of the Dufek Massif: *Ohio State Univ.*

Research Found. Rept. **825-1**(1) (IGY proj. 4.10, NSF grant Y/4.0/285), 164–208.

BAKER, B. H., 1967, Geology of the Mount Kenya area: *Kenya Geol. Survey Rept.* **79**. (78 pp.)

BAKLUND, O. O., 1911, Obshchiy obzor deyatelnosti ekspeditsii brat'ev Kuznetsovykh na Polyarnyy Ural letom 1909 goda: *Akad. Nauk SSSR, Fizikomatematicheskiy otdel, Zapiski, ser. 8*, **28**(1), 70–3.

BAKULIN, F. G., 1958, *L'distost' i osadki pri ottaivanii mnogoletnemerzlykh chetvertichnykh otlozheniy Vorkutskogo rayona*: Moskva, Akad. Nauk SSSR, Inst. Merzlotovedeniya im. V. A. Obrucheva. (95 pp.)

BAKULIN, F. G., SAVEL'YEV, B. A., and ZHUKOV, V. F., 1957, Fizicheskiye yavleniya i protsessy v ottaivayushchikh gruntakh: 72–83 (Glava IV) in Akad. Nauk SSSR, Institut Merzlotovedeniya im. V. A. Obrucheva, *Materialy po laboratornym issledovaniyam merzlykh gruntov* **3**: Moskva, Izdatel'stvo Akad. Nauk SSSR. (323 pp.)

1972, Physical processes in thawing ground [Fizicheskiye lavleniia i protsessy v ottaivaiushchikh gruntakh]: *US Army Corps of Engineers, Cold Regions Research and Engineering Laboratory Draft Translation* **325**. (13 pp.)

BALCH, E. S., 1900, *Glacières or freezing caverns*: Philadelphia, Allen, Lane & Scott. (337 pp.)

BALDUZZI, F., 1959, Experimentelle Untersuchungen über den Bodenfrost: *Versuchsanstalt für Wasserbau und Erdbau Mitt.* **44**. (17 pp.)

1960, Experimental investigation of soil freezing [Experimentelle Untersuchungen über den Bodenfrost]: *Canada Natl. Research Council Tech. Translation* **912**. (43 pp.)

BALKWILL, H. R., *et al.*, 1974, Glacial features and pingos, Amund Ringnes Island, Arctic Archipelago: *Canadian J. Earth Sci.* **11**, 1319–25.

BALL, D. F., 1976, Close-packed patterned arrangement of stones and shells on shore-line platforms: *Biuletyn Peryglacjalny* **25**, 5–7.

BALL, D. F., and GOODIER, R., 1970, Morphology and distribution of features resulting from frost-action in Snowdonia: *Field Studies* **3**, 193–217.

BALLANTYNE, C. K., 1978, The hydrologic significance of nivation features in permafrost areas: *Biuletyn Peryglacjalny* **27**, 5–10.

BALLARD, T. M., 1973, Soil physical properties in a sorted stripe field: *Arctic and Alpine Research* **5**, 127–31.

BALOBAYEV [BALOBAEV], V. T., 1964, Teploobmen merzlykh gornykh porod s atmosferoy pri nalichii rastitel'nogo pokrova: 39–53 in *Teplovyye protsessy v merzlykh gornykh porodakh*: Moskva, Izdatel'stvo Nauka.

1973, Heat exchange between permafrost and the atmosphere in the presence of a vegetation cover [Teploobmen merzlykh gornykh porod s atmosferoi pri nalichii rastitel'nogo pokrova]: *Canada Natl. Research Council Tech. Translation* **1670**. (18 pp.)

1978, Rekonstruktsiya paleoclimata po sovremennym geotermicheskim dannym – Reconstruction of paleoclimate from present-day geothermal data: 10–14 in *Third International Conference on Permafrost* (Edmonton, Alta., 10–13 July 1978), Proc. **1**: Ottawa, Canada Natl. Research Council. (947 pp.)

BARANOV, I. YA., 1959, Geograficheskoye rasprostraneniye sezonnopromerzayushchikh pochv i mnogoletnemerzlykh gornyk porod: 193–219 (Glava VII) in Inst. Merzlotovedeniya im. V. A. Obrucheva, *Osnovy geokriologii (merzlotovedeniya), Chast' pervaya, Obshchaya geokriologiya*: Moskva, Akad. Nauk SSSR. (459 pp.)

1964, Geographical distribution of seasonally frozen ground and permafrost [Geograficheskoye rasprostraneniye sezonnopromerzayushchikh pochv i mnogoletnemerzlykh gornyk porod]: *Canada Natl. Research Council Tech. Translation* **1121**. (85 pp.)

BARANOV, I. J. [YA.], and KUDRYAVTSEV, V. A., 1966, Permafrost in Eurasia: 98–102 in *Permafrost International Conference* (Lafayette, Ind., 11–15 Nov. 1963) *Proc.*: Natl. Acad. Sci.–Natl. Research Council Pub. **1287**. (563 pp.)

BARNES, P. W., and HOPKINS, D. M., eds., 1978, Geological sciences: 101–22 in *Environmental assessment of the Alaskan Continental Shelf – Interim synthesis: Beaufort-Chukchi*: US Dept. Commerce, National Oceanic and Atmospheric Administration (NOAA) Environmental Research Laboratories: Boulder, Col. (362 pp.)

BARNETT, D. M., 1966, Preliminary field investigations of movement on certain Arctic slope forms: *Geog. Bull.* **8**, 377–82.

BARR, D. J., and SWANSTON, D. N., 1970, Measurement of creep in a shallow, slide-prone till soil: *Am. J. Sci.* **269**, 467–80.

BARR, WILLIAM, and SYROTEUK, MERVYN, 1973, The pingos of the Tuktoyaktuk area: Univ. Saskatchewan, Inst. Northern Studies, *The Musk-Ox* **12**, 3–12.

BARSCH, DIETRICH, 1969a, Studien und Messungen an Blockgletschern in Macun, Unterengadin: *Zeitschr. Geomorphologie N.F., Supp.* **8**, 11–30.

1969b, Permafrost in der oberen subnivalen Stufe der Alpen: *Geog. Helvetica* **24**, 10–12.

1971, Rock glaciers and ice-cored moraines: *Geografiska Annaler* **53**A, 203–6.

1973, Permafrost in the upper subnival step of the Alps [Permafrost in der oberen subnivalen Stufe der Alpen]: 3–9 in Translation of two Swiss articles on permafrost in the Alps: *Canada Natl. Research Council Tech. Translation* **1657**. (29 pp.)

1977*a*, Nature and importance of mass-wasting by rock glaciers in alpine permafrost environments: *Earth Surface Processes* **2**, 231–45.

1977*b*, Ein Permafrostprofil aus Graubünden, Schweizer Alpen: *Zeitschr. Geomorphologie N. F.* **21**, 79–86.

1977*c*, Eine Abschätzung von Schuttproduktion und Schutttransport im Bereich activer Blockgletscher der Schweizer Alpen: *Zeitschr. Geomorphologie N. F., Supp.* **28**, 148–60.

1977*d*, Alpiner Permafrost – ein Beitrag zur Verbreitung, zum Charakter und zur Ökologie am Beispiel der Schweizer Alpen: 118–41 in Poser, Hans, ed., Formen, Formengesellschaften und Untergrenzen in den heutigen periglazialen Höhenstufen der Hochgebirge Europas und Afrikas zwischen Arktis und Äquator. Bericht über ein Symposium: *Akad. Wiss. Göttingen Abh., Math.-Phys. Kl. Folge 3*, **31**. (355 pp.)

1978, Active rock glaciers as indicators for discontinuous alpine permafrost. An example from the Swiss Alps: 348–53 in *Third International Conference on Permafrost* (Edmonton, Alta., 10–13 July 1978), Proc. **1**: Ottawa, Canada Natl. Research Council. (947 pp.)

BARSCH, DIETRICH, and HELL, GÜNTER, 1975, Photogrammetrische Bewegungsmessungen am Blockgletscher Murtèl I, Oberengadin, Schweizer Alpen: *Zeitschr. Gletscherkunde u. Glazialgeologie* **11**(2), 111–42.

BARSCH, DIETRICH, and TRETER, UWE, 1976, Zur Verbreitung von Periglazialphänomen in Rondane/Norwegen: *Geografiska Annaler* **58**A, 83–93.

BARSCH, DIETRICH, and UPDIKE, R. G., 1971, Periglaziale Formung am Kendrick Peak in Nord-Arizona während der letzten Kaltzeit: *Geog. Helvetica* **26**, 99–114.

BASHENINA, N. V., 1948, Proiskhozdeniye rel'efa Yuzhnogo Urala: Moskovskiy ordena Lenina Gosudarstvennyy universitet im. M. V. Lomonosova, *Nauchnoissledovatelskiy institut geografii, OPI* **2**, 1–232.

1967, *Formirovaniye sovremennogo rel'efa zemnoy poverkhnosti*: Moskva. (194 pp.)

BATES, R. E., and BILELLO, M. A., 1966, Defining the cold regions of the Northern Hemisphere: *US Army Cold Regions Research and Engineering Laboratory Tech. Rept.* **178**. (11 pp.)

BATTLE, W. R. B., 1960, Temperature observations in bergschrunds and their relationship to frost shattering: 83–95 in Lewis, W. V., ed., Norwegian cirque glaciers, London, *Royal Geog. Soc. Research Ser.* **4**. (104 pp.)

BAULIG, HENRI, 1956, Pénéplaines et pédiplaines: *Soc. belge études géog.* **25**(1), 25–58.

1957, Peneplains and pediplains: *Geol. Soc. America Bull.* **68**, 913–30.

BAULIN, V. V., 1962, Osnovnye etapy istorii razvitiya mnogoletnemerzlykh porod na territorii Zapadno-Sibirskoy nizmennosti: *Akademiya Nauk SSSR, Institut Merzlotovedeniya im. V. A. Obrucheva Trudy* **19**, 5–18.

BAULIN, V. V., *et al.*, 1967, *Geokriologicheskiye usloviya Zapadno-Sibirskoy Nizmennosti*: Moskva, Izdatel'stvo 'Nauka'. (314 pp.)

BAULIN, V. V., DUBIKOV, G. I., and UVARKIN, YU. T., 1973, Osnovnyye cherty stroyeniya i razvitiya vechnomerzlykh tolshch Zapadno-Sibirskoy ravniny: 10–17 in *Akademiya Nauk SSSR, Sektsiya Nauk o Zemle, Sibirskoye Otdeleniye, II Mezhdunarodnaya Konferentsiya po Merzlotovedeniyu, Doklady i soobshcheniya* **2** (Regional'naya geokriologiya): Yakutsk, Yakutskoye Knizhnoye Izdatel'stvo. (154 pp.)

1978, The main features of the structure and development of the permafrost of the west Siberian Plain [Osnovnyye cherty stroyeniya i razvitiya vechnomerzlykh tolshch Zapadno-Sibirskoy ravniny]: 63–7 in Sanger, F. J., ed., *USSR Contribution Permafrost Second International Conference* (Yakutsk, USSR, 13–28 July 1973): Washington, DC, Natl. Acad. Sci. (866 pp.)

BEATY, C. B., 1974, Needle ice and wind in the White Mountains of California: *Geology* **2**, 565–7.

BEHR, F. M., 1918, Über geologisch wichtige Frosterscheinungen in gemässigten Klimaten: *Deutsche Geol. Gesell. Zeitschr.* **70**, B. Monatsber., 95–117.

BEHRE, C. H., JR., 1933, Talus behavior above timber in the Rocky Mountains: *J. Geol.* **41**, 622–35.

BELKNAP, R. L., 1941, Physiographic studies in the Holstenborg District of southern Greenland: 199–255 in Hobbs, W. H., ed., Reports of the Greenland Expeditions of the University of Michigan (2), *Michigan Univ. Studies, sci. ser.* **6**. (287 pp.)

BELOPUKHOVA, E. B., 1973, Osobennosti sovremennogo razvitiya mnogoletnemerzlykh porod zapadnoy Sibiri: 84–6 in *Akademiya Nauk SSSR, Sektsiya Nauk o Zemle, Sibirskoye Otdeleniye, II*

Mezhdunarodnaya Konferentsiya po Merzlotovedeniyu, Doklady i soobshcheniya **2** (Regional'naya geokriologiya): Yakutsk, Yakutskoye Knizhnoye Izdatel'stvo. (154 pp.)

1978, Features of the contemporary development of permafrost in Western Siberia [Osobennosti sovremennogo razvitiya mnogoletnemerzlykh porod zapadnoy Sibiri]: 112–13 in Sanger, F. J., ed., *USSR Contribution Permafrost Second International Conference* (Yakutsk, USSR, 13–28 July 1973): Washington, DC, Natl. Acad. Sci. (866 pp.)

BENEDICT, J. B., 1966, Radiocarbon dates from a stone-banked terrace in the Colorado Rocky Mountains, USA: *Geografiska Annaler* **48**A, 24–31.

1970*a*, Frost cracking in the Colorado Front Range: *Geografiska Annaler* **52**A, 87–93.

1970*b*, Downslope soil movement in a Colorado alpine region: rates, processes, and climatic significance: *Arctic and Alpine Research* **2**, 165–226.

1973, Origin of rock glaciers [Letter]: *J. Glaciology* **12**(66), 520–2.

1976, Frost creep and gelifluction: A review: *Quaternary Research* **6**, 55–76.

1979, Fossil ice-wedge polygons in the Colorado Front Range: Origin and significance: *Geol. Soc. America Bull. Part 1* **90**, 173–80.

BENNINGHOFF, W. S., 1952, Interaction of vegetation and soil frost phenomena: *Arctic* **5**, 34–44.

1966, Relationships between vegetation and frost in soils: 9–13 in *Permafrost International Conference* (Lafayette, Ind., 11–15 Nov. 1963) *Proc.*: Natl. Acad. Sci.–Natl. Research Council Pub. **1287**. (563 pp.)

BENOIT, G. R., 1975, Effects of freeze-thaw cycles on soil hydraulic conductivity normal to the direction of freezing: *Eastern Snow Conference, 32d (Feb. 1975), Proc.*, 51–8.

BERG, T. E., 1969, Fossil sand wedges at Edmonton, Alberta, Canada: *Biuletyn Peryglacjalny* **19**, 325–33.

BERG, T. E., and BLACK, R. F., 1966, Preliminary measurements of growth of nonsorted polygons, Victoria Land, Antarctica: 61–108 in Tedrow, J. F. C., ed., Antarctic soils and soil-forming processes, *Am. Geophys. Union Antarctic Research. Ser.* **8** (*Natl. Acad. Sci.–Natl. Research Council Pub.* **1418**). (177 pp.)

BERGER, HERFRIED, 1967, Vorgänge und Formen der Nivation in den Alpen, 2 ed.: *Buchreihe des Landemuseums für Kärnten* **17**, Klagenfurt. (89 pp.)

BERTOUILLE, H., 1975, Gélifraction des roches humides: *Fondation Française d'Études Nordiques, 6ᵉ Congrès International, 'Les Problèmes Posés par la Gélifraction. Recherches Fondamentales et Appliquées'*

(Le Havre, 23–25 Apr. 1975), *Rept.* **306**. (11 pp.)

BESCHEL, R. L., 1966, Hummocks and their vegetation in the high arctic: 13–20 in *Permafrost International Conference* (Lafayette, Ind., 11–15 Nov. 1963) *Proc.*: Natl. Acad. Sci.–Natl. Research Council Pub. **1287**. (563 pp.)

BESKOW, G., 1930, Erdfliessen und Strukturböden der Hochgebirge im Licht der Frosthebung: *Geol. Fören. Stockholm, Förh.* **52**, 622–38.

1935, Tjälbildningen och tjällyftningen: *Sveriges Geol. Undersökning Avh. och uppsatser, ser. C* **375** [Årsbok **26**(3)]. (242 pp.)

1947, Soil freezing and frost heaving with special application to roads and railroads (Translated by J. O. Osterberg): *Northwestern Univ., Technol. Inst.* (145 pp.)

BEUF, BERNARD, *et al.*, 1971, Les grès du Paléozoic Inférieur au Sahara: *Inst. Français du Pétrole, Science et Technique du Pétrole* **18**, Paris, Éditions Technip. (464 pp.)

BIBUS, ERHARD, 1975, Geomorphologische Untersuchungen zur Hang- und Talentwicklung im zentralen West-Spitzbergen: *Polarforschung* **45**(2), 102–19.

BIBUS, ERHARD, NAGEL, G., and SEMMEL, A., 1976, Periglazialer Reliefformung im zentralen Spitzbergen: *Catena* **3**, 29–44.

BIERMANS, M. B. G. M., DIJKEMA, K. M., and DE VRIES, D. A., 1976, Water movement in porous media towards an ice front: *Nature* **264**(11), 166–7.

BIK, M. J. J., 1967, The periglacial origin of prairie mounds: 83–94 in Clayton, Lee, and Freers, T. F., eds., *Glacial geology of the Missouri Coteau and adjacent areas*: North Dakota Geol. Survey Misc. Ser. **30**. (170 pp.)

1969, The origin and age of the prairie mounds of southern Alberta, Canada: *Biuletyn Peryglacjalny* **19**, 85–130.

BIRD, J. B., 1951, The physiography of the middle and lower Thelon Basin: *Geog. Bull.* **1**, 14–29.

1967, *The physiography of arctic Canada, with special reference to the area south of Parry Channel*: Baltimore, The Johns Hopkins Press. (336 pp.)

BIRMAN, J. H., 1952, Pleistocene clastic dikes in weathered granite gneiss, Rhode Island: *Am. J. Sci.* **250**, 721–34.

BIULETYN PERYGLACJALNY, 1954–, **1**–[An international journal founded by Jan Dylik and published by Łodzkie Towarzystwo Naukowe, Łodz, Poland.]

BLACK, R. F., 1950, Permafrost: 247–73 in Trask, P. D., ed., *Applied sedimentation*: New York, John Wiley & Sons, Inc. (707 pp.)

1951, Eolian deposits of Alaska: *Arctic* **4**, 89–111.

1952, Polygonal patterns and ground conditions from aerial photographs: *Photogrammetrical Eng.* **18**, 123–34.

1953, Permafrost – A review: *New York Acad. Sci. Trans. Ser. 2*, **15**(5), 126–31.

1954, Permafrost – a review: *Geol. Soc. America Bull.* **65**, 839–55.

1963, Les coins de glace et le gel permanent dans le Nord de l'Alaska: *Annales Géog.* **72**(391), 257–71.

1965a, Ice-wedge casts of Wisconsin: *Wisconsin Acad. Sci., Arts and Letters, Trans.* **54**, 187–222.

1965b, Paleoclimatologic implications of ice-wedge casts of Wisconsin (abs.): 37 in *International Assoc. for Quaternary Research (INQUA) Cong., 7th (Boulder, Col., 1965), Abstracts General Sessions.* (532 pp.)

1969a, Slopes in southwestern Wisconsin, USA, periglacial or temperate?: *Biuletyn Peryglacjalny* **18**, 69–82.

1969b, Climatically significant fossil periglacial phenomena in northcentral United States: *Biuletyn Peryglacjalny* **20**, 225–38.

1969c, Thaw depressions and thaw lakes: A review: *Biuletyn Peryglacjalny* **19**, 131–50.

1969d, Geology, especially geomorphology, of northern Alaska: *Arctic* **22**(3), 283–99.

1973a, Growth of patterned ground in Victoria Land, Antarctica: 193–203 in *North American Contribution, Permafrost Second International Conference* (Yakutsk, USSR, 13–28 July 1973): Washington, DC, Natl. Acad. Sci. (783 pp.)

1973b, Cryomorphic processes and microrelief features, Victoria Land, Antarctica: 11–24 in Fahey, B. D., and Thompson, R. D., eds., *Research in polar and alpine geomorphology*: Guelph Symposium on Geomorphology, 1973, 3d, Proc.: Norwich, England, Geo Abstracts Ltd (Univ. East Anglia). (206 pp.)

1974a, Ice-wedge polygons of northern Alaska: 247–75 in Coates, D. R., ed., *Glacial geomorphology*: Annual Geomorphology Series, 1974, 5th, Proc.: Binghamton, New York, State Univ. New York. (398 pp.)

1974b, Periglacial processes and environments [Review]: *Am. J. Sci.* **274**, 666–9.

1976a, Periglacial features indicative of permafrost: Ice and soil wedges: *Quaternary Research* **6**, 3–26.

1976b, Features indicative of permafrost: 75–94 in Donath, F. A., ed., *Annual review of earth and planetary sciences* **4**, Palo Alto, Cal., Annual Reviews, Inc. (484 pp.)

1978a, Comparison of some permafrost features on Earth and Mars: 45–7 in *Colloquium on Planetary Water and Polar Processes, 2d* (Hanover, N.H., 16–18 Oct. 1978), Proc.: Hanover, N.H., US Army Corps of Engineers Cold Regions Research and Engineering Laboratory. (209 pp.)

1978b, Fabrics of ice wedges in central Alaska: 247–53 in *Third International Conference on Permafrost* (Edmonton, Alta., 10–13 July 1978), Proc. **1**: Ottawa, Canada Natl. Research Council. (947 pp.)

BLACK, R. F., and BARKSDALE, W. L., 1949, Oriented lakes of northern Alaska: *J. Geol.* **57**, 105–18.

BLACK, R. F., and BERG, T. E., 1964, Glacier fluctuations recorded by patterned ground, Victoria Land: 107–22 in Adie, R. J., ed., *Antarctic geology. First International Symposium on Antarctic Geology* (Cape Town, 16–21 Sept. 1963) Proc.: Amsterdam, North-Holland Publishing Co. (758 pp.)

BLACK, R. F., and HAMILTON, T. D., 1972, Mass-movement studies near Madison, Wisconsin: 121–79 in Morisawa, Marie, ed., Quantitative geomorphology: Some aspects and applications: Second Annual Geomorphology Symposia Series (Binghamton, NY, 15–16 Oct. 1971) Proc.: Binghamton, NY, State Univ. New York (Binghamton) Pubs. in Geomorphology. (315 pp.)

BLACKWELL, J. M., 1965, *Surficial geology and geomorphology of the Harding Lake area, Big Delta Quadrangle, Alaska*: Univ. Alaska, MS Thesis. (91 pp.)

BLAGBROUGH, J. W., 1976, Rock glaciers in the Capitan Mountains, south central New Mexico (abs.): *Geol. Soc. America Abs. with Programs* **8**(5), 570–1.

BLAGBROUGH, J. W., and BREED, W. J., 1967, Protalus ramparts on Navajo Mountain, southern Utah: *Am. J. Sci.* **265**, 759–72.

BLAGBROUGH, J. W., and FARKAS, S. E., 1968, Rock glaciers in the San Mateo Mountains, south-central New Mexico: *Am. J. Sci.* **266**, 812–23.

BLANCK, E., RIESER, A., and MORTENSEN, H., 1928, Die wissenschaftlichen Ergebnisse einer bodenkundlichen Forschungsreise nach Spitzbergen im Sommer 1926: *Chemie der Erde* **3**, 588–698.

BLEICH, K., and GROSCHOPF, P., 1959, Periglazialbildungen am Fusse der Schwäbischen Alb bei Bad Überkingen: *Oberrheinisch Geol. Ver. Jahrb. und Mitt. N.F.* **41**, 95–102.

BOARDMAN, JOHN, 1978, Grèzes litées near Keswick, Cumbria: *Biuletyn Peryglacjalny* **27**, 23–34.

BOBOV, N. G., 1960, Sovremennoye obrazovaniye gruntovykh zhil i melkopoligal'nogo rel'efa na Leno-Vilyuiskom mezdurech'e: *Akad. Nauk SSSR*,

Inst. Merzlotovedeniya im. V. A. Obrucheva, Trudy **16**, 24–9.

1970, The formation of beds of ground ice: *Soviet Geography, Review and Translation* **11**, 456–563. [From: *Akad. Nauk SSSR, Izvestiya, seriya geograficheskaya*, 1969 **6**, 63–8.]

BOCH, S. G., 1946, Snezhniki i snezhnaya eroziya v severnykh chastyakh Urala: *Vsesoyuznogo Geograficheskogo obshchestva, Izvestiya* **78**(2), 207–222.

1948, Eshche neskol'ko zamechaniy o prirode snegovoy erozii: *Vsesoyuznogo Geograficheskogo obshchestva, Izvestiya* **80**(6), 609–11.

BOCH, S. G., and KRASNOV, I. I., 1943, O nagornykh terraskh i drevnikh poverkhnostyakh vyravnivaniya na Urale i svyazannykh s nimi problemakh: *Vsesoyuznogo Geograficheskogo obshchestva, Izvestiya* **75**(1), 14–25.

1946, K voprosu o granitse maksimal'nogo chetvertichnogo oledeniya v predelakh Uralskogo khrebta v svyazi s nabludeniyami nad nagornymi terrasami: *Akad. Nauk SSSR, Chetvertichnoy komisii, Byulleten* **8**, 46–72.

BOGOLOMOV, N. S., and SKLYAREVSKAYA, A. N., 1969, O vzryvakh gidrolakkolitov v yuzhnoy chasti Chitinskoy oblasti: 127–30 in Alekseyev, V. R., *et al., Naledi Sibiri*: Akad. Nauk SSSR, Sibirskoye Otdeleniye, Inst. Merzlotovedeniye: Moskva, Izdatel'stvo 'Nauka'. (206 pp.)

1973, On explosion of hydrolaccoliths in southern part of Chitinskaya Oblast [O vzryvakh gidrolakkolitov v yuzhnoy chasti Chitinskoy oblasti]: 187–91 in Alekseyev, V. R., *et al.*, Siberian naleds [*Naledi Sibiri*]: *US Army Corps of Engineers, Cold Regions Research and Engineering Laboratory Draft Translation* **399**. (300 pp.)

BONES, J. G., 1973, Process and sediment size arrangement on high arctic talus, southwest Devon Island, NWT, Canada: *Arctic and Alpine Research* **5**, 29–40.

BORNS, H. W., JR., 1965, Late glacial ice-wedge casts in northern Nova Scotia, Canada: *Science* **148**, 1223–5.

BOSTOCK, H. S., 1948, Physiography of the Canadian Cordillera, with special reference to the area north of the fifty-fifth parallel: *Canada Geol. Survey Mem.* **247**. (106 pp.)

BOSTROM, R. C., 1967, Water expulsion and pingo formation in a region affected by subsidence: *J. Glaciology* **6**, 568–72.

BOULTON, G. S., *et al.*, 1977, A British ice-sheet model and patterns of glacial erosion and deposition in Britain: 231–46 in Shotton, F. W., ed., *British

Quaternary studies – Recent advances: Oxford, Clarendon Press. (298 pp.)

BOUT, PIERRE, 1953, *Etudes de géomorphologie dynamique en Islande: Expéditions Polaires Françaises* **3**: Paris, Hermann & Cie, *Actualités Scientifiques et Industrielles* **1197**. (218 pp.)

BOUT, PIERRE, *et al.*, 1955, Géomorphologie et glaciologie en Island Centrale: *Norois* **2**(8), 461–573.

BOUT, PIERRE, and GODARD, ALAIN, 1973, Aspects du modelé périglaciaire en Scandinavie du Nord: *Biuletyn Peryglacjalny* **22**, 49–79.

BOWLER, J. M., 1978, Glacial-age aeolian events at high and low latitudes: A Southern Hemisphere perspective: 149–72 in Van Zinderen Bakker, E. M., ed., *Antarctic glacial history and world paleoenvironments*: International Council of Scientific Unions – Scientific Committee on Antarctic Research, Symposium 10th INQUA Congress (Birmingham, UK, 17 Aug. 1977) Proc.: Rotterdam, A. A. Balkema. (172 pp.)

BOWLEY, W. W., and BURGHARDT, M. D., 1971, Thermodynamics and stones: *EOS, Am. Geophys. Union Trans.* **52**, 4–7.

BOYD, D. W., 1976, Normal freezing and thawing degree-days from normal monthly temperatures: *Canadian Geotech. J.* **13**, 176–80.

BOYÉ, MARC, 1950, *Glaciaire et périglaciaire de l'Ata Sund nord-oriental, Groenland: Expéditions Polaires Françaises* **1**: Paris, Hermann & Cie, *Actualités Scientifiques et Industrielles* **1111**. (176 pp.)

BOYER, S. J., 1975, Chemical weathering of rocks on the Lassiter Coast, Antarctic Peninsula, Antarctic: *New Zealand J. Geol. and Gephys.* **18**(4), 623–8.

BOYTSOV, M. N., 1963, Genezis i evolyutsiya treshchinno-poligonal'nogo rel'efa: 55–80 in Materialy po Chetvertichnoi geologii i geomorfologii: *Vsesoyuznyye Nauchno-issledovatel'skiy Geologicheskiy Institut, Trudy Novaya seriya* **90**.

BRADLEY, W. C., 1965, Glacial geology, periglacial features and erosion surfaces in Rocky Mountain National Park: *International Assoc. for Quaternary Research (INQUA) Cong., 7th (Boulder, Col., 1965), Guidebook Boulder area, Colorado, USA*, 27–33.

BRADLEY, W. C., HUTTON, J. T., and TWIDALE, C. R., 1978, Role of salts in development of granitic tafoni, South Australia: *J. Geol.* **86**, 647–54.

BRANDT, G. H., 1972, Chemical additives to reduce frost heave and water accumulation in soils: 30–42, and discussion 42–4 in Highway Research Board, *Frost action in soils*: Natl. Acad. Sci.–Natl. Acad. Eng., *Highway Research Record* **393**. (88 pp.)

BRAZEL, A. J., 1972, *Active layer thermal regimes in an

alpine pass, Alaska: Univ. Michigan, Ph.D. Thesis. (70 pp.)

BREMER, HANNA, 1965, Musterböden in tropisch-subtropischen Gebieten und Frostmusterböden: *Zeitschr. Geomorphologie N.F.* **9**, 222–36.

BRETZ, J H., 1935, Physiographic studies in East Greenland: 159–245 in Boyd, L. A., *The fiord region of East Greenland*: Am. Geog. Soc. Spec. Pub. **18**. (369 pp.)

BREWER, MAX, 1958, Some results of geothermal investigations of permafrost in northern Alaska: *Am. Geophys. Union Trans.* **39**, 19–26.

BREWER, R., and HALDANE, A. D., 1957, Preliminary experiments in the development of clay orientation in soils: *Soil Sci.* **84**, 301–9.

BRIDGMAN, P. W., 1912, Water, in the liquid and five solid forms, under pressure: *Am. Acad. Arts and Sci. Proc.* **47**(13), 441–558.

BRIGGS, GEOFFREY, *et al.*, 1977, Martian dynamical phenomena during June–November 1976: Viking Orbiter imaging results: *J. Geophys. Research* **82** (28), 4121–49.

BRINK, V. C., *et al.*, 1967, Needle ice and seedling establishment in southwestern British Columbia: *Canadian J. Plant Sci.* **47**, 135–9.

BRITTON, M. E., 1957, Vegetation of the Arctic Tundra: 26–61 in Hanson, H. P., ed., *Arctic biology* (18th Annual Biology Colloquium): Corvallis, Ore., Oregon State College. (134 pp.)

BROCHU, MICHEL, 1978, Disposition des fragments rocheux dans les dépôts de solifluxion, dans le éboulis de gravité et dans les dépôts fluviatiles: Mesure dans l'est de l'Arctique Nord-Américain et comparaison avec d'autres régions du globe: *Biuletyn Peryglacjalny* **27**, 35–51.

BROCKIE, W. J. 1968, A contribution to the study of frozen ground phenomena – preliminary investigations into a form of miniature stone stripes in East Otago: *New Zealand Geog. Conf., 5th, Proc.,* 191–201.

BROOKES, I. A., 1971 [1972], Fossil ice wedge casts in western Newfoundland: *Maritime Sediments* **7**(3), 118–22.

BROSCHE, K.-U., 1971, Beobachtungen an rezenten Periglazialerscheinungen in einigen Hochgebirgen der Iberischen Halbinsel (Sierra Segura, Sierra de Gredos, Serra da Estrêla, Sierra del Moncayo): *Die Erde* **102**(1), 34–52.

1977, Formen, Formengesellschaften und Untergrenzen in den heutigen periglazialen Höhenstufen der Hochgebirge der Iberischen Halbinsel: 178–202 in Poser, Hans, ed., Formen, Formengesellschaften und Untergrenzen in den heutigen periglazialen Höhenstufen der Hochgebirge Europas und Afrikas zwischen Arktis und Äquator. Bericht über ein Symposium: *Akad. Wiss. Göttingen Abh., Math.-Phys. Kl. Folge 3*, **31**. (355 pp.)

1978, Ergebnisse einer vergleichenden Studie zum rezenten und vorzeitlichen periglazialen Formenschatz auf der Iberischen Halbinsel: *Biuletyn Peryglacjalny* **27**, 53–104.

BROWN, JERRY, 1965, Radiocarbon dating, Barrow, Alaska: *Arctic* **18**, 37–48.

1966, Ice-wedge chemistry and related frozen ground processes, Barrow, Alaska: 94–8 in *Permafrost International Conference* (Lafayette, Ind., 11–15 Nov. 1963) *Proc.*: Natl. Acad. Sci.–Natl. Research Council Pub. **1287**. (563 pp.)

1967, An estimation of the volume of ground ice, coastal plain, Arctic Alaska: *US Army Cold Regions Research and Engineering Laboratory Tech. Note (Nov. 1967)*. (22 pp.)

1969, Buried soils associated with permafrost: 115–27 in Pawluk, S., ed., *Symposium on Pedology and Quaternary Research Proc.*: Edmonton, Univ. Alberta. (218 pp.)

1973, Environmental considerations for the utilization of permafrost terrain: 587–90 in *North American Contribution, Permafrost Second International Conference* (Yakutsk, USSR, 13–28 July 1973): Washington, DC, Natl. Acad. Sci. (783 pp.)

BROWN, JERRY, and GRAVE, N. A., 1979*a*, Physical and thermal disturbance and protection of permafrost: In *Third International Conference on Permafrost* (Edmonton, Alta., 10–13 July 1978), Proc. **2**: Ottawa, Canada Natl. Research Council. (In press.)

BROWN, JERRY, RICKARD, WARREN, and VIETOR, DONALD, 1969, The effect of disturbance on permafrost terrain: *US Army Corps of Engineers, Cold Regions Research and Engineering Laboratory Spec. Rept.* **138**. (13 pp.)

BROWN, JERRY, and SELLMAN, P. V., 1973, Permafrost and coastal plain history of Arctic Alaska: 31–47 in Britton, M. E., ed., Alaskan arctic tundra [Naval Arctic Research Laboratory, 25th anniversary celebration Proc.]: *Arctic Inst. North America Tech. Paper* **25**. (224 pp.)

BROWN, R. J. E., 1960, The distribution of permafrost and its relation to air temperature in Canada and the USSR: *Arctic* **13**, 163–77.

1966*a*, Influence of vegetation on permafrost: 20–5 in *Permafrost International Conference* (Lafayette, Ind., 11–15 Nov. 1963) *Proc.*: Natl. Acad. Sci.–Natl. Research Council Pub. **1287**. (563 pp.)

1966*b*, Relation between mean annual air and ground temperatures in the permafrost region of

Canada: 241–7 in *Permafrost International Conference* (Lafayette, Ind., 11–15 Nov. 1963) *Proc.*: Natl. Acad. Sci.–Natl. Research Council Pub. **1287**. (563 pp.)

1967a, Permafrost in Canada: *Canada Geol. Survey Map* **1246** A – Natl. Research Council Pub. NRC 9769.

1967b, Comparison of permafrost conditions in Canada and the USSR: *Polar Record* **13**, 741–51.

1968, Occurrence of permafrost in Canadian peatlands: *International Peat Cong., 3d, Quebec, 18–23 Aug. 1968, Proc.*, 174–81.

1969, Factors influencing discontinuous permafrost in Canada: 11–53 in Péwé, T. L., ed., *The periglacial environment*: Montreal, McGill–Queen's Univ. Press. (487 pp.)

1970, *Permafrost in Canada*: Toronto, Univ. Toronto Press. (234 pp.)

1972, Permafrost in the Canadian Arctic Archipelago: *Zeitschr. Geomophologie N.F., Supp.* **13**, 102–30.

1973a, Ground ice as an initiator of landforms in permafrost regions: 25–42 in Fahey, B. D., and Thompson, R. D., eds., *Research in polar and alpine geomorphology*: Guelph Symposium on Geomorphology, 1973, 3d, Proc.: Norwich, England, Geo Abstracts Ltd (Univ. East Anglia). (206 pp.)

1973b, Influence of climatic and terrain factors on ground temperature at three locations in the permafrost region of Canada: 27–34 in *North American Contribution, Permafrost Second International Conference* (Yakutsk, USSR, 13–28 July 1973): Washington, DC, Natl. Acad. Sci. (783 pp.)

1974a, Permafrost – Distribution and relation to environmental factors in the Hudson Bay Lowland: 35–68 in *Proceedings: Symposium on the physical environment of the Hudson Bay Lowland* (Univ. Guelph, 30–31 Mar. 1973): Guelph, Ont., Univ. Guelph. (126 pp.)

1974b, Some aspects of airphoto interpretation of permafrost in Canada: *Canada Natl. Research Council, Div. Building Research Tech. Paper* **409**. (20 pp.)

1975, Permafrost investigations in Quebec and Newfoundland (Labrador): *Canada Natl. Research Council, Div. Building Research Tech. Paper* **449**. (36+ pp.)

1978a, Influence of climate and terrain on ground temperature in the continuous permafrost zone of northern Manitoba and Keewatin District, Canada: 15–21 in *Third International Conference on Permafrost* (Edmonton, Alta., 10–13 July 1978),

Proc. **1**: Ottawa, Canada Natl. Research Council. (947 pp.)

1978b, Permafrost: Plate 32 in *Hydrological Atlas of Canada*: Ottawa, Fisheries and Environment Canada. (34 plates.)

BROWN, R. J. E., and KUPSCH, W. O., 1974, Permafrost terminology: *Canada Natl. Research Council, Assoc. Comm. on Geotechnical Research Tech. Mem.* **111**. (62 pp.)

BROWN, R. J. E., and PÉWÉ, T. L., 1973, Distribution of permafrost in North America and its relationship to the environment: A review, 1963–1973: 71–100 in *North American Contribution, Permafrost Second International Conference* (Yakutsk, USSR, 13–28 July 1973): Washington, DC, Natl. Acad. Sci. (783 pp.)

BROWN, R. J. E., and WILLIAMS, G. P., 1972, The freezing of peatland: *Canada Natl. Research Council, Div. Building Research Tech. Paper* **381**. (24 pp.)

BROWN, W. H., 1925, A probable fossil glacier: *J. Geol.* **33**, 464–6.

BRÜNING, HERBERT, 1964, Kinematische Phasen und Denudationsvorgänge bei der Fossilation von Eiskeilen: *Zeitschr. Geomorphologie N.F.* **8**, 343–50.

1966, Vorkommen und Entwicklungsrhythmus oberpleistozäner Periglazial–Erscheinungen und ihr Wert für pleistozäne Hangformung: *Forschungen zur Deutschen Landeskunde* **156**. (97 pp. + 52 plates.)

1973, Der Mainzer Raum und das nördliche Rheinhessen im Quartär: *Natur und Museum* **103**, 360–6.

BRUNNER, F. K., and SCHEIDEGGER, A. E., 1974, Kinematics of a scree slope: *Revista Italiana di Geofisica* **23**(1–2), 89–94.

BRUNNSCHWEILER, DIETER, 1962, The periglacial realm in North America during the Wisconsin glaciation: *Biuletyn Peryglacjalny* **11**, 15–27.

1964, Der pleistozäne Periglazialbereich in Nordamerika: *Zeitschr. Geomorphologie N.F.* **8**, 223–31.

BRYAN, KIRK, 1934, Geomorphic processes at high altitudes: *Geog. Rev.* **24**, 655–6.

1945, Glacial versus desert origin of loess: *Am. J. Sci. 5th ser.* **243**, 245–8.

1946, Cryopedology – the study of frozen ground and intensive frost-action with suggestions on nomenclature: *Am. J. Sci.* **244**, 622–42.

1951, The erroneous use of *tjaele* as the equivalent of perennially frozen ground: *J. Geol.* **59**, 69–71.

BRYAN, KIRK, and RAY, L. L., 1940, Geologic antiquity of the Lindenmeier site in Colorado: *Smithsonian Misc. Colln.* **99**(2). (76 pp.)

BRYSON, R. A., 1966, Air masses, streamlines, and the boreal forest: *Geog. Bull.* **8**(3), 228–69.

BUDD, W. F., JENSSEN, D., and RADOK, U., 1970, The extent of basal melting in Antarctica: *Polarforschung* **6**, 293–306.

1971, Derived physical characteristics of the Antarctic Ice Sheet: *Univ. Melbourne Meteorol. Dept. Pub.* **18**. (178 pp.)

BÜDEL, JULIUS, 1937, Eiszeitliche und rezente Verwitterung und Abtragung im ehemals nicht vereisten Teil Mitteleuropas: *Petermanns Mitt., Ergänzungsheft* **229**. (71 pp.)

1944, Die morphologischen Wirkungen des Eiszeitklimas im gletscherfreien Gebiet: *Geol. Rundschau* **34**, 482–519.

1948, Die klima-morphologischen Zonen der Polarländer: *Erdkunde* **2**, 22–53.

1949a, Die räumliche und zeitliche Gliederung des Eiszeitklimas: *Naturwissenschaften* **36**, 105–12, 133–40.

1949b, Neue Wege der Eiszeitforschung: *Erdkunde* **3**, 82–95.

1950, Das System der klimatischen Morphologie: *Deutscher Geographentag (München 1948) Verhandlungen* **27**(4), 1–36 (65–100 of whole volume).

1951, Die Klimazonen des Eiszeitalters: *Eiszeitalter und Gegenwart* **1**, 16–26.

1953, Die 'periglazial'-morphologischen Wirkungen des Eiszeitklimas auf der ganzen Erde: *Erdkunde* **7**, 249–66.

1959, Periodische und episodische Solifluktion im Rahmen der klimatischen Solifluktionstypen: *Erdkunde* **13**, 297–314.

1960, Die Frostschutt-Zone Südost Spitzbergens: *Colloquium Geographicum* **6**. (105 pp.)

1961, Die Abtragungsvorgänge auf Spitzbergen im Umkreis der Barentsinsel: 337–75 in *Deutscher Geographentag Köln, Tagungsbericht und wissenschaftliche Abhandlungen*: Wiesbaden, Franz Steiner Verlag G.m.b.H. (407 pp.)

1963, Klima-genetische Geomorphologie: *Geog. Rundschau* **15**, 269–85.

1969, Der Eisrinden-Effekt als Motor der Tiefenerosion in der excessiven Talbildungszone: *Würzburger Geog. Arbeiten* **25**. (41 pp.)

1977, *Klima-Geomorphologie*: Berlin and Stuttgart, Gebrüder Bornträeger. (304 pp.)

BULL, A. J., 1940, Cold conditions and land forms in the South Downs: *Geologist's Assoc., London, Proc.* **51**, 63–71.

BUNGE, ALEXANDER V., 1884, Naturhistorische Beobachtungen und Fahrten im Lena-Delta: *Akad. Imp. Sci. St Petersbourg Bull.*, 3d ser., **29**, 422 76.

1902, Einige Worte zur Bodeneisfrage: *Russisch-Kaiserlichen Mineral. Gesell. St Petersburg Verh.*, 2d Ser., **40**(1), 203–9.

BUNTING, B. T., 1961, The role of seepage moisture in soil formation, slope development, and stream initiation: *Am. J. Sci.* **257**, 503–18.

1977, The occurrence of vesicular structures in arctic and subarctic soils: *Zeitschr. Geomorphologie N.F.* **21**, 87–95.

BUNTING, B. T., and JACKSON, R. H., 1970, Studies of patterned ground on SW Devon Island, NWT: *Geografiska Annaler* **52**, 194–208.

BURDICK, J. L., and JOHNSON, PHILIP, eds., 1977, *Proceedings of the Second International Symposium on Cold Regions Engineering* (held at the University of Alaska, Fairbanks, 12–14 Aug. 1976): Cold Regions Engineers Prof. Assoc. (Univ. Alaska, Dept. Civil Engineering). (597 pp.)

BURROUS, C. M., 1977, *Experimental upfreezing of objects: Effects of object geometry*: Univ. Washington, Quaternary Research Center, Periglacial Laboratory Rept. (Ms.)

BUTRYM, J., et al., 1964, New interpretation of 'periglacial structures': *Polska Acad. Nauk Oddzial in Krakowie, Folia Quaternaria* **17**. (34 pp.)

BUTZER, K. W., 1964, *Environment and archeology*: Chicago, Aldine Publishing Co. (524 pp.)

1973, Pleistocene 'periglacial' phenomena in southern Africa: *Boreas* **2**(1), 1–11.

BYELORUSOVA, ZH. M., 1963, O nagornykh (soliflyuktsionnykh) terrasakh ostrova Bol'shogo Lyakhovskogo: 88–92 in Rutilevskiy, G. L., and Sisko, R. K., *Novosibirskiye ostrova*: Glavnogo upravleniya severnogo morskogo puti Ministerstva morskogo flota, Arkticheskogo i Antarcticheskogo nauchno-issledovatelskogo instituta, Trudy, Leningrad.

BYRNE, J., and TRENHAILE, A. S., 1977, Soil tongues in the Leamington Moraine, Ontario: *Canadian Geographer* **21**(3), 274–81.

CADY, W. M., et al., 1955, The Central Kuskokwim region, Alaska: *US Geol. Survey Prof. Paper* **268**. (132 pp.)

CAILLEUX, ANDRÉ, 1942, Les actions éoliennes périglaciaires en Europe: *Soc. Géol. France* **21**(1–2), (Mém. 46). (176 pp.)

1947, Caractères distinctifs des coulées de blocailles liées au gel intense: *Géol. Soc. France Comptes rendus*, 323–4.

1971, Lacs en ourson, cernes et thermokarst: *Cahiers Géog. Québec* **15**(34), 131–6.

1972, Les formes et dépôts nivéo-éoliens actuels en Antarctique et au Nouveau-Québec: *Cahiers Géog. Québec* **16**(39), 377–409.

1973*a*, Éolisations périglaciaires quaternaires au Canada: *Biuletyn Peryglacjalny* **22**, 81–115.

1973*b*, Répartition et signification des différents critères d'éolisation périglaciaire: *Biuletyn Peryglacjalny* **23**, 50–63.

1974, Formes précoces et albédos du nivéo-éolien: *Zeitschr. Geomorphologie N.F.* **18**, 437–59.

1976, Les pingos quaternaires de France: *Rev. Géog. Montréal* **30**(4), 374–9.

CAILLEUX, A., and CALKIN, P., 1963, Orientation of hollows in cavernously weathered boulders in Antarctica: *Biuletyn Peryglacjalny* **12**, 147–50.

CAILLEUX, A., GUILCHER, A., and TRICART, J., 1956, Phénomènes périglaciaires d'âge présumé Wurm: 34, Plate 3 in Tricart, Jean, Cartes des phénomènes périglaciaires quaternaires en France: *Carte géologique detaillée de la France, Mémoires*. (40 pp.)

CAILLEUX, ANDRÉ, and ROUSSET, CLAUDE, 1968, Présence de réseaux polygonaux de Fentes en coin en Basse-Provence occidentale et leur signification paléoclimatique: *Acad. Sciences Paris Comptes Rendus* **266**, 669–71.

CAILLEUX, A., and TAYLOR, G., 1954, *Cryopédologie, études des sols gelés: Expéditions Polaires Françaises, Missions Paul-Emile Victor* **IV**: Paris, Hermann & Cie, *Actualités Scientifiques et Industrielles* **1203**. (218 pp.)

CAINE, NEL, 1967, The tors of Ben Lomond, Tasmania: *Zeitschr. Geomorphologie N.F.* **11**, 418–29.

1968*a*, The log-normal distribution and rates of soil movement: An example: *Rev. Géomorphologie dynamique* **18**(1), 1–7.

1968*b*, The blockfields of northeastern Tasmania: *Australian Natl. Univ., Dept. Geog. Pub.* **G/6**. (127 pp.)

1969, A model for alpine talus slope development by slush avalanching: *J. Geol.* **77**(1), 92–100.

1972*a*, The distribution of sorted patterned ground in the English Lake District: *Rev. Géomorphologie dynamique* **21**, 49–56.

1972*b*, Air photo analysis of blockfield fabrics in Talus Valley, Tasmania: *J. Sed. Petrology* **42**(1), 33–48.

1974, The geomorphic processes of the alpine environment: 721–48 in Ives, J. D., and Barry, R. G., eds., *Arctic and alpine environments*: Methuen & Co. Ltd. (999 pp.)

CAINE, N. F., and JENNINGS, J. N., 1968, Some block-streams of the Toolong Range Kosciusko State Park, New South Wales: *Royal Soc. New South Wales, J. and Proc.* **101**, 93–103.

CAIRNES, L. D., 1912, Differential erosion and equiplanation in portions of Yukon and Alaska: *Geol. Soc. America Bull.* **23**, 333–48.

CALKIN, P. E., 1971, Glacial geology of the Victoria valley system, Southern Victoria Land, Antarctica: 363–412 in Crary, A. P., ed., Antarctic snow and ice studies—II, *Am. Geophysical Union, Antarctic Research Ser.* **16**. (412 pp.)

CALKIN, P. E., and NICHOLS, R. L., 1972, Quaternary studies in Antarctica: 625–43 in Adie, R. J., ed., *Antarctic geology and geophysics* (Internat. Union Geol. Sci., Ser. B, No. 1): Oslo, Universitetsforlaget. (876 pp.)

CALKIN, P. E., and RUTFORD, R. H., 1974, The sand dunes of Victoria Valley, Antarctica: *Geog. Rev.* **64**, 189–216.

CAMERON, R. E., 1969, Cold desert characteristics and problems relevant to other arid lands: 167–205 in McGinnies, W. G., and Goldman, B. J., eds., *Arid lands in perspective*: Tucson, Arizona, Univ. Arizona Press; Washington, DC, Am. Assoc. Advancement of Science. (421 pp.)

CAMPBELL, I. B., and CLARIDGE, G. G. C., 1975, Morphology and age relationships of antarctic soils: 83–8 in Suggate, R. P., and Cresswell, M. M., eds., Quaternary studies (Selected papers from IX INQUA Congress, Christchurch, New Zealand, 2–10 December 1973): *Royal Soc. New Zealand Bull.* **13**. (321 pp.)

CANADA, NATIONAL RESEARCH COUNCIL, 1978, *Third International Conference on Permafrost*: Proc. **1–2**: Ottawa. (1: 947 pp.) (2: In press.)

CAPPS, S. R., JR., 1910, Rock glaciers in Alaska: *J. Geol.* **18**, 359–75.

CAREY, K. L., 1970, Icing occurrence, control and prevention, an annotated bibliography: *US Army Corps of Engineers, Cold Regions Research and Engineering Laboratory Spec. Rept.* **151**. (151 pp.)

1973, Icings developed from surface water and ground water: *US Army Corps of Engineers, Cold Regions Research and Engineering Laboratory, Cold Regions Science and Engineering Mon.* **III**–D3. (65 pp.)

CARLETON UNIVERSITY and ÉCOLE NATIONALE DES PONTS ET CHAUSÉES, 1978, *Soil freezing and highway construction (Gels des sols et des chausées)*: Ottawa (Carleton Univ.) and Paris (École Nationale des Ponts et Chausées). (105 pp.)

CARLSON, R. F., 1977, Design considerations of a northern chilled gas pipeline stream crossing: 178–84 in *Frost action in soils*: Univ. Luleå, Internat. Symposium (Luleå, Sweden, 16–18 Feb. 1977) Proc. 1. (215 pp.)

CARR, M. H., and SCHABER, G. G., 1977, Martian

permafrost features: *J. Geophys. Research* **82**(28), 4039–54.

CARSON, C. E., and HUSSEY, K. M., 1962, The oriented lakes of Arctic Alaska: *J. Geol.* **70**, 417–39.

CARSON, M. A., 1976, Mass-wasting, slope development and climate: 101–36 in Derbyshire, Edward, ed., *Geomorphology and climate*: London, New York, Sydney, Toronto, John Wiley & Sons. (512 pp.)

CARSON, M. A., and KIRKBY, M. J., 1972, *Hillslope form and process*: Cambridge University Press. (475 pp.)

CASAGRANDE, ARTHUR, 1932, Discussion: 168–72 in Benkelman, A. C., and Olmstead, E. R., A new theory of frost heaving, *Natl. Research Council Highway Research Board, Proc. 11th Ann. Mtg.* (1), 152–77.

CASS, L. A., and MILLER, R. D., 1959, Role of the electric double layer in the mechanism of frost heaving: *US Army Snow Ice and Permafrost Research Establishment Research Rept.* **49**. (15 pp.)

CEGŁA, JERZY, 1973, Próba wyjaśnienia genezy gruntów strukturalnych SW Spitsbergenu w Świetle teorii układów niestatecznego warstwowania gęstościowego (Tentative explanation of origin of structural grounds in SW Spitsbergen in the light of the hypothesis of systems with reversed density gradient): *Czasopismo Geograficzne* **44**(2), 237–43.

CEGŁA, JERZY, and DŻUŁYŃSKI, STANISŁAW, 1970, Układy niestatecznie warstwowane i ich występowanie w srodowisku peryglacjalnym: *Acta Universitatis Wratislaviensis* **124** (*Studia geograficzne* **13**), 17–42.

CEGŁA, JERZY, and KOZARSKI, STEFAN, 1977, Sedimentary and geomorphological consequences of the occurrence of naled sheets on the outwash plain of the Gås Glacier, Sörkappland, Spitsbergen: *Acta Universitatis Wratislaviensis* **387**, 63–84.

CHAIX, ANDRÉ, 1923, Les coulées de blocs du Parc national suisse d'Engadine: *Le Globe, Mémoires* [*Société de géographie, Genève*] **42**. (38 pp.)

1943, Les coulées de blocs du Parc national suisse: *Le Globe, Mémoires* [*Société de géographie, Genève*] **82**, 121–8.

CHALMERS, BRUCE, and JACKSON, K. A., 1970, Experimental and theoretical studies of the mechanism of frost heaving: *US Army Corps of Engineers, Cold Regions Research and Engineering Laboratory Research Rept.* **199**. (22 pp.)

CHAMBERLAIN, E. J., and BLOUIN, S. E., 1976, Freeze-thaw enhancement of the drainage and consolidation of fine-grained dredged material in confined disposal areas: *US Army Corps of Engineers, Cold Regions Research and Engineering Laboratory*

(*CRREL*) *Final Rept. to US Army Engineer Waterways Experiment Station* [*under*] *Task 5A07*. (44 pp. + 61 Figs.)

1978, Densification by freezing and thawing of fine material dredged from waterways: 623–8 in *Third International Conference on Permafrost* (Edmonton, Alta., 10–13 July 1978), Proc. **1**: Ottawa, Canada Natl. Research Council. (947 pp.)

CHAMBERS, M. J. G., 1966a, Investigations of patterned ground at Signy Island, South Orkney Islands: I. Interpretation of mechanical analyses: *British Antarctic Survey Bull.* **9**, 21–40.

1966b, Investigations of patterned ground at Signy Island, South Orkney Islands: II. Temperature regimes in the active layer: *British Antarctic Survey Bull.* **10**, 71–83.

1967, Investigations of patterned ground at Signy Island, South Orkney Islands: III. Miniature patterns, frost heaving and general conclusions: *British Antarctic Survey Bull.* **12**, 1–22.

1970, Investigations of patterned ground at Signy Island, South Orkney Islands: IV. Long-term experiments: *British Antarctic Survey Bull.* **23**, 93–100.

CHENG, KUO-TUNG, 1975, *Tung t'u*: Peking, K'o Hsüeh Ch'u Pan She. (124 pp.)

CHERNOV, A. A., and CHERNOV, G. A., 1940, *Geologicheskoye stroyeniye bassey na reki Kos'-yu v Perhorshkom krayye*: Moskva.

CHIGIR, V. G., 1972, Fiziko-geograficheskiye usloviya moroznogo vyvetrivaniya v severnom Prikaspii, Predkavkaz'e i problema lëssof: 95–104 in Popov, A. I., ed., *Problemy kriolitologii* **2**: Moskva, Izdatel'stvo Moskovskogo Universiteta. (148 pp.)

1974, Physical-geographic conditions of frost weathering in the northern Caspian area and the North Caucasian Piedmont area, and the problem of loess [Fiziko-geograficheskiye usloviya moroznogo vyvetrivaniya v severnom Prikaspii, Predkavkaz'e i problema lëssof]: 95–105 in Popov, A. I., et al., Problems of cryolithology: *US Army Corps of Engineers, Cold Regions Research and Engineering Laboratory Draft Translation* **433**. (147 pp.)

CHOLNOKY, EUGEN V., 1911, Spitzbergen: *Földrajzi Közlemének* (*Nemzetközi kiadás* [International edition]) **39**, 93–134.

CHOU, YU-WU, and TU, YUNG-HUAN, 1963, Ch'ing tsang kao yüan tung t'u ch'u pu k'ao ch'a [Preliminary survey of the permafrost of the Ch'ing-hai Tibetan Plateau]: *Kexue Tongbao* [K'o Hsüeh T'ung Pao] **25**(2), 60–3.

CHRISTENSEN, LEIF, 1974, Crop-marks revealing large-scale patterned ground structures in cultivated area, southwestern Jutland, Denmark: *Boreas* **3**, 153–80.

CHURCH, M., 1972, Baffin Island sandurs; a study of arctic fluvial processes: *Canada Geol. Survey Bull.* **216**. (208 pp.)

CHURCH, R. E., PÉWÉ, T. L., and ANDRESEN, M. J., 1965, Origin and environmental significance of large-scale patterned ground, Donnelly Dome area, Alaska: *US Army Cold Regions Research and Engineering Laboratory Research Rept.* **159**. (71 pp.)

CLAPPERTON, C. M., 1975, Further observations on the stone runs of the Falkland Islands: *Biuletyn Peryglacjalny* **24**, 211–17.

CLARK, G. M., 1968, Sorted patterned ground: New Appalachian localities south of the glacial border: *Science* **161**, 355–6.

CLAYTON, LEE, and BAILEY, P. K., 1970, Tundra polygons in the northern Great Plains (abs.): *Geol. Soc. America, Abs. with Programs* **2**(6), 382.

CLAYTON, R. A. S., 1977, The geology of northwestern South Georgia: 1. Physiography: *British Antarctic Survey Bull.* **46**, 85–98.

CLIMAP, 1976, The surface of the ice-age earth: *Science* **191**, 1131–7.

COGLEY, J. G., 1972, Processes of solution in an Arctic limestone terrain: 201–11 in Price, R. J., and Sugden, D. E., compilers, *Polar geomorphology*: Inst. British Geographers Spec. Pub. **4**. (215 pp.)

COHEN, J. B., 1973, Solid waste disposal in permafrost areas: 590–8 in *North American Contribution, Permafrost Second International Conference* (Yakutsk, USSR, 13–28 July 1973): Washington, DC, Natl. Acad. Sci. (783 pp.)

COLLARD, MARCEL, 1973, Cryoturbations en forme de stries enracinées parallèls à la pente: *Biuletyn Peryglacjalny* **22**, 323–7.

CONANT, L. C., BLACK, R. F., and HOSTERMAN, J. W., 1976, Sediment-filled pots in upland gravels of Maryland and Virginia: *US Geol. Survey J. Research* **4**, 353–8.

CONRAD, T. A., 1839, Notes on American geology: *Am. J. Sci.* **35**, 237–51.

COOK, F. A., 1956, Additional notes on mud circles at Resolute Bay, Northwest Territories: *Canadian Geographer* **8**, 9–17.

1966, Patterned ground research in Canada: 128–30 in *Permafrost International Conference* (Lafayette, Ind., 11–15 Nov. 1963) *Proc.*: Natl. Acad. Sci–Natl. Research Council Pub. **1287**. (563 pp.)

COOK, F. A., and RAICHE, V. G., 1962, Freeze-thaw cycles at Resolute, NWT: *Geog. Bull.* **18**, 64–78.

COOKE, R. U., 1970, Stone pavements in deserts: *Assoc. Am. Geographers Annals* **60**, 560–77.

COOKE, R. U., and WARREN, ANDREW, 1973, *Geomorphology in deserts*: Berkeley and Los Angeles, Univ. California Press. (394 pp.)

COOPE, G. R., 1977, Fossil coleopteran assemblages as sensitive indicators of climatic changes during the Devensian (Last) cold stage: *Royal Soc. London Phil. Trans.* B**280**, 313–40.

COOPE, G. R., MORGAN, ANNE, and OSBORNE, P. J., 1971, Fossil Coleoptera as indicators of climatic fluctuations during the Last Glaciation in Britain: *Palaeogeog., Palaeoclimatol., and Palaeoecol.* **10**, 87–101.

CORBEL, J., 1961, Morphologie périglaciaire dans l'Arctique: *Annales Géog.* **70**, 1–24.

CORTE, A. E., 1959, Experimental formation of sorted patterns in gravel overlying a melting ice surface: *US Army Corps of Engineers, Snow Ice and Permafrost Research Establishment Research Rept.* **55**. (19 pp.)

1961, The frost behavior of soils: laboratory and field data for a new concept – I, Vertical sorting: *US Army Corps of Engineers, Cold Regions Research and Engineering Laboratory Research Rept.* **85**(1). (22 pp.)

1962a, Vertical migration of particles in front of a moving freezing plane: *J. Geophys. Research* **67**(3), 1085–90.

1962b, Relationship between four ground patterns, structure of the active layer, and type and distribution of ice in the permafrost: *US Army Corps of Engineers, Cold Regions Research and Engineering Laboratory Research Rept.* **88**. (79 pp.)

1962c, The frost behavior of soils: laboratory and field data for a new concept – II, Horizontal sorting: *US Army Corps of Engineers, Cold Regions Research and Engineering Laboratory Research Rept.* **85**(2). (20 pp.)

1962d, The frost behavior of soils – I, Vertical sorting: 9–34 in Soil behavior on freezing with and without additives: *Natl. Acad. Sci.–Natl. Research Council Highway Research Board Bull.* **317**. (34 pp.)

1962e, The frost behavior of soils – II, Horizontal sorting: 44–66 in Soil behavior associated with freezing: *Natl. Acad. Sci.–Natl. Research Council Highway Research Board Bull.* **331**. (115 pp.)

1963a, Vertical migration of particles in front of a moving freezing plane: *US Army Cold Regions Research and Engineering Laboratory Research Rept.* **105**. (8 pp.)

1963b, Particle sorting by repeated freezing and thawing: *Science* **142**, 499–501.

1963c, Relationship between four ground patterns, structure of the active layer, and type and distribution of ice in the permafrost: *Biuletyn Peryglacjalny* **12**, 7–90.

1966a, Experiments on sorting processes and the origin of patterned ground: 130–5 in *Permafrost International Conference* (Lafayette, Ind., 11–15 Nov. 1963) *Proc.*: Natl. Acad. Sci.–Natl. Research Council Pub. **1287**. (563 pp.)

1966b, Particle sorting by repeated freezing and thawing: *Biuletyn Peryglacjalny* **15**, 175–240.

1967a, Soil mound formation by multicyclic freeze-thaw: 1333–8 in Oura, Hirobumi, ed., *Physics of snow and ice, Internat. Conference Low Temperature Science* (Sapporo, Japan, 14–19 Aug. 1966) *Proc.* **1**(2): Sapporo, Hokkaido Univ., 713–1414.

1967b, Informe preliminar del progreso effectuado en el estudio de las estructuras de crioturbación Pleistocénas fósiles en la Provincia de Santa Cruz: *Terceras Jornadas Geol. Argentinas* **2**, 9–19.

1969a, Geocryology and engineering: 119–85 in Varnes, D. J., and Kiersch, George, eds., *Reviews in engineering geology* **2**: Boulder, Colo., Geol. Soc. America. (350 pp.)

1969b, Formacion en el laboratorio de estructuras como pliegues por congelamiento y descongelamiento multiple: *Cuartas Jornadas Geológicas Argentinas* **1**, 215–27.

1976, Rock glaciers: *Biuletyn Peryglacjalny* **26**, 175–97.

1978, Rock glaciers as permafrost bodies with a debris cover as an active layer. A hydrological approach. Andes of Mendoza, Argentine: 262–9 in *Third International Conference on Permafrost* (Edmonton, Alta., 10–13 July 1978), *Proc.* **1**: Ottawa, Canada Natl. Research Council. (947 pp.)

CORTE, A. E., and HIGASHI, AKIRA, 1964, Experimental research on desiccation cracks in soil: *US Army Cold Regions Research and Engineering Laboratory Research Rept.* **66**. (72 pp.)

COSTIN, A. B., 1955, A note on gilgaies and frost soils: *J. Soil Sci.* **6**, 32–4.

1972, Carbon-14 dates from the Snowy Mountains area, southeastern Australia, and their interpretation: *Quaternary Research* **2**, 579–90.

COSTIN, A. B., *et al.*, 1964, Snow action on Mount Twynam, Snowy Mountains, Australia: *J. Glaciology* **5**(38), 219–28.

1967, Nonsorted steps in the Mt Kosciusko area, Australia: *Geol. Soc. America Bull.* **78**, 979–91.

1973, Forces developed by snowpatch action, Mt Twynam, Snowy Mountains, Australia: *Arctic and Alpine Research* **5**, 121–6.

COSTIN, A. B., and POLACH, H. A., 1971, Slope deposits in the Snowy Mountains, south-eastern Australia: *Quaternary Research* **1**, 228–35.

1972, Age and significance of slope deposits, Black Mountain, Canberra: *Australian J. Soil Research* **11**, 13–25.

COSTIN, A. B., and WIMBUSH, D. J., 1973, Frost cracks and earth hummocks at Kosciusko, Snowy Mountains, Australia: *Arctic and Alpine Research* **5**, 111–120.

COTTON, C. A., and TE PUNGA, M. T., 1955, Solifluction and periglacially modified landforms at Wellington, New Zealand: *Royal Soc. New Zealand Trans.* **82**(5), 1001–31.

COUTARD, J.-P., *et al.*, 1970, Gélifraction expérimentale des calcaires de la Campagne de Caen; comparaison avec quelques dépôts périglaciaires de cette région: *CNRS, Centre de Géomorphologie de Caen Bull.* **6**, 7–44.

CRAIG, B. G., 1969, photograph, inside front cover: *Papers on Quaternary research in Canada* (*INQUA Congress, 8th, Paris, 30 Aug.–5 Sept. 1969*), Canada Geol. Survey.

CRAMPTON, C. B., 1977a, A study of the dynamics of hummocky microrelief in the Canadian north: *Canadian J. Earth Sci.* **14**, 639–49.

1977b, A note on asymmetric valleys in the central Mackenzie River catchment, Canada: *Earth Surface Processes* **2**, 427–9.

CRAWFORD, C. B., and JOHNSTON, G. H., 1971, Construction on permafrost: *Canadian Geotech. J.* **8**, 236–51.

CROSS, C. W., HOWE, ERNEST, and RANSOME, F. L., 1905, Description of the Silverton Quadrangle: *US Geol. Survey, Geologic Atlas of the United States*, Silverton Folio, **120**. (34 pp.)

CRUICKSHANK, J. G., and COLHOUN, E. A., 1965, Observations on pingos and other landforms in Schuchertdal, Northeast Greenland: *Geografiska Annaler* **47**A, 224–36.

CUNNINGHAM, F. F., 1969, The Crow Tors, Laramie Mountains, Wyoming, USA: *Zeitschr. Geomorphologie N.F.* **13**(1), 56–74.

CURREY, D. R., 1964, A preliminary study of valley asymmetry in the Ogotoruk Creek area, northwestern Alaska: *Arctic* **17**, 85–98.

CUTTS, J. A., *et al.*, 1976, North polar region of Mars: Imaging results from Viking 2: *Science* **194**, 1329–37.

CZAJKA, WILLI, 1955, Rezente und pleistozäne Verbreitung und Typen des periglazialen Denudationszyklus in Argentinien: *Soc. Geog. Fenniae Acta Geographica* **14**, 121–40.

CZEPPE, ZDZISŁAW, 1959, Uwagi o procesie wymarzania głazów: *Czasopismo Geograficzne* **30**, 195–202.

——— 1960, Thermic differentiation of the active layer and its influence upon the frost heave in periglacial regions (Spitsbergen): *Acad. polonaise sciences Bull.*, *sér. sci. géol. et géog.* **8**(2), 149–52.

——— 1961, Participation of flowing water in creation of the periglacial microrelief: 89 in *International Assoc. for Quaternary Research* (*INQUA*) *Cong., 6th* (*Warsaw, Poland, Aug.–Sept. 1961*), *Abstracts of Papers.* (210 pp.)

——— 1966, Przebieg głównych procesow morfogenetycznych w południowozachadnim Spitsbergenie: *Uniwersytetu Jagiellońskiego, Zeszyty Naukowe Prace Geograficzne, Zeszyt* **13**, *Prace Instytutu Geograficznego, Zeszylt* **35**. (129 pp.)

CZUDEK, TADÉÀS, 1964, Periglacial slope development in the area of the Bohemian Massif in Northern Moravia: *Biuletyn Peryglacjalny* **14**, 169–94.

——— 1973*a*, Zur klimatischen Talasymmetrie des Westteiles der Tschechoslowakei: *Zeitschr. Geomorphologie N.F., Supp.* **17**, 49–57.

——— 1973*b*, Die Talasymmetrie im Nordteil der Moravská brána (Mährische Pforte): *Acta Sci. Nat. Brno* **7**(3), 1–48.

CZUDEK, TADÉÀS, and DEMEK, JAROMÍR, 1961, Výzman pleistocenní kryoplanace na vývoj povrchových tvarů České vysočiny; Symposion o problémech pleistocénu: *Anthropos* **14** (*N.S.* **6**), 57–69.

——— 1970*a*, Pleistocene cryopedimentation in Czechoslovakia: *Acta Geographica Lodziensia* **24**, 101–8.

——— 1970*b*, Thermokarst in Siberia and its influence on the development of lowland relief: *Quaternary Research* **1**, 103–20.

——— 1971, Der Thermokarst im Ostteil des Mitteljakutischen Tieflandes: Scripta Facultatis Scientiarum Naturalium Univ. Purkynianse Brunensis, *Geographia* **1**, 1–19.

——— 1972, Present-day cryogenic processes in the mountains of eastern Siberia: *Geographia Polonica* **23**, 5–20.

——— 1973*a*, The valley cryopediments in Eastern Siberia: *Biuletyn Peryglacjalny* **22**, 117–30.

——— 1973*b*, Die Reliefentwicklung während der Dauerfrostbodendegradation: *Československé Akademie Věd Rozpravy, Řada Matematických a Přírodních Věd* **83**(2). (69 pp.)

CZUDEK, TADÉÀS, DEMEK, JAROMÍR, and STEHLÍK, O., 1961, Formy zvětrávání a odnosu pískovcův v Hostýnských vrších a Chřibech: *Časopis pro mineralogii a geologii* **6**, 262–9.

DAHL, EILIF, 1956, Rondane: Mountain vegetation in South Norway and its relation to the environment: *Norske Vidensk.-Akad. Oslo, I. Mat.-naturv. Kl.* **3**. (374 pp.)

DAHL, RAGNAR, 1966*a*, Block fields, weathering pits and tor-like forms in the Narvik mountains, Nordland, Norway: *Geografiska Annaler* **48**A, 55–85.

——— 1966*b*, Block fields and other weathering forms in the Narvik mountains: *Geografiska Annaler* **48**A, 224–7.

DALQUEST, W. W., and SCHEFFER, V. B., 1942, The origin of the Mima mounds of Western Washington: *J. Geol.* **50**, 68–84.

DALY, R. A., 1912, Geology of the North American Cordillera at the Forty-ninth Parallel: *Canada Geol. Survey Mem.* **38**. (857 pp.) (3 pts.)

DANILOV, I. D., 1972, Merzlotnyye i psevdomerzlotnyye klinovidnyye deformatsii v osadochnykh poradakh: 31–48 in Popov, A. I., ed., *Problemy kriolitologii* **2**: Moskva, Izdatel'stvo Moskovskogo Universiteta. (148 pp.)

——— 1973*a*, Subaqueous pseudomorphoses in Pleistocene deposits: *Biuletyn Peryglacjalny* **22**, 339–45.

——— 1973*b*, Plastic and ruptural deformations in basin sediments: *Biuletyn Peryglacjalny* **22**, 329–37.

——— 1974, Frost and pseudofrost wedge-shaped deformations in sedimentary rock [Merzlotnyye i psevdomerzlotnyye klinovidnyye deformatsii v osadochnykh poradakh]: 31–48 in Popov, A. I., *et al.*, Problems of cryolithology: *US Army Corps of Engineers, Cold Regions Research and Engineering Laboratory Draft Translation* **433**. (147 pp.)

DANILOVA, N. S., 1956, Gruntovyye zhily i ikh proiskhozhdeniye: 109–22 in Meyster, L. A., ed., *Materialy k osnovam ucheniya o merzlykh zonakh zemnoy kory, Vypusk – III*: Moskva, Akad. Nauk SSSR, Inst. Merzlotovedeniya im. V. A. Obrucheva. (229 pp.)

——— 1963, Soil wedges and their origin [Gruntovyye zhily i ikh proiskhozhdeniye]: 90–9 in Meister, L. A., ed., Data on the principles of the study of frozen zones in the earth's crust. Issue III: *Canada Natl. Research Council Tech. Translation* **1088**. (169 pp.)

DANSGAARD, WILLI, *et al.*, 1971, Climatic record revealed by the Camp Century ice core: 37–56 in Turekian, K. K., ed., *The Late Cenozoic glacial ages*: New Haven and London, Yale University Press. (606 pp.)

DAVIDSON, D. W., *et al.*, 1978, Natural gas hydrates in northern Canada: 937–43 in *Third International Conference on Permafrost* (Edmonton, Alta., 10–13

July 1978), Proc. **1**: Ottawa, Canada Natl. Research Council. (947 pp.)

DAVIES, J. L., 1969, *Landforms of cold climates*: Cambridge, Mass., MIT Press. (200 pp.)

DAVIES, W. E., 1961*a*, Polygonal features on bedrock, North Greenland: *US Geol. Survey Prof. Paper* **424**-D, D218–19.

1961*b*, Surface features of permafrost in arid areas: *Folia Geog. Danica* **9**, 48–56. (Internat. Geog. Cong., 19th, Norden 1960, Symposium SD 2, Physical geography of Greenland.)

DAVIS, J. L., *et al.*, 1976, Impulse radar experiments on permafrost near Tuktoyaktuk, Northwest Territories: *Canadian J. Earth Sci.* **13**, 1584–90.

DAVIS, W. M., 1909, The geographical cycle: 249–78 in Davis, W. M., *Geographical essays* (D. W. Johnson, ed.): Boston, Ginn and Company. (777 pp.)

DAVISON, CHARLES, 1888*a*, Note on the movement of scree-material: *Geol. Soc. London Quart. J.* **44**, 232–8.

1888*b*, Second note on the movement of scree-material: *Geol. Soc. London Quart. J.* **44**, 825–6.

1889, On the creeping of the soilcap through the action of frost: *Geol. Mag.* (Great Britain), decade 3, **6**, 255–61.

DAY, T. J., and GALE, R. J., 1976, Geomorphology of some arctic gullies: *Canada Geol. Survey Paper* **76–1**B, 173–85.

DE GEER, GERARD, 1912*a*, A geochronology of the last 12 000 years: *Internat. Geol. Congress, 11th (Stockholm, 1910), Compte rendu*, 241–57.

1912*b*, Exkursion Al: Spitsbergen: *Internat. Geol. Congress 11th (Stockholm, 1910), Compte rendu*, 1205–26.

1940, Geochronologia Suecica Principles: *K. Svenska Vetensk. Handl.*, ser. *3* **18**(6), Stockholm, Almqvist & Wiksells. (367 pp. and atlas.)

DEGRAFF, J. V., 1976, Relict patterned ground, Bear River Range, north-central Utah: *Utah Geology* **3**(2), 111–16.

DELORME, L. D., ZOLTAI, S. C., and KALAS, L. L., 1977, Freshwater shelled vertebrate indicators of paleoclimate in northwestern Canada during late glacial times: *Canadian J. Earth Sci.* **14**, 2029–46.

1978, Freshwater shelled vertebrate indicators of paleoclimate in northwestern Canada during late glacial times: Reply: *Canadian J. Earth Sci.* **15**, 462–3.

DEMEK, JAROMÍR, 1964*a*, Slope development in granite areas of Bohemian Massif: *Zeitschr. Geomorphologie N.F., Supp.* **5**, 82–106.

1964*b*, Altiplanation terraces in Czechoslovakia and their origin: *J. Czechoslovak Geog. Soc., Cong. Suppl.*, 55–65.

1964*c*, Castle koppies and tors on the Bohemian Highland (Czechoslovakia): *Biuletyn Peryglacjalny* **14**, 192–216.

1964*d*, Formy zvětrávání a odnosu granodioritu v Novohradských horách: *Zprávy Geografického ústavu ČSAV 1964* **9**, 6–15.

1965, Kryoplanační terasy v jihozápadní Anglii: *Sborník Československé společnosti zeměpisné* **70**, 272–7.

1967, Zpráva o studiu kryogenních jevů v Jakutsku: *Sborník Československé společnosti zeměpisné* **72**, 99–114.

1968*a*, Cryoplanation terraces in Yakutia: *Biuletyn Peryglacjalny* **17**, 91–116.

1968*b*, Beschleunigung der geomorphologischen Prozesse durch die Wirkung des Menschen: *Geol. Rundschau* **58**, 111–21.

1969*a*, Cryoplanation terraces, their geographical distribution, genesis and development: *Československé Akademie Věd Rozpravy, Řada Matematických a Přírodních Věd* **79**(4). (80 pp.)

1969*b*, Cryogene processes and the development of cryoplanation terraces: *Biuletyn Peryglacjalny* **18**, 115–25.

DEMEK, JAROMÍR, *et al.*, 1965, *Geomorfologie Českých zemí*: Praha. (335 pp.)

DEMEK, JAROMÍR, and KUKLA, JIŘI, eds., 1969, *Periglazialzone, Löss und Paläolithikum der Tschechoslowakei*: Brno, Tschechoslowakische Akad. Wissenschaften, Geog. Inst. (158 pp.)

DEMEK, JAROMÍR, and ŠMARDA, J., 1964, Periglaciální jevy v Bulharsku: *Zprávy Geografického ústavu ČSAV 1964* **3**, 1–4.

DENGIN, I. P., 1930, Sledy drevnego oledeneniya v Yablovom khrebte i problema goltsovykh terras: *Izvestiya Gosudarstvennogo Geograficheskogo Obshchestva* **62**(2), 153–87.

DENNY, C. S., 1936, Periglacial phenomena in southern Connecticut: *Am. J. Sci. 5th ser.* **32**, 322–42.

1951, Pleistocene frost action near the border of the Wisconsin drift in Pennsylvania: *Ohio J. Sci.* **51**, 116–25.

1956, Surficial geology and geomorphology of Potter County, Pennsylvania: *US Geol. Survey Prof. Paper* **288**. (72 pp.)

DERBYSHIRE, EDWARD, 1972, Tors, rock weathering and climate in southern Victoria Land, Antarctica: 93–105 in Price, R. J., and Sugden, D. E., compilers, *Polar geomorphology*: Inst. British Geographers Spec. Pub. **4**. (215 pp.)

1973, Periglacial phenomena in Tasmania: *Biuletyn Peryglacjalny* **22**, 131–48.

DERRUAU, M., 1956, Les formes périglaciaires du Labrador-Ungava Central comparées à celles de l'Island Centrale: *Rev. Géomorphologie dynamique* **7**, 12–16.

DEYNOUX, M., and TROMPETTE, R., 1976, Late Precambrian mixtites: Glacial and/or nonglacial? Dealing especially with the mixtites of West Africa: *Am. J. Sci.* **276**(10), 1302–15.

DIETRICH, R. V., 1977, Wind erosion by snow: *J. Glaciology* **18**(78), 148–9.

DINGMAN, S. L., 1975, Hydrologic effects of frozen ground – Literature review and synthesis: *US Army Corps of Engineers, Cold Regions Research and Engineering Laboratory Spec. Rept.* **218**. (55 pp.)

DINGWALL, P. R., 1973, Rock glaciers in the Canadian Rocky Mountains (abs.): 80–1 in *International Union for Quaternary Research (INQUA) Cong., 9th (Christchurch, New Zealand, 2–10 Dec. 1973), Abstracts.* (418 pp.)

DIONNE, J.-C., 1966a, Un type particulier de buttes gazonnées: *Rev. Géomorphologie dynamique* **16**, 97–100.

1966b, Formes de cryoturbation fossiles dans le sud-est du Québec: *Cahiers Géog. Québec* **10**(19), 89–100.

1966c, Fentes en coin fossiles dans le Québec méridional: *Acad. Sci. [Paris] Comptes rendus* **262**, 24–7.

1968a, Observations sur les tourbières reticulées du Lac Saint-Jean: *Mimeo.* (16 pp.)

1968b, Morphologie et sédimentologie glacielles, littoral sud du Saint-Laurent: *Zeitschr. Geomorphologie N.F., Supp.* **7**, 56–84.

1969, Nouvelles observations de fentes de gel fossiles sur la côte sud du Saint-Laurent: *Rev. Géog. Montréal* **23**, 307–16.

1970a, Aspects morpho-sédimentologiques du glaciel, en particulier des côtes du Saint-Laurent: *Service Canadien des Forêtes, Laboratoire de Recherches Forestières, Ste-Foy, Québec, Rapport d'Information* **Q-F-X-9**. (324 pp.)

1970b, Fentes en coin fossiles dans la région de Québec: *Rev. Géog. Montréal* **24**, 313–18.

1971a, Contorted structures in unconsolidated Quaternary deposits, Lake Saint-Jean and Saguenay regions, Quebec: *Rev. Géog. Montréal* **25**, 5–33.

1971b, Vertical packing of flat stones: *Canadian J. Earth Sci.* **8**, 1585–95.

1971c, Fentes de cryoturbation tardiglaciaires dans la région de Québec: *Rev. Géog. Montréal* **25**(3), 245–64.

1972, Caractéristiques des schorres des régions froides, en particulier de l'estuaire du Saint-Laurent: *Zeitschr. Geomorphologie N.F., Supp.* **13**, 131–62.

1973a, Distinction entre stries glacielles et stries glaciaires: *Rev. Géog. Montréal* **27**(2), 185–90.

1973b, Fentes de cryoturbation fossiles à Arthabaska, Québec: *Rev. Géog. Montréal* **27**(2), 190–6.

1974a, Cryosols avec triage sur rivage et fond de lacs, Québec central subarctique: *Rev. Géog. Montréal* **28**(4), 323–42.

1974b, Bibliographie annotée sur les aspects géologiques du glaciel (Annotated bibliography on the geological aspects of drift ice): *Service Canadien des Forêts, Centre de Recherches Forestières des Laurentides Rapport d'Information* **LAU-X-9**. (122 pp.)

1975, Paleoclimatic significance of late Pleistocene ice-wedge casts in southern Québec, Canada: *Palaeogeog., Palaeoclimatol., Palaeoecol.* **17**, 65–76.

1976a, Le glaciel de la région de la Grande Rivière, Québec subarctique: *Rev. Géog. Montréal* **30**(1–2) (*Le glaciel* – Premier colloque international sur l'action géologiques des glaces flottantes (Québec, Canada, 20–24 Apr. 1974), 133–53.

1976b, L'action glacielle dans les schores du littoral oriental de la Baie de James: *Cahiers Géog. Québec* **20**(50), 303–26.

ed., 1976c, *Le glaciel* – Premier colloque international sur l'action géologiques des glaces flottantes (Québec, Canada, 20–24 Apr. 1974): *Rev. Géog. Montréal* **30**(1–2). (236 pp.)

1978a, Le glaciel en Jamésie et en Hudsonie, Québec subarctique: *Géog. phys. Quat.* **32**(1), 3–70.

1978b, Formes et phénomènes périglaciaires en Jamésie Québec subarctique: *Géog. phys. Quat.* **32**(3), 187–247.

DIONNE, J.-C., and LAVERDIÈRE, CAMILLE, 1967, Sur la mise en place en milieu littoral de cailloux plats posés sur la tranche: *Zeitschr. Geomorphologie N.F.* **11**, 262–85.

1972, Ice formed beach features from Lake St Jean, Quebec: *Canadian J. Earth Sci.*, **9**, 979–90.

DIONNE, J.-C., and SHILTS, W. W., 1974, A Pleistocene clastic dike, Upper Chaudière Valley, Québec: *Canadian J. Earth Sci.* **11**, 1594–605.

DIRKSEN, C., and MILLER, R. D., 1966, Closed-system freezing of unsaturated soil: *Soil Sci. Soc. America Proc.* **30**(2), 168–73.

DOBROLYUBOVA, T. A., and SOCHKINA, E. D., 1935, Obshchaya geologicheskaya karta Evropeyskoy chasti SSSR (Severnyy Ural), list 123: *Leningradskogo geologo-gidrogeodezicheskogo upravleniya, Trudy* **8**.

DOLGUSHIN, L. D., 1961, *Geomorfologiya zapadnoy chasti Aldanskogo nagorya*: Moskva, 124–5.

DOMRACHEV, D. V., 1913, Dannyye o klimate, pochvakh i rastitelnosti verkhnego techeniya r Tungira Yakutskoy obl.: *St Petersburg, Amurskaya Ekspeditsiya, Trudy* **14**.

DOSTOVALOV, B. N., and KUDRYAVTSEV, V. A., 1967, *Obshcheye merzlotovedeniye*: Moskva, Izdatel'stvo Moskovskogo Universiteta. (403 pp.)

DOSTOVALOV, B. N., and POPOV, A. I., 1966, Polygonal systems of ice-wedges and conditions of their development: 102–5 in *Permafrost International Conference* (Lafayette, Ind., 11–15 Nov. 1963) *Proc.*: Natl. Acad. Sci.–Natl. Research Council Pub. **1287**. (563 pp.)

DOW, W. G., 1964, The effects of salinity on the formation of mud-cracks: *The Compass of Sigma Gamma Epsilon* **41**, 162–6.

DREW, J. V., *et al.*, 1958, Rate and depth of thaw in arctic soils: *Am. Geophys. Union Trans.* **39**, 697–701.

DREW, J. V., and TEDROW, J. C. F., 1962, Arctic soil classification and patterned ground: *Arctic* **15**, 109–16.

DRURY, W. H., JR., 1956, Bog flats and physiographic processes in the Upper Kuskokwim River Region, Alaska: *Harvard Univ. Gray Herbarium Contr.* **178**. (130 pp.)

1962, Patterned ground and vegetation on southern Bylot Island, Northwest Territories, Canada: *Harvard Univ. Gray Herbarium Contr.* **190**. (111 pp.)

DÜCKER, ALFRED, 1933, 'Steinsohle' oder 'Brodelpflaster'?: *Centralbl. Mineralogie*, Jahrg. **1933**(B), 264–7.

1940, Frosteinwirkung auf bindige Böden: *Strassenbau-Jahrb.*, 1939–1940, 111–26.

1954, Die Periglazial-Erscheinungen im holsteinischen Pleistozän: 5–54 in Poser, Hans, ed., Studien über die Periglazial-Erscheinungen in Mitteleuropa: *Göttinger Geog. Abh.* **16**. (96 pp.)

1956, Gibt es eine Grenze zwischen frostsicheren und frostempfindlichen Lockergesteinen?: *Strasse und Autobahn* **7**(3), 78–82.

1958, Is there a dividing line between non-frostsusceptible and frostsusceptible soils? [Gibt es eine Grenze zwischen frostsicheren und frostempfindlichen Lockergesteinen?]: *Canada Natl. Research Council Tech. Translation* **722**. (12 pp.)

DUMITRASHKO, N. V., 1938, Osnovnye momenty geomorfologii i molodye dvizheniya v rayone reki Verkhney Angary: *Akad. Nauk SSSR, otdeleniye matematicheskikh i estestvenykh nauk, Izvestiya*, 399–413.

1948, Osnovnyye problemy geomorfologii i paleogeografif Baykalskoy gornoy oblasti: *Instituta geografii AN SSSR, Trudy* **42**; *Materialy po geomorfologii i paleogeografii SSSR*, 75–141.

DUNN, J. R., and HUDEC, P. P., 1965, The influence of clays on water and ice in rock pores (2): *New York State, Department Public Works Phys. Research Rept.* RR **65**–5. (149 pp.)

1966, Frost deterioration: ice or ordered water? (abs.): *Geol. Soc. America Spec. Paper* **101**, 256.

1972, Frost and sorbtion effects in argillaceous rocks: 65–78 in Highway Research Board, *Frost action in soils: Natl. Acad. Sci.–Natl. Acad. Eng., Highway Research Record* **393**. (88 pp.)

DUPARC, L., and PEARCE, F., 1905, Sur la présence de hautes terrasses dans l'Oural du Nord: *Soc. Géog. Bull.* **11**, 384–5.

DUPARC, L., PEARCE, F., and TIKANOWICH, M., 1909, Recherches géologiques et pétrographiques sur l'Oural du Nord: *Soc. Physique et histoire naturelle de Genève Mem.* **36**, 149–65.

DURY, G. H., 1965, Theoretical implications of underfit streams: *US Geol. Survey Prof. Paper* **452**C. (43 pp.)

1970, General theory of meandering valleys and underfit streams: 264–75 in Dury, G. H., ed., *Rivers and river terraces*: London, Macmillan. (283 pp.)

DUTKIEWICZ, LEOPOLD, 1967, The distribution of periglacial phenomena in NW-Sörkapp, Spitsbergen: *Biuletyn Peryglacjalny* **16**, 37–83.

DYBECK, M. W., 1957, An investigation of some soil polygons in central Iceland: *J. Glaciology* **3**(22), 143–5.

DYKE, A. S., 1976, Tors and associated weathering phenomena, Somerset Island, District of Franklin: *Canada Geol. Survey Paper* **76**–1B, 209–16.

DYLIK, JAN, 1951, Some periglacial structures in Pleistocene deposits of middle Poland: *Soc. Sci. Math. et Lettres Łódź Bull.*, cl. 3 sci. math. et nat. **3**(2), 1–6.

1960, Rhythmically stratified slope waste deposits: *Biuletyn Peryglacjalny* **8**, 31–41.

1964a, Éléments essentiels de la notion de 'périglaciaire': *Biuletyn Peryglacjalny* **14**, 111–32.

1964b, The essentials of the meaning of the term of 'periglacial': *Soc. Sci. et Lettres Łódź Bull.* **15**(2), 1–19.

1966, Problems of ice-wedge structures and frost-fissure polygons: *Biuletyn Peryglacjalny* **15**, 241–91.

1967, Solifluxion, congelifluxion and related slope processes: *Geografiska Annaler* **49**A, 167–77.

1969, L'action du vent pendant le dernier âge froid

sur le territoire de la Pologne Centrale: *Biuletyn Peryglacjalny* **20**, 29–44.

1972, Rôle du ruissellement dans le modelé périglaciaire: *Göttinger Geog. Abh.* **60** (Hans Poser Festschrift), 169–80.

DYLIK, J., and MAARLEVELD, G. C., 1967, Frost cracks, frost fissures and related polygons: *Mededelingen van de Geol. Stichting, nieuwe ser.* **18**, 7–21.

DYLIKOWA, ANNA, 1969, Le problème des dunes intérieures en Pologne à la lumerière des études de structure: *Biuletyn Peryglacjalny* **20**, 45–80.

DŽUŁYŃSKI, STANISŁAW, 1963, Polygonal structures in experiments and their bearing on some 'periglacial phenomena': *Acad. polonaise sci. Bull. sér. sci. géol. et géog.* **11**(3), 145–50.

1965, Experiments on clastic wedges: *Acad. polonaise sci. Bull., sér. sci. géol. et géog.* **13**, 301–4.

1966, Sedimentary structures resulting from convection-like pattern of motion: *Soc. géol. Pologne Annales* **36**(1), 3–21.

EAGER, W. L., and PRYOR, W. T., 1945, Ice formation on the Alaska Highway: *Public Roads* **24**, 55–74, 82.

EAKIN, H. M., 1916, The Yukon–Koyukuk region, Alaska: *US Geol. Survey Bull.* **631**. (88 pp.)

1918, The Cosna–Nowitna region, Alaska: *US Geol. Survey Bull.* **667**, 50–4.

EDELMAN, C. H., FLORSCHÜTZ, F., and JESWIET, J., 1936, Ueber spätpleistozäne und frühholozäne kryoturbate Ablagerungen in den östlichen Niederlanden: *Geologisch-Mijnbouwkundid Genootschap voor Nederland en Koloniën, Verhandelingen Geol. Ser.* **11**, 301–60.

EDELMAN, C. H., and MAARLEVELD, G. C., 1949, De asymmetrische dalen van de Veluwe: *Kon. Nederlandsch Aardrijksk. Genoot. Tijdschr., 2d ser.*, **66**, 143–6.

EDELSHTEYN, J. S., 1936, *Instruktsiya dlya geomorfologicheskogo izucheniya i kartirovaniya Urala*: Leningrad, Vsesoyuznyy arkticheskiy institut. (91 pp.)

EISSMANN, L., 1978, Mollisoldiapirismus: *Zeitschr. Angewandte Geologie* **24**(3), 130–8.

EKBLAW, W. E., 1918, The importance of nivation as an erosive factor, and of soil flow as a transporting agency, in northern Greenland: *Natl. Acad. Sci. Proc.* **4**, 288–93.

ELKHORAIBI, M. C., 1975, *Volume changes of frozen soils*: Carleton Univ., Ph.D. Thesis. (72 pp. + appendices.)

ELLENBERG, LUDWIG, 1974a, Shimobashira – Kammeis in Japan: *Geog. Helvetica* **1974**(1), 1–5.

1974b, The periglacial stage in Europe (especially the Alps) and Japan – A comparison: *Tokyo Metropolitan Univ. Geog. Repts.* **9**, 53–65.

1976a, Zur Periglazialmorphologie von Ura Nippon, der schneereichen Seite Japans: *Geog. Helvetica* **1976**(3), 139–51.

1976b, Rezente Periglazialerscheinungen im Gebiet des Daisetsu San in Nordjapan: *Eiszeitalter und Gegenwart* **27**, 174–88.

ELTON, C. S., 1927, The nature and origin of soil-polygons in Spitsbergen: *Geol. Soc. London Quart. J.* **83**, 163–94.

EMBLETON, CLIFFORD, and KING, C. A. M., 1975, *Periglacial geomorphology: Glacial and periglacial geomorphology, 2d ed.*, **2**: London, Edward Arnold; New York, Halstead Press. (203 pp.)

EMMERT, U., 1965, Erläuterungen zur Geologischen Karte von Bayern 1:25 000 *Blatt* **6288** *Wiesentheid*. München.

ENVIRONMENT CANADA, 1974, *Permafrost hydrology – Workshop Seminar Proc.*: Can. Natl. Comm., Internat. Hydrological Decade: Ottawa, Environment Canada. (102 pp.)

EVANS, I. S., 1970, Salt crystallization and rock weathering: A review: *Rev. Géomorphologie dynamique* **19**, 153–77.

1971, Physical weathering in hot deserts; the processes of haloclasty and thermoclasty: Paper presented to *British Geomorphological Research Group Symposium on Geomorphology in the Arid Tropics (Durham, 25 Sept. 1977)*. (8 pp.)

EVANS, R., 1976, Observations on a stripe pattern: *Biuletyn Peryglacjalny* **25**, 9–22.

EVENSON, E. B., 1973, The ice-foot complex: Its morphology, classification, mode of formation, and importance as a sediment transporting agent: *The Michigan Academician* **6**(1), 43–57.

EVERETT, D. H., 1961, The thermodynamics of frost damage to porous solids: *Faraday Soc. Trans.* **57**(7), 1541–51.

EVERETT, K. R., 1963a, *Slope movement in contrasting environment*: Ohio State Univ., Ph.D. Thesis. (251 pp.)

1963b, Slope movement, Neotoma Valley, southern Ohio: *Ohio State Univ. Research Foundation, Inst. Polar Studies Rept.* **6**. (59 pp.)

1966, Slope movement and related phenomena: 175–220 in Wilimovsky, N. J., ed., *Environment of the Cape Thompson region, Alaska*: US Atomic Energy Commission, Division of Technical Information. (1250 pp.)

EYLES, N., 1977, Late Wisconsinan glacitectonic structures and evidence of postglacial permafrost in north-central Newfoundland: *Canadian J. Earth Sci.* **14**(12), 2797–2806.

FAHEY, B. D., 1973, An analysis of diurnal freeze-thaw and frost heave cycles in the Indian Peaks region of the Colorado Front Range: *Arctic and Alpine Research* **5**(3, Part 1), 269–81.

—— 1974, Seasonal frost heave and frost penetration measurements in the Indian Peaks region of the Colorado Front Range: *Arctic and Alpine Research* **6**, 63–70.

—— 1975, Nonsorted circle development in a Colorado alpine location: *Geografiska Annaler* **57**A, 153–64.

FAIRBRIDGE, R. W., 1977, Sea-ice transport and strandflats (abs.): *International Union for Quaternary Research (INQUA) Cong., 10th (Birmingham, England, 16–24 Aug. 1977), Abstracts*, 131.

FALCONER, ALLAN, 1969, Processes acting to produce glacial detritus: *Earth Sci. J.* **3**(1), 40–3.

FALCONER, G., 1966, Preservation of vegetation and patterned ground under a thin ice body in northern Baffin Island, NWT: *Geog. Bull.* **8**, 194–200.

FARRAND, W. R., 1961, Frozen mammoths and modern geology: *Science* **133**, 729–35.

FAUCHER, D., 1931, Note sur la dissymétrie des vallées de l'Armagnac: *Soc. d'Histoire Naturelle de Toulouse Bull.* **61**, 262–8.

FEDOROFF [FEDEROFF], N., 1966a, Les sols du Spitsberg Occidental: 111–228 in Audin, ed. *Spitsberg 1964 et premières observations 1965*. Lyon, Centre National Recherche Scientifique (CNRS).

—— 1966b, Les cryosols: *Science du Sol* **1966(2)** (Numéro Spécial), 77–110.

FEL'DMAN, G. M., 1967, Raschet migratsii vlagi i opredeleniye sloystoy tekstury grunta pri promerzanii: *Inzhenerno-Fizicheskiy Zhurnal* **13**(6), 812–20.

—— 1972, Moisture migration and stratified texture in freezing soils [Raschet migratsii vlagi i opredeleniye sloystoy tekstury grunta pri promerzanii]: *J. Engineering Physics (Inzhenerno-Fizicheskii Zhurnal)* **13**(6), 1967, 425–9. (New York, Consultants Bureau translation, 1972.)

FERRIANS, O. J., JR., 1965, Permafrost map of Alaska: *US Geol. Survey Misc. Geol. Inv. Map* **I–445**.

FERRIANS, O. J., JR., and HOBSON, G. D., 1973, Mapping and predicting permafrost in North America: A review, 1963–1973: 479–98 in *North American Contribution, Permafrost Second International Conference (Yakutsk, USSR, 13–28 July 1973)*: Washington, DC, Natl. Acad. Sci. (783 pp.)

FERRIANS, O. J., JR., KACHADOORIAN, REUBEN, and GREENE, G. W., 1969, Permafrost and related engineering problems in Alaska: *US Geol. Survey Prof. Paper* **678**. (37 pp.)

FIGURIN, A. E., 1823, Izvlecheniye iz zapisok medikokhirurga Figurina, vedennykh vo vremya opisi beregov Severo-Vostochnoy Sibiri: *Gosudarstv. Admiralt. Depart., Zapiski* **5**, 259–328.

FINK, J., 1976, Internationaler Lössforschungen – Bericht der INQUA-Lösskommission: *Eiszeitalter und Gegenwart* **27**, 220–35.

FINN, W. D., YONG, R. N., and LEE, K. K., 1978, Liquefaction of thawed layers in frozen soils: *Am. Soc. Civil Eng., J. Geotech. Eng. Div.* **104**(G10), 1243–55.

FISCH, W., SEN., FISCH, W., JR., and HAEBERLI, W., 1977, Electrical DC resistivity soundings with long profiles on rock glaciers and moraines in the Alps of Switzerland: *Zeitschr. Gletscherkunde u. Glazialgeologie* **13**(1–2), 239–60.

FISCHER, ERIC, and ELLIOTT, F. E., 1950, *A German and English glossary of geographical terms*: New York, Am. Geog. Soc. (111 pp.)

FITZE, P. F., 1969, *Untersuchungen von Solifluktionserscheinungen im Alpenquerprofil zwischen Säntis und Lago di Como*: Univ. Zürich, dissertation. (62 pp.)

—— 1971, Messungen von Bodenbewegungen auf West-Spitzbergen: *Geog. Helvetica* **26**, 148–52.

—— 1975, Nonsorted circles – ein Vergleich zwischen arktischen und alpinen Formen: *Geog. Helvetica* **30**, 75–82.

FLEISHER, P. J., and SALES, J., 1971, Clastic wedges of periglacial significance, central New York (abs.): *Geol. Soc. America Abs. with Programs* **3**(1), 28–9.

FLEMAL, R. C., 1972, Ice injection origin of the DeKalb mounds, north-central Illinois, USA: 130–5 in *Internat. Geol. Cong., 24th, Montreal, Sec. 12, Quaternary geology*. (226 pp.)

—— 1976, Pingos and pingo scars: Their characteristics, distribution, and utility in reconstructing former permafrost environments: *Quaternary Research* **6**, 37–53.

FLEMAL, R. C., HINKLEY, K. C., and HESLER, J. L., 1973, DeKalb mounds: A possible Pleistocene (Woodfordian) pingo field in north-central Illinois: 229–50 in Black, R. F., Goldthwait, R. P., and Willman, H. B., eds., *The Wisconsinan Stage: Geol. Soc. America Mem.* **136**. (334 pp.)

FLEMAL, R. C., ODOM, I. E., and VAIL, R. G., 1972, Stratigraphy and origin of the paha topography of northwestern Illinois: *Quaternary Research* **2**, 232–43.

FLEMING, R. W., and JOHNSON, A. M., 1975, Rates of seasonal creep of silty clay soil: *Quart. J. Eng. Geol.* **8**, A1–A29.

FLINT, R. F., 1948, Glacial geology and geomorphology: 91–210 in Boyd, L. A., *The coast of Northeast Greenland*: Am. Geog. Soc. Spec. Pub. **30**. (339 pp.)

1957, *Glacial and Pleistocene geology*: New York, John Wiley & Sons, Inc. (553 pp.)

1961, Geological evidence of cold climate: 140–55 in Nairn, A. E. M., ed., *Descriptive palaeoclimatology*: New York, Interscience. (300 pp.)

1971, *Glacial and Quaternary geology*: New York, John Wiley. (892 pp.)

FLINT, R. F., and DENNY, C. S., 1958, Quaternary geology of Boulder Mountain, Aquarius Plateau, Utah: *US Geol. Survey Bull.* **1061**-D. (164 pp.)

FLINT, R. F., and SKINNER, B. J., 1977, *Physical geology*, 2d ed.: New York, John Wiley & Sons. (594 pp. + appendices.)

FLOHR, E. F., 1935, Beobachtungen über die Bahnen der Schneesmelzwässer im Riesengebirge. Ein Beitrag zum Problem der Blockrinnen ('Steinstreifen'): *Gesell. Erdkunde Berlin Zeitschr.*, **1935**, 353–69.

FORSGREN, BERNT, 1968, Studies of palsas in Finland, Norway and Sweden, 1964–1966: *Biuletyn Peryglacjalny* **17**, 117–23.

FOSCOLOS, A. E., RUTTER, N. W., and HUGHES, O. L., 1977, The use of pedological studies in interpreting the Quaternary history of central Yukon Territory: *Canada Geol. Survey Bull.* **271**. (48 pp.)

FOTIYEV, S. M., DANILOVA, N. S., and SHEVELEVA, N. S., 1974, *Geokriologicheskiye usloviya Sredney Sibiri*: Moskva, 'Nauka'. (147 pp.)

FRÄNZLE, O., 1959, Glaziale und periglaziale Formbildung im östlichen Kastilischen Scheidegebirge (Zentralspanien): *Bonner Geog. Abh.* **26**. (80 pp.)

FRASER, J. K., 1959, Freeze-thaw frequencies and mechanical weathering in Canada: *Arctic* **12**, 40–53.

FRECHEN, J., and ROSAUER, E. A., 1959, Aufbau und Gliederung des Würm-Löss-Profils von Kärlich im Neuwieder Becken: *Fortschritte in der Geologie von Rheinland und Westfalen* **4**, 267–82.

FRENCH, H. M., 1971a, Ice cored mounds and patterned ground, southern Banks Island, western Canadian Arctic: *Geografiska Annaler* **53**A, 32–8.

1971b, Slope asymmetry of the Beaufort Plain, Northwest Banks Island, NWT, Canada: *Canadian J. Earth Sci.* **8**, 717–31.

1972a, The role of wind in periglacial environments, with special reference to northwest Banks Island, western Canadian Arctic: 82–4 in Adams, W. P., and Helleiner, F. M., eds., *International Geography*

1972, **1** (Internat. Geog. Cong., 22d, Montreal): Toronto, Univ. Toronto Press. (694 pp.)

1972b, Asymmetrical slope development in the Chiltern Hills: *Biuletyn Peryglacjalny* **21**, 51–73.

1973, Cryopediments on the chalk of southern England: *Biuletyn Peryglacjalny* **22**, 149–56.

1974a, Mass-wasting at Sachs Harbour, Banks Island, NWT, Canada: *Arctic and Alpine Research* **6**, 71–8.

1974b, Active thermokarst processes, eastern Banks Island, Western Canadian Arctic: *Canadian J. Earth Sci.* **11**, 785–94.

1975a, Pingo investigations and terrain disturbance studies, Banks Island, District of Franklin: *Canada Geol. Survey Paper* **75–1**, Part A, 459–64.

1975b, Man-induced thermokarst, Sachs Harbour airstrip, Banks Island, Northwest Territories: *Canadian J. Earth Sci.* **12**, 132–44.

1976a, *The periglacial environment*: London, Longman Group Limited; New York, Longman Inc. (309 pp.)

1976b, Geomorphological process and terrain disturbance studies, Banks Island, District of Franklin: *Canada Geol. Survey Paper* **76–1**A, 289–92.

1976c, Current field measurements concerning the nature and rate of periglacial processes – Results of a survey sponsored by IGU Co-ordinating Committee for Periglacial Research: *Biuletyn Peryglacjalny* **25**, 79–91.

1977, The pingos of Banks Island, Western Arctic (abs.): *International Union for Quaternary Research (INQUA) Cong., 10th (Birmingham, England, 16–24 Aug. 1977), Abstracts*, 148.

FRENCH, H. M., and DUTKIEWICZ, L., 1976, Pingos and pingo-like forms, Banks Island, Western Canadian Arctic: *Biuletyn Peryglacjalny* **26**, 211–22.

FRENCH, H. M., and EGGINTON, P., 1973, Thermokarst development, Banks Island, Western Canadian Arctic: 203–12 in *North American Contribution, Permafrost Second International Conference* (Yakutsk, USSR, 13–28 July 1973): Washington, DC, Natl. Acad. Sci. (783 pp.)

FRENZEL, BURKHARD, 1959, Die Vegetations- und Landschaftszonen Nord-Eurasiens während der letzten Eiszeit und während der postglazialen Wärmezeit – I, Teil: *Allgemeine Grundlagen, Akad. Wiss. u. Lit. Mainz Abh., Kl. Math.–Naturwiss.* **1959**(13), 937–1099.

1960, Die Vegetations- und Landschaftszonen Nord-Eurasiens während der letzten Eiszeit und während der postglazialen Wärmezeit – II, Teil: *Rekonstruktionsversuch der letzteiszeitlichen und wärmezeitlichen Vegetation Nord-Eurasiens, Akad. Wiss. u.*

Lit. Mainz Abh., Kl. Math.–Naturwiss. **1960**(6), 287–453.

1967, *Die Klimaschwankungen des Eiszeitalters*: Braunschweig, Friedr. Vieweg & Sohn. (296 pp.)

1968, Grundzüge der pleistozänen Vegetationsgeschichte Nord-Eurasiens: *Erdwissenschaftliche Forschung* **1**: Wiesbaden, Franz Steiner. (326 pp.)

1973a, *Climatic fluctuations of the Ice Age* (Translated by A. E. M. Nairn): Cleveland and London, The Press of Case Western Reserve University. (306 pp.)

1973b, Some remarks on the Pleistocene vegetation: 281–92 in Behre, K.-E., *et al.*, State of research on the Quaternary of the Federal Republic of Germany: *Eiszeitalter und Gegenwart* **23–24**, 219–270.

FRENZEL, BURKHARD, and TROLL, CARL, 1952, Die Vegetationszonen des nördlichen Eurasiens während der letzten Eiszeit: *Eiszeitalter und Gegenwart* **2**, 154–67.

FRIEDMAN, J. D., *et al.*, 1971, Observations on Icelandic polygon surfaces and palsa areas. Photo interpretation and field studies: *Geografiska Annaler* **53**A, 115–45.

FRIES, MAGNUS, WRIGHT, H. E., JR., and RUBIN, MEYER, 1961, A late Wisconsin buried peat at North Branch, Minnesota: *Am. J. Sci.* **259**, 679–93.

FRISTRUP, BØRGE, 1953, Wind erosion within the arctic deserts: *Geografisk Tidsskrift* **52**(1952–53), 51–65.

FRITZ, PETER, 1976, Gesteinsbedingte Standorts- und Formendifferenzierung rezenter Periglazialerscheinungen in den Ostalpen: *Österreichischen Geog. Gesell. Mitt.* **118**(2–3), 237–72.

FRÖDIN, JOHN, 1918, Über das Verhältnis zwischen Vegetation und Erdfliessen in den alpinen Regionen des schwedischen Lappland: *Lunds Univ. Årssk., N.F. Avd. 2* **14**(24). (32 pp.)

FROEHLICH, W., and SŁUPIK, J., 1978, Frost mounds as indicators of water transmission zones in the active layer of permafrost during winter season/ Khangay Mts., Mongolia: 188–93 in *Third International Conference on Permafrost* (Edmonton, Alta., 10–13 July 1978), Proc. **1**: Ottawa, Canada Natl. Research Council. (947 pp.)

FROLOV, A. D., 1976, *Elektricheskiye i uprugiye svoystva kriogennykh porod*: Moskva, 'NEDRA'. (254 pp.)

FUJII, YOSHIYUKI, and HIGUCHI, KEIJI, 1972, Fuji-san no eikyûtôdo (On the permafrost at the summit of Mt Fuji): *Seppyo* **34**(4), 173–86.

1976, Ground temperature and its relation to permafrost occurrences in the Khumbu region and

Hidden Valley: *Seppyo* **38** (Glaciers and climates of Nepal Himalayas), 125–8.

1978, Distribution of alpine permafrost in the northern hemisphere and its relation to air temperature: 366–71 in *Third International Conference on Permafrost* (Edmonton, Alta., 10–13 July 1978), Proc. **1**: Ottawa, Canada Natl. Research Council. (947 pp.)

FUKUDA, MASAMI, 1972, Ganseki nai no mizu no toketsu ni tsuite – II (Freezing-thawing process of water in pore space of rocks – II): *Low Temperature Sci., Ser. A* **30**, 183–9.

FUKUDA, MASAMI, and KINOSHITA, SEIICHI, 1974, Permafrost at Mt Taisetsu, Hokkaido and its climatic environment: *The Quaternary Research* **12**(4), 192–202.

FÜRBRINGER, WERNER, and HAYDN, RUPERT, 1974, Zur Frage der Orientierung Nordalaskischer Seen mit Hilfe des Satellitenbildes: *Polarforschung* **44**(1), 47–53.

FURRER, GERHARD, 1954, Solifluktionsformen im schweizerischen Nationalpark: *Schweizer. naturf. Gesell. Ergebnisse der wissenschaftlichen Untersuchungen des schweizerischen Nationalparks* **4** (Neue Folge) (29), 201–75.

1955, Bodenformen aus dem subnivalen Bereich: *Die Alpen (Les Alpes)* **31**, 146–51.

1965a, *Die Höhenlage von subnivalen Bodenformen*: Univ. Zürich, Habilitationsschrift, Philosophischen Fakultät – II. (89 pp.)

1965b, Die subnivale Höhenstufe und ihre Untergrenze in den Bündner und Walliser Alpen: *Geog. Helvetica* **20**(4), 185–92.

1966, Beobachtungen an rezenten und fossilen (kaltzeitlichen) Strukturböden: *Experientia* **22**(8), 489–96.

1968, Untersuchungen an Strukturböden in Ostspitzbergen, ihre Bedeutung für die Erforschung rezenter und fossiler Frostmusterformen in den Alpen bzw. im Alpenvorland: *Polarforschung* **6**, 202–6.

1972, Bewegungsmessungen auf Solifluktionsdecken: *Zeitschr. Geomorphologie N.F., Supp.* **13** (*Glazial- und Periglazialmorphologie*), 87–101.

FURRER, GERHARD, and BACHMANN, FRITZ, 1968, Die Situmetrie (Einregelungsmessung) als morphologische Untersuchungsmethode: *Geog. Helvetica* **23**, 1–14.

FURRER, GERHARD, and DORIGO, GUIDO, 1972, Abgrenzung und Gliederung der Hochgebirgsstufe der Alpen mit Hilfe von Solifluktionsformen: *Erdkunde* **26**, 98–107.

FURRER, GERHARD, and FITZE, PETER, 1970, Beitrag

zum Permafrostproblem in den Alpen: *Natur-forschenden Gesell. Zürich, Viertelsjahrsschr.* **115**(3), 353–68.

1973, Treatise on the permafrost problem in the Alps [Beitrag zum Permafrostproblem in den Alpen]: 10–29 in Translation of two Swiss articles on permafrost in the Alps: *Canada Natl. Research Council Tech. Translation* **1657**. (29 pp.)

FURRER, GERHARD, and FREUND, RALF, 1973, Beobachtungen zum subnivalen Formenschatz am Kilimanjaro: *Zeitschr. Geomorphologie N.F., Supp.* **16**, 180–203.

FYLES, J. G., 1963, Surficial geology of Victoria and Stefansson Islands, District of Franklin: *Canada Geol. Survey Bull.* **101**. (38 pp.)

FYODOROV, J. S., compiler, and IVANOV, N. S., ed., 1974, *English–Russian geocryological dictionary*: Yakutsk, Yakutsk State Univ. (127 pp.)

GAKKELYA, G. J., and KOROTKEVICH, E. S., 1962, Severnaya Yakutiya (Fizikogeograficheskaya kharakteristika): *Arkticheskogo i antarkticheskogo nauchno-issledovatel'skogo Instituta, Trudy* **236**, 79.

GALLOWAY, J. P., and CARTER, L. D., 1978, Preliminary map of pingos in National Petroleum Reserve in Alaska: *US Geol. Survey Open File Rept.* **78–795**.

GALLOWAY, R. W., 1961*a*, Periglacial phenomena in Scotland: *Geografiska Annaler* **43**, 348–53.

1961*b*, Solifluction in Scotland: *Scottish Geog. Mag.* **77**, 75–87.

1961*c*, Ice wedges and involutions in Scotland: *Biuletyn Peryglacjalny* **10**, 169–93.

1965, Late Quaternary climates in Australia: *J. Geol.* **73**, 603–18.

1970, The full-glacial climate in the southwestern United States: *Assoc. Am. Geographers Annals* **60**, 245–56.

GALON, RAJMUND, ed., 1969, Procesy i formy wydmowe w Polsce (Dunes processes and forms in Poland): *Polskiej Akad. Nauk, Instytut Geografii, Prace Geograficzne* **75**. (386 pp.)

GANESHIN, G. S., 1949, O nagornykh terrasakh v nizhnem Priamurye: *Vsesoyuznogo Geograficheskogo obshchestva, Izvestiya* **81**(2), 254–6.

GARDNER, J. S., 1969, Snowpatches: their influence on mountain wall temperatures and the geomorphic implications: *Geografiska Annaler* **51**A, 114–20.

1973, The nature of talus shift on alpine talus slopes: An example from the Canadian Rocky Mountains: 95–106 in Fahey, B. D., and Thompson, R. D., eds., *Research in polar and alpine geomorphology*: Guelph Symposium on Geomorphology,

1973, 3d, Proc.: Norwich, England, Geo Abstracts Ltd (Univ. East Anglia). (206 pp.)

GARLEFF, KARSTEN, 1970, Verbreitung und Vergesellschaftung rezenter Periglazialerscheinungen in Skandinavien: *Göttinger Geog. Abh.* **51**, 7–66.

GARY, MARGARET, MCAFEE, ROBERT, JR., and WOLF, C. L., eds., 1972, *Glossary of geology*: Washington, DC, American Geological Inst. (805 and A-52 pp.)

GASANOV, SH. SH., 1973, Mekhanizm samoregulirovaniya predel'nykh razmerov ledyanykh zhil: 65–9 in *Akademiya Nauk SSSR, Sektsiya Nauk o Zemle, Sibirskoye Otdeleniye, II Mezhdunarodnaya Konferentsiya po Merzlotovedeniyu, Doklady i soobshcheniya* **3** (Genezis, sostav i stroyeniye merzlykh tolshch i podzemnyye l'dy): Yakutsk, Yakutskoye Knizhnoye Izdatel'stvo. (102 pp.)

1978, Self-regulating mechanism of size-limited ice wedges [Mekhanizm samoregulirovaniya predel'nykh razmerov ledyanykh zhil]: 195–7 in Sanger, F. J., ed., *USSR Contribution Permafrost Second International Conference* (Yakutsk, USSR, 13–28 July 1973): Washington, DC, Natl. Acad. Sci. (866 pp.)

GATES, W. L., 1976, Modeling the ice-age climate: *Science* **191**, 1138–44.

GEIGER, RUDOLF, 1965, *The climate near the ground*: Cambridge, Mass., Harvard Univ. Press. (611 pp.)

GEIKIE, JAMES, 1894, *The great ice age and its relation to the antiquity of man*, 3 ed.: London, Edward Stanford. (850 pp.)

GELL, A., 1974*a*, Some observations on ice in the active layer and in massive ice bodies, Tuktoyaktuk coast, NWT: *Canada Geol. Survey Paper* **74–1**, Part A, 387.

1974*b*, A contact between massive ice and wedge ice, Tuktoyaktuk coast, District of Mackenzie: *Canada Geol. Survey Paper* **74–1**, Part B, 245–6.

1975, Tension-crack ice, icing mound ice, Tuktoyaktuk coast, District of Mackenzie: *Canada Geol. Survey Paper* **75–1**, Part A, 465–6.

1978*a*, Thermal contraction cracks in massive segregated ice, Tuktoyaktuk Peninsula, NWT, Canada: 277–81 in *Third International Conference on Permafrost* (Edmonton, Alta., 10–13 July 1978), Proc. **1**: Ottawa, Canada Natl. Research Council. (947 pp.)

1978*b*, Fabrics of icing-mound and pingo ice in permafrost: *J. Glaciology* **20**(84), 563–9.

1978*c*, Ice-wedge ice, Mackenzie Delta – Tuktoyaktuk Peninsula area, N.W.T., Canada: *J. Glaciology* **20**(84), 555–62.

GERASIMOV, I. P., and MARKOV, K. K., 1968, Permafrost and ancient glaciation: *Canada Defence Research Board Translation* T**499**R, 11–19.

GERDEL, R. W., 1969, Characteristics of the cold regions: *US Army Corps of Engineers, Cold Regions Research and Engineering Laboratory, Cold Regions Science and Engineering Mon.* **1**A. (53 pp.)

GERLACH, TADEUSZ, 1972, Contribution à la connaissance du développement actuel des buttes gazonées (thufurs) dans les Tatras Polonaises: 57–74 in Macar, P., and Pissart, A., eds., *Processus périglaciaires étudiés sur le terrain: Union Géographique Internationale, Symposium International de Géomorphologie* (Liège-Caen, 1–9 July 1971), *Part 1* (Séances tenues à Liège et excursions en Belgique): Les Congrès et Colloques de l'Université de Liège: Liège, Université de Liège. (339 pp.)

GEUKENS, F., 1947, De asymmetrie der droge dalen van Haspengouw: *Natuurwet. Tijdschr.* **29**, 13–18.

GILBERT, G. K., 1890, Lake Bonneville: *US Geol. Survey Mon.* **1**. (438 pp.)

1908, Lake ramparts: *Sierra Club Bull.* **6**(4), 225–34.

GILL, DON, 1972, Modification of levee morphology by erosion in the Mackenzie delta, Northwest Territories, Canada: 123–38 in Price, R. J., and Sugden, D. E., compilers, *Polar geomorphology*: Inst. British Geographers Spec. Pub. **4**. (215 pp.)

GLAZOVSKIY, A. F., 1978, Kamennye gletchery (Sostoyaniye problemy): 59–72 in Gorbunov, A. P., ed., *Kriogennye yavleniya vysokogoriy*: Akad. Nauk SSSR, Sibirskoye Otdeleniye: Novosibirsk, Izdatel'stvo 'Nauka'. (150 pp.)

GLEN, J. W., 1955, The creep of polycrystalline ice: *Royal Soc. Proc., Ser. A* **228**, 519–38.

1975, The mechanics of ice: *US Army Corps of Engineers, Cold Regions Research and Engineering Laboratory, Cold Regions Science and Engineering Mon.* **II**–C2b. (43 pp.)

GLORIOD, A., and TRICART, JEAN, 1952, Étude statistique des vallées asymétriques sur la feuille St-Pol au 1:50.000: *Rev. Géomorphologie dynamique* **3**, 88–98.

GOLD, L. W., *et al.*, 1972, Thermal effects in permafrost: 25–45 in Canadian Northern Pipeline Research Conference (Ottawa, 2–4 Feb. 1972), Proc.: *Canada Natl. Research Council, Assoc. Comm. on Geotechnical Research Tech. Mem.* **104**. (331 pp.)

GOLD, L. W., and LACHENBRUCH, A. H., 1973, Thermal conditions in permafrost – A review of North American literature: 3–23 in *North American Contribution, Permafrost Second International Conference* (Yakutsk, USSR, 13–28 July 1973): Washington, DC, Natl. Acad. Sci. (783 pp.)

GOLDTHWAIT, LAWRENCE, 1957, Ice action on New England lakes: *J. Glaciology* **3**(22), 99–103.

GOLDTHWAIT, R. P., 1940, Geology of the Presidential Range: *New Hampshire Acad. Sci. Bull.* **1**. (43 pp.)

1969, Patterned soils and permafrost on the Presidential Range (abs.): 150 in *International Union for Quaternary Research (INQUA) Cong., 8th (Paris, 1969), Résumés des Communications*. (389 pp.)

1976*a*, Frost sorted patterned ground: A review: *Quaternary Research* **6**, 27–35.

1976*b*, Past climates on 'the hill': Part 2. Permafrost fluctuations: *Mt Washington Observatory Bull.* **17**(2), 38–41.

GONZALEZ, M. A., and CORTE, A. E., 1976, Pleistocene geocryogenic structures at 38° S. L., 60° W. and 200 m above sea level, Gonzalez Chavez, Buenos Aires Province, Argentina: *Biuletyn Peryglacjalny* **25**, 23–33.

GOODRICH, L. E., 1978, Some results of a numerical study of ground thermal regimes: 30–4 in *Third International Conference on Permafrost* (Edmonton, Alta., 10–13 July 1978), Proc. **1**: Ottawa, Canada Natl. Research Council. (947 pp.)

GOODWIN, C. W., and OUTCALT, S. I., 1975, The development of a computer model of annual snow-soil thermal regime in arctic tundra terrain: 227–9 in Weller, Gunter, and Bowling, S. A., eds., *Climate of the Arctic*: Alaska Science Conference, 24th (Fairbanks, Alaska, 15–17 Aug. 1973): Fairbanks, Univ. Alaska Geophysical Institute. (436 pp.)

GORBUNOV, A. P., 1967, *Vechnaya merzlota Tyan'-Shanya*: Frunze, Akad. Nauk Kirgizskoy SSR, Izdatel'stvo 'Ilim'. (165 pp.)

1978, Permafrost investigations in high-mountain regions: *Arctic and Alpine Research* **10**, 283–94.

GORCHAKOVSKIY, P. L., 1954, Vysokogornaya vegetatsiya Yaman-Tau, samoy vysokoy gory yuzhnogo Urala: *Botanicheskiy zhurnal* **39**, 827–41.

GOUDIE, ANDREW, 1974, Further experimental investigation of rock weathering by salt and other mechanical processes: *Zeitschr. Geomorphologie N.F., Supp.* **21**, 1–12.

GOW, A. J., UEDA, H. T., and GARFIELD, D. E., 1968, Antarctic ice sheet: Preliminary results of first core hole to bedrock: *Science* **161**, 1011–13.

GOŹDZIK, JAN, 1973, Geneza i pozycja stratygraficzna struktur peryglacjalnych w środkowej Polsce [Origin and stratigraphical position of periglacial structures in middle Poland]: *Acta Geographica Lodziensia* **31**. (119 pp.)

1976, O szczelinowych strukturach pasowych w Polsce: *Acta Geographica Lodziensia* **37**, 7–23.

[Summary: stripe fissue structures in Poland, p. 21–3.]

GRADWELL, M. W., 1954, Soil frost studies at a high country station – 1: *New Zealand J. Sci. and Technol.*, sec. B **36**, 240–57.

1957, Patterned ground at a high-country station: *New Zealand J. Sci. and Technol.* **38**(B), 793–806.

1960, Soil frost action in snow-tussock grassland: *New Zealand J. Sci.* **3**, 580–90.

GRAF, K. J., 1971, *Beiträge zur Solifluktion in den Bündner Alpen (Schweiz) und in den Anden Perus und Boliviens*: Univ. Zürich, dissertation. (152 pp.)

1973, Vergleichende Betrachtungen zur Solifluktion in verschiedenen Breitenlagen: *Zeitschr. Geomorphologie N. F.*, Supp. **16**, 104–54.

1976, Zur Mechanik von Frostmusterungsprozessen in Bolivien und Ecuador: *Zeitschr. Geomorphologie N. F.* **20**, 417–47.

GRAHMANN, RUDOLF, 1932, Der Löss in Europa: *Gesell. Erdkunde Leipzig Mitt.* **51**, 5–24.

GRAVE, N. A., 1956, An archaeological determination of the age of some hydrolaccoliths (pingos) in the Chuckchee Peninsula: *Canada Defence Research Board Translation* T**218**R. (3 pp.)

1967, Temperature regime of permafrost under different geographical and geological conditions: 1339–43 in Oura, Hirobumi, ed., *Physics of snow and ice, Internat. Conference Low Temperature Science (Sapporo, Japan, 14–19 Aug. 1966) Proc.* **1**(2): Sapporo, Hokkaido Univ., 714–1414.

1968a, Merzlyye tolshchi zemli: *Priroda*, 1968 **1**, 46–53.

1968b, The earth's permafrost beds [Merzlyye tolshchi zemli]: *Canada Defense Research Board Translation* T**499**R, 1–10.

1977, Coordinator, Permafrost study during the period 1972–1975. A review: International Geographical Union, Coordinating Comm. for Periglacial Research. [Mimeo. rept. for Comm., 5 pp.]

GRAVE, N. A., and SUKHODROVSKIY [SUKHODROVSKII], V. L., 1978, Rel'yefoobrazuyushchiye protsessy oblasti vechnoy merzloty i printsipy ikh preduprezhdeniya i ogranicheniya na osvaivayemykh territoriyakh – Terrain-forming processes in the permafrost region and the principles of their prevention and limitation in territories under development: 467–71 in *Third International Conference on Permafrost* (Edmonton, Alta., 10–13 July 1978), Proc. **1**: Ottawa, Canada Natl. Research Council. (947 pp.)

GRAVIS, G. F., 1964, *Stadiynost' v razvitii nagornykh terras (na primere khrebta Udokan)*: Voprosy geografii Zabaykals'kogo Severa, 133–42.

1969, Fossil slope deposits in the northern Arctic asymmetrical valleys: *Biuletyn Peryglacjalny* **20**, 239–57.

1971, Sezonnyye bugry pucheniya kak indikator mnogoletnemerzlykh gornykh porod: In *Geokriologicheskiye issledovaniya*: Yakutsk, Yakutskoye Knizhnoye Izdatel'stvo.

1974, Paleogeokriologicheskiye usloviya Mongol'skoy Narodnoy Respubliki: 132 in Melnikov, P. I., ed., *Geokriologicheskiye usloviya Mongol'skoy Narodnoy Respubliki. Sovmestnaya Sovetsko-Mongol' skaya Nauchno-Issledovatel'skaya Geologicheskaya Ekspeditsiya, Trudy, vyp.* **10**: Moskva, Izdatel'stvo 'Nauka'. (200 pp.)

GRAVIS, G. F., *et al.*, 1973, Geokriologicheskaya kharakteristika Mongol'skoy Narodnoy Respubliki i nektoryye osobennosti razvitiya merzlykh tolshch v proshlom: 37–45 in *Akademiya Nauk SSSR, Sektsiya Nauk o Zemle, Sibirskoye Otdeleniye, II Mezhdunarodnaya Konferentsiya po Merzlotovedeniyu, Doklady i soobshcheniya* **2** (Regional'naya geokriologiya): Yakutsk, Yakutskoye Knizhnoye Izdatel'stvo. (154 pp.)

1978, The geocryological characteristics of the Mongolian People's Republic and some characteristics of permafrost development in the past [Geokriologicheskaya kharakteristika Mongol'skoy Narodnoy Respubliki i nektoryye osobennosti razvitiya merzlykh tolshch v proshlom]: 81–6 in Sanger, F. J., ed., *USSR Contribution Permafrost Second International Conference* (Yakutsk, USSR, 13–28 July 1973): Washington, DC, Natl. Acad. Sci. (866 pp.)

GRAVIS, G. F., and LISUN, A. M., 1974, Climatic changes and permafrost evolution in Mongolia: *Inter-Nord* **13–14**, 73–85.

GRAWE, O. R., 1936, Ice as an agent of rock weathering: A discussion: *J. Geol.* **44**, 173–82.

GRECHISHCHEV, S. YE. [S. E.], 1970, *K osnovam metodiki prognoza temperaturnykh napryazheniy i deformatsiy v merzlykh gruntakh*: Moskva, Ministerstvo Geologii SSSR, Vsesoyuznyy Nauchno-Issledovatel'skiy Institut Gidrogeologii i Inzhenernoy Geologii (VSEGINGEO). (53 pp.)

1973a, Osnovnyye zakonomernosti termoreologii i temperaturnogo rastreskivaniya merzlykh gruntov: 26–34 in *Akademiya Nauk SSSR, Sektsiya Nauk o Zemle, Sibirskoye Otdeleniye, II Mezhdunarodnaya Konferentsiya po Merzlotovedeniyu, Doklady i soobshcheniya* **4** (Fizika, fiziko-khimiya i mekhanika merzlykh gornykh porod i l'da): Yakutsk, Yakutskoye Knizhnoye Izdatel'stvo. (246 pp.)

1973b, Basic laws of thermorheology and thermal

cracking in frozen ground [Osnovnyye zakonomernosti termoreologii i temperaturnogo rastreskivaniya merzlykh gruntov]: 29–40 in Tsytovich, N. A., *et al.*, Physics, physical chemistry and mechanics of permafrost and ice: *US Army Corps of Engineers, Cold Regions Research and Engineering Laboratory Draft Translation* **439**. (326 pp.)

1975, Fundamental methodology of prognosis of temperature stresses [K osnovam metodiki prognoza temperaturnykh napryazheniy i deformatsiy v merzlykh gruntakh]: *US Army Corps of Engineers, Cold Regions Research and Engineering Laboratory Draft Translation* **462**. (48 pp.)

1976, Basis of method for predicting thermal stresses and deformations in frozen soils [K osnovam metodiki prognoza temperaturnykh napryazheniy i deformatsiy v merzlykh gruntakh]: *Canada Natl. Research Council Tech. Translation* **1886**. (52 pp.)

1978, Basic laws of thermorheology and temperature cracking of frozen ground [Osnovnyye zakonomernosti termoreologii i temperaturnogo rastreskivaniya merzlykh gruntov]: 228–34 in Sanger, F. J., ed., *USSR Contribution Permafrost Second International Conference* (Yakutsk, USSR, 13–28 July 1973): Washington, DC, Natl. Acad. Sci. (866 pp.)

GREGORY, K. J., 1966, Aspect and landforms in north east Yorkshire: *Biuletyn Peryglacjalny* **15**, 115–20.

GRIGORIEW, A. A., 1925, Die Typen des Tundren–Mikroreliefs von Polar–Eurasien, ihre geographische Verbreitung und Genesis: *Geog. Zeitschr.* **31**, 345–59.

GRIGOR'YEV [GRIGOR'EV], N. F., 1966, *Mnogoletnemerzlyye porody primorskoy zony Yakutii*: Akademiya Nauk SSSR, Sibirskoye Otdeleniye, Institut Merzlotovedeniya: Moskva, Izdatel'stvo 'Nauka'. (180 pp.)

1973, Rol' kriogennykh faktorov v protsessakh rossypeobrazovaniya v pribrezhnoy zone shel'fa morya Laptevykh: 19–22 in *Akademiya Nauk SSSR, Sektsiya Nauk o Zemle, Sibirskoye Otdeleniye, II Mezhdunarodnaya Konferentsiya po Merzlotovedeniyu, Doklady i soobshcheniya* **3** (Genezis, sostav i stroyeniye merzlykh tolshch i podzemnyye l'dy): Yakutsk, Yakutskoye Knizhnoye Izdatel'stvo. (102 pp.)

1976, Perennially frozen rocks of the coastal zone of Yakutia [Mnogoletnemerzlyye porody primorskoy zony Yakutii]: *US Army Corps of Engineers, Cold Regions Research and Engineering Laboratory Draft Translation* **512**. (192 pp.)

1978, The role of cryogenous factors in processes leading to the formation of placers in the coastal zone of the shelf of the Laptev Sea [Rol' kriogennykh faktorov v protsessakh rossypeobrazovaniya v pribrezhnoy zone shel'fa morya Laptevykh]: 165–7 in Sanger, F. J., ed., *USSR Contribution Permafrost Second International Conference* (Yakutsk, USSR, 13–28 July 1973): Washington, DC, Natl. Acad. Sci. (866 pp.)

GRIM, R. E., 1952, Relation of frost action to the clay-mineral composition of soil materials: 167–72 in Frost action in soils: a symposium: *Natl. Acad. Sci.–Natl. Research Council Highway Research Board Spec. Rept.* **2**. (385 pp.)

GRIMBERBIEUX, JEAN, 1955, Origine et asymétrie des vallées sèches de Hesbaye: *Soc. Géol. Belgique Annales* **78**, 267–86.

GRUBB, A. M., and BUNTING, B. T., 1976, Micromorphological studies of soil tonguing phenomena in the Burford loam, southern Ontario, Canada: *Biuletyn Peryglacjalny* **26**, 237–52.

GRUNER, P. M., 1912, Die Bodenkultur Islands: *Archiv für Biontologie* **3**(2). (213 pp.)

GUILCHER, A., 1950, Nivation, cryoplanation et solifluction quaternaires dans les colins de Bretagne Occidentale et du Nord de Devonshire: *Rev. Géomorphologie dynamique* **1**, 53–78.

GUILLIEN, YVES, 1951, Les grèzes litées de Charente: *Rev. Géog. Pyrénées Sud-Ouest* **22**, 154–62.

1954, Le litage des grèzes: *Acad. Sci.* [Paris], *Comptes rendus* **238**, 2250–1.

1964, Grèzes litées et bancs de neige: *Geologie en Mijnbouw* **43**, 103–12.

GUILLIEN, YVES, and LAUTRIDOU, J.-P., 1970, Recherches de gélifraction expérimentale du Centre de Géomorphologie – I. Calcaires des Charentes: *CNRS, Centre de Géomorphologie de Caen Bull.* **5**. (45 pp.)

1974, Conclusions des recherches de gélifraction expérimentale sur les calcaires des Charentes: 25–33 in Recherches de gélifraction expérimentale du Centre de Géomorphologie – VI. Nouveaux résultats sur faciès calcaires: *CNRS, Centre de Géomorphologie de Caen Bull.* **19**. (43 pp.)

GULLENTOPS, F., 1977, Fossil periglacial conditions in Western Europe: *International Union for Quaternary Research (INQUA) Cong., 10th (Birmingham, England, 16–24 Aug. 1977), Abstracts*, 186.

GULLENTOPS, F., *et al.*, 1966, Observations géologiques et palynologiques dans la Vallée de la Lienne: *Acta Geog. Lovanensia* **4**, 192–204.

GULLENTOPS, F., and PAULISSEN, E., 1978, The drop soil of the Eisden type: *Biuletyn Peryglacjalny* **27**, 105–15.

GUSSOW, W. C., 1954, Piercement domes in Canadian

Arctic: *Am. Assoc. Petroleum Geologists Bull.* **38**, 2225–6.

GUTHRIE, R. D., and MATTHEWS, J. V., JR., 1971, The Cape Deceit fauna – Early Pleistocene mammalian assemblage from the Alaskan Arctic: *Quaternary Research* **1**, 474–510.

HABRICH, WULF VON, 1968, Vegetationshöcker auf steilgeneigten Terrassenhängen in der Frostschutzzone Nordostkanadas: *Polarforschung* **6**, 212–15.

HACK, J. T., 1965, Geomorphology of the Shenandoah Valley Virginia and West Virginia and origin of the residual ore deposits: *US Geol. Survey Prof. Paper* **484**. (84 pp.)

HACK, J. T., and GOODLETT, J. C., 1960, Geomorphology and forest ecology of a mountain region in the Central Appalachians: *US Geol. Survey Prof. Paper* **347**. (66 pp.)

HAEBERLI, WILFRIED, 1975, Untersuchungen zur Verbreitung von Permafrost zwischen Flüelapass und Piz Grialetsch (Graubünden): *Eidgenössischen Technischen Hochschule Zürich, Versuchsanstalt für Wasserbau, Hydrologie und Glaziologie Mitt.* **17**. (221 pp.)

1978, Special aspects of high mountain permafrost methodology and zonation in the Alps: 378–84 in *Third International Conference on Permafrost* (Edmonton, Alta., 10–13 July 1978), Proc. **1**: Ottawa, Canada Natl. Research Council. (947 pp.)

HAEFELI, R., 1953, Creep problems in soils, snow and ice: *Intern. Conf. Soil Mechanics and Found. Eng., 3d, Switzerland 1953, Proc.* **3**, 238–51.

1954, Kriechprobleme im Boden, Schnee und Eis: *Wasser- und Energiewirtschaft* **46**(3), 51–67.

HAESAERTS, P., 1974, Séquence paléoclimatique du Pléistocène Supérieur du Bassin de la Haine (Belgique): *Soc. Géol. Belgique Annales* **97**, 105–37.

HAESAERTS, P., and VAN VLIET, B., 1974, Compte rendu de l'excursion du 25 Mai 1974 consacrée à la stratigraphie des limons aux environs de Mons: *Soc. Géol. Belgique Annales* **97**, 547–60.

HAGEDORN, HORST, 1961, Morphologische Studien in den Geestgebieten zwischen Unterelbe und Unterweser: *Göttinger Geog. Abh.* **26**. (80 pp.)

HAGEDORN, JÜRGEN, 1964, Geomorphologie des Uelzener Beckens: *Göttinger Geog. Abh.* **31**. (200 pp.)

1969, Beiträge zur Quartärmorphologie griechischer Hochgebirge: *Göttinger Geog. Abh.* **50**. (135 pp.)

1970, Zum Problem der Glatthänge: *Zeitschr. Geomorphologie N. F.* **14**, 103–13.

1975, Note on the occurrence of needle-ice phenom-

ena in the Southern Sinai Mountains: *Zeitschr. Geomorphologie N. F., Supp.* **21**, 35–8.

1977, Probleme der periglazialen Höhenstufung in Griechenland: 223–37 in Poser, Hans, ed., Formen, Formengesellschaften und Untergrenzen in den heutigen periglazialen Höhenstufen der Hochgebirge Europas und Afrikas zwischen Arktis und Äquator. Bericht über ein Symposium: *Akad. Wiss. Göttingen Abh., Math.-Phys. Kl. Folge 3,* **31**. (355 pp.)

HALICKI, BRONISŁAW, 1957a, Kongeliflukcja i soliflukcja w Karpatach: *Biuletyn Peryglacjalny* **5**, 89–90.

1957b, Congelifluction and solifluction in the Carpathians: *Biuletyn Peryglacjalny* **5**, 225–6.

HALLET, B., 1978, Solute redistribution in freezing ground: 85–91 in *Third International Conference on Permafrost* (Edmonton, Alta., 10–13 July 1978), Proc. **1**: Ottawa, Canada Natl. Research Council. (947 pp.)

HALLSWORTH, E. G., ROBERTSON, G. K., and GIBBONS, R. F., 1955, Studies in pedogenesis in New South Wales – VII, The 'gilgai' soils: *J. Soil Sci.* **6**, 1–31.

HAMBERG, AXEL, 1915, Zur Kenntnis der Vorgänge im Erdboden beim Gefrieren und Auftauen sowie Bemerkungen über die erste Kristallisation des Eises in Wasser: *Geol. Fören. Stockholm, Förh.* **37**, 583–619.

HAMELIN, L.-E., 1957, Les tourbières réticulées du Québec–Labrador subarctique: interprétation morphoclimatique: *Cahiers Géog. Québec* **2**(3), 87–106.

1959, Dictionnaire franco-anglais des glaces flottantes: *Univ. Laval (Québec), Inst. Géographie Travaux* **9**. (Mimeo, 64 pp.)

1961, Périglaciaire du Canada: idées nouvelles et perspective globales: *Cahiers Géog. Québec* **5**(10), 141–203.

1969, Le glaciel de Iakoutie, en Sibérie nordique: *Cahiers Géog. Québec* **13**(29), 205–16.

1976, La famille du mot 'glaciel': *Rev. Géog. Montréal* **30**(1–2) (*Le glaciel* – Premier colloque international sur l'action géologique des glaces flottantes, Québec, 20–24 Apr. 1974), 233–6.

HAMELIN, L.-E., and COOK, F. A., 1967, *Le périglaciaire par l'image; Illustrated glossary of periglacial phenomena*: Quebec, Les Presses de l'Université Laval. (237 pp.)

HAMILTON, A. B., 1966, Freezing shrinkage in compacted clays: *Canadian Geotech. J.* **3**, 1–17.

HAMILTON, J. J., 1963, Volume changes in undisturbed clay profiles of western Canada: *Canadian Geotech. J.* **1**, 27–42.

HAMMEN, T. VAN DER, *et al.*, 1967, Stratigraphy, climatic succession, and radiocarbon dating of the Last Glacial in the Netherlands: *Geologie en Mijnbouw* **46**, 79–95.

HANNELL, F. G., 1973, The thickness of the active layer on some of Canada's Arctic slopes: *Geografiska Annaler* **55**A, 177–84.

HANSEN, B. L., and LANGWAY, C. C., JR., 1966, Deep core drilling in ice and core analysis at Camp Century, Greenland, 1961–1966: *Antarctic J. US* **1**(5), 207–8.

HANSEN, SIGURD, 1940, Varvighed i danske og skaanske senglaciale Aflejringer. Med. saerlig Hensyntagen til Egernsund Issøsystemet: *Danmarks Geol. Undersøgelse, 2d Raekke* **63**. (478 pp.)

HANSON, H. C., 1950, Vegetation and soil profiles in some solifluction and mound areas in Alaska: *Ecology* **31**, 606–30.

HARDEN, DEBORAH, BARNES, PETER, and REIMNITZ, ERK, 1977, Distribution and character of naleds in northeastern Alaska: *Arctic* **30**, 28–40.

HARRIS, CHARLES, 1972, Processes of soil movement in turf-banked solifluction lobes, Okstindan, northern Norway: 155–174 in Price, R. J., and Sugden, D. E., compilers, *Polar geomorphology*: Inst. British Geographers Spec. Pub. **4**. (215 pp.)

1973, Some factors affecting the rates and processes of periglacial mass movements: *Geografiska Annaler* **55**A, 24–8.

1977, Engineering properties, groundwater conditions, and the nature of soil movement on a solifluction slope in north Norway: *Quart. J. Eng. Geology* **10**, 27–43.

HARRIS, H. J. H., and CARTWRIGHT, KEROS, 1977, Pressure fluctuations in an Antarctic aquifer: The freight-train response to a moving rock glacier: Special Committee on Antarctic Research (SCAR), Third Symposium on Antarctic Geology and Geophysics (Madison, Wisc., 22–27 Aug. 1977), preprint. (18 pp.)

HARRIS, S. A., 1964, Seasonal density changes in the alluvial soils of northern Iraq: *Internat. Cong. Soil Science, 8th (Bucharest, 1964) Trans.* **2**, 291–303.

1973, Studies of soil creep, western Alberta, 1970 to 1972: *Arctic and Alpine Research* **5**(3, Part 2) (International Geog. Union, Comm. on High Altitude Geoecology Symposium, Calgary, Alberta, 1–8 Aug. 1972, Proc.), 171–80.

1975, Petrology and origin of stratified scree in New Zealand: *Quaternary Research* **5**, 199–214.

1976, Loess wedges on the Banks Peninsula, South Island, New Zealand: *Canadian Assoc. Geographers*

Ann. Meeting (Univ. Laval, 23–27 May 1976), Prog. and Résumés, 46–9.

HARRISON, S. S., 1970, Note on the importance of frost weathering in the disintegration and erosion of till in east-central Wisconsin: *Geol. Soc. America Bull.* **81**, 3407–9.

HARTSHORN, J. H., and SCHAFER, J. P., organizers, 1965, New England: 5–38 in *International Assoc. for Quaternary Research (INQUA) Cong., 7th (Boulder, Col., 1965): Guidebook for Field Conference A, New England – New York State.* (92 pp.)

HASTENRATH, STEFAN, 1973, Observations on the periglacial morphology of Mts Kenya and Kilimanjaro, East Africa: *Zeitschr. Geomorphologie N. F., Supp.* **16**, 161–79.

1974, Glaziale und periglaziale Formbildung in Hoch-Semyen, Nord-Äthiopien: *Erdkunde* **28**, 176–86.

1977, Observations on soil frost phenomena in the Peruvian Andes: *Zeitschr. Geomorphologie N. F.* **21**, 357–62.

HASTENRATH, STEFAN, and WILKINSON, J., 1973, A contribution to the periglacial morphology of Lesotho, southern Africa: *Biuletyn Peryglacjalny* **22**, 157–67.

HAUNSCHILD, H., 1967, Erläuterungen zur Geologischen Karte von Bayern 1 : 25 000 *Blatt* **6829** *Ornbau.* München.

HAY, THOMAS, 1936, Stone stripes: *Geog. J.* **87**, 47–50.

HEGINBOTTOM, J. A., 1973, Some effects of surface disturbance on the permafrost active layer at Inuvik, NWT, Canada: 649–57 in *North American Contribution, Permafrost Second International Conference (Yakutsk, USSR, 13–28 July 1973)*: Washington, DC, Natl. Acad. Sci. (783 pp.)

1974, The effects of surface disturbance on ground ice content and distribution: *Canada Geol. Survey Paper* **74-1**, Part A, 273.

1978, An active retrogressive thaw flow slide on eastern Melville Island, District of Franklin: *Canada Geol. Survey Paper* **78-1**A, 525–6.

HEINE, KLAUS, 1975, Permafrost am Pico de Orizaba/Mexiko: *Eiszeitalter und Gegenwart* **26**, 212–17.

1977*a*, Zur morphologischen Bedeutung des Kammeises in der subnivalen Zone randtropischer semihumider Hochgebirge: *Zeitschr. Geomorphologie N. F.* **21**, 57–78.

1977*b*, Beobachtungen und Überlegungen zur eiszeitlichen Depression von Schneegrenze und Strukturbodengrenze in den Tropen und Subtropen: *Erdkunde* **31**, 161–77.

HEINTZ, A. E., and GARUTT, V. E., 1965, Determination of the absolute age of the fossil remains of mammoth and woolly rhinoceros from the permafrost in Siberia by the help of radiocarbon (C_{14}): *Norsk geol. tidsskr.* **45**, 73–9.

HELAAKOSKI, A. R., 1912, Havaintoja jäätymisilmiöiden geomorfologisista vaikutuksista: *Finland Geografiska Fören. Medd., Julkaisuja* **9**, 1–108.

HELBIG, KLAUS, 1965, Asymmetrische Eiszeittäler in Süddeutschland und Ostösterreich: *Würzburger Geog. Arbeiten* **14**. (108 pp.)

HEMPEL, LUDWIG, 1958, Zur geomorphologischen Höhenstufung der Sierra Nevada Spaniens: *Erdkunde* **12**(4), 270–7.

HENDERSON, E. P., 1959a, Surficial geology of Sturgeon Lake map-area, Alberta: *Canada Geol. Survey Mem.* **303**. (108 pp.)

1959b, A glacial study of central Quebec–Labrador: *Canada Geol. Survey Bull.* **50**. (94 pp.)

1968, Patterned ground in southeastern Newfoundland: *Canadian J. Earth Sci.* **5**, 1443–53.

HENDERSON, J., and HOEKSTRA, P., 1977, Electromagnetic methods for mapping shallow permafrost: 16–24 in Proceedings of a symposium on permafrost geophysics, 12 October 1976: *Canada Natl. Research Council, Assoc. Comm. on Geotechnical Research Tech. Mem.* **119**. (144 pp.)

HENKE, J.-H., 1964, Über eine interessante Froststruktur im episodisch-solifluidal bewegten Boden während der Würmeiszeit: *Eiszeitalter und Gegenwart* **15**, 221 3.

HENOCH, W.-E.-S., 1960, String-bogs in the Arctic 400 miles north of the tree-line: *Geog. J.* **126**, 335–9.

HERMESH, RHEINHARD, 1972, *A study of the ecology of the Athabasca sand dunes with emphasis on the phytogenic aspects of dune formation*: Univ. Saskatchewan, M. S. Thesis. (158 pp.)

HERZ, KARL, 1964, Ergebnisse mikromorphologischer Untersuchungen im Kingsbay-Gebiet (Westspitzbergen): *Petermanns Mitt.* **108**, 45–53.

HERZ, KARL, and ANDREAS, GOTTFRIED, 1966, Untersuchungen zur Morphologie der periglazialen Auftauschicht im Kongsfjordgebiet (Westspitzbergen): *Petermanns Mitt.* **110**(3), 190–8.

HEY, R. W., 1963, Pleistocene screes in Cyrenaica (Libya): *Eiszeitalter und Gegenwart* **14**, 77–84.

HEYWOOD, W. W., 1957, Isachsen area, Ellef Ringnes Island, District of Franklin, Northwest Territories: *Canada Geol. Survey Paper* **56–8**. (36 pp.)

HIGASHI, A., 1977, Kanchi-kogaku Kisoron, sono 7 [Fundamentals of cold region engineering (7)]: *Seppyo* **39**(1), 15–25.

HIGASHI, AKIRA, and CORTE, A. E., 1971, Solifluction: A model experiment: *Science* **171**, 480–2.

1972, Growth and development of perturbations on the soil surface due to the repetition of freezing and thawing: *Hokkaido Univ., Faculty of Eng. Mem.* **13** suppl., 49–63.

HIGGINBOTTOM, I. A., and FOOKES, P. G., 1971, Engineering aspects of periglacial features in Britain: *Quart. J. Eng. Geol.* **3**(for 1970), 85–117.

HIGHWAY RESEARCH BOARD, 1948, Bibliography on frost action in soils, annotated: *Natl. Acad. Sci.–Natl. Research Council Highway Research Board Bibliography* **3**. (57 pp.)

1952a, Frost action in roads and airfields (*Highway Research Board Spec. Rept.* **1**): *Natl. Acad. Sci.–Natl. Research Council Pub.* **211**. (287 pp.)

1952b, Frost action in soils, a symposium (*Highway Research Board Spec. Rept.* **2**): *Natl. Acad. Sci.–Natl. Research Pub.* **213**. (385 pp.)

1957, Fundamental and practical concepts of soil freezing (*Highway Research Board Bull.* **168**): *Natl. Acad. Sci.–Natl. Research Council Pub.* **528**. (205 pp.)

1959, Highway pavement design in frost areas, a symposium: Part 1, Basic considerations (*Highway Research Board Bull.* **225**): *Natl. Acad. Sci.–Natl. Research Council Pub.* **685**. (131 pp.)

1962, Soil behavior associated with freezing (*Highway Research Board Bull.* **331**): *Natl. Acad. Sci.–Natl. Research Council Pub.* **1013**. (115 pp.)

1963, Pavement design in frost areas: II. Design considerations (*Highway Research Record* **33**): *Natl. Acad. Sci.–Natl. Research Council Pub.* **1153**. (270 pp.)

1969, Effects of temperature and heat on engineering behavior of soils (*Highway Research Board Spec. Rept.* **103**): *Natl. Acad. Sci.–Natl. Research Council Pub.* **1641**. (300 pp.)

1970, Frost action: Bearing, thrust, stabilization, and compaction: *Natl. Acad. Sci.–Natl. Acad. Eng., Highway Research Record* **304**. (51 pp.)

1972, Frost action in soils: *Natl. Acad. Sci.–Natl. Acad. Eng., Highway Research Record* **393**. (88 pp.)

HOBBS, P. V., 1974, *Ice physics*: Oxford, Clarendon Press. (837 pp.)

HOBBS, W. H., 1911, Requisite conditions for the formation of ice ramparts: *J. Geol.* **19**, 157–60.

1931, Loess, pebble bands, and boulders from glacial outwash of the Greenland continental glacier: *J. Geol.* **39**, 381–5.

HOBSON, G. D., et al., 1977, Permafrost distribution in the southern Beaufort Sea as determined from seismic measurements: 91–8 in Proceedings of a symposium on permafrost geophysics, 12 October 1976: *Canada Natl. Research Council, Assoc. Comm. on Geotechnical Research Tech. Mem.* **119**. (144 pp.)

HODDER, A. P. W., 1976, Cavitation-induced nucleation of ice: A possible mechanism for frost-cracking in rocks: *New Zealand J. Geol. and Geophys.* **19**, 821–6.

HOEKSTRA, PIETER, 1969, Water movement and freezing pressures: *Soil Sci. Soc. Amer. Proc.* **33**, 512–18.

HOEKSTRA, PIETER, and MILLER, R. D., 1965, Movement of water in a film between glass and ice: *US Army Cold Regions Research Engineering Laboratory Research Rept.* **153**. (8 pp.)

HÖGBOM, BERTIL, 1910, Einige Illustrationen zu den geologischen Wirkungen des Frostes auf Spitzbergen: *Uppsala Univ., Geol. Inst. Bull.* **9**, 41–59.

1914, Über die geologische Bedeutung des Frostes: *Uppsala Univ., Geol. Inst. Bull.* **12**, 257–389.

HÖGBOM, IVAR, 1923, Ancient inland dunes of northern and middle Europe: *Geografiska Annaler* **5**, 113–243.

HÖLLERMANN, P. W., 1964, Rezente Verwitterung, Abtragung und Formenschatz in den Zentralalpen am Beispiel des oberen Suldentales (Ortlergruppe): *Zeitschr. Geomorphologie N. F., Supp.* **4**. (257 pp.)

1967, Zur Verbreitung rezenter periglazialer Kleinformen in den Pyrenäen und Ostalpen: *Göttinger Geog. Abh.* **40**. (198 pp.)

1972, Beiträge zur Problematik der rezenten Strukturbodengrenze: *Göttinger Geog. Abh.* **60** (Hans-Poser-Festschrift), 235–60.

1977, Die periglaziale Höhenstufe der Gebirge in einem West-Ost-Profil von Nordiberien zum Kaukasus: 238–60 in Poser, Hans, ed., Formen, Formengesellschaften und Untergrenzen in den heutigen periglazialen Höhenstufen der Hochgebirge Europas und Afrikas zwischen Arktis und Äquator. Bericht über ein Symposium: *Akad. Wiss. Göttingen Abh., Math.–Phys. Kl. Folge 3,* **31**. (355 pp.)

HÖLLERMANN, P. W., and POSER, HANS, 1977, Grundzüge der räumlichen Ordnung in der heutigen periglazialen Höhenstufe der Gebirge Europas und Afrikas. Rückblick und Ausblick: 333–54 in Poser, Hans, ed., Formen, Formengesellschaften und Untergrenzen in den heutigen periglazialen Höhenstufen der Gebirge Europas und Afrikas zwischen Arktis und Äquator. Bericht über ein Symposium: *Akad. Wiss. Göttingen Abh., Math.–Phys. Kl. Folge 3,* **31**. (355 pp.)

HOLMES, C. D., and COLTON, R. B., 1960, Patterned ground near Dundas (Thule Air Force Base), Greenland: *Medd. om Grønland* **158**(6). (15 pp.)

HOLMES, G. E., HOPKINS, D. M., and FOSTER, H. L.,

1968, Pingos in central Alaska: *US Geol. Survey Bull.* **1241**-H. (40 pp.)

HOLMQUIST, P. J., 1898, Ueber mechanische Störungen und chemische Umsetzungen in dem Bänderthon Schwedens: *Uppsala Univ., Geol. Inst. Bull.* **3**, 412–32.

HOPKINS, D. M., 1949, Thaw lakes and thaw sinks in the Imuruk Lake area, Seward Peninsula, Alaska: *J. Geol.* **57**, 119–31.

1972, The paleogeography and climatic history of Beringia during late Cenozoic time: *Inter-Nord* **12**, 121–50.

HOPKINS, D. M., *et al.*, 1955, Permafrost and ground water in Alaska: *US Geol. Survey Prof. Paper* **264**-F, 113–46.

HOPKINS, D. M., and EINARSSON, THORLEIFUR, 1966, Pleistocene glaciation on St George, Pribilof Islands: *Science* **152**, 343–5.

HOPKINS, D. M., MACNEIL, F. S., and LEOPOLD, E. B., 1960, The coastal plain at Nome, Alaska: A late Cenozoic type section for the Bering Strait region: 46–57 in *Internat. Geol. Cong., 21st (Copenhagen, 1960) Proc., pt.* **4**, *Chronology and climatology of the Quaternary.* (157 pp.)

HOPKINS, D. M., and SIGAFOOS, R. S., 1951, Frost action and vegetation patterns on Seward Peninsula, Alaska: *US Geol. Survey Bull.* **974**-C, 51–100.

1954, Role of frost thrusting in the formation of tussocks: *Am. J. Sci.* **252**, 55–9.

HORBERG, LELAND, 1949, A possible fossil ice wedge in Bureau County, Illinois: *J. Geol.* **57**, 132–6.

1951, Intersecting minor ridges and periglacial features in the Lake Agassiz Basin, North Dakota: *J. Geol.* **59**, 1–18.

HORN, M., 1971, Erläuterungen zur Geologischen Karte von Hessen 1 : 25 000 *Blatt* **4721** *Naumburg.* Wiesbaden.

HÖVERMANN, JÜRGEN, 1953, Die Periglazial-Erscheinungen im Harz: *Göttinger Geog. Abh.* **14**, 1–39.

1954, Die Periglazial-Erscheinungen im Tegernseegebiet (bayerische Voralpen): *Göttinger Geog. Abh.* **15**, 91–124.

1960, Über Strukturböden im Elburs (Iran) und zur Frage des Verlaufs der Strukturbodengrenze: *Zeitschr. Geomorphologie N.F.* **4**, 173–4.

1962, Über Verlauf und Gesetzmässigkeit der Strukturbodengrenze: *Biuletyn Peryglacjalny* **11**, 201–7.

HÖVERMANN, JÜRGEN, and KUHLE, MATTHIAS, 1978, Verbreitung und Bildung von Frostmusterböden in den Gebieten des Vorderen Orients und der Sahara: 321–31 in Association Géographique d'Alsace, *Colloque sur le Périglaciaire d'Altitude du Domaine Méditerranéen et Abords* (Strassbourg – Université Louis Pasteur, 12–14 May 1977): Strassbourg, Assoc. Géog. d'Alsace. (366 pp.)

HOWARD, R. B., COWEN, BARRY, and INOUYE, DAVID, 1977, Reappraisal of desert pavement formation (abs.): *Geol. Soc. America Abs. with Programs* **9**(4), 438–9.

HOWE, ERNEST, 1909, Landslides in the San Juan Mountains, Colorado: including a consideration of their causes and their classification: *US Geol. Survey Prof. Paper* **67**. (58 pp.)

HOWE, JOHN, 1971, Temperature readings in test bore holes: *Mt Washington Observatory News Bull.* **12**(2), 37–40.

HOWITT, FRANK, 1971, Permafrost in Prudhoe Bay field: Geology and physical characteristics (abs.): *Internat. Symposium on Arctic Geology, 2d* (*San Francisco, Cal., 1–4 Feb. 1971*), *Program abstracts*, 27.

HRÁDEK, M., 1967, Drobné tvary v pegmatitu Čertových kamenů v Hrubém Jeseníku: *Zprávy Geografichkého ústavu ČSAV* **3**, 1–8.

HSIEH, TZU-CH'U, *et al.*, 1975, Basic features of the glaciers of the Mt Jolmo Lungma region, southern part of the Tibet autonomous region, China: *Scientia Sinica* **38**(1), 106–30.

HUDEC, P. P., 1974, Weathering of rocks in Arctic and sub-Arctic environment: 313–35 in Aitken, J. D., and Glass, D. J., eds., *Canadian Arctic geology*: Geol. Assoc. Canada – Canadian Soc. Petrol. Geol., Symposium on the geology of the Canadian Arctic (Saskatoon, 24–26 May 1973), Proc. (368 pp.)

HUGHES, O. L., 1969, Distribution of open-system pingos in central Yukon Territory with respect to glacial limits: *Canada Geol. Survey Paper* **69–34**. (8 pp.)

HUGHES, T. M., 1884, On some tracks of terrestrial and freshwater animals: *Geol. Soc. London Quart. J.* **40**, 178–86.

HUME, J. D., and SCHALK, MARSHALL, 1976, The effects of ice on the beach and nearshore, Point Barrow, Arctic Alaska: *Rev. Géog. Montréal* **30**(1–2) (*Le glaciel* – Premier colloque international sur l'action géologique des glaces flottantes, Québec, Canada, 20–24 Apr. 1974), 105–14.

HUME, W. F., 1925, *Geology of Egypt*, v. 1: Cairo, Government Press. (408 pp.)

HUNT, C. B., and WASHBURN, A. L., 1966, Patterned ground: B104–33 in Hunt, C. B., *et al.*, Hydrologic basin, Death Valley, California: *US Geol. Survey Prof. Paper* **494**-B. (138 pp.)

HUNTER, J. A. M., *et al.*, 1976, The occurrence of permafrost and frozen sub-seabottom materials in the southern Beaufort Sea: *Canada, Dept. of Environment Beaufort Sea Project* (*Victoria, BC*), *Beaufort Sea Tech. Rept.* **22**. (174 pp.)

HUSSEY, K. M., 1962, Ground patterns as keys to photo-interpretation of arctic terrain: *Iowa Acad. Sci.* **69**, 332–41.

HUSSEY, K. M., and MICHELSON, R. W., 1966, Tundra relief features near Point Barrow, Alaska: *Arctic* **19**, 162–84.

HUTCHINSON, J. N., 1974, Periglacial solifluxion: an approximate mechanism for clayey soils: *Géotechnique* **24**(3), 438–43.

HUXLEY, J. S., and ODELL, N. E., 1924, Notes on surface markings in Spitsbergen: *Geog. J.* **63**, 207–29.

IL'IN, R. S., 1934, Nagornyye terasy i kurumy: *Gosudarstvennogo GeografichESKOGO obshchestva, Izvestiya* **66**(4), 621–5.

INGERSOLL, L. R., ZOBEL, O. J., and INGERSOLL, A. C., 1954, *Heat conduction with engineering, geological, and other applications*: Madison, University of Wisconsin Press. (325 pp.)

INGLIS, D. R., 1965, Particle sorting and stone migration by freezing and thawing: *Science* **148**, 1616–17.

INSTITUT MERZLOTOVEDENIYA im. V. A. OBRUCHEVA, 1956, *Osnovnyye ponyatiya i terminy geokriologii* (*merzlotovedeniya*): Moskva, Akad. Nauk SSSR. (16 pp.)

1960, Fundamental concepts and terms in geocryology (permafrost studies) [Osnovnyye ponyatiya i terminy geokriologii (merzlotovedeniya)]: *US Army Corps of Engineers, Arctic Construction and Frost Effects Laboratory Translation* **28**. (11 pp.)

ISHERWOOD, D. J., 1975, *Soil chemistry and rock weathering in an Arctic environment*: Univ. Colorado, Ph.D. Thesis. (173 pp.)

IVAN, A., 1965, Zpráva o výzkumu kryoplanačních teras v severozápadni části Rychlebských hor: *Zprávy Geografického ústavu ČSAV* **7**(**146**-B), 1–3.

IVERONOVA, M. I., 1964, Stationary studies of the recent denudation processes on the slopes of the R. Tchon–Kizilsu Bazin, Tersky Alatau ridge, Tien-Shan: *Zeitschr. Geomorphologie N.F., Supp.* **5** (*Fortschritte der internationalen Hangforschung*), 206–12.

IVES, J. D., 1966, Block fields, associated weathering forms on mountain tops and the Nunatak hypothesis: *Geografiska Annaler* **48**A, 220–3.

1973, Permafrost and its relationship to other environmental parameters in a midlatitude, high-altitude setting, Front Range, Colorado Rocky

Mountains: 121–5 in *North American Contribution, Permafrost Second International Conference* (Yakutsk, USSR, 13–28 July 1973): Washington, DC, Natl. Acad. Sci. (783 pp.)

1974, Permafrost: 159–94 in Ives, J. D., and Barry, R. G., eds., *Arctic and alpine environments*: London, Methuen & Co. Ltd. (999 pp.)

IVES, J. D., and FAHEY, B. D., 1971, Permafrost occurrence in the Front Range, Colorado Rocky Mountains, USA: *J. Glaciology* **10**, 105–11.

IVES, R. L., 1940, Rock glaciers in the Colorado Front Range: *Geol. Soc. America Bull.* **51**, 1271–94.

JÄCKLI, HEINRICH, 1957, Gegenwartsgeologie des bündnerischen Rheingebietes: *Beiträge zur Geologie der Schweiz, Geotechnische Ser.* **36**. (136 pp.)

JACKSON, K. A., and CHALMERS, BRUCE, 1957, Study of ice formation in soils: *US Army Corps of Engineers New England Division, Arctic Construction and Frost Effects Laboratory Tech. Rept.* **65** (Revised). (29 pp.)

JACKSON, K. A., and UHLMANN, D. R., 1966, Particle sorting and stone migration due to frost heave: *Science* **152**, 545–6.

JAHN, ALFRED, 1948a, Badania nad strukturą i temperaturą gleb w Zachodniej Grenlandii: *Polska Akad. Umiejętności, Rozprawy Wydz., matem.-przyr.* **72**A, (6). (121 pp., 63–183 of whole volume.)

1948b, Badania nad struktura i temperaturą gleb w Grenlandii zachodniej – Research on the structure and temperature of the soils in western Greenland: *Acad. polonaise sci. et lettres Bull., cl. sci. math. et nat.– Ser. A, sci. math. N° sommaire A*, 1940–1946, 50–9.

1958, Periglacial microrelief in the Tatras and on the Babia Góra: *Biuletyn Peryglacjalny* **6**, 227–49.

1960, Some remarks on evolution of slopes on Spits- bergen: *Zeitschr. Geomorphologie N.F., Supp.* **1** (*Morphologie des versants*), 49–58.

1961a, Quantitative analysis of some periglacial processes in Spitsbergen: *Uniwersytet Wrocławski im. Bolesława Bieruta, zeszyty naukowe, nauki przyrodnicze, ser. B* **5** (Nauk o Ziemi II), 1–34.

1961b, Problemy geograficzne Alaski wo świetle podrózi naukowej odbytej v 1960 roku: *Czasopismo Geograficzne* **32**(2), 115–81.

1965, Formy i procesy stokowe w Karkonoszach: *Opera Corcontica* **2**, 7–16.

1966, *Alaska*: Warszawa, Państwowe Wydawn- ictwo Naukowe. (498 pp.)

1967, Some features of mass movement on Spits- bergen slopes: *Geografiska Annaler* **49**A, 213–25.

1970, *Zagadnienia strefy peryglacjalnej*: Warszawa, Państwowe Wydawnictwo Naukowe. (202 pp.)

1972a, Tundra polygons in the Mackenzie Delta area: *Göttinger Geog. Abh.* **60** (Hans Poser Festschrift), 285–92.

1972b, Niveo-eolian processes in the Sudetes Moun- tains: *Geographia Polonica* **23**, 93–110.

1972c, Some regularities in thermokarst develop- ment: 167–76 in Macar, P., and Pissart, A., eds., *Processus périglaciaires étudies sur le terrain*: Union Géographique Internationale, Symposium Inter- national de Géomorphologie (Liège-Caen, 1–9 July 1971), *Part* **1** (Séances tenues à Liège et excursions en Belgique): Les Congrès et Colloques de l'Université de Liège: Liège, Université de Liège. (339 pp.)

1975, *Problems of the periglacial zone (Zagadnienia strefy peryglacjalnej)*: Warsaw, Państwowe Wy- dawnictwo Naukowe. (223 pp.)

1976a, Geomorphological modelling and nature protection in Arctic and Subarctic environments: *Geoforum* **7**, 121–37.

1976b, Pagórki mrozowe typu palsa (Palsa-type frost mounds): 123–39 in Problemy geografii fizycznej: *Studia Soc. Scientiarum Torunensis, Sec. C (Geog. et Geol.)* **8**(2–4). (312 pp.)

1977a, Periglacial forms produced by shore ice at Hornsund (Spitsbergen): *Acta Universitatis Wratis- laviensis* **387**, 19–29.

1977b, The permafrost active layer in the Sudety Mountains during the last glaciation: *Quaestiones Geographicae* **4**, 29–41.

1978, Mass wasting in permafrost and non- permafrost environments: 295–300 in *Third Inter- national Conference on Permafrost* (Edmonton, Alta., 10–13 July 1978), Proc. **1**: Ottawa, Canada Natl. Research Council. (947 pp.)

JAHN, ALFRED, and CIELIŃSKA, MARIA, 1974, The rate of soil movement in the Sudety Mountains: 86–101 in Poser, Hans, ed., Geomorphologische Prozesse und Prozesskombinationen in der Gegenwart unter verschiedenen Klimabedin- gungen (Rept. Commission Present-day Geo- morphological Processes, International Geo- graphical Union): *Akad. Wiss. Göttingen Abh., Math.–Phys. Kl. Folge 3*, **29**. (440 pp.)

JAHN, ALFRED, and CZERWIŃSKI, JANUSZ, 1965, The rôle of impulses in the process of periglacial soil structure formation (Rola impulsow w procesie formowania sie peryglacjalnych struktur gle- bowych): *Acta Universitatis Wratislaviensis* **44** (*Studia Geograficzne* **7**), 1–24.

JAMES, P. A., 1971, The measurement of soil frost- heave in the field: *British Geomorphological Research Group Tech. Bull.* **8**. (43 pp.)

JENNESS, J. L., 1952, Erosive forces in the physiography of western Arctic Canada: *Geog. Rev.* **42**, 238–52.

JENNINGS, J. N., 1978, The geomorphic role of stone movement through snow creep, Mount Twynam, Snowy Mountains, Australia: *Geografiska Annaler* **60**A, 1–8.

JENNINGS, J. N., and COSTIN, A. B., 1978, Stone movement through snow creep, 1963–1975, Mount Twynam, Snowy Mountains, Australia: *Earth Surface Processes* **3**(1), 3–22.

JERSAK, JÓSEF, 1975, Frost fissures in loess deposits: *Biuletyn Peryglacjalny* **24**, 245–58.

JESSBERGER, H. L., 1973, Frost-susceptibility criteria: 40–6 in Highway Research Board, *Soils: Loess, suction, and frost action: Natl. Acad. Sci. – Natl. Acad. Eng., Highway Research Record* **429**. (63 pp.)

ed., 1978, *International symposium on ground freezing* (Bochum, Germany, 8–10 Mar. 1978), Proc.: Bochum, Ruhr Univ. (363 pp.)

JESSOP, A. M., 1971, The distribution of glacial perturbation of heat flow in Canada: *Canadian J. Earth Sci.* **8**, 162–6.

1973, Terrestrial heat flow and permafrost [summary]: 51–3 in Brown, R. J. E., ed., Proceedings of a seminar on the thermal regime and measurements in permafrost, 2 and 3 May 1972: *Canada Natl. Research Council, Assoc. Comm. on Geotechnical Research Tech. Mem.* **108**. (85 pp.)

JESSUP, R. W., 1960, The stony tableland soils of the southeastern portion of the Australian arid zone and their evolutionary history: *J. Soil Sci.* **11**(2), 188–96.

JOHANSSON, SIMON, 1914, Die Festigkeit der Bodenarten bei verschiedenem Wassergehalt nebst Vorschlag zu einer Klassifikation: *Sveriges Geol. Undersökning Årsbok* **7**(3) (avh. och uppsatser, ser. C, 256). (110 pp.)

JOHNSON, D. L., and HANSEN, K. L., 1974, The effects of frost-heaving on objects in soil: *Plains Anthropologist* **19**(64), 81–98.

JOHNSON, D. L., MUHS, D. R., and BARNHARDT, M. L., 1977, The effects of frost heaving on objects in soils, II: Laboratory experiments: *Plains Anthropologist* **22–76**(1), 133–47.

JOHNSON, J. H., 1961, Sea surface temperature and monthly average and anomaly charts, northeastern Pacific ocean, 1947–58: *US Dept. Interior, Fish and Wildlife Service Special Sci. Rept. – Fisheries* **385**. (56 pp.)

JOHNSON, J. P., 1973, Some problems in the study of rock glaciers: 84–94 in Fahey, B. D., and

Thompson, R. D., eds., *Research in polar and alpine geomorphology*: Guelph Symposium on Geomorphology, 1973, 3d, Proc.: Norwich, England, Geo Abstracts Ltd (Univ. East Anglia). (206 pp.)

JOHNSON, P. G., 1975, Mass movement processes in Metalline Creek, southwest Yukon Territory: *Arctic* **28**, 130–9.

1978, Rock glacier types and their drainage systems, Grizzly Creek, Yukon Territory: *Canadian J. Earth Sci.* **15**, 1496–507.

JOHNSON, T. C., et al., 1975, Roadway design in seasonal frost areas: *US Army Corps of Engineers, Cold Regions Research and Engineering Laboratory Tech. Rept.* **259**. (104 pp.)

JOHNSON, W. H., 1978, Patterned ground, wedge-shaped bodies, and circular landforms in central and eastern Illinois (abs.): *Geol. Soc. America Abs. with Programs* **10**(6), 257.

JOHNSSON, GUNNAR, 1958, Periglacial wind and frost erosion at Klågerup, S. W. Scania: *Geografiska Annaler* **40**, 232–43.

1959, True and false ice-wedges in southern Sweden: *Geografiska Annaler* **41**, 15–33.

1960, Cryoturbation at Zaragoza, northern Spain: *Zeitschr. Geomorphologie N.F.* **4**, 74–80.

1962, Periglacial phenomena in southern Sweden: *Geografiska Annaler* **44**, 378–404.

JOHNSTON, J. H., 1973, Salt weathering processes in the McMurdo Dry Valley regions of South Victoria Land, Antarctica: *New Zealand J. Geol. and Geophys.* **16**(2), 221–4.

JORRÉ, GEORGES, 1933, Problème des 'terrasses goletz' sibériennes: *Rev. Géog. Alpine* **21**, 347–71.

JOURNAUX, ANDRÉ, 1973, Action of running water and solifluction in Canadian semi-arid Arctic (abs.): 176–7 in *International Union for Quaternary Research (INQUA) Cong., 9th (Christchurch, New Zealand, 2–10 Dec. 1973), Abstracts.* (418 pp.)

1976a, Alternances du ruisselement et de la solifluction dans les milleiux périglaciaires: Exemples Canadiens et experimentations: *Biuletyn Peryglacjalny* **26**, 269–73.

1976b, Les grèzes litées du Châtillonnais: *Assoc. française pour l'Étude du Quaternaire* **1976**(3–4), 123–38.

JUDGE, A. S., 1973a, Deep temperature observations in the Canadian North: 35–40 in *North American Contribution, Permafrost Second International Conference* (Yakutsk, USSR, 13–28 July 1973): Washington DC, Natl. Acad. Sci. (783 pp.)

1973b, The thermal regime of the Mackenzie Valley: Observations of the natural state: *Environmental-*

Social Comm. Northern Pipelines [Canada], *Task Force on Northern Oil Development Rept.* **73–38**. (177 pp.)

1974, Occurrence of offshore permafrost in northern Canada: 427–37 in Reed, J. C., Sater, J. E., and Gunn, W. W., eds., *The coast and shelf of the Beaufort Sea* (*Beaufort Sea Coast and Shelf Research Symposium Proc.*): Arctic Inst. North America. (750 pp.)

1975, The occurrence of permafrost beneath the sea-bottom of Kugmallit Bay, Beaufort Sea, Canada (abs.): *Geol. Soc. America Abs. with Programs* **7**(6), 793–4.

1977, Permafrost, hydrates and the offshore thermal regime: 99–113 in Proceedings of a symposium on permafrost geophysics, 12 October 1976: *Canada Natl. Research Council, Assoc. Comm. on Geotechnical Research Tech. Mem.* **119**. (144 pp.)

JUDSON, SHELDON, 1949, Rock-fragment slopes caused by past frost action in the Jura Mountains (Ain), France: *J. Geol.* **57**, 137–42.

1965, Quaternary processes in the Atlantic Coastal Plain and Appalachian Highlands: 133–6 in Wright, H. E., and Frey, D. G., eds., *The Quaternary of the United States*: Princeton, NJ, Princeton Univ. Press. (922 pp.)

JUMIKIS, A. R., 1973, Effect of porosity on amount of soil water transferred in a freezing silt: 305–10 in *North American Contribution, Permafrost Second International Conference* (Yakutsk, USSR, 13–28 July 1973): Washington, DC, Natl. Acad. Sci. (783 pp.)

1977, *Thermal geotechnics*: New Brunswick, NJ, Rutgers Univ. Press. (375 pp.)

JÜNGST, H., 1934, Zur geologischen Bedeutung der Synärese: *Geol. Rundschau* **25**, 312–25.

KACHADOORIAN, REUBEN, and FERRIANS, O. J., JR., 1973, Permafrost-related engineering geology problems posed by the trans-Alaska pipeline: 684–7 in *North American Contribution, Permafrost Second International Conference* (Yakutsk, USSR, 13–28 July 1973): Washington, DC, Natl. Acad. Sci. (783 pp.)

KACHURIN, S. P., 1959, Kriogennyye fiziko-geologicheskiye yavleniya v rayonakh s mnogoletnemerzlymi porodami: 365–98 (Glava XI) in Inst. Merzlotovedeniya im. V. A. Obrucheva, *Osnovy geokriologii* (*merzlotovedeniya*), *Chast' pervaya, Obshchaya geokriologiya*: Moskva, Akad. Nauk SSSR. (459 pp.)

1961, *Termokarst na territorii SSSR* (Thermokarst on the territory of the USSR): Moscow, Akad. Nauk. (291 pp.)

1962, Thermokarst within the territory of the USSR: *Biuletyn Peryglacjalny* **11**, 49–55.

1964, Cryogenic physico-geological phenomena in permafrost regions [Kriogennyye fiziko-geologicheskiye yavleniya v rayonakh s mnogoletnemerzlymi porodami]: *Canada Natl. Research Council Tech. Translation* **1157**. (91 pp.)

KACZOROWSKI, R. T., 1976, Origin of the Carolina Bays: II-16 – II-36 in Hayes, M. O., and Kana, T. W., eds., *Terrigenous clastic depositional environments. Some modern examples* (AAPG Field course): Univ. South Carolina Coastal Research Div. (Dept. Geology) Tech. Rept. 11-CRD. (Part I, 131 pp.; Part II, 185 pp.)

1977, The Carolina Bays and their relationship to modern oriented lakes (abs.): *Geol. Soc. America Abs. with Programs* **9**(2), 151–2.

KAISER, KARLHEINZ, 1958, Wirkungen des pleistozänen Bodenfrostes in den Sedimenten der Niederrheinischen Bucht: *Eiszeitalter und Gegenwart* **9**, 110–29.

1960, Klimazeugen des periglazialen Dauerfrostbodens in Mittel- und West-europa: *Eiszeitalter und Gegenwart* **11**, 121–41.

1965, Die Ausdehnung der Vergletscherungen und ‚periglazialen‘ Erscheinungen während der Kaltzeiten des quärtaren Eiszeitalters innerhalb der syrisch-libanesischen Gebirge und die Lage der klimatischen Schneegrenze zur Würmeiszeit im östlichen Mittelmeergebiet: 127–48 in *International Assoc. for Quarternary Research* (*INQUA*) *Cong., 6th* (*Warsaw, Poland, Aug.–Sept. 1961*), *Report* **3**: Lódź. (500 pp.)

1969, The climate of Europe during the Quaternary Ice Age: 10–37 in Wright, H. E., Jr., ed., Quaternary geology and climate: *International Assoc. for Quarternary Research* (*INQUA*) *Cong., 7th* (*Boulder, Col., 1965*), *Proc.* **16**: Washington, DC, Natl. Acad. Sci. (162 pp.)

1970, Über Konvergenzen arider und 'periglazialer' Oberflächenformung und zur Frage einer Trockengrenze solifluidaler Wirkungen am Beispiel des Tibesti-Gebirges in der zentralen Ostsahara: *Freie Univ. Berlin, 1. Geog. Inst. Abh.* **13** (*Aktuelle Probleme geographischer Forschung – J. H. Schultze Festschrift*), 147–88.

KALETSKAYA, M. W., and MIKLUKHO-MAKLAY, A. D., 1958, Nekotorye cherty chetvertichnoy istorii vostochnoy chasti pechorskogo basseyna i zapadnogo sklona Poliarnogo Urala: *Instituta geografii AN SSSR, Trudy* **76**; *Materialy po geomorfologii i paleografii SSSR* **20**, 1–67.

KAPLAR, C. W., 1965, Stone migration by freezing of soil: *Science* **149**, 1520–1.

—— 1969, Phenomena and mechanism of frost heaving: *Highway Research Board Annual Mtg., 49th Washington, DC 1970, preprint.* (44 pp.)

—— 1970, Phenomenon and mechanism of frost heaving: 1–13 in Highway Research Board, *Frost action: Bearing, thrust, stabilization, and compaction: Natl. Acad. Sci. – Natl. Acad. Eng., Highway Research Record* **304**. (51 pp.)

—— 1971, Experiments to simplify frost susceptibility testing of soils: *US Army Corps of Engineers, Cold Regions Research Engineering Laboratory Tech. Rept.* **223**. (23 pp.)

—— 1974, Freezing test for evaluating relative frost susceptibility of various soils: *US Army Corps of Engineers, Cold Regions Research and Engineering Laboratory Tech. Rept.* **250**. (37 pp.)

KAPLINA, T. N., 1965, *Kriogennyye sklonovyye protsessy*: Moskva, Izdatel'stvo 'Nauka'. (296 pp.)

KARPOV, V. M., and PUZANOV, I. I., 1970, *Stroitel'stvo i vechnaya merzlota*: Leningrad, Izdatel'stvo Literatury po Stroitel'stvu. (95 pp.)

KARRASCH, HEINZ, 1970, Das Phänomen der klimabedingten Reliefasymmetrie in Mitteleuropa: *Göttinger Geog. Abh.* **56**. (299 pp.)

—— 1972, Flächenbildung unter periglazialen Klimabedingungen?: *Göttinger Geog. Abh.* **60** (Hans-Poser-Festschrift), 155–68.

—— 1974, Hangglättung und Kryoplanation an Beispielen aus den Alpen und kanadischen Rocky Mountains: 287–300 in Poser, Hans, ed., Geomorphologische Prozesse und Prozesskombinationen in der Gegenwart unter verschiedenen Klimabedingungen (Symposium and Report of Commission on Present-day Geomorphological Processes, Intl. Geog. Union): *Akad. Wiss. Göttingen Abh., Math.-Phys. Kl. Folge 3*, **29**. (440 pp.)

—— 1977, Die klimatischen und aklimatischen Varianzfaktoren der periglazialen Höhenstufe in den Gebirgen West- und Mitteleuropas: 157–77 in Poser, Hans, ed., Formen, Formengesellschaften und Untergrenzen in den heutigen periglazialen Höhenstufen der Hochgebirge Europas und Afrikas zwischen Arktis und Äquator. Bericht über ein Symposium: *Akad. Wiss. Göttingen Abh., Math.-Phys. Kl. Folge 3*, **31**. (355 pp.)

KATASONOV, E. M., 1972, Regularities in cryogenic phenomena development: 34–5 in Adams, W. P., and Helleiner, F. M., eds., *International Geography 1972* **1** (Internat. Geog. Cong., 22d, Montreal): Toronto, Univ. Toronto Press. (694 pp.)

—— 1973a, Classification of frost-caused phenomena with references to the genesis of the sediments in Central Yakutia: *Biuletyn Peryglacjalny* **23**, 71–80.

—— 1973b, Present-day ground- and ice veins in the region of the middle Lena: *Biuletyn Peryglacjalny* **23**, 81–9.

—— 1975, Sovremennyye zemlyanyye i ledyanyye zhily v srednem techenii r. Leny: 9–15 in Markov, K. K., and Spasskaya, I. I., eds., *Paleogeografiya i periglyatsial'nyye yavleniya pleistotsena*: Moskva, Akad. Nauk SSSR, Komissiya po izucheniyu chetvertichnogo perioda, Izdatel'stvo 'Nauka'. (224 pp.)

—— 1977, Cryolithogenic deposits of Yakutia and their importance for understanding the Quaternary history of permafrost area (abs.): *International Union for Quaternary Research (INQUA) Cong., 10th (Birmingham, England, 16–24 Aug. 1977), Abstracts,* 234.

KATASONOV, E. M., and IVANOV, M. S., 1973, *Cryolithology of Central Yakutia*: Internat. Permafrost Conf., 2d (Yakutsk, USSR, 13–28 July 1973), Guidebook, Excursion on the Lena and Aldan Rivers: Yakutsk, USSR Acad. Sci., Sec. Earth Sci., Siberian Div. (38 pp.)

KATASONOV, E. M., and PUDOV, G. G., 1972, Kriolitologicheskiye issledovaniya v rayone Van'kinoy guby morya Laptevykh: 130–6 in *Merzlotnyye issledovaniya vyp.* **12**: Moskva, Izdatel'stvo Moskovskogo Universiteta. (224 pp.)

KATASONOV, E. M., and SOLOV'EV, P. A., 1969, *Guide to trip round Central Yakutia*: Internat. Symposium, Paleogeography and Periglacial Phenomena of Pleistocene: Yakutsk, USSR. (88 pp.)

KATASONOVA, E. G., 1963, Rol' termokarsta v razvitii delli: 91–100 in Katasonov, E. M., ed., *Usloviya i osobennosti razvitiya merzlykh tolshch v Sibiri i na Severo-Vostoke*: Moskva, Akademiya Nauk SSSR, Sibirskoye Otdeleniye, Inst. Merzlotovedeniya, Izdatel'stvo Akad. Nauk SSSR. (119 pp.)

KATZ, D. L., 1971, Depths to which frozen gas fields (gas hydrates) may be expected: *J. Petroleum Technology*, April 1971, 419–23.

KAYE, C. A., and POWER, W. R., JR., 1954, A flow cast of very recent date from northeastern Washington: *Am. J. Sci.* **252**, 309–10.

KELLER, B., 1910, Po dolinam i goram Altaya: *Pochvennobotanicheskoy ekspeditsii po issledovaniyam Aziatskoy Rosii, Trudy, Chast II., Botanicheskiye issledovaniya,* 233–4.

KELLETAT, DIETER, 1969, Verbreitung und Vergesellschaftung rezenter Periglazialerscheinungen im Appenin: *Göttinger Geog. Abh.* **48**. (114 pp.)

1970a, Rezente Periglazialerscheinungen im Schottischen Hochland. Untersuchungen zu ihrer Verbreitung und Vergesellschaftung: *Göttinger Geog. Abh.* **51**, 67–140.

1970b, Zum Problem der Verbreitung, des Alters und der Bildungsdauer alter (inaktiver) Periglazialerscheinungen im Schottischen Hochland: *Zeitschr. Geomorphologie N.F.* **14**, 510–19.

1977a, Die rezente periglaziale Höhenstufe in den Gebirgen der nördlichen Mittelbreiten Europas: 105–17 in Poser, Hans, ed., Formen, Formengesellschaften und Untergrenzen in den heutigen periglazialen Höhenstufen der Hochgebirge Europas und Afrikas zwischen Arktis und Äquator. Bericht über ein Symposium: *Akad. Wiss. Göttingen Abh., Math.-Phys. Kl. Folge 3,* **31**. (355 pp.)

1977b, Die rezente periglaziale Höhenstufe des Appenin: geomorphologische Ausstattung, gegenwärtige Formungsprozesse und Probleme der Abgrenzung: 203–22 in Poser, Hans, ed., Formen, Formengesellschaften und Untergrenzen in den heutigen periglazialen Höhenstufen der Hochgebirge Europas und Afrikas zwischen Arktis und Äquator. Bericht über ein Symposium: *Akad. Wiss. Göttingen Abh., Math.-Phys. Kl. Folge 3,* **31**. (355 pp.)

KELLETAT, DIETER, and GASSERT, DETLEF, 1974, Spätglaziale Eiskeil-Spaltennetze in der Umgebung der Sellagruppe, Grödner Dolomiten: *Zeitschr. Geomorphologie N.F.* **18**, 307–15.

KENDREW, W. G., 1941, *The climates of the continents*, 3 ed.: London, Oxford Univ. Press. (473 pp.)

KENNEDY, B. A., 1976, Valley-side slopes and climate: 171–201 in Derbyshire, Edward, ed., *Geomorphology and climate*: London, New York, Sydney, Toronto, John Wiley & Sons. (512 pp.)

KENNEDY, B. A., and MELTON, M. A., 1972, Valley asymmetry and slope forms of a permafrost area in the Northwest Territories, Canada: 107–121 in Price, R. J., and Sugden, D. E., compilers, *Polar geomorphology*: Inst. British Geographers Spec. Pub. **4**. (215 pp.)

KERFOOT, D. E., 1972, Thermal contraction cracks in an arctic tundra environment: *Arctic* **25**, 142–50.

1973, Thermokarst features produced by man-made disturbances to the tundra terrain: 60–72 in Fahey, B. D., and Thompson, R. D., eds., *Research in polar and alpine geomorphology*: Guelph Symposium on Geomorphology, 1973, 3d, Proc.: Norwich, England, Geo Abstracts Ltd (Univ. East Anglia). (206 pp.)

KERFOOT, D. E., and MACKAY, J. R., 1972, Geomorphological process studies, Garry Island, NWT: 115–30 in Kerfoot, D. E., ed., *Mackenzie Delta area monograph*: Internat. Geog. Cong., 22d, Montreal (1972): St Catherines, Ont., Brock Univ. (174 pp.)

KERSTEN, M. S., 1949, *Laboratory research for the determination of the thermal properties of soils*: Final report to U.S. Army Corps of Engineers, St. Paul District: Univ. Minnesota, Engineering Experiment Station. (225 pp.)

KESSELI, J. E., 1941, Rock streams in the Sierra Nevada, California: *Geog. Rev.* **31**, 203–27.

KESSLER, PAUL, 1925, *Das eiszeitliche Klima und seine geologischen Wirkungen im nicht vereisten Gebiet*: Stuttgart, E. Schweizerbart'sche. (204 pp.)

KHMYZNIKOV, P. K., 1937, [Erosion of shores in the Laptev Sea. Northern Sea route]: *Sbornik Statey po Gidrografii i Moreplavaniyu* **7** (Leningrad, Izd-vo. Glavsevmorputi).

KIM, DO-JONG, 1967, *Die dreidimensionale Verteilung der Strukturböden auf Island in ihrer klimatischen Abhängigkeit*: Univ. Bonn, Dissertation. (227 pp.)

1970, Über die Strukturbödenerscheinungen auf dem Halla-San: *Geog. Naksan (Nakusan Siri) (Seoul Natl. Univ., Dept. Geog.)* **1**, 9. (Korean text, 3–10.)

KINDLE, E. M., 1917, Some factors affecting the development of mud-cracks: *J. Geol.* **25**, 135–44.

KING, C. A. M., and HIRST, R. A., 1964, The boulder-fields of the Åland Islands: *Fennia* **89**(2). (41 pp.)

KING, LESTER, 1958, The problem of tors [Correspondence]: *Geog. J.* **124**, 289–91.

KING, LORENZ, 1977, Permafrost Untersuchungen in Tarfala (Schwedisch Lappland) mit hilfe der Hammerschlagseismic: *Zeitschr. Gletscherkunde u. Glazialgeologie* **12**, 187–204.

KING, R. B., 1971, Boulder polygons and stripes in the Cairngorm Mountains, Scotland: *J. Glaciology* **10**, 375–86.

KINOSITA, S., et al., 1978, The comparison of frozen soil structure beneath a high center polygon, trough and center, at Barrow, Alaska: 301–4 in *Third International Conference on Permafrost* (Edmonton, Alta., 10–13 July 1978), Proc. **1**: Ottawa, Canada Natl. Research Council. (947 pp.)

1979, Core samplings of the uppermost layer in a tundra area: 17–44 in Kinosita, S., ed., *Joint studies on physical and biological environments in the permafrost, Alaska and North Canada*: Hokkaido, Hokkaido Univ. Institute of Low Temperature Science. (149 pp.)

KIRKBY, M. J., 1967, Measurement and theory of soil creep: *J. Geol.* **75**, 359–78.

KLAER, W., 1956, Verwitterungsformen im Granit auf Korsika: *Petermanns Mitt. Ergänzungsheft* **261**. (146 pp.)

KLASSEN, R. W., 1979, Thermokarst terrain near Whitehorse, Yukon Territory: *Canada Geol. Survey Paper* **79-1**A, 385–8.

KLATKA, TADEUSZ, 1961*a*, Problèmes des sols striés de la partie septentrionale de la presqu'île de Sorkapp (Spitsbergen): *Biuletyn Peryglacjalny* **10**, 291–320.

1961*b*, Indices de structure et de texture des champs de pierres des Lysógory: *Soc. Sci. et Lettres Łodz Bull.* **12**(10), 1–21.

1962, Geneza i wiek globorzy Łysogorskich: *Acta Geographica Lodziensia* **12**. (129 pp.)

1968, Microrelief of slopes in the coastal area south of Hornsund, Vestspitsbergen: 265–81 in Birkenmajer, K., ed., *Polish Spitsbergen Expeditions 1957–1960*: Warsaw, Polish Academy of Sciences. (466 pp.)

KLATKOWA, HALINA, 1965, Niecki i doliny denudacyjne w okolicach Łodzi: *Acta Geographica Lodziensia* **19**. (141 pp.)

KLIEWER, R. M., 1973, A general solution for the two-dimensional, transient heat conduction problem in permafrost, using implicit, finite difference methods: 41–51 in *North American Contribution, Permafrost Second International Conference* (Yakutsk, USSR, 13–28 July 1973): Washington, DC, Natl. Acad. Sci. (783 pp.)

KLIMASZEWSKI, MIECZYSŁAW, 1960, Geomorphological studies of the western part of Spitsbergen between Kongsfjord and Eidembukta: *Uniwersytetu Jagiellońskiego, Zeszytu Naukowe, Prace Geograficzne, Seria Nowa, Zeszyt* **1**, *Prace Instytutu Geograficznego Zeszyt* **23**, 91–167.

KNECHTEL, M. M., 1951, Giant playa-crack polygons in New Mexico compared with arctic tundra-crack polygons (abs.): *Geol. Soc. America Bull.* **62**, 1455.

1952, Pimpled plains of eastern Oklahoma: *Geol. Soc. America Bull.* **63**, 689–700.

KNIGHT, G. R., 1971, Ice wedge cracking and related effects on buried pipelines: 384–95 in Burdick, J. L., ed., Symposium on Cold Regions Engineering (Univ. Alaska, 17–19 Aug. 1970) Proc. **1**: College, Alaska Section, Am. Soc. Civil Eng., and Univ. Alaska. (395 pp.)

KOAZE, TAKASHI, NOGAMI, MICHIO, and IWATA, SHUJI, 1974*a*, Paleoclimatic significance of fossil periglacial phenomena in Hokkaidô, northern Japan:

The Quaternary Research **12**(4), 177–91.

1974*b*, Ice-wedge casts in eastern Hokkaido, Japan: *Tokyo Geog. Soc. J. Geography* **83**(1), 48–60.

KOERNER, R. M., 1961, Glaciological observations in Trinity Peninsula, Graham Land, Antarctica: *J. Glaciology* **3**, 1063–74.

KOJAN, EUGENE, 1967, Mechanics and rate of natural soil creep: *Idaho Dept. Highways, 5th Annual Engineering Geology and Soils Engineering Symposium (Pocatello, Idaho, 1967) Proc.*, 233–53.

KOKKONEN, P., 1930, Beobachtungen über die durch den Bodenfrost verursachte Hebung der Erdoberfläche und in der Frostschicht befindlicher Gegenstände: *Maataloustieteellinen Aikakauskirja* **3**, 84–100.

KONISHCHEV, V. N., 1973, Kriogennoye vyvetrivaniye: 38–45 in *Akademiya Nauk SSSR, Sektsiya Nauk o Zemle, Sibirskoye Otdeleniye, II Mezhdunarodnaya Konferentsiya po Merzlotovedeniyu, Doklady i soobshcheniya* **3** (Genezis, sostav i stroyeniye merzlykh tolshch i podzemnyye l'dy): Yakutsk, Yakutskoye Knizhnoye Izdatel'stvo. (102 pp.)

1978, Frost weathering [Kriogennoye vyvetrivaniye]: 176–81 in Sanger, F. J., ed., *USSR Contribution Permafrost Second International Conference* (Yakutsk, USSR, 13–28 July 1973): Washington, DC, Natl. Acad. Sci. (866 pp.)

KONISHCHEV [KONIŠČEV], V. N., FAUSTOVA, M. A., and ROGOV, V. V., 1973, Cryogenic processes as reflected in ground microstructure: *Biuletyn Peryglacjalny* **22**, 213–19.

KONISHCHEV [KONISHTCHEV], V. N., ROGOV, V. V., and SHCHURINA, G. N., 1975, Cryogenic transformation of clayey sediment rocks: *Fondation Française d'Études Nordiques, 6e Congrès International, 'Les Problèmes Posés par la Gélifraction. Recherches Fondamentales et Appliquées'* (Le Havre, 23–25 Apr. 1975) *Rept.* **105**. (Preprint.)

KÖPPEN, W., 1936, Das geographische System der Klimate: C1–44 in Köppen, W., and Geiger, R., eds., *Handbuch der Klimatologie* **1**: Berlin, Gebrüder Borntraeger.

KÖPPEN-GEIGER, 1954, *Klima der Erde – Climate of the earth* (map – 1 : 16,000,000): Darmstadt, Germany, Justus Perthes.

KORNILOV, B. A., 1962, *Rel'ef yugovostochnoy okrainy Aldanskogo nagorya*: Moskva. (96 pp.)

KOROFEYEV, N. V., 1939, K voprosu genezisa nagornykh terras: *Problemy Arktiky* **6**, 89–91.

KORZHUYEV, S. S., 1959, Geomorfologiya severo-zapadnoy chasti Stanovogo khrebta i yeye yuzhnogo obramleniya: *Instituta geografii AN SSSR, Trudy* **78**, 74–123.

KOSTYAYEV [KOST'JAEV, KOSTYAEV], A. G., 1965, Ledanyye zhily i konvektivnaya neustoychivost' gruntov: 133–40 in Popov, A. I., ed., *Podzemnyy led (Underground ice – Issue 1*). K VII Mezhdunarodnomu Kongressu Assotsiatsii po Izucheniyu Chetvertichnogo Perioda (INKVA) v SSHA, 1965: Moskva, Izdatel'stvo Moskovskogo Universiteta. (217 pp.)

1966, Über die Grenze der unterirdischen Vereisung und die Periglazialzone im Quartär: *Petermanns Mitt.* **110**(4), 253–61.

1969, Wedge- and fold-like diagenetic disturbances in Quaternary sediments and their paleographic significance: *Biuletyn Peryglacjalny* **19**, 231–70.

1973, Some rare varieties of stone circles: *Biuletyn Peryglacjalny* **22**, 347–52.

KOZARSKI, STEFAN, 1974, Evidences of Late-Würm permafrost occurrence in North-West Poland: *Quaestiones Geographicae* **1**, 65–86.

KOZARSKI, STEFAN, and SZUPRYCZYŃSKI, JAN, 1973, Glacial forms and deposits in the Sidujökull deglaciation area: 255–311 in Galon, Rajmund, ed., Scientific results of the Polish Geographical Expedition to Vatnajökull (Iceland): *Geographia Polonica* **26**. (312 pp.)

KOZIAR, ANDREW, and STRANGWAY, D. W., 1978, Permafrost mapping by audiofrequency magnetotellurics: *Canadian J. Earth Sci.* **15**, 1539–46.

KOZLOV, M. T., 1966, K voprosu ob obrazovanii nagornykh terras: *Formirovaniye rel'efa i chetvertichnykh otlozheniy Kolskogo Poluostrova*, 126–32.

KOZMIN, N. M., 1890, O lednikovykh yavleniyakh v Olekminsko-Vitimskoy gornoy strane i o svyazi ikh s obrazovaniyem zolotnoshykh rossypey: *Imperatorskogo Russkogo Geograficheskogo obshchestva, Vostochno – Sibirskogo otdela, Izvestiya* **21**(1).

KRAJEWSKI, KAZIMIERZ, 1977, Późnoplejstoceńskie i Holoceńskie procesy wydmotwórcze w Pradolinie Warszawsko-Berlińskiej w widłach Warty i Neru: *Acta Geographica Lodziensia* **39**. (87 pp.)

KRÁL, V., 1968, Geomorfologie vrcholové oblasti Krušných hor a problém paroviny: *Československé Akademie Věd Rozpravy, Řada Matematických a Přírodních Věd* **78**(9), 1–65.

KRASNOV, I. I., and KOZLOVSKAYA, S. F., 1966, *Geologiya Sibirskoy platformy*, Moskva, 399–419.

KRINSLEY, DAVID, and CAVALLERO, LILLIAN, 1970, Scanning electron microscopic examination of periglacial eolian sands from Long Island: *J. Sed. Petrology* **40**, 1345–50.

KROPACHEV, A. M., and KROPACHEVA, T. S., 1956, Nagornyye terrasy odnogo iz rayonov vostochnogo Zapolyarya: *Ministerstvo vysshego obrazovaniya*

SSSR, Molotovskiy Gornyy institut, Nauchnyye Trudy, Sbornik **1**, 126–35.

KRUMME, OSKAR, 1935, Frost und Schnee in ihrer Wirkung auf den Boden im Hochtaunus: *Rhein-Mainische Forschungen* **13**. (73 pp.)

KRYUCHKOV, V. V., 1968, Pochvy kraynego severa nado berech': *Priroda* **12**, 72–4.

1969, Soils of the far north should be conserved: 11–13 in Brown, Jerry, Rickard, Warren, and Vietor, Donald, The effect of disturbance on permafrost terrain: *US Army Corps of Engineers, Cold Regions Research and Engineering Laboratory Spec. Rept.* **138**. (13 pp.)

1976a, *Chutkaya subarktika*: Moskva, Izdatel'stvo 'Nauka'. (137 pp.)

1976b, Sensitive subarctic [Chutkaya subarktika]: *US Army Corps of Engineers, Cold Regions Research and Engineering Laboratory Draft Translation* **556**. (129 pp.)

KUDRYAVTSEV, V. A., 1966, Theory of the development of frozen rock masses: 106–8 in *Permafrost International Conference* (Lafayette, Ind., 11–15 Nov. 1963) *Proc.*: Natl. Acad. Sci.–Natl. Research Council Pub. **1287**. (563 pp.)

ed., 1974, *Osnovy merzlotnogo prognoza pri inzhenerno-geologicheskikh issledovaniyakh*: Moskva, Izdatel'stvo Moskovskogo Universiteta. (431 pp.)

ed., 1978, *Obshcheye merzlotovedeniya (Geokriologiya)*, Izd. 2: Moskva, Izdatel'stvo Moskovskogo Universiteta. (464 pp.)

KUDRYAVTSEV, V. A., et al., 1977, Fundamentals of frost forecasting in geological engineering investigations [Osnovy merzlotnogo prognoza pri inzhenerno-geologicheskikh issledoviyakh]: *US Army Corps of Engineers, Cold Regions Research and Engineering Laboratory Draft Translation* **606**. (489 pp.)

KUDRYAVTSEV, V. A., KONDRAT'YEVA, K. A., and ROMANOVSKIY, N. N., 1978, Zonal'nyye i regional'nyye zakonomernosti formirovaniya kriolitozony SSSR – Zonal and regional patterns of formation of the permafrost region in the U.S.S.R.: 419–26 in *Third International Conference on Permafrost* (Edmonton, Alta., 10–13 July 1978), Proc. **1**: Ottawa, Canada Natl. Research Council. (947 pp.)

KUENEN, P. H., 1960, Experimental abrasion 4: Eolian action: *J. Geol.* **68**, 427–49.

KUENEN, P. H., and PERDOK, W. G., 1962, Experimental abrasion 5. Frosting and defrosting of quartz grains: *J. Geol.* **70**, 648–58.

KUHLE, MATTHIAS, 1974, Vorläufige Ausführungen morphologischer Feldarbeitsergebnisse aus dem

S/E-Iranischen Hochgebirge am Beispiel des Kuh-I-Jupar: *Zeitschr. Geomorphologie N.F.* **18**, 472–83.

1978a, Über Periglazialerscheinungen im Kuh-I-Jupar (SE-Iran) und im Dhaulagiri-Himalaya (Nepal) sowie zum Befund einer Solifluktions-obergrenze: 289–309 in Association Géographique d'Alsace, *Colloque sur le Périglaciaire d'Altitude du Domaine Méditerranéen et Abords* (Strassbourg – Université Louis Pasteur, 12–14 May 1977): Strassbourg, Assoc. Géog. d'Alsace. (366 pp.)

1978b, Obergrenze von Frostbodenerscheinungen: *Zeitschr. Geomorphologie N. F.* **22**, 350–6.

1978c, Über Solifluktion und Strukturböden in S.E.-Iranischen Hochgebirgen: *Biuletyn Peryglacjalny* **27**, 117–31.

KUNSKÝ, J., and LOUČEK, D., 1956, Stone stripes and thufurs in the Krkonoše: *Biuletyn Peryglacjalny* **4**, 345–9.

KUSHEV, S. L., 1957, Rel'ef i prirodnyye usloviya Tuvinskoy avtonomnoy oblasti: *Tuvinskoy kompleksnoy ekspedicii, Trudy* **3**, 11–45.

KUZ'MIN, R. O., 1977, K voprosu o stroyenii kriolitosfery Marsa: 7–25 in Popov, A. I., ed., *Problemy kriolitologii* **6**: Izdatel'stvo Moskovskogo Universiteta. (232 pp.)

LABA, J. T., 1970, Lateral thrust in frozen granular soils caused by temperature change: 27–37 in Highway Research Board, *Frost action: Bearing, thrust, stabilization, and compaction: Natl. Acad. Sci.-Natl. Acad. Eng., Highway Research Record* **304**, (51 pp.)

LACHENBRUCH, A. H., 1957, Thermal effects of the ocean on permafrost: *Geol. Soc. America Bull.* **68**, 1515–29.

1959, Periodic heat flow in a stratified medium with application to permafrost problems: *US Geol. Survey Bull.* **1083**-A, 1–36.

1961, Depth and spacing of tension cracks: *J. Geophys. Res.* **66**(12), 4273–92.

1962, Mechanics of thermal contraction cracks and ice-wedge polygons in permafrost: *Geol. Soc. America Spec. Paper* **70**. (69 pp.)

1966, Contraction theory of ice wedge polygons: A qualitative discussion: 63–71 in *Permafrost International Conference* (Lafayette, Ind., 11–15 Nov. 1963) *Proc.*: Natl. Acad. Sci.-Natl. Research Council Pub. **1287**. (563 pp.)

1970a, Some estimates of the thermal effects of a heated pipeline in permafrost: *US Geol. Survey Circ.* **632**. (23 pp.)

1970b, Thermal considerations in permafrost: J1–2, and discussion J2–5 in Adkison, W. L., and

Borsge, M. M., eds. *Geological Seminar on the North Slope of Alaska Proc.*: Los Angeles, Am. Assoc. Petroleum Geologists, Pacific Sec., A1–R10.

LACHENBRUCH, A. H., *et al.*, 1962, Temperatures in permafrost: 791–803 in *Temperature – Its measurement and control in science and industry* **3**, Part 1: New York, Reinhold Publishing Corp.

LACHENBRUCH, A. H., and MARSHALL, B. V., 1969, Heat flow in the Arctic: *Arctic* **22**, 300–11.

1977, Sub-sea temperatures and a simple tentative model for offshore permafrost at Prudhoe Bay, Alaska: *US Geol. Survey Open-File Rept.* **77–395**. (54 pp.)

LADURIE, E. L. R., 1971, *Times of feast, times of famine. A history of climate since the year 1000*: Garden City, NY, Doubleday & Company, Inc. (426 pp.)

LAGAREC, DANIEL, 1973, Postglacial permafrost features in eastern Canada: 126–31 in *North American Contribution, Permafrost Second International Conference* (Yakutsk, USSR, 13–28 July 1973): Washington, DC, Natl. Acad. Sci. (783 pp.)

LAMAKINI, V. V., and LAMAKINI. N. V., 1930, Sayano-Dzhidinskoye nagorye: *Zemlevedeniye* **32**(1–2), 21–54.

LAMBE, T. W., and KAPLAR, C. W., 1971, Additives for modifying the frost susceptibility of soils (1): *US Army Corps of Engineers, Cold Regions Research and Engineering Laboratory Tech. Rept.* **123**(1). (41 pp.)

LAMBE, T. W., KAPLAR, C. W., and LAMBIE, T. J., 1969, Effect of mineralogical composition of fines on frost susceptibility of soils: *US Army Corps of Engineers, Cold Regions Research and Engineering Laboratory Tech. Rept.* **207**. (31 pp.)

1971, Additives for modifying the frost susceptibility of soils (2): *US Army Corps of Engineers, Cold Regions Research and Engineering Laboratory Tech. Rept.* **123**(2). (41 pp.)

LAMONTHE, CLAUDE, and ST-ONGE, DENIS, 1961, A note on a periglacial erosional process in the Isachsen area, NWT: *Geog. Bull.* **16**, 104–13.

LANG, W. B., 1943, Gigantic drying cracks in Animas Valley, New Mexico: *Science*, New Ser. **98**, 583–4.

LASCA, N. P., 1969, The surficial geology of Skeldal, Mesters Vig, Northeast Greenland: *Medd. om Grønland* **176**(3). (56 pp.)

LAURSEN, DAN, 1950, The stratigraphy of the marine Quaternary deposits in West Greenland: *Medd. om Grønland* **151**(1). (142 pp.)

LAUTRIDOU, J. P., 1971, Conclusions générales des expériences de gélifraction expérimentale: 63–84 in Recherches de gélifraction expérimentale du

Centre de Géomorphologie – V: *CNRS, Centre de Géomorphologie de Caen Bull.* **10**. (84 pp.)

1975, Les recherches de gélifraction expérimentale du Centre de Géomorphologie du C.N.R.S.: *Fondation Française d'Études Nordiques, 6ᵉ Congrès International, 'Les Problèmes Posés par la Gélifraction. Recherches Fondamentales et Appliquées'* (Le Havre, 23–25 Apr. 1975), *Rept.* **106**. (6 pp.)

LAUTRIDOU, D. P., and RAGOT, J. P., 1977, Essais de gel sur les echantillons de Monsieur Ragot: *Centre de Géomorphologie du CNRS, Laboratoire de Cryoclastie Rept.* **6**. (15 pp.)

LAWSON, D. E., and BROWN, JERRY, 1978, Disturbance of permafrost, massive ground ice, and surficial materials: 14–29 in Lawson, D. E., *et al.*, Tundra disturbance and recovery following the 1949 exploratory drilling, Fish Creek, Northern Alaska: *US Army Corps of Engineers, Cold Regions Research and Engineering Laboratory Report* **78–28**. (81 pp.)

LAZAREV, P. A., 1961, Kratkiy geomorfologicheskiy ocherk khrebta Tuora-Sis: *Voprosy geografii Yakutii* **1**, 5–11.

LECONTE, JOSEPH, 1873, On the great lava-flood of the Northwest, and on the structure and age of the Cascade Mountains: *California Acad. Sci. Proc.* **5**, 214–20.

1874, On the great lava-flood of the Northwest; and on the age of the Cascade Mountains: *Am. J. Sci. and Arts, 3d Ser.*, **7** (Whole Ser. **107**), 167–80, 259–67.

LEE, H. A., 1957, Surficial geology of Fredericton, York and Sunbury counties, New Brunswick: *Canada Geol. Survey Paper* **56–2**. (11 pp.)

LEFFINGWELL, E. DE K., 1915, Ground-ice wedges; the dominant form of ground-ice on the north coast of Alaska: *J. Geol.* **23**, 635–54.

1919, The Canning River Region, northern Alaska: *US Geol. Survey Prof. Paper* **109**. (251 pp.)

LEGGET, R. F., 1966, Permafrost in North America: 2–6 in *Permafrost International Conference* (Lafayette, Ind., 11–15 Nov. 1963) *Proc.*: Natl. Acad. Sci.–Natl. Research Council Pub. **1287**. (563 pp.)

1973, Soil: *Geotimes* **18**(9), 38–9.

LEGGET, R. F., BROWN, R. J. E., and JOHNSTON, G. H., 1966, Alluvial fan formation near Aklavik, Northwest Territories, Canada: *Geol. Soc. America Bull.* **77**, 15–29.

LESHCHIKOV, F. N., and RYASHCHENKO, T. G., 1973, Izmeneniye sostava i svoystv glinistykh gruntov pri promerzanii: 76–9 in *Akademiya Nauk SSSR, Sektsiya Nauk o Zemle, Sibirskoye Otdeleniye, II Mezhdunarodnaya Konferentsiya po Merzlotovedeniyu, Doklady i soobshcheniya* **3** (Genezis, sostav i stroy-eniye merzlykh tolshch i podzemnyye l'dy): Yakutsk, Yakutskoye Knizhnoye Izdatel'stvo. (102 pp.)

1978, Changes in the composition and properties of clay soils during freezing [Izmeneniye sostava i svoystv glinistykh gruntov pri promerzanii]: 201–3 in Sanger, F. J., ed., *USSR Contribution Permafrost Second International Conference* (Yakutsk, USSR, 13–28 July 1973): Washington, DC, Natl. Acad. Sci. (866 pp.)

LEWELLEN, R. I., 1970, *Permafrost erosion along the Beaufort Sea coast*: [Published by author] (25 pp.)

1973, The occurrence and characteristics of near-shore permafrost, northern Alaska: 131–6 in *North American Contribution, Permafrost Second International Conference* (Yakutsk, USSR, 13–28 July 1973): Washington, DC, Natl. Acad. Sci. (783 pp.)

1974, Offshore permafrost of Beaufort Sea, Alaska: 417–26 in Reed, J. C., Sater, J. E., and Gunn, W. W., eds., *The coast and shelf of the Beaufort Sea (Beaufort Sea Coast and Shelf Research Symposium Proc.)*: Arctic Inst. North America. (750 pp.)

LEWIS, C. A., 1966, The nivational landforms and the reconstructed snowline of Slaettaratindur, Faeroe Islands: *Biuletyn Peryglacjalny* **15**, 293–302.

LEWIS, W. V., 1939, Snowpatch erosion in Iceland: *Geog. J.* **94**, 153–61.

LEWKOWICZ, A. G., DAY, T. J., and FRENCH, H. M., 1978, Observations on slopewash processes in an arctic tundra environment, Banks Island, District of Franklin: *Canada Geol. Survey Paper* **78–1**A, 516–20.

LIEDTKE, HERBERT, 1975, 76–87 in *Die nordischen Vereisungen in Mitteleuropa*: Forschungen zur deutschen Landeskunde **204**. (160 pp.)

LIESTØL, OLAV, 1977, Pingos, springs, and permafrost in Spitsbergen: *Norsk Polarinstitutt Årbok* **1975**, 7–29.

LIMBERT, D. W. S., 1977*a*, Climatological summary for 1973: *British Antarctic Survey Bull.* **45**, 131–7.

1977*b*, Climatological summary for 1974: *British Antarctic Survey Bull.* **45**, 139–45.

LINDSAY, J. F., 1973, Ventifact evolution in Wright Valley, Antarctica: *Geol. Soc. America Bull.* **84**, 1791–7.

LINELL, K. A., 1973, Long-term effects of vegetative cover on permafrost stability in an area of discontinuous permafrost: 688–93 in *North American Contribution, Permafrost Second International Conference* (Yakutsk, USSR, 13–28 July 1973): Washington, DC, Natl. Acad. Sci. (783 pp.)

LINELL, K. A., and JOHNSTON, G. H., 1973, Engineering design and construction in permafrost regions: A review: 553–75 in *North American Contribution, Permafrost Second International Conference* (Yakutsk, USSR, 13–28 July 1973): Washington, DC, Natl. Acad. Sci. (783 pp.)

LINELL, K. A., and KAPLAR, C. W., 1959, The factor of soil and material type in frost action: 81–128 in Highway pavement design in frost areas, a symposium: Part 1. Basic considerations: *Natl. Acad. Sci.–Natl. Research Council Highway Research Board Bull.* **225**. (131 pp.)

1966, Description and classification of frozen soils: 71–7 (Appendix A) in Stearns, S. R., Permafrost (perennially frozen ground): *US Army Cold Regions Research and Engineering Laboratory, Cold Regions Science and Engineering* [Mon.] **I**(A2). (77 pp.)

LINTON, D. L., 1955, The problem of tors: *Geog. J.* **121**, 470–81 (and discussion, 482–6).

1958, The problem of tors [Correspondence]: *Geog. J.* **124**, 289–91.

1964, The origin of Pennine tors – an essay in analysis: *Zeitschr. Geomorphologie N. F.* **8**, Sonderheft, 5–24.

1969, The abandonment of the term 'periglacial': 65–70 in van Zinderen Bakker, E. M., ed., *Palaeoecology of Africa and of the surrounding islands and Antarctica* **5**: Cape Town, Balkema. (240 pp.)

LLIBOUTRY, LOUIS, 1955, L'origine des sols striés et polygonaux des Andes de Santiago (Chili): *Acad. Sci. Paris Comptes Rendus* **240**, 1793–4.

1956, *Nieves y glaciales de Chile. Fundamentos de glaciologia*: Santiago, Universidad de Chile. (471 pp.)

1961, Phénomènes·cryonivaux dans les Andes de Santiago (Chili): *Biuletyn Peryglacjalny* **10**, 209–24.

1965, *Traité de glaciologie* **2**, *Glaciers, variations du climat, sols gelés*: Paris, Masson & Cie. (1040 pp.)

1974, Microstriated ground in the Andes: *J. Glaciology* **13**(68), 322.

LÖHR, H., and BRUNNACKER, K., 1974, Metternicher und Eltviller Tuffhorizont im Würmlöss am Mittel- und Niederrhein: *Hessisches Landesamt für Bodenforschung Notizblatt* **102**, 168–90.

LONGWELL, C. R., 1928, Three common types of desert mud-cracks: *Am. J. Sci.*, 5th ser. **15**, 136–45.

LONGWELL, C. R., FLINT, R. F., and SANDERS, J. E., 1969, *Physical geology*: New York, John Wiley. (685 pp.)

LONGWELL, C. R., KNOPF, A., and FLINT, R. F., 1939, *A textbook of geology, part I–Physical geology*, 2d (revised) ed.: New York, John Wiley. (543 pp.)

LORENZO, J. L., 1969, Minor periglacial phenomena among the high volcanoes of Mexico: 161–75 in Péwé, T. L., ed., *The periglacial environment*: Montreal, McGill–Queen's Univ. Press. (487 pp.)

LÖSCHE, HEINRICH, 1930, Lassen sich die diluvialen Breitenkreise aus klimabedingten Vorzeitformen rekonstruieren?: *Deutsche Seewarte (Hamburg), Archiv der Deutschen Seewarte* **48**(7). (39 pp.)

LÖTSCHERT, WILHELM, 1972, Über die Vegetation frostgeformter Böden auf Island: *Natur u. Museum* **102**(1), 1–12.

LOUIS, HERBERT, 1930, Morphologische Studien in Südwest-Bulgarian: *Geog. Abh.* [*Pencks*], 3d ser. **2**. (119 pp.)

LÖVE, DORIS, 1970, Subarctic and subalpine: Where and what?: *Arctic and Alpine Research* **2**, 63–73.

LOVELL, C. W., JR., 1957, Temperature effects on phase composition and strength of partially-frozen soil: 74–95 in Fundamental and practical concepts of soil freezing: *Natl. Acad. Sci.–Natl. Research Council Highway Research Board Bull.* **168**. (205 pp.)

LOW, P. F., HOEKSTRA, PIETER, and ANDERSON, D. M., 1967, Some thermodynamic relationships for soils at or below the freezing point **2**: Effects of temperature and pressure on unfrozen soil water: *US Army Cold Regions Research and Engineering Laboratory Research Rept.* **222**. (5 pp.)

ŁOZIŃSKI, W., 1909, Über die mechanische Verwitterung der Sandsteine im gemässigten Klima: *Acad. sci. cracovie Bull. internat., cl. sci. math. et naturelles* **1**, 1–25.

1912, Die periglaziale Fazies der mechanischen Verwitterung: *Internat. Geol. Congress, 11th, Stockholm 1910, Compte rendu*, 1039–53.

LUCKMAN, B. H., 1971, The role of snow avalanches in the evolution of alpine talus slopes: 93–110 in Brunsden, D., compiler, *Slopes – form and process*: Inst. Brit. Geographers Spec. Pub. **3**. (178 pp.)

1976, Rockfalls and rockfall inventory data: Some observations from Surprise Valley, Jasper National Park, Canada: *Earth Surface Processes* **1**, 287–98.

1977, The geomorphic activity of snow avalanches: *Geografiska Annaler* **59**A(1–2), 31–48.

1978, Geomorphic work of snow avalanches in the Canadian Rocky Mountains: *Arctic and Alpine Research* **10**, 261–76.

LUCKMAN, B. H. and CROCKETT, J. K., 1978, Distribution and characteristics of rock glaciers in the southern part of Jasper National Park, Alberta: *Canadian J. Earth Sci.* **15**, 540–50.

LUNDQVIST, G., 1949, The orientation of the block material in certain species of earth flow: 335–47 in *Glaciers and Climate: Geografiska Annaler 1949* **1–2**.

LUNDQVIST, JAN, 1962, Patterned ground and related frost phenomena in Sweden: *Sveriges Geol. Under-*

sökning *Årsbok* **55**(7) (*Avh. och uppsatser, ser C* **583**). (101 pp.)

1969, Earth and ice mounds: a terminological discussion: 203–15 in Péwé, T. L., ed., *The periglacial environment*: Montreal, McGill–Queen's Univ. Press. (487 pp.)

LÜTTIG, GERD, 1960, Neue Ergebnisse quartärgeologischer Forschung im Raume Alfeld-Hameln-Elze: *Geol. Jahrb.* **77**, 337–90.

LYFORD, W. H., GOODLETT, J. C., and COATES, W. H., 1963, Landforms, soils with fragipans, and forest on a slope in the Harvard Forest: *Harvard Forest Bull.* **30**. (68 pp.)

LYUBIMOVA, E. L., 1955, Botaniko-geograficheskiye issledovaniya yuzknoy chasti Pripolyarnogo Urala: *Instituta geografii AN SSSR, Trudy* **64**, 201–41.

MAARLEVELD, G. C., 1960, Wind directions and cover sands in the Netherlands: *Biuletyn Peryglacjalny* **8**, 49–58.

1965, Frost mounds, a summary of the literature of the past decade: *Mededelingen van de Geologische Stichting, Nieuwe Serie* **17**. (16 pp.)

1976, Periglacial phenomena and the mean annual temperature during the last glacial time in the Netherlands: *Biuletyn Peryglacjalny* **26**, 57–78.

MACCARTHY, G. R., 1952, Geothermal investigations on the arctic slope of Alaska: *Am. Geophys. Union Trans.* **33**(4), 589–93.

MACKAY, J. R., 1953, Fissures and mud circles on Cornwallis Island, NWT: *The Canadian Geographer* **3**, 31–7.

1956, Notes on oriented lakes of the Liverpool Bay area, Northwest Territories: *Rev. Canadienne Géographie* **10**(4), 169–73.

1958a, Arctic 'vegetation arcs': *Geog. J.* **124**, 294–5.

1958b, The Anderson River Map-Area, NWT: *Canada, Dept. Mines and Technical Surveys, Geog. Branch Mem.* **5**. (137 pp.)

1962, Pingos of the Pleistocene Mackenzie River delta area: *Geog. Bull.* **18**, 21–63.

1963a, The Mackenzie Delta area, NWT: *Canada, Dept. Mines and Tech. Surveys, Geog. Branch Mem.* **8**. (202 pp.)

1963b, Origin of the pingos of the Pleistocene Mackenzie Delta area: *Canada Natl. Research Council, Associate Comm. Soil and Snow Mechanics Tech. Memo.* **76**, 79–83.

1963c, Progress of break-up and freeze-up along the Mackenzie River: *Geog. Bull.* **19**, 103–16.

1965, Gas-domed mounds in permafrost, Kendall Island, NWT: *Geog. Bull.* **7**, 105–15.

1966a, Pingos in Canada: 71–6 in *Permafrost International Conference* (Lafayette, Ind., 11–15 Nov. 1963) *Proc.*: Natl. Acad. Sci.–Natl. Research Council Pub. **1287**. (563 pp.)

1966b, Segregated epigenetic ice and slumps in permafrost, Mackenzie Delta area, NWT: *Geog. Bull.* **8**, 59–80.

1968, Discussion of the theory of pingo formation by water expulsion in a region affected by subsidence [Letter]: *J. Glaciology* **7**(50), 346–50.

1970, Disturbances to the tundra and forest tundra environment of the western Arctic: *Canadian Geotech. J.* **7**, 420–32.

1971a, Ground ice in the active layer and the top portion of permafrost: 26–30 in Brown, R. J. E., ed., *Proceedings of a seminar on the permafrost active layer, 4 and 5 May 1971: Canada Natl. Research Council Tech. Memo.* **103**. (63 pp.)

1971b, The origin of massive icy beds in permafrost, western arctic coast, Canada: *Canadian J. Earth Sci.* **8**, 397–422.

1972a, The world of underground ice: *Assoc. Am. Geographers Annals* **62**, 1–22.

1972b, Some observations on growth of pingos: 141–7 in Kerfoot, D. E., ed., *Mackenzie Delta area monograph*: Internat. Geog. Cong., 22d, Montreal (1972): St Catherines, Ont., Brock Univ. (174 pp.)

1972c, Some observations on ice-wedges, Garry Island, NWT: 131–9 in Kerfoot, D. E., ed., *Mackenzie Delta area monograph*: Internat. Geog. Cong., 22d, Montreal (1972): St Catherines, Ont., Brock Univ. (174 pp.)

1972d, Offshore permafrost and ground ice, southern Beaufort Sea, Canada: *Canadian J. Earth Sci.* **9**, 1550–61.

1973a, Problems in the origin of massive icy beds, western Arctic, Canada: 223–8 in *North American Contribution, Permafrost Second International Conference* (Yakutsk, USSR, 13–28 July 1973): Washington, DC, Natl. Acad. Sci. (783 pp.)

1973b, The growth of pingos, Western Arctic Coast, Canada: *Canadian J. Earth Sci.* **10**, 979–1004.

1973c, Winter cracking (1967–1973) of ice-wedges, Garry Island, NWT: *Canada Geol. Survey Paper* **73-1**, Part B, 161–3.

1973d, A frost tube for the determination of freezing in the active layer above permafrost: *Canadian Geotech. J.* **10**, 392–6.

1974a, Reticulate ice veins in permafrost, northern Canada: *Canadian Geotech. J.* **11**, 230–7.

1974b, The rapidity of tundra polygon growth and destruction, Tuktoyaktuk Peninsula – Richards Island area, NWT: *Canada Geol. Survey Paper* **74-1**, Part A, 391–2.

1974*c*, Ice-wedge cracks, Garry Island, Northwest Territories: *Canadian J. Earth Sci.* **11**, 1336–83.

1974*d*, Measurement of upward freezing above permafrost with a self-positioning thermister probe: *Canada Geol. Survey Paper* **74–1**, Part B, 250–1.

1974*e*, Seismic shot holes and ground temperatures, Mackenzie Delta area, Northwest Territories: *Canada Geol. Survey Paper* **74–1**, Part A, 389–90.

1975*a*, Relict ice wedges, Pelly Island, NWT: *Canada Geol. Survey Paper* **75–1**, Part A, 131–2.

1975*b*, Freezing processes at the bottom of permafrost, Tuktoyaktuk Peninsula area, District of Mackenzie: *Canada Geol. Survey Paper* **75–1**, Part A, 471–4.

1975*c*, The closing of ice-wedge cracks in permafrost, Garry Island, Northwest Territories: *Canadian J. Earth Sci.* **12**, 1668–74.

1975*d*, Reticulate ice veins in permafrost, northern Canada: Reply: *Canadian Geotech. J.* **12**, 163–5.

1975*e*, Some resistivity surveys of permafrost thickness, Tuktoyaktuk Peninsula, NWT: *Canada Geol. Survey Paper* **75–1**B, 177–80.

1975*f*, The stability of permafrost and recent climatic change in the Mackenzie Valley, NWT: *Canada Geol. Survey Paper* **75–1**B, 173–6.

1976*a*, Pleistocene permafrost, Hooper Island, Northwest Territories: *Canada Geol. Survey Paper* **76–1**A, 17–18.

1976*b*, Ice wedges as indicators of recent climatic change, Western Arctic Coast: *Canada Geol. Survey Paper* **76–1**A, 233–4.

1976*c*, Ice segregation at depth in permafrost: *Canada Geol. Survey Paper* **76–1**A, 287–8.

1976*d*, The growth of ice wedges (1966–1975), Garry Island, NWT, Canada: 180–2 in International Geographical Congress, 23d (Moscow, 27 July–3 Aug. 1976), *International Geography* **1** (*Geomorphology and Paleogeography*). (409 pp.)

1976*e*, The age of Ibuyuk Pingo, Tuktoyaktuk Peninsula, District of Mackenzie: *Canada Geol. Survey Paper* **76–1**B, 59–60.

1976*f*, On the origin of pingos – A comment: *J. Hydrology* **30**, 295–8.

1977*a*, The widths of ice wedges: *Canada Geol. Survey Paper* **77–1**A, 43–4.

1977*b*, Permafrost growth and subpermafrost pore water expulsion, Tuktoyaktuk Peninsula, District of Mackenzie: *Canada Geol. Survey Paper* **77–1**A, 323–6.

1977*c*, Probing for the bottom of the active layer: *Canada Geol. Survey Paper* **77–1**A, 327–8.

1977*d*, Pulsating pingos, Tuktoyaktuk Peninsula, NWT: *Canadian J. Earth Sci.* **14**, 209–22.

1977*e*, Changes in the active layer from 1968 to 1976 as a result of the Inuvik fire: *Canada Geol. Survey Paper* **77–1**B, 273–5.

1978*a*, The surface temperature of an ice-rich melting permafrost exposure, Garry Island, Northwest Territories: *Canada Geol. Survey Paper* **78–1**A, 521–2.

1978*b*, Contemporary pingos: A discussion: *Biuletyn Peryglacjalny* **27**, 133–54.

1978*c*, Freshwater shelled invertebrate indicators of paleoclimate in northwestern Canada during late glacial times: Discussion: *Canadian J. Earth Sci.* **15**, 461–3.

1978*d*, Sub-pingo water lenses, Tuktoyaktuk Peninsula, Northwest Territories: *Canadian J. Earth Sci.* **15**, 1219–27.

1979, An equilibrium model for hummocks (nonsorted circles), Garry Island, Northwest Territories: *Canada Geol. Survey Paper* **79–1**A, 165–7.

MACKAY, J. R., and BLACK, R. F., 1973, Origin, composition, and structure of perennially frozen ground and ground ice: A review: 185–92 in *North American Contribution, Permafrost Second International Conference* (Yakutsk, USSR, 13–28 July 1973): Washington, DC, Natl. Acad. Sci. (783 pp.)

MACKAY, J. R., KONISHCHEV, V. N., and POPOV, A. I., 1979, Geological control of the origin, characteristics, and distribution of ground ice: In *Third International Conference on Permafrost* (Edmonton, Alta., 10–13 July 1978), Proc. **2**: Ottawa, Canada Natl. Research Council. (In Press.)

MACKAY, J. R., and LAVKULICH, L. M., 1974, Ionic and oxygen isotopic fractionation in permafrost growth: *Canada Geol. Survey Paper* **74–1**, Part B, 255–6.

MACKAY, J. R., and MACKAY, D. K., 1976, Cryostatic pressures in nonsorted circles (mud hummocks), Inuvik, Northwest Territories: *Canadian J. Earth Sci.* **13**, 889–97.

1977, The stability of ice-push features, Mackenzie River, Canada: *Canadian J. Earth Sci.* **14**, 2213–25.

MACKAY, J. R., and MATHEWS, W. H., 1967, Observations on pressures exerted by creeping snow, Mount Seymour, British Columbia, Canada: 1185–97 in Oura, Hirobumi, ed., *Physics of snow and ice*, Internat. Conference Low Temperature Science (*Sapporo, Japan, 14–19 Aug. 1966*) Proc. **1**(2): Sapporo, Hokkaido Univ., 713–1414.

1974*a*, Needle ice striped ground: *Arctic and Alpine Research* **6**, 79–84.

1974*b*, Movement of sorted stripes, the Cinder Cone,

Garibaldi Park, BC, Canada: *Arctic and Alpine Research* **6**, 347–59.

1975, Orientation of soil stripes caused by needle ice: *J. Glaciology* **14**(71), 329–31.

MACKAY, J. R., RAMPTON, V. N., and FYLES, J. G., 1972, Relic Pleistocene permafrost, Western Arctic, Canada: *Science* **176**, 1321–3.

MACKAY, J. R., and STAGER, J. K., 1966, The structure of some pingos in the Mackenzie Delta area, NWT: *Geog. Bull.* **8**(4), 360–8.

MACKIN, J. H., 1947, Altitude and local relief of Bighorn area during the Cenozoic: *Wyoming Geol. Assoc. Field Conf., Bighorn Basin, Guidebook*, 103–20.

MACNAMARA, E. E., 1973, Macro- and microclimates of the Antarctic coastal oasis, Molodezhnaya: *Biuletyn Peryglacjalny* **23**, 201–36.

MAKEROV, J., 1913, Nagornyye terrasy v Sibiri i proiskhozhdeniye yikh: *Geologicheskogo komiteta, Izvestiya* **32**(8), 761–801.

MALAURIE, JEAN, 1952, Sur l'asymétrie des versants dans l'île de Disko, Groenland: *Acad. Sci. [Paris] Comptes rendus* **234**, 1461–2.

1968, Thèmes de recherche géomorphologique dans le Nord-Ouest du Groenland: *Centre de Recherches et Documentation Cartographiques et Géographiques [Paris], Mémoires et Documents, Numéro hors série.* (495 pp.)

MALAURIE, JEAN, and GUILLIEN, YVES, 1953, Le modelé cryo-nival des versants meubles de Skansen (Disko, Groenland). Interprétation générale des grèzes litées: *Soc. Géol. France Bull.*, ser. 6, **3**, 703–21.

MALDE, H. E., 1961, Patterned ground of possible solifluction origin at low altitude in the western Snake River Plain, Idaho: *US Geol. Survey Prof. Paper* **424**-B, B-170–3.

1964, Patterned ground in the western Snake River Plain, Idaho, and its possible cold-climate origin: *Geol. Soc. America Bull.* **75**, 191–208.

MANGERUD, JAN, and SKREDEN, S. A., 1972, Fossil ice wedges and ground wedges in sediments below the till at Voss, western Norway: *Norsk Geol. Tidsskr.* **52**, 73–96.

MANSIKKANIEMI, HANNU, 1970, Ice-push action on sea shores, southeastern Finland: *Turun Yliopiston Maantieteen Laitoksen Julkaisuja – Publicationes Instituti Geographici Universitatis Turkuensis* **50**. (30 pp.)

1976, Ice action on the seashore, southern Finland: observations and experiments: *Fennia* **148**, 1–17.

MARBUT, C. F., and WOODWORTH, J. B., 1896, The clays about Boston: 989–98 in Shaler, N. S.,

Woodworth, J. B., and Marbut, C. F., Glacial brick clays of Rhode Island and southeastern Massachusetts: *US Geol. Survey 17th Ann. Rept.* (1). (1076 pp.)

MARCINEK, JOACHIM, and NITZ, BERNHARD, 1973, *Das Tiefland der Deutschen Demokratischen Republik*: Gotha/Leipzig, VEB Herman Haack, Geographisch-Kartographische Anstalt. (288 pp.)

MARKOV, K. K., 1961, Sur les phénomènes périglaciaires du Pléistocène dans le territoire de l'URSS: *Biuletyn Peryglacjalny* **10**, 75–85.

MARKOV, K. K., and BODINA, E. L., 1961, Karta periglayatsial'ykh obrazovaniy Antarktiky: 53–60 in *Antarktika, Doklady Komissii 1960, Akad. Nauk SSSR, Mezhduvedomstvennaya Komissiya po Izucheniyu Antarktiki.* (88 pp.)

1966, Map of periglacial formations in Antarctica: 49–59 in *Antarctica, Commission Reports 1960, Acad. Sci. USSR. Interdepartmental Commission on Antarctic Research.* (103 pp.) (Israel Program for Scientific Translations, Jerusalem.)

MARKUSE, GERHARD, 1976, Der Dauerfrostboden in der UdSSR und Probleme seiner Nutzung: *Geographische Berichte* **79**(2), 118–31.

MARNEZY, ALAIN, 1977*a*, Glaciers rocheux et phénomènes périglaciaires dans le vallon de la Rocheure (Massif de la Vanoise): *Rev. Géog. Alpine* **65**(2), 147–65.

1977*b*, Aspects du modèle périglaciaire dans le Vallon de la Rocheure (Massif de la Vanoise): *Rev. Géog. Alpine* **65**(4), 367–84.

MARTIN, R. T., 1959, Rhythmic ice banding in soil: 11–23 in Frost effects in soils and on pavement surfaces (*Highway Research Board Bull.* **218**): *Natl. Acad. Sci.–Natl. Research Council Pub.* **671**. (48 pp.)

MARTINI, ANDRZEJ, 1967, Preliminary experimental studies on frost weathering of certain rock types from the West Sudetes: *Biuletyn Peryglacjalny* **16**, 147–94.

1973, Experimental investigations of frost weathering on granites: 61–5 in Hrádek, Mojmír, ed., Symposium of the INQUA Commission on genesis and lithology of Quaternary deposits (Poland–Czechoslovakia, 2–7 Oct. 1972): *Československá Akad. Věd, Geografický Ústav Brno Studia Geographica* **33**. (128 pp.)

MARUSZCZAK, H., 1961, Phénomènes périglaciaires dans le Pirin et sur la Vitocha (Bulgarie): *Biuletyn Peryglacjalny* **10**, 225–34.

MATHER, K. F., GOLDTHWAIT, R. P., and THIESMEYER, L. R., 1942, Pleistocene geology of western Cape Cod, Massachusetts: *Geol. Soc. America Bull.* **53**, 1127–74.

MATHEWS, W. H., 1955, Permafrost and its occurrence in the southern Coast Mountains of British Columbia: *Canadian Alpine J.* **38**, 94–8.

MATHEWS, W. H., and MACKAY, J. R., 1963, Snow-creep studies, Mount Seymour, BC: Preliminary field investigations: *Geog. Bull.* **20**, 58–75.

1975, Snow creep: Its engineering problems and some techniques and results of its investigation: *Canadian Geotech. J.* **12**, 187–98.

MATHYS, HANS, 1974, Klimatische Aspekte zur Frost-verwitterung in der Hochgebirgsregion: *Natur-forschenden Gesell. Bern, Mitt.* **31**, 49–62.

MATTHES, F. E., 1900, Glacial sculpture of the Bighorn Mountains, Wyoming: *US Geol. Survey 21st Ann. Rept.* (2), 173–90.

MATTHEWS, B., 1962, Frost-heave cycles at Scheffer-ville, October 1960–June 1961 with a critical examination of methods used to determine them: 112–24 in Field Research in Labrador–Ungava: *McGill Sub-Arctic Research Lab. Ann. Rept. 1960–1961, McGill Sub-Arctic Research Papers* **12**. (137 pp.)

MAYO, L. R., 1970, Classification and distribution of aufeis deposits, Brooks Range and Arctic Slope, Alaska (abs.): *Alaska Sci. Conf., 20th, College, Alaska 1969, Proc.*, 310.

MCARTHUR, D. S., and ONESTI, L. J., 1970, Contorted structures in Pleistocene sediments near Lansing, Michigan: *Geografiska Annaler* **52**A, 186–93.

MCCABE, L. H., 1939, Nivation and corrie erosion in west Spitsbergen: *Geog. J.* **94**, 447–65.

MCCALLIEN, W. J., RUXTON, B. P., and WALTON, B. J., 1964, Mantle rock tectonics. A study in tropical weathering at Accra, Ghana: [*Great Britain*] *Over-seas Geol. Surveys, Overseas Geol. and Mineral Resources* **9**(3), 257–94.

MCCANN, S. B., and COGLEY, J. G., 1973, The geomorphic significance of fluvial activity at high latitudes: 118–35 in Fahey, B. D., and Thompson, R. D., eds., *Research in polar and alpine geomorphology*: Guelph Symposium on Geomorphology, 1973, 3d, Proc.: Norwich, England, Geo Abstracts Ltd (Univ. East Anglia). (206 pp.)

MCCANN, S. B., HOWARTH, P. J., and COGLEY, J. G., 1972, Fluvial processes in a periglacial environment, Queen Elizabeth Islands, NWT, Canada: *Inst. British Geographers Trans.* **55**, 69–82.

MCCRAW, J. D., 1967, Some surface features of McMurdo Sound region, Victoria Land, Antarctica: *New Zealand J. Geol. and Geophys.* **10**, 394–417.

MCCULLOCH, D. S., and HOPKINS, DAVID, 1966, Evidence for an early Recent warm interval in northwestern Alaska: *Geol. Soc. America Bull.* **77**, 1089–107.

MCCULLOCH, D. S., TAYLOR, D. W., and RUBIN, MEYER, 1965, Stratigraphy, non-marine mollusks, and radiometric dates from Quaternary deposits in the Kotzebue Sound area, western Alaska: *J. Geol.* **73**, 442–53.

MCDOWALL, I. C., 1960, Particle size reduction of clay minerals by freezing and thawing: *New Zealand J. Geol. and Geophys.* **3**(3), 337–43.

MCGINNIS, L. D., NAKAO, K., and CLARK, C. C., 1973, Geophysical identification of frozen and unfrozen ground, Antarctica: 136–46 in *North American Contribution, Permafrost Second International Conference* (Yakutsk, USSR, 13–28 July 1973): Washington, DC, Natl. Acad. Sci. (783 pp.)

MCINTYRE, A., RUDDIMAN, W. F., and JANTZEN, R., 1972, Southward penetration of the North Atlantic Polar Front: faunal and floral evidence of large-scale surface water mass movements over the last 225 000 years: *Deep-Sea Research* **19**, 61–77.

MCLAREN, PATRICK, 1977, *The coasts of eastern Melville and western Byam Martin Islands: Coastal processes and related geology of a high arctic environment*: Univ. South Carolina, Ph.D. Thesis. (304 pp.)

MCROBERTS, E. C., and MORGENSTERN, N. R., 1974*a*, The stability of thawing slopes: *Canadian Geotech. J.* **11**, 447–69.

1974*b*, Stability of slopes in frozen soil, Mackenzie Valley, NWT: *Canadian Geotech. J.* **11**, 554–73.

1975, Pore water expulsion during freezing: *Canadian Geotech. J.* **12**, 130–41.

MCROBERTS, E. C., and NIXON, J. F., 1975*a*, Reticulate ice veins in permafrost, northern Canada: Discussion: *Canadian Geotech. J.* **12**, 159–62.

1975*b*, Some geotechnical observations on the role of surcharge pressure in soil freezing: *Am. Geophys. Union, Conference on Soil-Water Problems in Cold Regions (Calgary, Alberta, Canada, 6–7 May 1975), Proc.*, 42–57.

MCSAVENEY, E. R., 1972, The surficial fabric of rock-fall talus: 181–97 in Morisawa, Marie, ed., *Quantitative geomorphology: Some aspects and applications*: Second Annual Geomorphology Symposia Series (Binghamton, NY, 15–16 Oct. 1971), Proc.: Binghamton, NY, State Univ. New York (Binghamton) Pubs. in Geomorphology. (315 pp.)

MCVEE, C. V., 1973, Permafrost considerations in land use planning management: 146–51 in *North American Contribution, Permafrost Second International Conference* (Yakutsk, USSR, 13–28 July 1973): Washington, DC, Natl. Acad. Sci. (783 pp.)

MEARS, BRAINERD, JR., 1966, Ice-wedge pseudomorphs in the Laramie Basin, Wyoming (abs.): *Geol. Soc. America Spec. Paper* **87**, 295.

1973, Were Rocky Mountain intermontane basins Pleistocene tundras? (abs.): 233 in *International Union for Quaternary Research (INQUA) Cong., 9th (Christchurch, New Zealand, 2–10 Dec. 1973), Abstracts*. (418 pp.)

MECKELEIN, WOLFGANG, 1959, *Forschungen in der zentralen Sahara: 1. Klimageomorphologie*: Braunschweig, Georg Westermann Verlag. (181 pp.)

1965, Beobachtungen und Gedanken zu geomorphologischen Konvergenzen in Polar- und Wärmewüsten: *Erdkunde* **19**, 31–9.

1974, Aride Verwitterung in Polargebieten im Vergleich zum subtropischen Wüstengürtel: *Zeitschr. Geomorphologie N.F., Supp.* **20**, 178–88.

MEIER, M. F., 1965, Glaciers and climate: 795–805 in Wright, H. E., and Frey, D. G., eds., *The Quaternary of the United States*: Princeton, NJ, Princeton University Press. (922 pp.)

MEINARDUS, WILH., 1912*a*, Beobachtungen über Detritussortierung und Strukturboden auf Spitzbergen: *Gesell. Erdkunde Berlin Zeitschr.*, **1912**, 250–59.

1912*b*, Über einige charakteristische Bodenformen auf Spitzbergen: *Naturh. Ver. Preuss. Rheinlande u. Westfalens, Medizinisch-naturwiss. Gesell. Münster Sitzungsber., Sitzung* **26**, 1–42.

1923, Meteorologische Ergebnisse der Kerguelenstation 1902–1903: 341–435 in Drygalski, E. v., *Deutsche Südpolar-Expedition 1901–1903* **3** (Meteorologie, Teilbd. 1, 1 Hälfte): Berlin and Leipzig. (578 pp.)

1930, Arktische Böden: 27–96 in Blanck, E., *Handbuch der Bodenlehre* **3**: Berlin, Julius Springer. (550 pp.)

MELLOR, MALCOLM, 1970, Phase composition of pore water in cold rocks: *US Army Corps of Engineers, Cold Regions Research and Engineering Laboratory Research Rept.* **292**. (61 pp.)

1971, Strength and deformability of rocks at low temperatures: *US Army Corps of Engineers, Cold Regions Research and Engineering Laboratory Research Rept.* **294**. (73 pp.)

MELNIKOV, P. I., 1963, Itogi geokriologicheskikh i gidrogeologicheskikh inzhenerno–geokriologicheskikh issledovaniy v Tsentral'noy i yuzhnoy Yakutii: Inst. Merzlotovedeniya im. V. A. Obrucheva, Doklady o rabotakh, predstavl. na soiskaniye uchen. st. d-ra Nauk.

1977, Environmental protection task and development of permafrost region [Zadachi okhrany okruzhayushchei sredy i osvoyeniye oblasti mnogoletnemerzlykh porod]: *US Army Corps of Engineers, Cold Regions Research and Engineering Laboratory Draft Translation* **617**. (18 pp.)

MELNIKOV, P. I., and BALOBAYEV, V. T., 1977, Permafrost at the territory of the USSR: 47–8 in Brown, Jerry, ed., Symposium: Geography of polar countries (Selected papers and summaries – 23d International Geographical Congress, Leningrad, USSR, 22–26 July 1976): *US Army Corps of Engineers, Cold Regions Research and Engineering Laboratory Spec. Rept.* **77–6**. (61 pp.)

MELNIKOV, P. I., and TOLSTIKHIN, N. I., eds., 1974, *Obshcheye merzlotovedeniye*: Novosibirsk, Sibirskoye Otdeleniye, Izdatel'stvo 'Nauka'. (291 pp.)

1979, Merzlotno-gidrogeologicheskoye rayonirovaniye Vostochnoy Sibiri – Permafrost and hydrogeological provinces of eastern Siberia: In *Third International Conference on Permafrost* (Edmonton, Alta., 10–13 July 1978), Proc. **2**: Ottawa, Canada Natl. Research Council. (In press.)

MELTON, F. A., 1954, 'Natural mounds' of northeastern Texas, southern Arkansas, and northern Louisiana: *Oklahoma Geol. Survey, The Hopper* **14**(7), 87–121.

MENSCHING, HORST, 1977, Bemerkungen zum Problem einer 'periglazialen' Höhenstufe in den Gebirgen der ariden Zone im nördlichen Afrika: 290–99 in Poser, Hans, ed., Formen, Formengesellschaften und Untergrenzen in den heutigen periglazialen Höhenstufen der Hochgebirge Europas und Afrikas zwischen Arktis und Äquator. Bericht über ein Symposium: *Akad. Wiss. Göttingen Abh., Math.–Phys. Kl. Folge 3,* **31**. (355 pp.)

MERRILL, R. K., and PÉWÉ, T. L., 1977, Late Cenozoic geology of the White Mountains, Arizona: *Arizona Bureau Geology and Mineral Technology Spec. Paper* **1**. (65 pp.)

MERTIE, J. B., JR., 1937, The Yukon–Tanana Region, Alaska: *US Geol. Survey Bull.* **872**. (276 pp.)

MESSENGER, J. A., and PÉWÉ, T. L., 1977, Paleogeographic maps of Alaska for the Wisconsinan (abs.): *Geol. Soc. America Abs. with Programs* **9**(6), 748–9.

MESSERLI, BRUNO, 1965, *Beiträge zur Geomorphologie der Sierra Nevada (Andalusien)*: Univ. Bern, dissertation. (178 pp.)

MICHALEK, D. D., 1969, *Fanlike features and related periglacial phenomena of the southern Blue Ridge*: Univ. North Carolina (Chapel Hill), Ph.D. Thesis. (198 pp.)

MICHAUD, JEAN, 1950, Emploi de marques dans

l'étude des mouvements du sol: *Rev. Géomorphologie dynamique* **1**, 180–9.

MICHAUD, JEAN, and CAILLEUX, ANDRÉ, 1950, Vitesses des mouvements du sol au Chambeyron (Basses-Alpes): *Acad. Sci. Paris Comptes rendus* **230**, 314–15.

MICHEL, J. P., 1975, Périglaciaire des environs de Paris: *Biuletyn Peryglacjalny* **24**, 259–352.

MIDDENDORFF, A. T. V., 1853, Zusatz: 312–16 in Ditmar, C. v., 1853, Ueber die Eismulden im östlichen Sibirien (Nakipni der Sibirischen Russen): *Akademie Impériale des Sciences de St.-Pétersbourg, Bulletin (Classe Phys.-Math.)* **11**(19–20), 306–16.

1861, Das Klima Sibiriens: 333–523 in *Dr. A. v. Middendorff's Sibirische Reise*, **4**, Part 1 (Uebersicht der Natur Nord- und Ost-Sibiriens): St Petersburg, Kaiserlichen Akademie der Wissenschaften. (783 pp.)

MIDDLETON, CHRISTOPHER, 1743, The effects of cold; together with observations of the longitude, latitude, and declination of the magnetic needle, at Prince of Wales Fort, upon Churchill-River in Hudson's Bay, North America [Read 1742]: *Royal Soc. London Phil. Trans.* **42**, 157–71.

MILLER, R., COMMON, R., and GALLOWAY, R. W., 1954, Stone stripes and other surface features of Tinto Hill: *Geog. J.* **80**(2), 216–19.

MILLER, R. D., 1966, Phase equilibria and soil freezing: 193–7 in *Permafrost International Conference* (Lafayette, Ind., 11–15 Nov. 1963) *Proc.*: Natl. Acad. Sci.–Natl. Res. Council Pub. **1287**. (563 pp.)

1972, Freezing and heaving of saturated and unsaturated soils: 1–11 in Highway Research Board, *Frost action in soils*: *Natl. Acad. Sci.–Natl. Acad. Eng., Highway Research Record* **393**. (88 pp.)

1977, Lens initiation in secondary heaving: 68–74 in *Frost action in soils*: Univ. Luleå, Internat. Symposium (Luleå, Sweden, 16–18 Feb. 1977) Proc. 2. (119 pp.)

1978, Frost heaving in non-colloidal soils: 707–13 in *Third International Conference on Permafrost* (Edmonton, Alta., 10–13 July 1978), Proc. **1**: Ottawa, Canada Natl. Research Council. (947 pp.)

MILLER, R. D., LOCH, J. P. G., and BRESLER, E., 1975, Transport of water and heat in a frozen permeameter: *Soil Sci. Soc. America Proc.* **39**(6), 1029–36.

MILORADOVICH, B. V., 1936, Geologicheskiy ocherk severo-vostochnogo poberezhya severnogo ostrova Novoy Zemli: *Arkticheskogo instituta, Trudy* **38**, 51–121.

MILTON, D. J., 1974, Mysterious Martian flood channels linked to 'dry ice': *US Geol. Survey News release 23 June*. (Mimeo., 3 pp.)

MITCHELL, G. F., 1971, Fossil pingos in the south of Ireland: *Nature* **230**, 43–4.

1973, Fossil pingos in Camaross Townland, Co. Wexford: *Royal Irish Acad. Proc.* **73**(B)(16), 269–82.

1977, Periglacial Ireland: *Royal Soc. London Phil. Trans.* **B280**, 199–209.

MOHAUPT, WILLI, 1932, *Beobachtungen über Bodenversetzungen und Kammeisbildungen aus dem Stubai und dem Grödener Tal*: Univ. Hamburg Thesis, Hamburg, Hans Christians Druckerei und Verlag. (54 pp.)

MOIGN, ANNIK, 1976, L'action des glaces flottantes sur le littoral et les fonds marins du Spitsberg central et nord-occidental: *Rev. Géog. Montréal* **30**(1–2) (*Le glaciel* – Premier colloque international sur l'action géologique des glaces flottantes, Québec, Canada, 20–24 Apr. 1974), 51–64.

MOLOCHUSHKIN, E. N., 1973, Vliyaniye termoabraznii na temperaturu mnogoletnemerzlykh porod v pribrezhnoy zone morya Laptevykh: 52–8 in *Akademiya Nauk SSSR, Sektsiya Nauk o Zemle, Sibirskoye Otdeleniye, II Mezhdunarodnaya Konferentsiya po Merzlotovedeniyu, Doklady i soobshcheniya* **2** (Regional'naya geokriologiya): Yakutsk, Yakutskoye Knizhnoye Izdatel'stvo. (154 pp.)

1978, The effect of thermal abrasion on the temperature of the permafrost in the coastal zone of the Laptev Sea [Vliyaniye termoabraznii na temperaturu mnogoletnemerzlykh porod v pribrezhnoy zone morya Laptevykh]: 90–3 in Sanger, F. J., ed., *USSR Contribution Permafrost Second International Conference* (Yakutsk, USSR, 13–28 July 1973): Washington, DC, Natl. Acad. Sci. (866 pp.)

MOORE, P. D., and BELLAMY, D. J., 1975, *Peatlands*: New York, Springer-Verlag. (221 pp.)

MORAWETZ, S. O., 1932, Beobachtungen an Schutthalden, Schuttkegeln und Schuttflecken: *Zeitschr. Geomorphologie* **7** (1932–3), 25–43.

1973, Permafrost-Schneegrenze-Periglaziales: *Univ. Salzburg Geog. Inst. Beiträge zur Klimatologie, Meteorologie und Klimamorphologie* **3** (*Festschrift Hanns Tollner*), 37–44.

MORGAN, A. V., 1969, Intraformational periglacial structures in the Nose Hill gravels and sands, Calgary, Alberta, Canada: *J. Geol.* **77**, 358–64.

1972, Late Wisconsinan ice-wedge polygons near Kitchener, Ontario, Canada: *Canadian J. Earth Sci.* **9**, 607–17.

MORGAN, A. V., and MORGAN, ANNE, 1977, The age and distribution of fossil ice-wedge polygon networks in south-western Ontario, Canada (abs.): *International Union for Quaternary Research (INQUA) Cong., 10th (Birmingham, England, 16–24 Aug. 1977), Abstracts*, 308.

MORGENSTERN, N. R., and NIXON, J. F., 1971, One-dimensional consolidation of thawing soils: *Canadian Geotech. J.* **8**, 558–65.

MORTENSEN, HANS, 1930, Einige Oberflächenformen in Chile und auf Spitzbergen im Rahmen einer vergleichenden Morphologie der Klimazonen: *Petermanns Mitt., Ergänzungsheft* **209**, 147–56.

—— 1932, Über die physikalische Möglichkeit der 'Brodel'-Hypothese: *Centralbl. Mineralog.* **1932**(B), 417–22.

—— 1933, Die 'Salzsprengung' und ihre Bedeutung für die regionalklimatische Gliederung der Wüsten: *Petermanns Mitt.* **79**, 130–5.

MÜCKENHAUSEN, E., 1960, Eine besondere Art von Pingos am Hohen Venn/Eifel: *Eiszeitalter und Gegenwart* **11**, 5–11.

MULLENDERS, WILLIAM, and GULLENTOPS, FRANS, 1969, The age of the pingos of Belgium: 321–35 in Péwé, T. L., ed., *The periglacial environment*: Montreal, McGill–Queen's Univ. Press. (487 pp.)

MULLER, E. H., 1973, Boulder concentrations in till at the close of a glacial cycle (abs.): 258 in *International Union for Quaternary Research (INQUA) Cong., 9th (Christchurch, New Zealand, 2–10 Dec. 1973), Abstracts*. (418 pp.)

MÜLLER, FRITZ, 1954, *Frostbodenerscheinungen in NE- und N-Groeland*: Zürich Univ., Philos. Fakultät II, Diplomarbeit für das Höhere Lehramt in Geographie. (221 pp.) (Thesis.)

—— 1959, Beobachtungen über Pingos. Detailuntersuchungen in Ostgrönland und in der kanadischen Arktis: *Medd. om Grønland* **153**(3). (127 pp.)

—— 1962, Analysis of some stratigraphic observations and radiocarbon dates from two pingos in the Mackenzie Delta area, NWT: *Arctic* **15**, 279–88.

—— 1963, Observations on pingos [Beobachtungen über Pingos]: *Canada Natl. Research Council Tech. Translation* **1073**. (117 pp.)

—— 1968, Pingos, modern: 845–7 in Fairbridge, R. W., ed., *The encyclopedia of geomorphology*: New York, Reinhold Book Corp. (1295 pp.)

MÜLLER, H.-M., 1978, Weichselzeitliche Eiskeilsysteme im Emsgebiet bei Münster (Westf.) – Ein Beitrag zur Datierung der Emsterrassen: *Neues Jahrbuch Geologie und Paläontologie, Monatshefte* **1978**, 117–28.

MULLER, S. W., 1947, *Permafrost or permanently frozen ground and related engineering problems*: Ann Arbor, Mich., J. W. Edwards. (231 pp.)

MURRMANN, R. P., 1973, Ionic mobility in permafrost: 352–9 in *North American Contribution, Permafrost Second International Conference* (Yakutsk, USSR, 13–28 July 1973): Washington, DC, Natl. Acad. Sci. (783 pp.)

NAKANO, YOSHISUKE, and BROWN, JERRY, 1972, Mathematical modeling and validation of the thermal regimes in tundra soils, Barrow, Alaska: *Arctic and Alpine Research* **4**, 19–38.

NANGERONI, GIUSEPPE, 1962, Les phénomènes périglaciaire en Italie: *Biuletyn Peryglacjalny* **11**, 57–63.

NANSEN, FRIDTJOF, 1904, The bathymetrical features of the North Polar Seas: In Nansen, Fridtjof, *The Norwegian North Polar Expedition 1893–1896, Scientific results* **4**, *part 13*: Christiania, Jacob Dybwad; New York, London, Bombay, Longmans, Green, and Co.; Leipzig, F. A. Brockhaus. (231 pp.)

—— 1922a, *Spitzbergen*, 3 ed.: Leipzig, F. A. Brockhaus. (327 pp.)

—— 1922b, The strandflat and isostasy: *Videnskapsselskapet i Kristiania [Norges Videnskaps Akad.] Skrifter 1921 I Mat.-Nat. Kl.* **2**(11). (313 pp.)

NATIONAL ACADEMY OF SCIENCES, 1973, *North American Contribution, Permafrost Second International Conference* (Yakutsk, USSR, 13–28 July 1973): Washington, DC, Natl. Acad. Sci. (783 pp.)

—— 1974, *Priorities for basic research on permafrost*: Washington, DC, Natl. Acad. Sci. (54 pp.)

—— 1975, *Opportunities for permafrost-related research associated with the trans-Alaska pipeline system*: Washington, DC, Natl. Acad. Sci. (37 pp.)

—— 1976, *Problems and priorities in offshore permafrost research*: Washington, DC, Natl. Acad. Sci. (43 pp.)

—— 1978, Sanger, F. J., ed., *USSR Contribution Permafrost Second International Conference* (Yakutsk, USSR, 13–28 July 1973): Washington, DC, Natl. Acad. Sci. (866 pp.)

NATIONAL ACADEMY OF SCIENCES–NATIONAL RESEARCH COUNCIL, 1966, *Permafrost International Conference Proceedings: Natl. Acad. Sci.–Natl. Research Council Pub.* **1287**. (563 pp.)

NATIONAL OCEANIC AND ATMOSPHERIC ADMINISTRATION [NOAA] (OFFICIALS OF), 1974, *Climates of the states* (in two volumes): Port Washington, NY, Water Information Center, Inc. (975 pp.)

NEAL, JAMES, 1965, Giant desiccation polygons of Great Basin playas: *Air Force Cambridge Research Laboratories, Environmental Research Paper* **123**. (30 pp.)

—— 1966, Giant desiccation stripes (abs.): *Geol. Soc. America Spec. Paper* **87**, 117.

NEAL, J. T., LANGER, A. M., and KERR, P. F., 1968,

Giant desiccation polygons of Great Basin playas: *Geol. Soc. America Bull.* **79**, 69–90.

NEIZVESTNOV, YA. V., *et al.*, 1971, Gidrogeologicheskoye rayonirovaniye i gidrogeologicheskiye usloviya sovetskogo sektora Arktiki: *V. kn.: Geologiya i poleznye iskopayemye severa Sibirskoy platformy, L.*, 92–105.

NEKRASOV, I. A., and GORDEYEV, P. P., 1973, *The north-east coast of Yakutia*: Internat. Permafrost Conf. 2 (Yakutsk, USSR, 13–28 July 1973), Guidebook: Yakutsk, USSR, Acad. Sci., Sec. Earth Sci., Siberian Div., 1973. (46 pp.)

NEWCOMB, R. C., 1952, Origin of the Mima Mounds, Thurston County region, Washington: *J. Geol.* **60**, 461–72.

NICHOLS, R. L., 1953, Marine and lacustrine ice-pushed ridges: *J. Glaciology* **2**, 172–5.

——— 1966, Geomorphology of Antarctica: 1–59 in Tedrow, J. C. F., ed., *Antarctic soils and soil forming processes*: Am. Geophysical Union, Antarctic Research Ser. **8**. (117 pp.)

——— 1969, Geomorphology of Inglefield Land, North Greenland: *Medd. om Grønland* **188**(1). (109 pp.)

——— 1971, Glacial geology of the Wright Valley, McMurdo Sound: 293–340 in Quam, L. O., ed., *Research in the Antarctic*: Am. Assoc. Advancement Science Pub. **93**. (768 pp.)

NICHOLSON, F. H., 1976, Patterned ground formation and description as suggested by Low Arctic and Subarctic examples: *Arctic and Alpine Research* **8**(4), 329–42.

——— 1978, Permafrost distribution and characteristics near Schefferville, Quebec: Recent studies: 428–33 in *Third International Conference on Permafrost* (Edmonton, Alta., 10–13 July 1978), Proc. **1**: Ottawa, Canada Natl. Research Council. (947 pp.)

NICHOLSON, F. H., and GRANBERG, H. B., 1973, Permafrost and snowcover relationships near Schefferville: 151–8 in *North American Contribution, Permafrost Second International Conference* (Yakutsk, USSR, 13–28 July 1973): Washington, DC, Natl. Acad. Sci. (783 pp.)

NICHOLSON, F. H., and THOM, B. G., 1973, Studies at the Timmins 4 Experimental Site: 159–66 in *North American Contribution, Permafrost Second International Conference* (Yakutsk, USSR, 13–28 July 1973): Washington, DC, Natl. Acad. Sci. (783 pp.)

NIKIFOROFF, C. C., 1952, Origin of microrelief in the Lake Agassiz Basin: *J. Geol.* **60**, 91–103.

NIKOL'SKAYA, V. V., and CHICHAGOV, V. P., 1962, Drevniye periglacial'nyye yavleniya v basseyne Amura: *Voprosy kriologii pri izuchenii chetvertichnykh otlozheniy*, 45–52.

NIKOL'SKAYA, V. V., and SHCHERBAKOV, I. N., 1956, Znaki byvshego oledeniya gor Tukuringra-Dzhagdy: *Akad. Nauk SSSR, Izvestiya, seriya geograficheskaya* **2**, 58–65.

NIKOL'SKAYA, V. V., TIMOFEYEV, D. A., and CHICHAGOV, V. P., 1964, Zonalnyye tipy pedimentov basseyna Amura: *Geograficheskogo obshchestva SSSR, Zabaykal'skogo otdel, Zapiski* **24**, 67–86.

NIXON, J. F., and MCROBERTS, E. C., 1973, A study of some factors affecting the thawing of frozen soils: *Canadian Geotech. J.* **10**, 439–52.

NOBLES, L. H., 1961, Surface features of the ice-cap margin, northwestern Greenland: 752–67 in Raasch, G. O., ed., *Geology of the Arctic*, 2 vols.: Toronto, Univ. Toronto press. (1196 pp.)

——— 1966, Slush avalanches in northern Greenland and the classification of rapid mass movements: *Internat. Assoc. Sci. Hydrol. pub.* **69**, 267–72.

NORGES TEKNISK-NATURVITENSKAPELIGE FORSKNINGSRÅD OG STATENS VEGVESENS UTVALG FOR FROST I JORD, 1970–, *Frost i Jord* **1**–.

NØRVANG, AKSEL, 1946, Nogle Forekomster af Arktisk Strukturmark (Brodelboden) bevarede i danske Istidsaflejringer: *Danmarks Geol. Undersøgelse – II, Raekke* **74**. (65 pp.)

OBERMAN, N. G., 1974, Regional'nyye osobennosti merzloy zony Timano-Ural'skoy oblasti: *Vysshikh uchebn. zavedenii, Geologiya i razvedka Izv.* **11**, 98–103.

OBERMAN, N. G., and KAKUNOV, B. B., 1973, Opredeleniye moshchnosti poyasa otritsatel'nykh temperatur gornykh porod na poberezh'ye arktiki: 130–7 in *Akademiya Nauk SSSR, Sektsiya Nauk o Zemle, Sibirskoye Otdeleniye, II Mezhdunarodnaya Konferentsiya po Merzlotovedeniyu, Doklady i soobshcheniya* **2** (Regional'naya geokriologiya); Yakutsk, Yakutskoye Knizhnoye Izdatel'stvo. (154 pp.)

——— 1978, Determination of the thickness of permafrost on the arctic coast [Opredeleniye moshchnosti poyasa otritsatel'nykh temperatur gornykh porod na poberezh'ye arktiki]: 143–7 in Sanger, F. J., ed., *USSR Contribution Permafrost Second International Conference* (Yakutsk, USSR, 13–28 July 1973): Washington, DC, Natl. Acad. Sci. (866 pp.)

O'BRIEN, ROBERT, 1971, Observations on pingos and permafrost hydrology in Schuchert Dal, N. E. Greenland: *Medd. om Grønland* **195**(1). (20 pp.)

O'BRIEN, R., ALLEN, C. R., and DODSON, B., 1968, Geomorphology: 3–16 in *Scoresby Land Expedition, 1968*: Univ. Dundee. (61 pp.)

OBRUCHEV, S. V., 1937, Soliflyuktsionnyye (nagornyye) terrasy i ikh genezis na osnovanii rabot v Chukotskom krayye: *Problemy Arktiki* **3**, 27–48; **4**, 57–83.

OBRUCHEV, V. A., 1945, Loess types and their origin: *Am. J. Sci. 5th ser.* **243**, 256–62.

OHLSON, BIRGER, 1964, Frostakitivität, Verwitterung und Bodenbildung in den Fjeldgegenden von Enontekiö, Finnisch–Lappland: *Fennia* **89**(3), 1–180.

OLLIER, C. D., and THOMASSON, A. J., 1957, Asymmetrical valleys of the Chiltern Hills: *Geog. J.* **123**, 71–80.

ORGANISATION FOR ECONOMIC CO-OPERATION AND DEVELOPMENT, 1973, *Symposium on Frost Action on Roads* (Oslo, Norway, 1–3 Oct. 1973), *Repts.* **1–3**: Paris, Organisation de Coopération et de Développement Economiques. (385, 450, and 71 pp.)

OSBORN, G. D., 1975, Advancing rock glaciers in the Lake Louise area, Banff National Park, Alberta: *Canadian J. Earth Sci.* **12**, 1060–2.

OSTERKAMP, T. E., and HARRISON, W. D., 1976, Subsea permafrost at Prudhoe Bay, Alaska: Drilling report and data analysis: *Univ. Alaska Geophysical Inst. Rept. UAG R-245* (Sea Grant Rept. 76–5). (67 pp. + appendices.)

1977, Sub-sea permafrost regime at Prudhoe Bay, Alaska, USA: *J. Glaciology* **19**(81), 627–37.

ØSTREM, GUNNAR, 1971, Rock glaciers and ice-cored moraines, a reply to D. Barsch: *Geografiska Annaler* **53**A, 207–13.

OUTCALT, S. I., 1970, *A study of needle ice events at Vancouver, Canada, 1961–68*: Univ. British Columbia, Ph.D. Thesis. (135 pp.)

1971, An algorithm for needle ice growth: *Water Resources Research* **7**, 394–400.

OUTCALT, S. I., and BENEDICT, J. B., 1965, Photo-interpretation of two types of rock glacier in the Colorado Front Range, USA: *J. Glaciology* **5**(42), 849–56.

OWENS, E. H., and HARPER, J. R., 1977, Frost-table and thaw depths in the littoral zone near Peard Bay, Alaska: *Arctic* **30**, 155–68.

OWENS, E. H., and MCCANN, S. B., 1970, The role of ice in the arctic beach environment with special reference to Cape Ricketts, southwest Devon Island, Northwest Territories, Canada: *Am. J. Sci.* **268**, 397–414.

PADALKA, G., 1928, O vysokykh terrasakh na Severnom Urale: *Geologicheskogo komiteta, Vestnik III* **4**, 9–15.

PAEPE, R., and PISSART, A., 1969, Periglacial structures in the Late-Pleistocene stratigraphy of Belgium: *Biuletyn Peryglacjalny* **20**, 321–36.

PALMER, A. C., 1967, Ice lensing, thermal diffusion and water migration in freezing soil: *J. Glaciology* **6**(47), 681–94.

PALMER, JOHN, and RADLEY, JEFFREY, 1961, Gritstone tors of the English Pennines: *Zeitschr. Geomorphologie N. F.* **5**, 37–52.

PANOŠ, V., 1960, Příspěvek k poznání geomorfologie krasové oblasti 'Na Pomezi' v Rychlebských horách: *Sborník Vlastivědného ústavu v Olomouci, oddíl A, Přírodní vědy* **4**, 33–88.

PANOV, D. G., 1937, Geomorfologicheskiy ocherk Polyarnykh Uralid i zapadnoy chasti polyarnogo shelfa: *Instituta geografii AN SSSR, Trudy* **26**.

PARIZEK, R. R., 1969, Glacial ice-contact rings and ridges: 49–102 in Schumm, S. A., and Bradley, W. C., eds., *United States contributions to Quaternary research: Geol. Soc. America Spec. Paper* **123**. (305 pp.)

PATALEYEV, A. V., 1955, Morozoboynyye treshchiny v gruntakh: *Priroda* **44**(12), 84–5.

PATERSON, W. S. B., 1969, *The physics of glaciers*: Oxford, Pergamon Press. (250 pp.)

PAVLOV, A. V., 1978, Teplofizika merzlotnykh landshaftov – Thermal physics of permafrost terrain: 68–74 in *Third International Conference on Permafrost* (Edmonton, Alta., 10–13 July 1978), Proc. **1**: Ottawa, Canada Natl. Research Council. (947 pp.)

PECK, R. B., HANSON, W. E., and THORNBURN, T. H., 1953, *Foundation engineering*: New York, John Wiley. (410 pp.)

PÉCSI, M., 1963, Die periglazialen Erscheinungen in Ungarn: *Petermanns Mitt.* **107**, 161–82.

1964, *Ten years of physicogeographic research in Hungary*: Akad. Kiadó, Budapest, Studies in Geography **1**. (132 pp.)

1965, Les principaux problèmes des recherches géomorphologiques dans les montagnes hongroises moyennes: *Geographia Polonica* **9**, 87–99.

PELTIER, L. C., 1950, The geographic cycle in periglacial regions as it is related to climatic geomorphology: *Assoc. Am. Geog. Annals* **40**, 214–36.

PENNER, E., 1968, Particle size as a basis for predicting frost action in soils: *Soils and Foundations* **8**(4), 21–9.

1972, Influence of freezing rate of frost heaving: 56–64 in Highway Research Board, *Frost action in soils: Natl. Acad. Sci.–Natl. Acad. Eng. Highway Research Record* **393**. (88 pp.)

1973, Frost heaving pressures in particulate materials: 379–85 in Organisation for Economic Co-operation and Development: *Symposium on Frost Action on Roads* (Oslo, Norway, 1–3 Oct. 1973), *Rept.* **1**: Paris, Organisation de Coopération et de Développement Économiques. (385 pp.)

1977, Fundamental aspects of frost action: 17–28 in *Frost action in soils*: Univ. Luleå, Internat. Symposium (Luleå, Sweden, 16–18 Feb. 1977) Proc. **2**. (119 pp.)

PENNER, EDWARD, and UEDA, TAKAO, 1977, The dependence of frost heaving on load application – Preliminary results: 92–101 in *Frost action in soils*: Univ. Luleå, Internat. Symposium (Luleå, Sweden, 16–18 Feb. 1977) Proc. **1**. (215 pp.)

PERLA, R. I., and MARTINELLI, M., JR., 1976, *Avalanche handbook*: US Dept. Agriculture, Forest Service, Agriculture Handbook 489. (238 pp.)

PEROV, V. F., 1959, O nablyudeniyakh nad protsessami nivatsii i soliflyuktsii v Khibinskikh gorakh: *Voprosy fizicheskoy geografii polyarnych stran* **2**, 83–6.

PESSL, FRED, JR., 1969, Formation of a modern ice-push ridge by thermal expansion of lake ice in southeastern Connecticut: *US Army Corps of Engineers, Cold Regions Research and Engineering Laboratory Research Rept.* **259**. (15 pp.)

PETERSON, K. M., and BILLINGS, W. D., 1978, Geomorphic processes and vegetational change along the Meade River sand bluffs in northern Alaska: *Arctic* **31**, 7–23.

PÉWÉ, T. L., 1948, Origin of the Mima Mounds: *Sci. Monthly* **66**, 293–6.

1951, An observation of wind-blown silt: *J. Geol.* **59**, 399–401.

1954, Effect of permafrost on cultivated fields, Fairbanks area, Alaska: *US Geol. Survey Bull.* **989**-*F*, 315–51.

1955, Origin of the upland silt near Fairbanks, Alaska: *Geol. Soc. America Bull.* **66**, 699–724.

1958, Geology of the Fairbanks (D-2) quadrangle: *US Geol. Survey Map GQ* **110**.

1959, Sand-wedge polygons (Tesselations) in the McMurdo Sound Region, Antarctica – A progress report: *Am. J. Sci.* **257**, 545–52.

1964, New type large-scale sorted polygons near Barrow, Alaska (abs.): *Geol. Soc. America Spec. Paper* **76**, 301.

1965a, Fairbanks area: 6–36 in *International Assoc. for Quaternary Research (INQUA) Cong., 7th (Boulder, Col., 1965): Guidebook for field conference F, central and south central Alaska.* (141 pp.)

1965b, Middle Tanana Valley: 36–54 in *International Assoc. for Quaternary Research (INQUA) Cong., 7th (Boulder, Col., 1965): Guidebook for field conference F, central and south central Alaska.* (141 pp.)

1966a, Ice-wedges in Alaska – classification, distribution, and climatic significance: 76–81 in *Perma-frost International Conference* (Lafayette, Ind., 11–15 Nov. 1963) *Proc.*: Natl. Acad. Sci.–Natl. Research Council Pub. **1287**. (563 pp.)

1966b, Paleoclimatic significance of fossil ice wedges: *Biuletyn Peryglacjalny* **15**, 65–73.

1966c, *Permafrost and its effect on life in the North*: Corvallis, Oregon State Univ. Press. (40 pp.)

1969, The periglacial environment: 1–9 in Péwé, T. L., ed., *The periglacial environment*: Montreal, McGill–Queen's Univ. Press. (487 pp.)

1970, Altiplanation terraces of early Quaternary age near Fairbanks, Alaska: *Acta Geographica Lodziensia* **24**, 357–63.

1971, Permafrost and environmental-engineering problems in Arctic (abs.): *Internat. Symposium on Arctic Geology, 2d (San Francisco, Cal., 1–4 Feb. 1971), Program abstracts,* 44.

1973a, Ancient altiplanation terraces near Fairbanks, Alaska: *Biuletyn Peryglacjalny* **23**, 99–100.

1973b, Ice wedge casts and past permafrost distribution in North America: *Geoforum* **15**, 15–26.

1974, Geomorphic processes in polar deserts: 33–52 in Smiley, T. L., and Zumberge, James, eds., *Polar deserts and modern man*: Tucson, Univ. Arizona Press. (173 pp.)

1975, Quaternary geology of Alaska: *US Geol. Survey Prof. Paper* **835**. (145 pp.)

1978, Tyndall figures in ice crystals of ground-ice in permafrost near Fairbanks, Alaska: 312–17 in *Third International Conference on Permafrost* (Edmonton, Alta., 10–13 July 1978), Proc. **1**: Ottawa, Canada Natl. Research Council. (947 pp.)

PÉWÉ, T. L., CHURCH, R. E., and ANDRESEN, M. J., 1969, Origin and paleoclimatic significance of large-scale patterned ground in the Donnelly Dome Area, Alaska: *Geol. Soc. America Special Paper* **103**. (87 pp.)

PÉWÉ, T. L., and SELLMANN, P. V., 1973, Geochemistry of permafrost and Quaternary stratigraphy: 166–9 in *North American Contribution, Permafrost Second International Conference* (Yakutsk, USSR, 13–28 July 1973): Washington, DC, Natl. Acad. Sci. (783 pp.)

PÉWÉ, T. L., and UPDIKE, R. G., 1976, *San Francisco Peaks – A guidebook to the geology*, 2nd ed.: Flagstaff, Ariz., Museum of Northern Arizona. (80 pp.)

PHILBERTH, KARL, 1960, Sur une explication de la régularité dans des sols polygonaux: *Acad. Sci. [Paris] Comptes rendus* **251**, 3004–6.

1961, Sols polygonaux et striés dans les Pyrénées: *Soc. géol. France, Compte rendu sommaire des séances,* 7th ser., **3**, 88–90.

1964, Recherches sur les sols polygonaux et striés: *Biuletyn Peryglacjalny* **13**, 99–198.

PICARD, KARL, 1956, Eiskeile bei Kellinghusen (Mittelholstein): *Neues Jahrbuch für Geologie und Paläontologie Monatshefte* **1956**, 365–73.

PIERCE, W. G., 1961, Permafrost and thaw depressions in a peat deposit in the Beartooth Mountains, northwestern Wyoming: *US Geol. Survey Prof. Paper* **424**-B, B154–6.

PIHLAINEN, J. A., 1962, An approximation of probable permafrost occurrence: *Arctic* **15**, 151–4.

PIHLAINEN, J. A., BROWN, R. J. E., and LEGGET, R. F., 1956, Pingo in the Mackenzie Delta, Northwest Territories, Canada: *Geol. Soc. America Bull.* **67**, 1119–22.

PIHLAINEN, J. A., and JOHNSTON, G. H., 1963, Guide to a field description of permafrost for engineering purposes: *Canada, Natl. Research Council, Assoc. Comm. Snow and Soil Mechanics Tech. Mem.* **79**. (21 pp.)

PIIROLA, JOUKO, 1969, Frost-sorted block concentrations in western Inari, Finnish Lapland: *Fennia* **99**(2). (35 pp.)

PISSART, A., 1956, L'origine périglaciare des viviers des Hautes Fagnes: *Soc. Géol. Belgique Annales* **79**, 1955–6, B119–31.

1958, Les dépressions fermées dans la région parisienne: Le probleme de leur origine: *Rev. Géomorphologie dynamique* **9**, 73–83.

1963a, Les traces de 'pingos' du Pays de Galles (Grande-Bretagne) et du plateau des Hautes Fagnes (Belgique): *Zeitschr. Geomorphologie N.F.* **7**, 147–65.

1963b, Des replats de cryoturbation au Pays de Galles: *Biuletyn Peryglacjalny* **12**, 119–35.

1964a, Contribution expérimentale à la genèse des sols polygonaux: *Soc. Géol. Belgique Annales* **87** (1963–4), *Bull.* **7**, B214–23.

1964b, Vitesse des mouvements du sol au Chambeyron (Basses Alpes): *Biuletyn Peryglacjalny* **14**, 303–9.

1965, Les pingos des Hautes Fagnes: Les problèmes de leur genèse: *Soc. Géol. Belgique Annales* **88**, 277–89.

1966a, Expériences et observations à propos de la genèse des sols polygonaux triés: *Rev. Belge Géographie* **90**(1), 55–73.

1966b, Le role géomorphologique du vent dans la région de Mould Bay (Ile Prince Patrick–NWT–Canada): *Zeitschr. Geomorphologie N. F.* **10**, 226–36.

1966c, Étude de quelques pentes de l'île Prince Patrick: *Soc. Géol. Belgique Annales* **89**, 377–402.

1967, Les pingos de l'île Prince-Patrick (76°N–120°W): *Geog. Bull.* **9**, 189–217.

1968, Les polygones de fente de gel de l'île Prince Patrick (Arctique Canadien – 76° lat. N.): *Biuletyn Peryglacjalny* **17**, 171–80.

1969, Le méchanism périglaciaire dressant les pierres dans le sol. Resultats d'expériences: *Acad. Sci. [Paris] Comptes rendus* **268**, 3015–17.

1970a, Les phénomènes physiques essentielles liés au gel, les structures périglaciaires qui en résultent et leur signification climatique: *Soc. Géol. Belgique Annales* **93**, 7–49.

1970b, The pingos of Prince Patrick Island (76°N–120°W) [Les pingos de l'île Prince Patrick (76°N–120°W]: *Canada Natl. Research Council Tech. Translation* **1401**. (46 pp.)

1972a, Vitesse des mouvements de pierres dans les sols et sur des versants périglaciaires au Chambeyron (Basses Alpes): 251–68 in Macar, P., and Pissart, A., eds., *Processus périglaciaires étudiés sur le terrain: Union Géographique Internationale, Symposium International de Géomorphologie* (Liège–Caen, 1–9 July 1971), *Part* **1** (Séances tenues à Liège et excursions en Belgique): Les Congrès et Colloques de l'Université de Liège: Liège, Université de Liège. (339 pp.)

1972b, Mouvements de sols gelés subissant des variations de température sous 0°: résultats de mesures dilatométriques: 124–6 in Adams, W. P., and Helleiner, F. M., eds., *International Geography 1972* **1** (Internat. Geog. Cong., 22d, Montreal): Toronto, Univ. Toronto Press. (694 pp.)

1973a, Résultats d'expériences sur l'action du gel dans le sol: *Biuletyn Peryglacjalny* **23**, 101–13.

1973b, L'origine des sols polygonaux et striés du Chambeyron (Basses–Alpes): *Soc. Géog. Liège Bull.* **9**, 33–53.

1974a, Détermination expérimentale des processus responsables des petits sols polygonaux triés de haute montagne: 241–8 in Poser, Hans, ed., *Geomorphologische Prozesse und Prozesskombinationen in der Gegenwart unter verschiedenen Klimabedingungen* (Symposium and Report of Commission on Present-day Geomorphological Processes, Intl. Geog. Union): *Akad. Wiss. Göttingen Abh., Math.–Phys. Kl. Folge 3*, **29**. (440 pp.)

1974b, Les viviers des Hautes Fagnes sont des traces de buttes périglaciaires. Mais s'agissait-il réellement de pingos?: *Soc. Géol. Belgique Annales* **97**, 359–81.

1975, Glace de ségrégation, soulèvement du sol et phénomènes thermokarstique dans les regions à pergélisol: *Soc. Géog. Liège Bull.* **11**, 89–96.

1976a, L'origine des sols polygonaux et striés décimétrique: 211–13 in International Geographical Congress, 23d (Moscow, 27 July–3 Aug. 1976), *International geography* **1** (*Geomorphology and Paleogeography*). (409 pp.)

1976*b*, Les dépôts et la morphologie périglaciaires de la Belgique: 115–35 in Pissart, A., coord., *Géomorphologie de la Belgique* (Hommage au Professeur Paul Macar): Liège, Université de Liège. (224 pp.)

1976*c*, Sols à buttes, cercles non triés et sols striés, non triés de l'île de Banks (Canada, NWT): *Biuletyn Peryglacjalny* **26**, 275–85.

1977*a*, Apparition et évolution des sols structuraux périglaciaires de haute montagne. Expériences de terrain au Chambeyron (Alpes, France): 142–56 in Poser, Hans, ed., Formen, Formengesellschaften und Untergrenzen in den heutigen periglazialen Höhenstufen der Hochgebirge Europas und Afrikas zwischen Arktis und Äquator. Bericht über ein Symposium: *Akad. Wiss. Göttingen Abh., Math.–Phys. Kl. Folge 3*, **31**. (355 pp.)

1977*b*, The origin of pingos in regions of thick permafrost (Canadian Arctic) (abs.): *International Union for Quaternary Research (INQUA) Cong., 10th* (*Birmingham, England, 16–24 Aug. 1977*), *Abstracts*, 361.

PISSART, A., *et al.*, 1972, Les cicatrices de pingos de la Brackvenn (Hautes Fagnes). Compte rendu de l'excursion du 3 Juillet 1971: 281–94 in Macar, P., and Pissart, A., eds., *Processus périglaciaire étudiés sur le terrain: Union Géographique Internationale, Symposium International de Géomorphologie* (Liège–Caen, 1–9 July 1971), *Part I* (Séances tenues à Liège et excursions en Belgique): Les Congrès et Colloques de l'Université de Liège: Liège, Université de Liège. (339 pp.)

1976, Reports of the laboratory experiments – Activity of working group of the Periglacial Committee of the IGU: *Biuletyn Peryglacjalny* **26**, 113–61.

PISSART, A., and FRENCH, H. M., 1976, Pingo investigations, north-central Banks Island, Canadian Arctic: *Canadian J. Earth Sci.* **13**, 937–46.

1977, The origin of pingos in regions of thick permafrost, western Canadian Arctic: *Quaestiones Geographicae* **4**, 149–60.

PISSART, A., VINCENT, J.-S., and EDLUND, S. A., 1977, Dépôts et phénomènes éoliens sur l'île de Banks, Territoires du Nord-Ouest, Canada: *Canadian J. Earth Sci.* **14**(11), 2462–80.

PITTY, A. F., 1971, *Introduction to geomorphology*: London, Methuen & Co., Ltd. (526 pp.)

PLASCHEV, A. V., 1956, Vzryv ledyanogo bugra: *Priroda* **45**(9), 113.

POPOV, A. I., 1961, Cartes des formations périglaciaires actuelles et Pléistocènes en territoire de l'URSS: *Biuletyn Peryglacjalny* **10**, 87–96.

1967, *Merzlotnyye yavleniya v zemnoy kore (kriolitologiya)*: Moskva, Izdatel'stvo Moskovskogo Universiteta. (304 pp.)

ed., 1969–, *Problemy kriolitologii (Problems of cryolithology)* **1**–: Moskva, Izdatel'stvo Moskovskogo Universiteta.

1973, *Al'bom kriogennykh obrazovananiy v zemnoy kore i rel'efe*: Moskva, Izdatel'stvo Moskovskogo Universiteta. (55 pp.)

1976, O geograficheskoy zonal'nosti kriolitogeneza: *Moskovskogo Universiteta, Vestnik Ser. 5 Geografiya* **4**, 55–60.

1977, Geographical zonality of frozen ground formation [O geograficheskoy zonal'nosti kriolitogeneza]: *Polar geography* **1**(1), 50–5.

1978, Cryolithogenesis, the composition and structure of frozen rocks, and ground ice (The current state of the problem): *Biuletyn Peryglacjalny* **27**, 155–70.

POPOV, A. I., ROZENBAUM, G. E., and VOSTOKOVA, A. V., 1978, Kriolitologicheskaya karta SSSR – Cryolithological map of the U.S.S.R. (Principles of compilation): 434–7 in *Third International Conference on Permafrost* (Edmonton, Alta., 10–13 July 1978), *Proc.* **1**: Ottawa, Canada Natl. Research Council. (947 pp.)

POPPE, V., and BROWN, R. J. E., 1976, Russian–English glossary of permafrost terms: *Canada, Natl. Research Council, Assoc. Comm. on Geotechnical Research Tech. Mem.* **117**. (25 pp.)

PORSILD, A. E., 1938, Earth mounds in unglaciated arctic northwestern America: *Geog. Rev.* **28**, 46–58.

PORTER, S. C., 1966, Pleistocene geology of Anaktuvuk Pass, Central Brooks Range, Alaska: *Arctic Inst. No. America Tech. Paper* **18**. (100 pp.)

1977, Present and past glaciation threshold in the Cascade Range, Washington, USA: Topographic and climatic controls, and paleoclimatic implications: *J. Glaciology* **18**(78), 101–16.

PORTER, S. C., and DENTON, G. H., 1967, Chronology of Neoglaciation in the North American Cordillera: *Am. J. Sci.* **265**, 177–210.

POSER, HANS, 1931, Beiträge zur Kenntnis der arktischen Bodenformen: *Geol. Rundschau* **22**, 200–31.

1932, Einige Untersuchungen zur Morphologie Ostgrönlands: *Medd. om Grønland* **94**(5). (55 pp.)

1936, Talstudien aus Westspitzbergen und Ostgrönland: *Zeitschr. Gletscherkunde* **24**, 43–98.

1947*a*, Dauerfrostboden und Temperaturverhältnisse während der Würm-Eiszeit im nicht vereisten Mittel- und Westeuropa: *Naturwissenschaften* **34**, 10–18.

1947*b*, Auftautiefe und Frostzerrung im Boden Mitteleuropas während der Würm-Eiszeit: *Naturwissenschaften* **34**, 232–8, 262–7.

1948*a*, Boden- und Klimaverhältnisse in Mittel- und Westeuropa während der Würmeiszeit: *Erdkunde* **2**, 53–68.

1948*b*, Äolische Ablagerungen und Klima des Spätglazials in Mittel- und Westeuropa: *Naturwissenschaften* **9**, 269–75, 307–12.

1950, Zur Rekonstruktion der spätglazialen Luftdruckverhältnisse in Mittel- und Westeuropa auf Grund der vorzeitlichen Binnendünen: *Erdkunde* **4**, 81–8.

1951, Die nördliche Lössgrenze in Mitteleuropa und das spätglaziale Klima: *Eiszeitalter und Gegenwart* **1**, 27–55.

1954, Die Periglazial–Erscheinungen in der Umgebung der Gletscher des Zemmgrundes (Zillertaler Alpen): *Göttinger Geog. Abh.* **15**, 125–80.

1957, Klimamorphologische Probleme auf Kreta: *Zeitschr. Geomorphologie N. F.* **1**, 113–42.

ed., 1974, Geomorphologische Prozesse und Prozesskombinationen in der Gegenwart unter verschiedenen Klimabedingungen. Bericht über ein Symposium: *Akad. Wiss. Göttingen Abh., Math.-Phys. Kl. Folge 3*, **29**. (440 pp.)

ed., 1977, Formen, Formengesellschaften und Untergrenzen in den heutigen periglazialen Höhenstufen der Hochgebirge Europas und Afrikas zwischen Arktis und Äquator. Bericht über ein Symposium: *Akad. Wiss. Göttingen Abh., Math.-Phys. Kl. Folge 3*, **31**. (355 pp.)

1978, Glatthänge als Konvergenzformen. Beobachtungen am Joúchtas auf Kreta: *Braunschweig. Wiss. Gesell. Abh.* **29**, 33–46.

POSER, HANS, and MÜLLER, THEODOR, 1951, Studien an den asymmetrischen Tälern des Niederbayerischen Hügellandes: *Akad. Wiss. Göttingen Nachrichten, Math.–Phys. Klasse IIb* **1951**(1), 1–32.

POTTER, NOEL, JR., 1969, *Rock glaciers and masswastage in the Galena Creek area, northern Absaroka Mountains*: Univ. Minnesota, Ph.D. Thesis. (150 pp.)

1972, Ice-cored rock glacier, Galena Creek, northern Absaroka Mountains, Wyoming: *Geol. Soc. America Bull.* **83**, 3025–57.

POTTER, NOEL, JR., and MOSS, J. H., 1968, Origin of the Blue Rocks block field and adjacent deposits, Berks County, Pennsylvania: *Geol. Soc. America Bull.* **79**, 255–62.

POTTS, A. S., 1970, Frost action in rocks: Some experimental data: *Inst. British Geographers Trans.* **49**, 109–24.

POWERS, W. E., 1936, The evidences of wind abrasion: *J. Geol.* **44**, 214–19.

PRECHTL, HANS, 1965, Geomorphologische Strukturen: *Tübinger Geog. Studien* **17**. (144 pp.)

PRENTICE, J. E., and MORRIS, P. G., 1959, Cemented screes in the Manifold Valley, North Staffordshire: Univ. Nottingham, Dept. Geography, *The East Midland Geographer* **11**, 16–19.

PREOBRAZHENSKIY, V. S., 1959, Alpiyskie i goltsovyye yavleniya v prirode khrebtov Stanovogo nagorya (Kodar i Udokan): *Akad. Nauk SSSR, Izvestiya, seriya geograficheskaya* **4**, 67–72.

1962, *Rel'ef i istoriya ego razvitiya*: Moskva, Prirodnyye usloviya osvoyeniya Severa Chitinskoy oblasti, 1–126.

PREST, V. K., 1969, Retreat of Wisconsin and recent ice in North America: *Canada Geol. Survey Map* **1257**A.

PREUSSER, HUBERTUS, 1973, Hypsometrischer Formenwandel der Polygone in Island: *Zeitschr. Geomorphologie N. F., Supp.* **16**, 155–60.

PRICE, L. W., 1970, Up-heaved blocks: A curious feature of instability in the tundra: *Assoc. Am. Geographers Proc.* **2**, 106–10.

1971, Vegetation, microtopography, and depth of active layer on different exposures in sub-arctic alpine tundra: *Ecology* **52**, 638–47.

1972, The periglacial environment, permafrost, and man: *Assoc. Am. Geographers, Comm. on College Geog. Resource Paper* **14**. (88 pp.)

1973, Rates of mass-wasting in the Ruby Range, Yukon Territory: 235–45 in *North American Contribution, Permafrost Second International Conference* (Yakutsk, USSR, 13–28 July 1973): Washington, DC, Natl. Acad. Sci. (783 pp.)

1974, The developmental cycle of solifluction lobes: *Assoc. Am. Geographers Annals* **64**(3), 430–8.

PRICE, W. A., 1968, Oriented lakes: 784–96 in Fairbridge, R. W., ed., *The encyclopedia of geomorphology*: New York, Reinhold Book Corp. (1295 pp.)

PRIESNITZ, K., and SCHUNKE, E., 1978, An approach to the ecology of permafrost in central Iceland: 473–9 in *Third International Conference on Permafrost* (Edmonton, Alta., 10–13 July 1978), Proc. **1**: Ottawa, Canada Natl. Research Council. (947 pp.)

PRINDLE, L. M., 1905, The gold placers of the Fortymile, Birch Creek and Fairbanks regions: *US Geol. Survey Bull.* **251**, 89.

PROKOPOVICH, N. P., 1969, Pleistocene permafrost in California's Central Valley? (abs.): *Geol. Soc. America Abs. with Programs for 1969* **5**, 66.

PROTAS'YEVA, I. V., 1967, *Aerometody v geokriologii*: Moskva, Akad. Nauk SSSR, Sibirskoye Otdeleniye, Inst. Merzlotovedeniya. (196 pp.)

1975, Aeromethods in geocryology [Aerometody v geokriologii]: *US Army Corps of Engineers, Cold Regions Research and Engineering Laboratory Draft Translation* **482**. (184 pp.)

PRYALUKHINA, A. F., 1958, O rastitelnosti goltsov i podgoltsovoy polosy Bikino-Imanskogo vodorazdela: *Botanicheskiy zhurnal* **43**, 92–6.

PULINA, MARIAN, 1968, Gleby poligonalne w jaskini Czarnej, Tatry Zachodnie (Les sols polygonaux dans la grotte Czarna, les Tatras Occidentales): *Speleologia* **3**(2), 99–104.

PULLAN, R. A., 1959, Tors: *Scottish Geog. Mag.* **75**, 51–5.

P'YAVCHENKO [PLYAVCHENKO], N. I., 1963, *Zabolochennyye lesa i bolota Sibiri*: Moskva, Akad. Nauk SSSR, Sibirskoye Otdeleniye, Inst. Lesa i Drevesiny. (217 pp.)

1969, Swampy forests and bogs of Siberia [Zabolochennyye lesa i bolota Sibiri]: *US Army Foreign Science and Technology Center Technical Translation FSTC–HT–23–310–70*. (215 pp.)

PYRCH, J. B., 1973, *The characteristics and genesis of stone stripes in north-central Oregon*: Portland State Univ., Oregon, M.Sc. Thesis. (134 pp.)

RABOTNOV, T. A., 1937, Vegetatsiya vysokogornogo poyassa basseyna verkhovyev rek Aldana i Timptona: *Gosudarstvennogo Geograficheskogo Obshchestva, Izvestiya* **69**, 585–605.

RADD, F. J., and OERTLE, D. H., 1973, Experimental pressure studies of frost heave mechanisms and the growth-fusion behavior of ice: 377–84 in *North American Contribution, Permafrost Second International Conference* (Yakutsk, USSR, 13–28 July 1973): Washington, DC, Natl. Acad. Sci. (783 pp.)

RAESIDE, J. D., 1949, The origin of schist tors in Central Otago: *New Zealand Geographer* **5**, 72–6.

RAILTON, J. B., and SPARLING, J. H., 1973, Preliminary studies on the ecology of palsa mounds in northern Ontario: *Canadian J. Botany* **51**(5), 1037–44.

RAMPTON, V. N., 1973, The influence of ground ice and thermokarst upon the geomorphology of the Mackenzie–Beaufort region: 43–59 in Fahey, B. D., and Thompson, R. D., eds., *Research in polar and alpine geomorphology*: Guelph Symposium on Geomorphology, 1973, 3d, Proc.: Norwich, England, Geo Abstracts Ltd (Univ. East Anglia). (206 pp.)

RAMPTON, V. N., and BOUCHARD, M., 1975, Surficial geology of Tuktoyaktuk, District of Mackenzie: *Canada Geol. Survey Paper* **74–53**. (17 pp.)

RAMPTON, V. N., and DUGAL, J. B., 1974, Quaternary stratigraphy and geomorphic processes on the Arctic Coastal Plain and adjacent areas, Demarcation Point, Yukon Territory, to Malloch Hill, District of Mackenzie: *Canada Geol. Survey Paper* **74–1**, Part A, 283.

RAMPTON, V. N., and MACKAY, J. R., 1971, Massive ice and icy sediments throughout the Tuktoyaktuk Peninsula, Richards Island, and nearby areas, District of Mackenzie: *Canada Geol. Survey Paper* **71–21**. (16 pp.)

RAMPTON, V. N., and WALCOTT, R. I., 1974, Gravity profiles across ice-cored topography: *Canadian J. Earth Sci.* **11**, 110–22.

RAPP, ANDERS, 1959, Avalanche boulder tongues in Lappland: *Geografiska Annaler* **41**(1), 34–48.

1960a, Recent development of mountain slopes in Kärkevagge and surroundings, northern Scandinavia: *Geografiska Annaler* **42**(2–3), 65–200.

1960b, Talus slopes and mountain walls at Tempelfjorden, Spitsbergen: *Norsk Polarinstitutt Skr.* **119**. (96 pp.)

1967, Pleistocene activity and Holocene stability of hillslopes, with examples from Scandinavia and Pennsylvania: 229–44 in L'évolution des versants, *Congrès et Colloques de l'Université de Liège* **40**. (384 pp.)

1975, Studies of mass wasting in the Arctic and in the Tropics: 79–103 in Yatsu, E., Ward, A. J., and Adams, F., eds., *Mass wasting*: Guelph Symposium on Geomorphology, 1975, 4th, Proc.: Norwich, England, Geo Abstracts Ltd (Univ. East Anglia). (202 pp.)

RAPP, ANDERS, and ANNERSTEN, LENNART, 1969, Permafrost and tundra polygons in northern Sweden: 65–91 in Péwé, T. L., ed., *The periglacial environment*: Montreal, McGill–Queen's Univ. Press. (487 pp.)

RAPP, ANDERS, and CLARK, G. M., 1971, Large nonsorted polygons in Padjelanta National Park, Swedish Lappland: *Geografiska Annaler* **53**A, 71–85.

RAPP, ANDERS, GUSTAFSSON, KJELL, and JOBS, PER, 1962, Iskilar i Padjelanta?: *Ymer* **82**, 188–202.

RAPP, ANDERS, and RUDBERG, STEN, 1964, Studies on periglacial phenomena in Scandinavia 1960–1963: *Biuletyn Peryglacjalny* **14**, 75–89.

RASMUSSEN, W. C., 1953, Periglacial frost-thaw basins in New Jersey: A discussion: *J. Geol.* **61**, 473–474.

RAUP, H. M., 1951, Vegetation and cryoplanation: *Ohio J. Sci.* **51**(3), 105–16.

1965, The structure and development of turf hummocks in the Mesters Vig district, Northeast Greenland: *Medd. om Grønland* **166**(3). (113 pp.)

1969, Observations on the relation of vegetation to mass-wasting processes in the Mesters Vig district, Northeast Greenland: *Medd. om Grønland* **176**(6). (216 pp.)

1971a, The vegetational relations of weathering, frost action, and patterned ground processes, in the Mesters Vig district, Northeast Greenland: *Medd. om Grønland* **194**(1). (92 pp.)

1971b, Miscellaneous contributions on the vegetation of the Mesters Vig district, Northeast Greenland: *Medd. om Grønland* **194**(2). (105 pp.)

REGER, R. D., 1975, *Cryoplanation terraces of interior and western Alaska*: Arizona State Univ., Ph.D. Thesis. (326 pp.)

REGER, R. D., and PÉWÉ, T. L., 1976, Cryoplanation terraces: Indicators of a permafrost environment: *Quaternary Research* **6**, 99–109.

REID, CLEMENT, 1887, On the origin of dry chalk valleys and of Coombe rock: *Geol. Soc. London Quart. J.* **43**, 364–73.

REID, J. R., 1970a, Report on formation of a 'frost boil', Umiat, Alaska, 1953: Written communication.

1970b, Ground patterns and frost contraction polygons in North Dakota: Written communication.

REITER, E. R., 1961, *Meteorologie der Strahlströme* (*Jet streams*): Wien, Springer-Verlag. (473 pp.)

1963, *Jet-stream meteorology*: Chicago and London, Univ. Chicago Press. (515 pp.)

RETZER, J. L., 1965, Present soil-forming factors and processes in Arctic and Alpine regions: *Soil Science* **99**(1), 38–44.

REYNOLDS, R. C., 1971, Clay mineral formation in an alpine environment: *Clays and Clay Minerals* **19**, 361–74.

REYNOLDS, R. C., and JOHNSON N. M., 1972, Chemical weathering in the temperate glacial environment of the Northern Cascade Mountains: *Geochimica et Cosmochimica Acta* **36**, 537–54.

RICH, E. E., and JOHNSON, A. M., eds. 1949, *James Isham's observations on Hudsons Bay, 1743 and notes and observations on a book entitled 'A voyage to Hudsons Bay in the Dobbs Galley, 1749'*: Toronto, The Champlain Society. (352 pp.)

RICHMOND, G. M., 1949, Stone nets, stone stripes, and soil stripes in the Wind River Mountains, Wyoming: *J. Geol.* **57**, 143–53.

1952, Comparison of rock glaciers and block streams in the La Sal Mountains, Utah (abs.): *Geol. Soc. America Bull.* **63**, 1292–3.

1962, Quaternary stratigraphy of the La Sal Mountains, Utah: *US Geol. Survey Prof. Paper* **324**. (135 pp.)

1964, Glaciation of Little Cottonwood and Bells Canyons, Wasatch Mountains, Utah: *US Geol. Survey Prof. Paper* **454**D. (41 pp.)

RICHTER, H., 1965, Die periglazialen Zonen ausserhalb des Jungmoränengebietes: 230–42 in Gellert, J. F., ed., *Die Weichsel-Eiszeit im Gebiet der DDR*, Berlin.

RICHTER, H., HAASE, G., and BARTHEL, H., 1963, Die Golesterrassen: *Petermanns Mitt.* **3**, 183–92.

RICHTER, KONRAD, 1951, Die stratigraphische Bewertung periglazialer Umlagerungen im nördlichen Niedersachsen: *Eiszeitalter und Gegenwart* **1**, 132–42.

RICKARD, W. E., JR., and BROWN, JERRY, 1974, Effects of vehicles on arctic tundra: *Environmental Conservation* **1**(1), 55–62.

RIEGER, SAMUEL, 1974, Arctic soils: 749–69 in Ives, J. D., and Barry, R. G., eds., *Arctic and alpine environments*: London, Methuen & Co. Ltd. (999 pp.)

RITCHIE, A. M., 1953, The erosional origin of the Mima Mounds of southwest Washington: *J. Geol.* **61**, 41–50.

ROBITAILLE, B., 1960, Géomorphologie du Sud-Est de l'Île Cornwallis, Territoires du Nord-Ouest: *Cahiers Géog. Québec* **4**, 359–65.

ROCHETTE, JEAN-CLAUDE, and CAILLEUX, ANDRÉ, 1971, Dépôts nivéo-éoliens annuels à Poste-de-la-Baleine, Nouveau-Québec: *Rev. Géog. Montréal* **25**, 35–41.

ROGERS, J. C., 1976, Beaufort Seacoast permafrost studies: 257–83 in US Departments of Commerce and Interior: *Environmental assessment of the Alaskan Continental Shelf* **13**. (427 pp.)

1977, Seismic investigation of offshore permafrost near Prudhoe Bay, Alaska: 71–7 in Proceedings of a symposium on permafrost geophysics, 12 October 1976: *Canada Natl. Research Council, Assoc. Comm. on Geotechnical Research Tech. Mem.* **119**. (144 pp.)

ROGERS, J. C., and MORACK, J. L., 1977a, *Beaufort Seacoast permafrost studies – Annual report (1 April 1977), Contract #03-5-022-55, Research Unit #271*: Fairbanks, Alaska, Univ. Alaska Geophysical Inst. (44 pp.)

1977b, *Beaufort Seacoast permafrost studies – Quarterly report (30 September 1977), Contract #03-5-022-55, Research Unit #271*: Fairbanks, Alaska, Univ.

Alaska Geophysical Inst. (3 pp.)

1978, Geophysical investigation of offshore permafrost, Prudhoe Bay, Alaska: 560–6 in *Third International Conference on Permafrost* (Edmonton, Alta., 10–13 July 1978), Proc. **1**: Ottawa, Canada Natl. Research Council. (947 pp.)

ROHDENBURG, H., 1967, Eiskeilhorizonte in südniedersächsischen und nordhessischen Lössprofilen: *Biuletyn Peryglacjalny* **16**, 225–45.

ROHDENBURG, H., and MEYER, B., 1969, Zur Deutung pleistozäner Periglazialformen in Mitteleuropa: *Göttinger Bodenkindliche Berichte* **7**, 49–70.

ROMANOVSKIY [ROMANOVSKIJ], N. N., 1973*a*, Zakonomernosti razvitiya poligonal'no-zhil'nykh obrazovaniy i ispol'zovaniye ikh dlya paleogeograficheskikh rekonstruktsiy: 50–9 in *Komissiya po Izucheniyu Chetvertichnogo Perioda, Paleokriologiya v Chetvertichnoy stratigrafii i paleogeografii* (Doklady v Kongresse INQUA 9th): Moskva, Akad. Nauk SSSR, Izdatel'stvo 'Nauka'. (143 pp.)

1973*b*, Regularities in formation of frost-fissures and development of frost-fissure polygons: *Biuletyn Peryglacjalny* **23**, 237–77.

1976, The scheme of correlation of polygonal structures: *Biuletyn Peryglacjalny* **26**, 287–94.

1977, *Formirovaniye poligonal'no-zhil'nykh struktur:* Novosibirsk, Akad. Nauk SSSR, Sibirskoye Otdeleniye, Izdatel'stvo 'Nauka'. (216 pp.)

RÖMKENS, M. J. M., 1969, *Migration of mineral particles in ice with a temperature gradient*: Cornell Univ., Ph.D. Thesis. (109 pp.)

RÖMKENS, M. J. M., and MILLER, R. D., 1973, Migration of mineral particles in ice with a temperature gradient: *J. Colloid and Interface Sci.* **42**(1), 103–11.

ROOT, J. D., 1975, Ice-wedge polygons, Tuktoyaktuk area, NWT: *Canada Geol. Survey Paper* **75–1**B, 181.

ROSE, J., and ALLEN, P., 1977, Middle Pleistocene stratigraphy in south-east Suffolk: *J. Geol. Soc. London* **133**(1), 83–102.

ROSENKRANTZ, ALFRED, 1940, Den danske Nûgssuaq Exspedition 1939: *Medd. Dansk Geol. Foren.* **9**, 653–63.

1942, The marine, Cretaceous sediments at Umîvik: 37–42 in Rosenkrantz, Alfred, *et al.*, A geological reconnaissance of the southern part of the Svartenhuk Peninsula West Greenland: *Medd. om Grønland* **135**(3). (72 pp.)

ROSS, JOHN, 1835, *Narrative of a second voyage in search of a North-West Passage and of a residence in the arctic regions during the years 1829, 1830, 1831, 1832, 1833*: London, A. W. Webster. (740 pp.)

ROTNICKI, KAROL, 1977, Oxbow-lake pingos in continental permafrost conditions, Mongolia (abs.): *International Union for Quaternary Research (INQUA) Cong., 10th (Birmingham, England, 16–24 Aug. 1977), Abstracts*, 386.

ROWE, J. S., and HERMESH, REINHARD, 1974, Saskatchewan's Athabasca sand dunes: *Nature Canada* **3**(3), 19–23.

RUDAVIN, V. V., 1967, Kriogennyye obrazovaniya i protsessy v Yuzhno-muyskom khrebte: *Geokriologicheskyye usloviya Zabaykal'ya i Pribaykal'ya*, 175–82.

RUDBERG, STEN, 1958, Some observations concerning mass movement on slopes in Sweden: *Geol. Fören. Stockholm, Förh.* **80**, 114–25.

1962, A report on some field observations concerning periglacial geomorphology and mass movement on slopes in Sweden: *Biuletyn Peryglacjalny* **11**, 311–23.

1964, Slow mass movement processes and slope development in the Norra Storfjäll area, southern Swedish Lappland: *Zeitschr. Geomorphologie N. F., Supp.* **5** (*Fortschritte der internationalen Hangforschung*), 192–203.

1970, Naturgeografiska uppsatser vid Göteborgs universitet höstterminen 1959 – vårterminen 1969: *Geografiska Fören. i Göteborg Medd.* **10**.

1972, Periglacial zonation – a discussion: *Göttinger Geog. Abh.* **60** (Hans–Poser–Festschrift), 221–33.

1977, Periglacial zonation in Scandinavia: 92–104 in Poser, Hans, ed., Formen, Formengesellschaften und Untergrenzen in den heutigen periglazialen Höhenstufen der Hochgebirge Europas und Afrikas zwischen Arktis und Äquator. Bericht über ein Symposium: *Akad. Wiss. Göttingen Abh., Math.–Phys. Kl. Folge 3*, **31**. (355 pp.)

RUDDIMAN, W. F., 1977, Late Quaternary deposition of ice-rafted sand in the subpolar North Atlantic (lat 40° to 65°N): *Geol. Soc. America Bull.* **88**, 1813–27.

RUHE, R. V., 1969, *Quaternary landscapes in Iowa*: Ames, Iowa, Iowa State Univ. Press. (255 pp.)

RUSANOV, B. S., *et al.*, 1967, *Geomorfologiya Vostochnoy Yakutii*: Yakutsk. (375 pp.)

RUSSELL, R. J., 1933, Alpine land forms in western United States: *Geol. Soc. America Bull.* **44**, 927–50.

1943, Freeze-thaw frequencies in the United States: *Am. Geophys. Union Trans.* **24**, 125–33.

RUTTEN, M. G., 1951, Polygon soils in Iceland: *Geologie en Mijnbouw* **13**(5), 161–7.

RUUHIJÄRVI, RAUNO, 1960, Über die regionale Einteilung der Nordfinnischen Moore: *Annales Botanici Societatis Zoologicae Botanicae Fennicae 'Vanamo'* **31**(1). (360 pp.)

RYCKBORST, H., 1975, On the origin of pingos: *J. Hydrology* **26**, 303–14.

1976, On the origin of pingos – A reply: *J. Hydrology* **30**, 299–301.

RYDQUIST, FOLKE, 1960, Studier inom öländska polygonmarker: *Stockholm Högskola, Geografiska Inst. Medd.* **125**, 50–74.

RYZHOV, B. V., 1961, K voprosu o geomorfologii i stroyenii chetvertichnogo pokrova verkhovyev seti Shilkinskogo-Argunskogo mezhdurechya v svyazi s usloviami zaleganiya kassiteronosnykh rossypey: *Vsesoyuznogo soveshchaniya po izuchenii chetvertichnogo perioda, Materialy* **3**, 277–82.

SABELBERG, UDO, *et al.*, 1976, Quartärgliederung und Aufbau von Warmzeit-Kaltzeit-Zyklen in Bereichen mit Dominanz periglazialer Hangsedimente, dargestellt am Quartärprofil Dreihausen/Hessen: *Eiszeitalter und Gegenwart* **27**, 93–120.

SABELBERG, UDO, ROHDENBURG, HEINRICH, and HAVELBERG, GERD, 1974, Bodenstratigraphische und geomorphologische Untersuchungen an den Lössprofilen Ostheim (Kreis Hanau) und Dreihausen (Kreis Marburg) und ihre Bedeutung für die Gliederung des Quartärs in Mitteleuropa: 101–120 in Semmel, Arno, ed., Das Eiszeitalter im Rhein-Main-Gebiet: *Rhein-Mainische Forschungen* **78**.

SALMI, M., 1968, Development of palsas in Finnish Lapland: International Peat Congress, 3d (Quebec, Canada, 19–23 Aug. 1968), Proc., 182–9.

SALVIGSEN, OTTO, 1977, An observation of palsa-like forms in Nordauslandet, Svalbard: *Norsk Polarinstitutt Årbok* **1976**, 364–7.

SAMUELSSON, CARL, 1927, Studien über die Wirkungen des Windes in den kalten und gemässigten Erdteilen: *Uppsala Univ., Geol. Inst. Bull.* **20**, 57–230.

SANCETTA, CONSTANCE, IMBRIE, JOHN, and KIPP, N. G., 1973, The climatic record of the past 130 000 years in North Atlantic deep-sea core V23–82; correlation with terrestrial record: 62–5 in *Mapping the atmospheric and oceanic circulations and other climatic parameters at the time of the last glacial maximum about 17 000 years ago.* Collected abstracts international (CLIMAP) Conference (Norwich, England, May 1973): Norwich, Univ. East Anglia, School of Environmental Sciences, Climatic Research Unit Pub. **2** (CRURP 2). (123 pp.)

SANDBERG, GUSTAF, 1938, Redogöelser for undersökningar utförda med understöd av sällskapets stipendier: *Ymer* **58**, 333–7.

SANGER, F. J., 1966, Degree-days and heat conduction in soils: 253–62 in *Permafrost International Conference* (Lafayette, Ind., 11–15 Nov. 1963) *Proc.*: Natl. Acad. Sci.–Natl. Research Council Pub. **1287**. (563 pp.)

1969, Foundations of structures in cold regions: *US Army Cold Regions Research and Engineering Laboratory, Cold Regions Science and Engineering Mon.* **III**-C4. (93 pp.)

SARTZ, R. S., 1973, Snow and frost depths on north and south slopes: *US Dept. Agriculture, Forest Service, North Central Experiment Station Research Note* **NC-157**. (2 pp.)

SAVEL'EV, B. A., 1960, Peculiarities of the ice-thawing process of the ice cover and in frozen ground: 160–7 in *Problems of the North* (Translation of *Problemy Severa* **1**, 1958), Canada Natl. Research Council. (376 pp.)

SCHAFER, J. P., 1949, Some periglacial features in central Montana: *J. Geol.* **57**, 154–74.

1965, Periglacial frost action in southern New England (Les effets du gel périglaciaire dans la partie méridonale de la Nouvelle-Angleterre) (abs.): 407 in *International Assoc. for Quaternary Research (INQUA) Cong., 7th (Boulder, Col., 30 Aug.–5 Sept. 1965), Abstracts General Sessions.* (532 pp.)

SCHAFER, J. P., and HARTSHORN, J. H., 1965, The Quaternary of New England: 113–28 in Wright, H. E., Jr., and Frey, D. G., eds., *The Quaternary of the United States*: Princeton, NJ, Princeton Univ. Press. (922 pp.)

SCHEIDEGGER, A. E., 1970, *Theoretical geomorphology*, 2d (revised) ed.: Berlin–Heidelberg–New York, Springer-Verlag. (435 pp.)

SCHENK, ERWIN, 1955a, Die Mechanik der periglazialen Strukturböden: *Hessischen Landesamtes für Bodenforschung Abh.* **13**. (92 pp.)

1955b, Die periglazialen Strukturbodenbildungen als Folgen der Hydratationsvorgänge im Boden: *Eiszeitalter und Gegenwart* **6**, 170–84.

1966, Origin of string bogs: 155–9 in *Permafrost International Conference* (Lafayette, Ind., 11–15 Nov. 1963) *Proc.*: Natl. Acad. Sci.–Natl. Research Council Pub. **1287**. (563 pp.)

SCHERMERHORN, L. J. G., 1976, Reply: *Am. J. Sci.* **276**(10), 1315–24.

SCHILLING, WERNER, and WIEFEL, HEINZ, 1962, Jungpleistozäne Periglazialbildungen und ihre regionale Differenzierung in einigen Teilen Thüringens und des Harzes: *Geologie* **2**, 428–60.

SCHIRMER, W., 1967, Ein Pleistozän-Profil nordöstlich Aschaffenburgs: *Geschicht- und Kunstverein Aschaffenburg Veröffentlichungen* **10**, 201–8.

1970, Das jüngere Pleistozän in der Tongrube Kärlich am Mittelrhein: *Mainzer Naturwissenschaftliches Archiv* **9**, 257–84.

SCHMERTMANN, J. H., and TAYLOR, R. S., 1965, Quantitative data from a patterned ground site over permafrost: *US Army Cold Regions Research and Engineering Laboratory Research Rept.* **96**. (76 pp.)

SCHMID, JOSEF, 1955, Der Bodenfrost als morphologischer Faktor: Heidelberg, Dr Alfred Hüthig Verlag. (144 pp.)

1958, Rezente und fossile Frosterscheinungen im Bereich der Gletscherlandschaft der Gurgler Ache (Ötztaler Alpen): *Schlern-Schriften* **190**, 255–64.

SCHMITTHENNER, HEINRICH, 1923, Die Oberflächenformen der Stufenlandschaft zwischen Maas und Mosel: *Geog. Abh. 2d Ser.* **1**. (89 pp.)

SCHÖNHAGE, W., 1969, Notes on the ventifacts in the Netherlands: *Biuletyn Peryglacjalny* **20**, 355–60.

SCHOSTAKOWITSCH, W. B., 1927, Der ewig gefrorene Boden Siberiens: *Gesell. Erdkunde Berlin Zeitschr.*, **1927**, 394–427.

SCHOTT, CARL, 1931, Die Blockmeere in den deutschen Mittelgebirgen: *Forschungen zur Deutschen Landes- und Volkskunde* **29**(1), 1–78.

SCHRAMM, J. R., 1958, The mechanism of frost heaving of tree seedlings: *Am. Philosophical Soc. Proc.* **102**(4), 333–50.

SCHUBERT, CARLOS, 1973, Striated ground in the Venezuelan Andes: *J. Glaciology* **12**(66), 461–8.

1975, Glaciation and periglacial morphology in the northwestern Venezuelan Andes: *Eiszeitalter und Gegenwart* **26**, 196–211.

SCHULTZ, C. B., and FRYE, J. C., eds., 1968, Loess and related eolian deposits of the world: *International Assoc. for Quaternary Research (INQUA) Cong., 7th (Boulder, Col., 1965), Proc.* **12**. (367 pp.)

SCHUNKE, EKKEHARD, 1973, Palsen und Kryokarst in Zentral-Island: *Akad. Wiss. Göttingen Nachrichten, II Math.-Phys. Kl.*, **1973**(4), 65–102.

1974a, Formungsvorgänge an Schneeflecken im isländischen Hochland: 274–86 in Poser, Hans, ed., Geomorphologische Prozesse und Prozesskombinationen in der Gegenwart unter verschiedenen Klimabedingungen (Symposium and Report of Commission on Present-day Geomorphological Processes, Internat. Geog. Union): *Akad. Wiss. Göttingen Abh., Math.-Phys. Kl. Folge 3*, **29**. (440 pp.)

1974b, Frostspaltenpolygone im westlichen Zentral-Island, ihre klimatischen und edaphischen Bedingungen: *Eiszeitalter und Gegenwart* **25**, 157–65.

1975a, Die Periglazialerscheinungen Islands in Abhängigkeit von Klima und Substrat: *Akad.*

Wiss. Göttingen Abh., Math.-Phys. Kl. Folge 3, **30**. (273 pp.)

1975b, Neue Beobachtungen zur Periglazialmorphologie Islands (Ein Forschungsbericht): *Gesell. Erdkunde Berlin Zeitschr.* **106**(1–2), 47–56.

1977a, Zur Ökologie der Thufur Islands (The ecology of Thufurs in Iceland): *Forschungstelle Neðri Ás, Hveragerði (Island), Berichte* **26**. (69 pp.)

1977b, Zur Genese der Thufur Islands und Öst-Grönlands: *Erdkunde* **31**, 279–87.

1977c, Periglazialformen und -formengesellschaften in der europäisch-atlantischen Arktis und Subarktis: 39–62 in Poser, Hans, ed., Formen, Formengesellschaften und Untergrenzen in den heutigen periglazialen Höhenstufen der Hochgebirge Europas und Afrikas zwischen Arktis und Äquator. Bericht über ein Symposium: *Akad. Wiss. Göttingen Abh., Math.-Phys. Kl. Folge 3*, **31**. (355 pp.)

SCHWARZBACH, MARTIN, 1963, Zur Verbreitung der Strukturböden und Wüsten in Island: *Eiszeitalter und Gegenwart* **14**, 85–95.

SCOTT, B. W., 1965, The ecology of the alpine tundra on Trail Ridge: *International Assoc. for Quaternary Research (INQUA) Cong., 7th (Boulder, Col., 1965), Guidebook Boulder area*, 13–16.

SCOTT, I. D., 1927, Ice-push on lake shores: *Michigan Acad. Sci. Arts and Letters Papers* **7**, 107–23.

SCOTT, K. M., 1978, Effects of permafrost on stream channel behavior in arctic Alaska: *U.S. Geol. Survey Prof. Paper* **1068**. (19 pp.)

SCOTT, R. F., 1969, The freezing process and mechanics of frozen ground: *US Army Corps of Engineers, Cold Regions Research and Engineering Laboratory, Cold Regions Science and Engineering Mon* **II**-D1. (67 pp.)

SCOTT, W. J., SELLMANN, P. V., and HUNTER, J. A., 1979, A review of applications of geophysics in permafrost regions: In *Third International Conference on Permafrost* (Edmonton, Alta., 10–13 July 1978), *Proc.* **2**: Ottawa, Canada Natl. Research Council. (In press.)

SCOTTER, G. W., 1975, Permafrost profiles in the Continental Divide region of Alberta and British Columbia: *Arctic and Alpine Research* **7**, 93–5.

SEGERSTROM, KENNETH, 1950, Erosion studies at Parícutin, State of Michoacán, Mexico: *US Geol. Survey Bull.* **965**-A, 1–151.

SEGUIN, M. K., 1977, Détermination de la géométrie et des propriétés physiques du pergélisol discontinu de la région de Schefferville: *Canadian J. Earth Sci.* **14**, 431–43.

SEIDENFADEN, GUNNAR, 1931, Moving soil and vegetation in East Greenland: *Medd. om Grønland* **87**(2). (21 pp.)

SEKYRA, JOSEF, 1956, The development of cryopedology in Czechoslovakia: *Biuletyn Peryglacjalny* **4**, 351–69.

1960, Působeni mrazu na pŭdu; Kryopedologie se zvláštním zřetelem k ČSR (Frost action on the ground; Cryopedology with special reference to Czechoslovakia): *Geotechnica* **27**. (164 pp.)

1964, Kvarterně-geologické a geomorfologické problémy krkonošského krystalinika: *Opera Corcontica* **1**, 7–24.

1972, Forms of mechanical weathering and their significance in the stratigraphy of the Quaternary in Antarctica: 669–74 in Adie, R. J., ed., *Antarctic geology and geophysics* (Internat. Union Geol. Sci., Ser. B, no. 1): Oslo, Universitetsforlaget. (876 pp.)

SELBY, M. J., 1972, Antarctic tors: *Zeitschr. Geomorphologie N.F., Supp.* **13**, 73–86.

1977, Transverse erosional marks on ventifacts from Antarctica: *New Zealand J. Geol. and Geophys.* **20**(5), 949–69.

SELBY, M. J., et al., 1973, Ventifact distribution and wind directions in the Victoria Valley, Antarctica (*Note*): *New Zealand J. Geol. and Geophys.* **16**(2), 303–6.

SELBY, M. J., RAINS, R. B., and PALMER, W. P., 1974, Eolian deposits of the ice-free Victoria Valley, southern Victoria Land, Antarctica: *New Zealand J. Geol. and Geophys.* **17**(3), 543–62.

SELLMANN, P. V., 1967, Geology of the USA CRREL permafrost tunnel Fairbanks, Alaska: *US Army Cold Regions Research and Engineering Laboratory Tech. Rept.* **199**. (22 pp.)

1972, Geology and properties of materials exposed in the USA CRREL permafrost tunnel: *US Army Corps of Engineers, Cold Regions Research and Engineering Laboratory Spec. Rept.* **177**. (16 pp.)

SELLMANN, P. V., et al., 1972, Terrain and coastal conditions on the Arctic Alaskan Coastal Plain: Arctic environmental data package supplement 1: *US Army Corps of Engineers, Cold Regions Research and Engineering Laboratory Spec. Rept.* **165**. (74 pp.)

1975, The classification and geomorphic implications of thaw lakes on the Arctic Coastal Plain, Alaska: *US Army Corps of Engineers, Cold Regions Research and Engineering Laboratory Research Rept.* **344**. (21 pp.)

1976, Delineation and engineering characteristics of permafrost beneath the Beaufort Sea: 391–408 in US Departments of Commerce and Interior: *Environmental assessment of the Alaskan Continental Shelf* **12**. (676 pp.)

SELZER, GEORG, 1959, 'Erdkegel' als heutige Frost-boden-Bildungen an Rutschhangen im Saarland: *Eiszeitalter und Gegenwart* **10**, 217–23.

SEMMEL, ARNO, 1968, Studien uber den Verlauf jungpleistozäner Formung in Hessen: *Frankfurter Geog. Hefte* **45**.

1969, Verwitterungs- und Abtragungserscheinungen in rezenten Periglazialgebieten (Lappland und Spitzbergen): *Würzburger Geog. Arbeiten* **26**. (82 pp.)

1973a, Periglacial sediments and their stratigraphy: 293, 305 in Behre, K.-E., et al., State of research on the Quaternary of the Federal Republic of Germany: *Eiszeitalter und Gegenwart* **23–4**.

1973b, Periglazialer Umlagerungszonen auf Moränen und Schotterterrassen der letzten Eiszeit im deutschen Alpenvorland: *Zeitschr. Geomorphologie N.F., Supp.* **17**, 118–32.

1974, Das Eiszeitalter im Rhein–Main–Gebiet: *Rheinmainische Forschungen* **78**.

1976, Aktuelle subnivale Hang- und Talentwicklung im zentralen West-Spitzbergen: *Deutscher Geographentag (Innsbruck, 1975) Verhandlungen* **40**, 396–400.

1977, Untersuchungen zur periglazialen Formung auf Spitsbergen: 34–8 in Poser, Hans, ed., Formen, Formengesellschaften und Untergrenzen in den heutigen periglazialen Höhenstufen der Hochgebirge Europas und Afrikas zwischen Arktis und Äquator. Bericht über ein Symposium: *Akad. Wiss. Göttingen Abh., Math.–Phys. Kl. Folge 3*, **31**. (355 pp.)

SEPPÄLÄ, MATTI, 1972a, The term 'palsa': *Zeitschr. Geomorphologie N. F.* **16**, 463.

1972b, Pingo-like remnants in the Peltojärvi area of Finnish Lapland: *Geografiska Annaler* **54**A, 38–45.

1973, On the formation of periglacial sand dunes in northern Fennoscandia (abs.): 318–19 in International Union for Quaternary Research (*INQUA*) Cong., 9th (Christchurch, New Zealand, 2–10 Dec. 1973), Abstracts. (418 pp.)

1974, Some quantitative measurements of the present-day deflation on Hietatievat, Finnish Lapland: 208–20 in Poser, Hans, ed., Geomorphologische Prozesse und Prozesskombinationen in der Gegenwart unter verschiedenen Klimabedingungen (Symposium and Report of Commission on Present-day Geomorphological Processes, Internat. Geog. Union): *Akad. Wiss. Göttingen Abh., Math.–Phys. Kl. Folge 3*, **29**. (440 pp.)

1976, Periglacial character of the climate of the Kevo region (Finnish Lapland) on the basis of meteorological observations 1962–71: *Kevo Subarctic Res. Stat. Rept.* **13**, 1–11.

1977, Distribution and character of palsas in Finnish Lapland (abs.): *International Union for Quaternary*

Research (*INQUA*) *Cong.*, *10th* (*Birmingham, England, 16–24 Aug. 1977*), *Abstracts*, 411.

SEVON, W. D., 1972, Late Wisconsinan periglacial boulder deposits in northeastern Pennsylvania (abs.): *Geol. Soc. America Abs. with Programs* **4**(1), 43–4.

SHARBATYAN, A. A., 1974a, Mnogoletnyaya kriolitozona: 370–2 in *Bol'shaya Sovetskaya Entsiklopediya* **16**: Moskva, Izdatel'stvo 'Sovetskaya Entsiklopediya'. (616 pp.)

1974b, *Ekstremal'nyye otsenki v geotermii i geokriologii*: Akad. Nauk SSSR, Institut Vodnykh Problem: Moskva, Izdatel'stvo 'Nauka'. (122 pp.)

1975, Perennial cryolithic zone [Mnogoletnyaya kriolitozona]: *US Army Corps of Engineers, Cold Regions Research and Engineering Laboratory Draft Translation* **484**. (5 pp.)

SHARBATYAN, A. A., and SHUMSKIY, P. A., 1974, Extreme estimations in geothermy and geocryology [Ekstremal'nyye otsenki v geotermii i geokriologii]: *US Army Corps of Engineers, Cold Regions Research and Engineering Laboratory Draft Translation* **465**. (140 pp.)

SHARP, R. P., 1942a, Soil structures in the St Elias Range, Yukon Territory: *J. Geomorphology* **5**, 274–301.

1942b, Periglacial involutions in northeastern Illinois: *J. Geol.* **50**, 113–33.

1942c, Multiple Pleistocene glaciation on San Francisco Mountain, Arizona: *J. Geol.* **50**, 481–503.

1949, Pleistocene ventifacts east of the Bighorn Mountains, Wyoming: *J. Geol.* **57**, 174–95.

1974, Ice on Mars: *J. Glaciology* **13**(68), 173–85.

SHARPE, C. F. S., 1938, *Landslides and related phenomena*: New York, Columbia Univ. Press. (137 pp.) (Also: New Jersey, Pageant Books, 1960.)

SHAW, JOHN, and HEALY, T. F., 1977, Rectilinear slope development in Antarctica: *Assoc. Am. Geographers Annals* **67**(1), 46–54.

SHEARER, J. M., 1972, Beaufort Sea, east of Mackenzie Bay, submarine pingo-like features: *Ice* **38**, 6.

SHEARER, J. M., *et al.*, 1971, Submarine pingos in the Beaufort Sea: *Science* **174**, 816–18.

SHERMAN, R. G., 1973, A groundwater supply for an oil camp near Prudhoe Bay, Arctic Alaska: 469–72 in *North American Contribution, Permafrost Second International Conference* (Yakutsk, USSR, 13–28 July 1973): Washington, DC, Natl. Acad. Sci. (783 pp.)

SHILTS, W. W., 1973, Drift prospecting; geochemistry of eskers and till in permanently frozen terrain: District of Keewatin; Northwest Territories: *Canada Geol. Survey Paper* **72–45**. (34 pp.)

1974, Physical and chemical properties of unconsolidated sediments in permanently frozen terrain, District of Keewatin: *Canada Geol. Survey Paper* **74–1**, Part A, 229–35.

1977, Geochemistry of till in perennially frozen terrain of the Canadian Shield – application to prospecting: *Boreas* **6**(2), 203–12.

1978, Nature and genesis of mudboils, central Keewatin, Canada: *Canadian J. Earth Sci.* **15**, 1053–68.

SHILTS, W. W., and DEAN, W. E., 1975, Permafrost features under arctic lakes, District of Keewatin, Northwest Territories: *Canadian J. Earth Sci.* **12**, 649–62.

SHORT, A. D., and WISEMAN, W. J., JR., 1973, Freezing effects on arctic beaches: *Louisiana State Univ., Coastal Studies Inst. Bull.* **7**, 23–31.

SHOTTON, F. W., 1960, Large scale patterned ground in the valley of the Worcestershire Avon: *Geol. Mag.* **97**, 404–8.

SHREVE, R. L., 1968a, The Blackhawk landslide: *Geol. Soc. America Spec. Paper* **108**. (47 pp.)

1968b, Sherman landslide: 395–401 in The great Alaska earthquake of 1964. *Hydrology* A: *Natl. Acad. Sci. Pub.* **1603**. (441 pp.)

SHROCK, R. R., 1948, *Sequence in layered rocks*: New York, McGraw-Hill. (507 pp.)

SHUMSKIY, P. A., 1959, Podzemnyye l'dy: 274–327 (Glava IX) in Inst. Merzlotovedeniya im. V. A. Obrucheva, *Osnovy geokriologii (merzlotovedeniya), Chast' pervaya, Obshchaya geokriologiya*: Moskva, Akad. Nauk SSSR. (459 pp.)

1964a, Ground (subsurface) ice [Podzemnyye l'dy]: *Canada Natl. Research Council Tech. Translation* **1130**. (118 pp.)

1964b, *Principles of structural glaciology* (Translated by David Kraus): New York, Dover Publications. (497 pp.)

SHUSHERINA, E. P., RACHEVSKIY, B. S., and OTROSHCHENKO, O. P., 1970, Issledovaniye temperaturnykh deformatsiy merzlykh gornykh porod: *Moskovskiy Gosudarstvennyy Universitet im. M. V. Lomonosova, Geologicheskiy Fakul'tet, Merzlotnyye Issledovaniya* **10**, 273–83.

SHUSHERINA, E. P., and ZAITSEV, V. N., 1976, Temperaturnyye deformatsiy mnogoletnemerzlykh dispersnykh porod i povtorno-zhil'nykh l'dov: *Moskovskiy Gosudarstvennyy Universitet im. M. V. Lomonosova, Geologicheskiy Fakul'tet, Merzlotnyye Issledovaniya* **15**, 187–97.

SHVETSOV, P. F., 1959, Obshchiye zakonomernosti vozniknoveniya i razvitiya mnogoletney kriolitozony: 77–107 (Glava IV) in Inst. Merzlotovedeniya im. V. A. Obrucheva, *Osnovy geokriologii* (*merzlotovedeniya*), *Chast' pervaya, Obshchaya geo-kriologiya:* Moskva, Akad. Nauk SSSR. (459 pp.)

1964, General mechanisms of the formation and development of permafrost [Obshchiye zakonomernosti vozniknoveniya i razvitiya mnogoletney kriolitozony]: *Canada Natl. Research Council Tech. Translation* **1117**. (91 pp.)

SIGAFOOS, R. S., and HOPKINS, D. M., 1952, Soil instability on slopes in regions of perennially-frozen ground: 176–92 in *Frost action in soils: a symposium: Natl. Acad. Sci.–Natl. Research Council Highway Research Board Spec. Rept.* **2**. (385 pp.)

SISSONS, J. B., 1965, Quaternary: 467–503 in Craig, G. Y., ed., *The geology of Scotland*: Edinburgh and London, Oliver & Boyd, Ltd. (556 pp.)

1976, A remarkable protalus rampart complex in Wester Ross: *Scottish Geog. Mag.* **92**(3), 182–90.

SLAYMAKER, OLAV, 1974, Rates of operation of geomorphological processes in the Canadian Cordillera: 319–32 in Poser, Hans, ed., Geomorphologische Prozesse und Prozesskombinationen in der Gegenwart unter verschiedenen Klimabedingungen (Symposium and Report of Commission on Present-day Geomorphological Processes, Internat. Geog. Union): *Akad. Wiss. Göttingen Abh., Math.–Phys. Kl. Folge 3*, **29**. (440 pp.)

SLOAN, C. E., ZENONE, CHESTER, and MAYO, L. R., 1976, Icings along the trans-Alaska pipeline route: *US Geol. Survey Prof. Paper* **979**. (31 pp.)

SMALLEY, I. J., 1972, The interaction of great rivers and large deposits of primary loess: *New York Acad. Sci. Trans. Ser. 2* **34**, 534–42.

SMALLEY, I. J., and KRINSLEY, D. H., 1978, Loess deposits associated with deserts: *Catena* **5**, 53–66.

SMITH, D. I., 1961, Operation Hazen – The geomorphology of the Lake Hazen region, NWT: *McGill Univ., Geog. Dept. Misc. Paper* **2**. (100 pp.)

1972, The solution of limestone in an Arctic environment: 187–200 in Price, R. J., and Sugden, D. E., compilers, *Polar geomorphology*: Inst. British Geographers Spec. Pub. **4**. (215 pp.)

SMITH, H. T. U., 1949a, Physical effects of Pleistocene climatic changes in nonglaciated areas: Eolian phenomena, frost action, and stream terracing: *Geol. Soc. America Bull.* **60**, 1485–516.

1949b, Periglacial features in the driftless area of southern Wisconsin: *J. Geol.* **57**, 196–215.

1962, Periglacial frost features and related phenomena in the United States: *Biuletyn Peryglacjalny* **11**, 325–42.

1964, Periglacial eolian phenomena in the United States: 177–86 in *International Assoc. for Quaternary*

Research (*INQUA*) *Cong., 6th* (*Warsaw, 1961*), *Rept.* **4**, Łódź, Poland. (596 pp.)

1965, Dune morphology and chronology in central and western Nebraska: *J. Geol.* **73**, 557–78.

1966, Wind-formed pebble ripples in Antarctica: *Geol. Soc. America Spec. Paper* **87**, 160.

1968, 'Piping' in relation to periglacial boulder concentrations: *Biuletyn Peryglacjalny* **17**, 195–204.

SMITH, JEREMY, 1960, Cryoturbation data from South Georgia: *Biuletyn Peryglacjalny* **8**, 73–9.

SMITH, M. W., 1975, Microclimatic influences on ground temperatures and permafrost distribution, Mackenzie Delta, Northwest Territories: *Canadian J. Earth Sci.* **12**, 1421–38.

1976, Permafrost in the Mackenzie Delta, Northwest Territories: *Canada Geol. Survey Paper* **75-28**. (34 pp.)

SOCHAVA, V. B., 1930, Gora Standukhina na kraynem severovostotske Azii: *Priroda* **11-12**.

SOERGEL, WOLFGANG, 1919, *Lösse, Eiszeiten, and palaolithische Kulturen*: Jena, G. Fischer. (177 pp.)

1936, Diluviale Eiskeile: *Deutsche Geol. Gesell. Zeitschr.* **88**, 223–47.

SOFRONOV, G. P., 1945, K geomorfologii Voykarskogo rayona-Polyarnyy Ural: *Akad. Nauk SSSR, Izvestiya, seriya geologicheskaya* **4**.

SOIL SURVEY STAFF, 1975, *Soil taxonomy. A basic system of soil classification for making and interpreting soil surveys*: US Dept. Agriculture, Soil Conservation Service Agriculture Handbook 436. (754 pp.)

SOLOV'YEV [SOLOVIEV, SOLOVYEV], P. A., 1962, Alasnyy rel'yef Centralnoy Yakutii i ego proiskhozhdeniye: 38–53 in *Mnogoletnemerzlyye porody i soputstvuyushchiye im yavleniya na territorii Yakutskoy SSR*: Moskva, Izdatel'stvo AN SSSR.

1973a, Alass thermokarst relief of Central Yakutia: Internat. Permafrost Conf., 2 (Yakutsk, USSR, 13–28 July 1973), Guidebook: Yakutsk, USSR, Acad. Sci., Sec. Earth Sci., Siberian Div. (48 pp.)

1973b, Thermokarst phenomena and landforms due to frost heaving in Central Yakutia: *Biuletyn Peryglacjalny* **23**, 135–55.

1975, Termokarstovyye yavleniya i formy mnogoletnego kriogennogo (moroznogo) pucheniya v Tsentral'noy Yakutii: 23–36 in Markov, K. K., and Spasskaya, I. I., eds., *Paleogeografiya i periglyatsial'nyye yavleniya pleystotsena*: Moskva, Akad. Nauk SSSR, Komissiya po izucheniyu Chetvertichnogo Perioda, Izdatel'stvo 'Nauka'. (224 pp.)

SOLOV'YEV, V. V., 1961, Sledy drevnego oledeniya i periglyacialnykh usloviy v Yuzhnom Primorye:

Vsesoyuznogo geologicheskogo Instituta, Trudy, novaya seriya **64**, 141–8.

SOONS, J. M., 1962, A survey of periglacial features in New Zealand: 74–87 in McCaskill, Murray, ed., *Land and Livelihood – Geographical essays in honor of George Jobberns*: Christchurch, New Zealand Geographical Society. (280 pp.)

SOONS, J. M., and GREENLAND, D. E., 1970, Observations on the growth of needle ice: *Water Resources Research* **6**, 579–93.

SØRENSEN, THORVALD, 1935, Bodenformen und Pflanzendecke in Nordostgrönland: *Medd. om Grønland* **93**(4). (69 pp.)

SOUCHEZ, R., 1967, Gélivation et évolution des versants en bordure de l'Inlandsis d'Antarctide orientale: 291–8 in L'évolution des versants: *Congrès et Colloques de l'Université de Liège* **40**. (384 pp.)

1969, Rate of frost shattering and slope development in dolomitic limestone, southwestern Ellesmere Island, Arctic Canada (abs.): 50 in *International Union for Quaternary Research (INQUA) Cong., 8th (Paris, 1969), Résumés des communications*. (389 pp.)

SPENCER, A. C., 1900, A peculiar form of talus (abs.): *Science* **11**, 188.

SPENCER, A. M., 1971, Late Pre-Cambrian glaciation in Scotland: *Geol. Soc. London Mem.* **6**. (100 pp.)

SPETHMANN, H., 1912, Über Bodenbewegungen auf Island: *Gesell. Erdkunde Berlin Zeitschr.*, **1912**, 246–8.

SPÖNEMANN, JÜRGEN, 1966, Geomorphologische Untersuchungen an Schicht-kämmen des Niedersächsischen Berglandes: *Göttinger Geog. Abh.* **36**. (167 pp.)

1974, Spülstreifen – eine besondere Erscheinung der Flächenspülung. Beobachtungen aus Kenia: 137–46 in Poser, Hans, ed., Geomorphologische Prozesse und Prozesskombinationen in der Gegenwart unter verschiedenen Klimabedingungen (Symposium and Report of Commission on Present-day Geomorphological Processes, Internat. Geog. Union): *Akad. Wiss. Göttingen Abh., Math.-Phys. Kl. Folge 3*, **29**. (440 pp.)

1977, Die periglaziale Höhenstufe Ostafrikas: 300–32 in Poser, Hans, ed., Formen, Formengesellschaften und Untergrenzen in den heutigen periglazialen Höhenstufen der Hochgebirge Europas und Afrikas zwischen Arktis und Äquator. Bericht über ein Symposium: *Akad. Wiss. Göttingen Abh., Math.-Phys. Kl. Folge 3*, **31**. (355 pp.)

SPRINGER, M. E., 1958, Desert pavement and vesicular layer of some soils of the desert of the Lahontan Basin, Nevada: *Soil Sci. Soc. America Proc.* **22**, 63–6.

ST-ONGE, DENIS, 1965, La géomorphologie de l'Île Ellef Ringnes, Territoires du Nord-Ouest, Canada: *Canada, Ministère des Mines et des Relevés techniques, Direction de la Géographie Étude géog.* **38**. (46 pp.)

1969, Nivation landforms: *Canada Geol. Survey Paper* **69–30**. (12 pp.)

STÄBLEIN, GERHARD, 1977, Periglaziale Formengesellschaften und rezente Formungsbedingungen in Grönland: 18–33 in Poser, Hans, ed., Formen, Formengesellschaften und Untergrenzen in den heutigen periglazialen Höhenstufen der Hochgebirge Europas und Afrikas zwischen Arktis und Äquator. Bericht über ein Symposium: *Akad. Wiss. Göttingen Abh., Math.–Phys. Kl. Folge 3*, **31**. (355 pp.)

STAGER, J. K., 1956, Progress report on the analysis of the characteristics and distribution of pingos east of the Mackenzie Delta: *Canadian Geographer* **7**, 13–20.

STALKER, A. M., 1960, Ice-pressed drift forms and associated deposits in Alberta: *Canada Geol. Survey Bull.* **57**. (38 pp.)

STARKEL, LESZEK, 1960, Periglacial covers in the Beskid Wyspowy (Carpathians): *Biuletyn Peryglacjalny* **8**, 155–69 (cf. 329–40).

1963, Der Stand der Forschungen über die morphogenetischen Prozesse im Quatär in den Karpathen: *Akad. Wiss. Göttingen Nachrichten, Math.–Phys. Kl. II*, **10**, 139–61.

STEARNS, S. R., 1966, Permafrost (perennially frozen ground): *US Army Cold Regions Research and Engineering Laboratory, Cold Regions Science and Engineering* [Mon.] **1**–A2. (77 pp.)

STECHE, HANS, 1933, Beiträge zur Frage der Strukturböden: *Sächsischen Akad. Wiss. Leipzig Berichte, Math.–phys. Kl.* **85**, 193–272.

STEFANSSON, VILHJALMUR, 1921, *The friendly Arctic*: New York, The Macmillan Company. (784 pp.)

STEFFENSEN, ESTHER, 1969, The climate and its recent variations at the Norwegian Arctic stations: *Norske Meteorologiske Institutt, Meteorologiske Annaler* **5**(8), 214–349.

STEHLÍK, O., 1960, Skalní tvary ve východní části Moravskoslezských Beskyd: *Dějepis a zeměpis ve škole* **3**, 46–7.

STEINEMANN, SAMUEL, 1953, Kammeis, eine anomale Wachstumsform der Eiskristalle: *Zeitschr. angew. Mathematik und Physik (ZAMP)* **4**, 500–6.

1955, Mushfrost, an anomalous growth form of the ice crystal [Kammeis, eine anomale Wachstumsform der Eiskristalle]: *Canada Natl. Research Council Tech. Translation* **528**. (11 pp.)

STEPHENSON, P. J., 1961, Patterned ground in Antarctica [correspondence]: *J. Glaciology* **3**, 1163–4.

STINGL, HELMUT, 1969, Ein periglazial–morphologisches Nord-Süd-Profil durch die Ostalpen: *Göttinger Geog. Abh.* **49**. (115 pp.)

1971, Zur Verteilung von Gross- und Miniaturformen von Strukturböden in den Ostalpen: *Akad. Wiss Göttingen Nachrichten, Math.–Phys. Kl. II*, **2**, 25–40.

1974, Zur Genese und Entwicklung von Strukturbodenformen: 249–62 in Poser, Hans, ed., Geomorphologische Prozesse und Prozesskombinationen in der Gegenwart unter verschiedenen Klimabedingungen (Symposium and Report of Commission on Present-day Geomorphological Processes, Internat. Geog. Union): *Akad. Wiss. Göttingen Abh., Math.–Phys. Kl. Folge 3*, **39**. (440 pp.)

STINGL, HELMUT, and HERRMANN, REIMER, 1976, Untersuchungen zum Strukturbodenproblem auf Island, Geländebeobachtungen und statistische Auswertung: *Zeitschr. Geomorphologie N. F.* **20**, 205–26.

STOCKER, ERICH, 1973, Bewegungsmessungen und Studien an Schrägterrassen an einem Hangausschnitt in der Kreuzeckgruppe (Kärnten): *Univ. Salzburg Geog. Inst. Beiträge zur Klimatologie, Meteorologie und Klimamorphologie* **3** (*Festschrift Hanns Tollner*), 193–203.

STONE, P., 1975, An unusual form of patterned ground, Cooper Bay, South Georgia: *British Antarctic Survey Bull.* **41–42**, 195–7.

STRAHLER, A. N., 1969, *Physical geography*, **3** ed.: New York, John Wiley. (733 pp.)

STREIFF-BECKER, RUDOLPH, 1946, Strukturböden in den Alpen: *Geog. Helvetica* **1**, 150–7.

STRIGIN, V. M., 1960, Vysotnaya fiziko-geograficheskaya poyasnost Denezhkina Kamnya na Severnom Urale: *Voprosy fizicheskoy geografii Urala* (*Moskovskoye obshchestvo ispytateley prirody, Geograficheskaya sektsiya*), 113–18.

STROCK, CLIFFORD, and KORAL, R. L., eds., 1959, *Handbook of air conditioning, heating and ventilating*, 2 ed.: New York, The Industrial Press. (1472 pp.)

STRÖMQUIST, LENNART, 1973, Geomorfologiska studier av blockhav och blockfält i norra Skandinavien (Geomorphological studies of block-fields in northern Scandinavia): *Uppsala Univ., Naturgeografiska Inst., Avdelningen för Naturgeografi, UNGI Rapport* **22**. (161 pp.)

SUGDEN, D. E., 1971, The significance of periglacial activity on some Scottish mountains: *Geog. J.* **137**, 388–92.

1977, Reconstruction of the morphology, dynamics, and thermal characteristics of the Laurentide ice sheet at its maximum: *Arctic and Alpine Research* **9**, 21–47.

SUGDEN, D. E., and WATTS, S. H., 1977, Tors, felsenmeer, and glaciation in northern Cumberland Peninsula, Baffin Island: *Canadian J. Earth Sci.* **14**(12), 2817–23.

SUKACHËV, V., 1910, Rastitelnost verkhney chasti basseyna reki Tungira, Olekminskogo okruga Yakutskòy oblasti: *Amurskoy ekspedicii, Trudy* **16** (Botanicheskiye issledovaniya), 265.

1911, K voprosu o vliyanii merzloty na pochvu: *Akad. Nauk Izv., ser. 6* **5**, 51–60.

[SUKATSCHEW, W.], 1912, Die Vegetation des oberen Einzugsgebietes des Flusses Tungir im Kreise Olekminsk im Bezirk Jakutsk: *Arbeiten der auf Allerhochsten Befehl ausgefuhrten Amur-Expedition* **1**(16), St Petersburg.

SUZDALSKIY, O. V., 1952, Po povodu nagornykh terras Visherskogo Urala: *Vsesoyuznogo Geograficheskogo obshchestva, Izvestiya* **84**, 102–3.

SVENSSON, HARALD, 1967, A tetragon patterned block field: 8–23 in Svensson, Harald, *et al.*, Polygonal ground and solifluction features: *Lund Studies in Geography, Ser. A. Physical Geography* **40**. (67 pp.)

1969, A type of circular lake in northernmost Norway: *Geografiska Annaler* **51**A(1–2), 1–12.

1970, Pingos i yttre delen av Adventdalen: *Norsk Polarinstitutt Årbok* **1969**, 168–74.

1973, Distribution and chronology of relict polygon patterns on the Laholm plain, the Swedish west coast: *Geografiska Annaler* **55**A(3–4), 159–75.

1976a, Relict ice-wedge polygons revealed in aerial photographs from Kaltenkirchen, northern Germany: *Geografisk Tidsskrift* **75**, 8–12.

1976b, Pingo problems in the Scandinavian countries: *Biuletyn Peryglacjalny* **26**, 33–40.

1976c, Iskilar som klimatindikator: *Svensk Geografisk Årsbok* **52**, 46–57.

1977, Observations on polygonal fissuring in non-permafrost areas of the northern countries: 63–76 in Poser, Hans, ed., Formen, Formengesellschaften und Untergrenzen in den heutigen periglazialen Höhenstufen der Hochgebirge Europas und Afrikas zwischen Arktis und Äquator. Bericht über ein Symposium: *Akad. Wiss. Göttingen Abh., Math.–Phys. Kl. Folge 3*, **31**. (355 pp.)

SVENSSON, HARALD, *et al.*, 1967, Polygonal ground and solifluction features: *Lund Studies in Geography, Ser. A. Physical Geography* **40**. (67 pp.)

SWINZOW, G. K., 1969, Certain aspects of engineering geology in permafrost: *Eng. Geol.* **3**, 177–215.

TABER, STEPHEN, 1918, Ice forming in clay soils will lift surface weights: *Eng. News-Rec.* **80**(6), 262–3.

1929, Frost heaving: *J. Geol.* **37**, 428–61.

1930*a*, The mechanics of frost heaving: *J. Geol.* **38**, 303–17.

1930*b*, Freezing and thawing of soils as factors in the destruction of road pavements: *Public Roads* **11**, 113–32.

1943, Perennially frozen ground in Alaska: its origin and history: *Geol. Soc. America Bull.* **54**, 1433–548.

1950, Intensive frost action along lake shores: *Am. J. Sci.* **248**, 784–93.

1952, Geology, soil mechanics, and botany: *Science* **115**, 713–14.

1953, Origin of Alaska silts: *Am. J. Sci.* **251**, 321–36.

TAILLEFER, F., 1944, La dissymétrie des vallées Gasconnes: *Rev. Géog. Pyrénées Sud-Ouest* **15**, 153–81.

TAKAGI, SHUNSUKI, 1977, Segregation freezing temperature as the cause of suction force: 59–66 in *Frost action in soils*: Univ. Luleå, Internat. Symposium (Luleå, Sweden, 16–18 Feb. 1977) Proc. 1. (215 pp.)

1978, Segregation freezing as the cause of suction force for ice lens formation: *US Army Corps of Engineers, Cold Regions Research and Engineering Laboratory Rept.* **78–6**. (13 pp.)

TANTTU, ANTTI, 1915, Ueber die Enstehung der Bülten und Stränge der Moore: *Acta Forestalia Fennica* **4**, 1–24.

TARNOCAL., C., and ZOLTAI, S. C., 1978, Earth hummocks of the Canadian Arctic and Subarctic: *Arctic and Alpine Research* **10**, 581–94.

TARNOGRADSKIY, G. S., 1963, Reliktovyye nagornyye terrasy na zapadnom sklone Severnogo Urala: *Vsesoyuznogo Geograficheskogo obshchestva, Izvestiya* **95**(4), 358–60.

TARR, R. S., 1897, Rapidity of weathering and stream erosion in the arctic latitudes: *Am. Geologist* **19**, 131–6.

TAYLOR, A. E., and JUDGE, A. S., 1977, Canadian geothermal data collection – Northern wells 1976–77: *Canada, Energy, Mines and Resources, Geothermal Service of Canada Geothermal Ser. No.* **10**, 10–12.

TAYLOR, GRIFFITH, 1922, *The physiography of the McMurdo Sound and Granite Harbour region: British Antarctic (Terra Nova) Expedition 1910–1913*: London, Harrison. (246 pp.)

TAYLOR, R. B., 1978, The occurrence of grounded ice ridges and shore ice piling along the northern coast of Somerset Island, NWT: *Arctic* **31**, 133–49.

TAYLOR, R. B., and MCCANN, S. B., 1976, The effect of sea and nearshore ice on coastal processes in Canadian Arctic Archipelago: *Rev. Géog. Montréal* **30**(1–2) (*Le glaciel* – Premier colloque international sur l'action géologique des glaces flottantes, Québec, Canada, 20–24 Apr. 1974), 123–32.

TEDROW, J. C. F., 1969, Thaw lakes, thaw sinks and soils in northern Alaska: *Biuletyn Peryglacjalny* **20**, 337–44.

1970, Soil investigations in Inglefield Land, Greenland: *Medd. om Grønland* **188**(3). (93 pp.)

1977, *Soils of the polar landscapes*: New Brunswick, NJ, Rutgers University Press. (638 pp.)

TEICHERT, C., 1935, Bedeutung des Windes in arktischen Gegenden: *Natur und Volk* **65**, 619–28.

1939, Corrasion by wind-blown snow in polar regions: *Am. J. Sci.* **237**, 146–8.

TE PUNGA, M. T., 1956*a*, Altiplanation terraces in Southern England: *Biuletyn Peryglacjalny* **4**, 331–8.

1956*b*, Fossil ice wedges near Wellington: *New Zealand J. Sci. and Tech.* **38B**(2), 97–102.

1957, Periglaciation in southern England: *Kon. Nederlandsch Aardrijks. Genoot. Tijdschr.*, 2d ser., **74**, 400–12.

TERZAGHI, KARL, 1952, Permafrost: *Boston Soc. Civil Engineers J.* **39**, 1–50.

THIBODEAU, ÉVA, and CAILLEUX, ANDRÉ, 1973, Zonation en latitude de structures de thermokarst et de tourbières vers 75° ouest, Québec: *Rev. Géog. Montréal* **27**, 117–38.

THIE, J., 1974, Distribution and thawing of permafrost in the southern part of the discontinuous permafrost zone in Manitoba: *Arctic* **27**, 189–200.

THOM, B. G., 1972, The role of spring thaw in string bog genesis: *Arctic* **25**, 236–9.

THOM, GORDON, 1978, Disruption of bedrock by the growth and collapse of ice lenses: *J. Glaciology* **20**(84), 571–5.

THOMAS, W. N., 1938, Experiments on the freezing of certain building materials: *Great Britain, Department Scientific and Industrial Research, Building Research Tech. Paper* **17**. (146 pp.)

THOMASSON, A. J., 1961, Some aspects of the drift deposits and geomorphology of south-east Hertfordshire: *Geologist's Assoc., London, Proc.* **72**, 287–302.

THOMPSON, W. F., 1962, Preliminary notes on the nature and distribution of rock glaciers relative to true glaciers and other effects of the climate on the ground in North America: *Internat. Assoc. Sci.*

Hydrology, Symposium of Obergurgl, Pub. **58**, 212–19.

1968, New observations on alpine accordances in the western United States: *Assoc. Am. Geographers Annals* **58**, 650–69.

THORARINSSON, SIGURDUR, 1951, Notes on patterned ground in Iceland, with particular reference to the Icelandic 'flás': *Geografiska Annaler* **33**, 144–56.

1964, Additional notes on patterned ground in Iceland with a particular reference to ice-wedge polygons: *Biuletyn Peryglacjalny* **14**, 327–36.

THORN, C. E., 1974*a*, *An analysis of nivation processes and their geomorphic significance, Niwot Ridge, Colorado Front Range*: Univ. Colorado, Ph.D. Thesis. (351 pp.)

1974*b*, Nivation, a reappraisal (abs.): *Geol. Soc. America Abs. with Programs* **6**(7), 987.

1976*a*, A model of stony earth circle development, Schefferville, Quebec: *Assoc. Am. Geographers Proc.* **8**, 19–23.

1976*b*, A quantitative evaluation of nivation: 226–8 in International Geographical Congress, 23d (Moscow, 27 July–3 Aug. 1976), *International Geography* **1** (*Geomorphology and Paleogeography*). (409 pp.)

1976*c*, Quantitative evaluation of nivation in the Colorado Front Range: *Geol. Soc. America Bull.* **87**, 1169–78.

THORODDSEN, TH., 1913, Polygonboden und 'thufur' auf Island: *Petermanns Mitt.* **59**(2), 253–5.

1914, An account of the physical geography of Iceland with special reference to the plant life: 187–343 in Kolderup-Rosenvinge, L., and Warming, Eugene, eds., 1912–18, *The botany of Iceland* **1**: Copenhagen, J. Frimodt; London, John Weldon. (675 pp.)

THORSTEINSSON, R., 1961, The history and geology of Meighen Island, Arctic Archipelago: *Canada Geol. Survey Bull.* **75**, 19.

THOSTE, V., 1974, *Die Niederterrassen des Rheins vom Neuwieder Becken bis in die Niederrheinische Bucht*: Univ. Köln Thesis.

TICE, A. R., ANDERSON, D. M., and BANIN, AMOS, 1973, The prediction of unfrozen water contents in frozen soils from liquid limit determinations: 329–42 in Organisation for Economic Co-operation and Development: *Symposium on Frost Action on Roads, Rept.* **1**: Paris, Organisation de Coopération et de Développement Economiques. (385 pp.)

1976, The prediction of unfrozen water contents in frozen soils from liquid limit determinations: *US Army Corps of Engineers, Cold Regions Research and Engineering Laboratory CRREL Rept.* **76–8**. (9 pp.)

TICE, A. R., BURROUS, C. M., and ANDERSON, D. M., 1978, Determination of unfrozen water in frozen soil by pulsed nuclear magnetic resonance: 149–55 in *Third International Conference on Permafrost* (Edmonton, Alta., 10–13 July 1978), Proc. **1**: Ottawa, Canada Natl. Research Council. (947 pp.)

TIMOFEYEV, D. A., 1965, *Srednaya i nizhnaya Olekma*: Moskva. (137 pp.)

TOLL, EDUARD V., 1895, Die fossilen Eislager und ihre Beziehungen zu den Mammuthleichen: *Wissenschaftliche Resultate der von der Kaiserlichen Akademie der Wissenschaften zur Erforschung des Janalandes und der Neusibirischen Inseln in den Jahren 1885 und 1886 ausgesandten Expedition—Abt. III, L'Acad. Impériale des Sciences de St.-Pétersbourg, VII ser.* **42**(13). (86 pp.)

TOLMACHEV, I. P., 1903, Geologicheskaya poyezdka v Kuznetski Alatau letom 1902 goda: *Imperatorskogo Russkogo Geograficheskogo obshchestva, Izvestiya* **39**, 390–436.

TOLSTIKHIN, N. I., and TOLSTIKHIN, O. N., 1974, Podzemnyye i poverkhnostnyye vody territorii rasprostraneniya merzloy zony: 192–229 (Glava IX) in Melnikov, P. I., and Tolstikhin, N. I., *Obshcheye merzlotovedeniye*: Novosibirsk, Sibirskoye Otdeleniye Izdatel'stvo 'Nauka'. (291 pp.)

TOTTEN, S. M., 1973, Glacial geology of Richland County, Ohio: *Ohio Geol. Survey Rept. of Investigations* **88**. (55 pp.)

TRAINER, F. W., 1961, Eolian deposits of the Matanuska Valley agricultural area, Alaska: *US Geol. Survey Bull.* **1121**–C. (34 pp.)

TRICART, JEAN, 1956*a*, Étude expérimentale du problème de la gélivation: *Biuletyn Peryglacjalny* **4**, 285–318.

1956*b*, Les actions périglaciaires du Quaternaire recent dans les Alpes du Sud: 189–97 in International Assoc. for Quaternary Research (INQUA) Cong., 4th (Rome–Pisa, Aug.–Sept. 1953), *Actes* **1**. (463 pp.)

1963, *Géomorphologie des régions froides*: Paris, Presses Universitaires de France. (289 pp.)

1966, Un chott dans le désert Chilien: La Pampa del Tamarugal: *Rev. Géomorphologie dynamique* **16**, 12–22.

1967, Le modelé des régions périglaciaires: Tricart, J., and Cailleux, A., *Traité de Géomorphologie* **2**, Paris, SEDES. (512 pp.)

1969, *Geomorphology of cold environments* (Translated by Edward Watson): London, Macmillan; New York, St Martin's Press. (320 pp.)

1970, Convergence de phénomènes entre l'action du

gel et celle du sel: *Acta Geographica Lodziensia* **24**, 425–36.

TROELSEN, J. C., 1952, An experiment on the nature of wind erosion, conducted in Peary Land, North Greenland: *Dansk. geol. Foren. Medd.* **12**, 221–2.

TROLL, CARL, 1944, Strukturböden, Solifluktion und Frostklimate der Erde: *Geol. Rundschau* **34**, 545–694.

—— 1947, Die Formen der Solifluktion und die periglazialer Bodenabtragung: *Erdkunde* **1**, 162–75.

—— 1958, Structure soils, solifluction, and frost climates of the earth [Strukturböden, Solifluktion, und Frostklimate der Erde]: *US Army Corps of Engineers, Snow Ice and Permafrost Research Establishment Translation* **43**. (121 pp.)

—— 1969, Inhalt, Probleme und Methoden geomorphologischer Forschung (mit besonderer Berücksichtigung der klimatischen Fragestellung): *Geol. Jahrbuch Beihandlung* **80**, 225–57.

—— 1973, Rasenabschälung (Turf Exfoliation) als periglaziales Phänomen der subpolaren Zonen und der Hochgebirge: *Zeitschr. Geomorphologie N.F., Supp.* **17**, 1–32.

TSKHURBAYEV, F. J., 1966, Geomorfologiya, chetvertichnyye otlozheniya i zolotonosnyye rossypy Nerskogo ploskogorya: *Geologiya rossypey zolota i zakonomernosti ikh razmeshcheniya v centralnoy chasti Yanokolymskogo skladchatogo poyasa*, 129–60.

TSVETAYEV, A. A., 1960, Klimaticheskiye osobennosti gornogo rayona Iremel: *Voprosy fizicheskoy geografii Urala*, 101–12.

TSYTOVICH, N. A., 1957, The fundamentals of frozen ground mechanics: 116–19 in *Internat. Conf. Soil Mech. and Found. Eng., 4th, London 1957, Proc.* **1**(28). (466 pp.)

—— 1958, Comments: 92–3 in *Internat. Conf. Soil Mech. and Found. Eng., 4th, London 1957, Proc.* **3**. (291 pp.)

—— 1966, Permafrost problems: 7 in *Permafrost International Conference* (Lafayette, Ind., 11–15 Nov. 1963) *Proc.*: Natl. Acad. Sci.–Natl. Research Council Pub. **1287**. (563 pp.)

—— 1975, *The mechanics of frozen ground*: Washington, DC, Scripta Book Company; New York [etc.], McGraw-Hill Book Company. (426 pp.)

TSYTOVICH, N. A., *et al.*, 1959, O fizicheskikh yavleniyakh i protsessakh v promerzayushchikh, merzlykh i protaivayushchikh gruntakh: 108–52 (Glava V) in Inst. Merzlotovedeniya im. V. A. Obrucheva, *Osnovy geokriologii (merzlotovedeniya), Chast' pervaya, Obshchaya geokriologiya*: Moskva, Akad. Nauk SSSR. (459 pp.)

—— 1964, Physical phenomena and processes in freezing, frozen and thawing soils [O fizicheskikh yav-leniyakh i protsessakh v promerzayushchikh, merzlykh i protaivayushchikh gruntakh]: *Canada Natl. Research Council Tech. Translation* **1164**. (109 pp.)

TUCK, RALPH, 1938, The loess of the Matanuska Valley, Alaska: *J. Geol.* **46**, 647–53.

TUCKER, M. E., 1978, Gypsum crusts (gypcrete) and patterned ground from northern Iraq: *Zeitschr. Geomorphologie N.F.* **22**, 89–100.

TUFNELL, LANCE, 1969, The range of periglacial phenomena in northern England: *Biuletyn Peryglacjalny* **19**, 291–323.

—— 1972, Ploughing blocks with special reference to north-west England: *Biuletyn Peryglacjalny* **21**, 237–70.

—— 1976, Ploughing block movements on the Moor House Reserve (England), 1965–75: *Biuletyn Peryglacjalny* **26**, 311–17.

TYRRELL, J. B., 1910, Ice on Canadian Lakes: *Canadian Inst. Trans.* **9**(20), 13–21.

TYRTIKOV, A. P., 1973, Vechnaya merzlota i rastitel'nost': 68–74 in *Akademiya Nauk SSSR, Sektsiya Nauk o Zemle, Sibirskoye Otdeleniye, II Mezhdunarodnaya Konferentsiya po Merzlotovedeniyu, Doklady i soobshcheniya* **2** (Regional'naya geokriologiya): Yakutsk, Yakutskoye Knizhnoye Izdatel'stvo. (154 pp.)

—— 1978, Permafrost and vegetation [Vechnaya merzlota i rastitel'nost']: 100–4 in Sanger, F. J., ed., *USSR Contribution Permafrost Second International Conference* (Yakutsk, USSR, 13–28 July 1973): Washington, DC, Natl. Acad. Sci. (866 pp.)

TYULINA, L. N., 1936, O lesnoy rastitelnosti Anadyrskogo kraya i eye vzaimootnoshenii s tundroy: *Arkticheskogo instituta, Trudy* **40**, 7–212.

—— 1948, O sledakh oledeneniya na severovostochnom poberezhye Baykala: *Problemy fizicheskoy geografii* **13**, 77–90.

TYULINA, L. O., 1931, O yavleniyakh svyazannykh s pochvennoy merzlotoy i moroznym vyvetrivaniyem na gore Iremel (Yuzhnyy Ural): *Gosudarstvennogo Geograficheskogo obshchestva, Izvestiya* **63**(2–3), 124–44.

TYUTYUNOV, I. A., 1964, *An introduction to theory of the formation of frozen rocks* (Translated from the Russian by J. O. H. Muhlhaus: translation edited by N. Rast): Oxford, Pergamon; New York, Macmillan. (94 pp.)

UGOLINI, F. C., 1966, Soils of the Mesters Vig district, Northeast Greenland. 1. The Arctic Brown and related soils: *Medd. om Grønland* **176**(1). (22 pp.)

—— 1979, Polar soil classification: In Fairbridge, R. W.,

and Finkl, C. W., eds., *Soil science – Morphology, genesis, and classification: Encyclopedia of earth science* **12**: Stroudsberg, Pa., Dowden, Hutchinson, & Ross, Inc. (In press.)

UGOLINI, F. C., and ANDERSON, D. M., 1973, Ionic migration and weathering in frozen antarctic soils: *Soil Science* **115**(6), 461–70.

UGOLINI, F. C., BOCKHEIM, J. G., and ANDERSON, D. M., 1973, Soil development and patterned ground evolution in Beacon Valley, Antarctica: 246–54 in *North American Contribution, Permafrost Second International Conference* (Yakutsk, USSR, 13–28 July 1973): Washington, DC, Natl. Acad. Sci. (783 pp.)

UNDERWOOD, J. R., JR., 1974, Contrasting types of patterned ground in southeast Asia (abs.): *Geol. Soc. America Abs. with Programs* **5**(2), 125–6.

US ARMY ARCTIC CONSTRUCTION AND FROST EFFECTS LABORATORY, 1958, Cold room studies, third interim report of investigations **1**: *US Army Corps of Engineers, New England Div. Tech. Rept.* **43**. (46 pp.)

US ARMY CORPS OF ENGINEERS COLD REGIONS RESEARCH AND ENGINEERING LABORATORY, 1951–, Bibliography on snow, ice, and permafrost: *CRREL Rept.* **12–**.

1973, II International Conference on Permafrost – II Mezhdunarodnaya Konferentsiya po Merzlotovedeniyu, *Draft Translations* **436–39**. (121 + 9, 172, 360, 326 pp.) (*Doklady i soobshcheniya* **6**, **5**, **7**, **4**, respectively.)

US ARMY WATERWAYS EXPERIMENT STATION, 1948, Trafficability of soils. Laboratory tests to determine effects of moisture content and density variations: *US Army Corps of Engineers Tech. Memo.* **3–240**. (First supp. 28 pp.)

1953, The unified soil classification system: *US Army Corps of Engineers Tech. Memo.* **3–357**. (30 pp.)

US DEPARTMENT OF COMMERCE, 1969–1970, *Monthly climatic data for the world* **22**: Asheville, N.C., Environmental Science Services Administration, Environmental Data Service. (532 pp.)

1975, *Monthly climatic data for the world* **28**: Asheville, N.C., National Oceanic and Atmospheric Administration (NOAA), Environmental Data Service, National Climatic Center.

VAN CAMPO, M., 1969, Végétation würmienne en France. Données bibliographiques. Hypothèse: 104–11 in *Études Françaises sur le Quaternaires – Présentées à l'occasion de VIIIe Congrès International de l'INQUA*: Paris, Suppl. au Bull. de l'AFEQ. (276 pp.)

VAN EVERDINGEN, R. O., 1976, Geocryological terminology: *Canadian J. Earth Sci.* **13**, 862–7.

1978, Frost mounds at Bear Rock, near Fort Norman, Northwest Territories, 1975–1976: *Canadian J. Earth Sci.* **15**, 263–76.

VANNEY, J.-R., and DANGEARD, LOUIS, 1976, Les dépôts glacio-marins actuels et anciens: *Rev. Géog. Montréal* **30**(1–2) (*Le glaciel* – Premier colloque international sur l'action géologique des glaces flottantes, Québec, Canada, 20–24 Apr. 1974), 9–50.

VAN ZUIDAM, R. A., 1976, Periglacial-like features in the Zaragoza region, Spain: *Zeitschr. Geomorphologie N. F.* **20**, 227–34.

VARNES, D. B., 1958, Landslide types and processes: 20–47 in Eckel, E. G., ed., Landslides and engineering practice: *Natl. Acad. Sci.–Natl. Research Council Pub.* **544** (*Highway Research Board Spec. Rept.* **29**). (232 pp.)

VARSANOFEVA, V. A., 1929, Geomorfologicheskiy ocherk basseyna reki Ylykha: *Instituta po izuchenii severa, Trudy* **42**. (120 pp.)

1932, Geomorfologicheskiye nablyudeniya v severnom Urale: *Gosudarstvennogo Geograficheskogo Obshchestva, Izvestiya* **64**(2–3), 105–71.

VELICHKO [VELIČKO], A. A., ed., 1969, *Lëss – peryglyatsial – paleolit na territorii sredney i vostochnoy Evropy* (*Loess – periglaciaire – paleolithique sur le territoire de l'Europe moyenne et orientale*) (Pour le VIII Congrès de l'INQUA): [Moskva] Akademii Nauk. (742 pp.)

1972, La morphologie cryogène relicte: caractères fondamentaux et cartographie: *Zeitschr. Geomorphologie N.F. Supp.* **13**, 59–72.

1973, Osnovnyye cherty kriogeneza ravninnykh territoriy Evropy v verkhnem pleystotsene: 17–25 in *Akademiya Nauk SSSR, Sektsiya Nauk o Zemle, Sibirskoye Otdeleniye, II Mezhdunarodnaya Konferentsiya po Merzlotovedeniyu, Doklady i soobshcheniya* **2** (Regional'naya geokriologiya): Yakutsk, Yakutskoye Knizhnoye Izdatel'stvo. (154 pp.)

1975, Paragenesis of a cryogenic (periglacial) zone: *Biuletyn Peryglacjalny* **24**, 89–110.

1978, Basic features in the cryogenesis of the European plains in the upper Pleistocene [Osnovnyye cherty kriogeneza ravninnykh territoriy Evropy v verkhnem pleystotsene]: 67–72 in Sanger, F. J., ed., *USSR Contribution Permafrost Second International Conference* (Yakutsk, USSR, 13–28 July 1973): Washington, DC, Natl. Acad. Sci. (866 pp.)

VELICHKO, A. A., and BERDNIKOV, V. V., 1969, Kriogennyye obrazovaniya: 551–63 in Velichko, A. A., ed., *Lëss – periglyatsial – paleolit na territorii sredney i vostochnoy Evropy* (*Loess – periglaciaire – paleo-lithique sur le territoire de l'Europe moyenne et orientale*) (Pour le VIII Congrès de l'INQUA): [Moskva] Akademii Nauk. (742 pp.)

VERESHCHAGIN, N. K., 1974, The mammoth 'cemeteries' of north-east Siberia: *Polar Record* **17**(106), 3–12.

VERNON, PETER, and HUGHES, O. L., 1966, Surficial geology, Dawson, Larsen Creek, and Nash Creek map-areas, Yukon Territory: *Canada Geol. Survey Bull.* **136**. (25 pp.)

VIERECK, L. A., 1973, Ecological effects of river flooding and forest fires on permafrost in the taiga of Alaska: 60–7 in *North American Contribution, Permafrost Second International Conference* (Yakutsk, USSR, 13–28 July 1973): Washington, DC, Natl. Acad. Sci. (783 pp.)

VIETORIS, L., 1972, Über den Blockgletscher des Äusseren Hochebenkars: *Zeitschr. Gletscherkunde und Glazialgeologie* **8**(1–2), 169–88.

VIGDORCHIK, MICHAEL, 1977a, A geographic based information management system for permafrost in the Beaufort and Chukchi Seas – Ann. Rept. (Oct. 1976–Mar. 1977): *US Dept. Commerce, National Oceanic and Atmospheric Administration* [*NOAA*], *Environmental Research Laboratories, Outer Continental Environmental Assessment Program* [*OCSEAP*], *Research Unit* **516**, *Contract* **3–7–022–35127**. (23 + 51 pp.)

1977b, A geographic based information management system for permafrost in the Beaufort and Chukchi Seas – Quart. Rept. (Apr.–June 1977): *U.S. Dept. Commerce, National Oceanic and Atmospheric Administration* [*NOAA*], *Environmental Research Laboratories, Outer Continental Environmental Assessment Program* [*OSCEAP*], *Research Unit* **516**, *Contract* **3–7–022–35127**. (84 pp.)

VILBORG, L., 1955, The uplift of stones by frost: *Geografiska Annaler* **37**, 164–9.

VINCENT, P. J., and CLARKE, J. V., 1976, The terracette enigma – A review: *Biuletyn Peryglacjalny* **25**, 65–77.

VISCHER, ANDREAS, 1943, Die postdevonische Tektonik von Ostgrönland zwischen 74° und 75° N. Br. – Kuhn Ø. Wollaston Forland, Clavering Ø und angrenzende Gebiete: *Medd. om Grønland* **133**(1). (194 pp.)

VOLKOVA, E. V., 1973a, O nekotorykh osobennostyakh migratsii vody pri promerzanii tonkodispersnykh gruntov: 176–80 in *Akademiya Nauk SSSR, Sektsiya Nauk o Zemle, Sibirskoye Otdeleniye, II Mezhdunarodnaya Konferentsiya po Merzlotovedeniyu, Doklady i soobshcheniya* **4** (Fizika, fiziko-khimiya i mekhanika merzlykh gornykh porod i

l'da): Yakutsk, Yakutskoye Knizhnoye Izdatel'-stvo. (246 pp.)

1973b, Some characteristic features of water migration during freezing of fine soil [O nekotorykh osobennostyakh migratsii vody pri promerzanii tonkodispersnykh gruntov]: 235–40 in Tsytovich, N. A., *et al.*, Physics, physical chemistry and mechanics of permafrost and ice: *US Army Corps of Engineers, Cold Regions Research and Engineering Laboratory Draft Translation* **439**. (326 pp.)

1978, Some characteristic features of water migration during freezing of fine soils [O nekotorykh osobennostyakh migratsii vody pri promerzanii tonkodispersnykh gruntov]: 324–7 in Sanger, F. J., ed., *USSR Contribution Permafrost Second International Conference* (Yakutsk, USSR, 13–28 July 1973): Washington, DC, Natl. Acad. Sci. (866 pp.)

VORNDRANG, GERHARD, 1972, Kryopedologische Untersuchungen mit Hilfe von Bodentemperaturmessungen (an einem zonalen Strukturbodenvorkommen in der Silvrettagruppe): *Münchener Geog. Abh.* **6**. (70 pp.)

VOSTOKOVA, A. V., 1973, Regional'noye kompleksnoye kartografirovaniye mnogoletnemerzlykh porod (pa primere Tiomenskoy oblasti): 26–37 in *Akademiya Nauk SSSR, Sektsiya Nauk o Zemle, Sibirskoye Otdeleniye, II Mezhdunarodnaya Konferentsiya po Merzlotovedeniyu, Doklady i soobshcheniya* **2** (Regional'naya geokriologiya): Yakutsk, Yakutskoye Knizhnoye Izdatel'stvo. (154 pp.)

1978, Comprehensive regional mapping of permafrost [Regional'noye kompleksnoye kartografirovaniye mnogoletnemerzlykh porod (pa primere Tiomenskoy oblasti)]: 73–81 in Sanger, F. J., ed., *USSR Contribution Permafrost Second International Conference* (Yakutsk, USSR, 13–28 July 1973): Washington, DC, Natl. Acad. Sci. (866 pp.)

VOTYAKOV, I. N., 1966, Obemnyye izmeneniya merzlykh dispersnykh gruntov v svyazi s fazovymi perekhodami vody pri temperaturnykh kolebaniyakh: *Materialy VIII Mezhduvedomstvennogo Soveshchaniya po Geokriologii, Vyp.* **5**, 11–21. (Yakutsk, 1966.)

1973a, Strukturnyye preobrazovaniya v merzlykh gruntakh pri izmenenii ikh temperatury: 78–82 in *Akademiya Nauk SSSR, Sektsiya Nauk o Zemle, Sibirskoye Otdeleniye, II Mezhdunarodnaya Konferentsiya po Merzlotovedeniyu, Doklady i soobshcheniya* **4** (Fizika, fiziko-khimiya i mekhanika merzlykh gornykh porod i l'da): Yakutsk, Yakutskoye Knizhnoye Izdatel'stvo. (246 pp.)

1973b, Structural transformations in frozen ground on variation of the ground temperature [Strukturnyye preobrazovaniya v merzlykh gruntakh pri izmenenii ikh temperatury]: 98–103 in Tsytovich, N. A., *et al.*, Physics, physical chemistry and mechanics of permafrost and ice: *US Army Corps of Engineers, Cold Regions Research and Engineering Laboratory Draft Translation* **439**. (326 pp.)

1978, Structural transformations in frozen soils on variation of the ground temperature [Strukturnyye preobrazovaniya v merzlykh gruntakh pri izmenenii ikh temperatury]: 264–7 in Sanger, F. J., ed., *USSR Contribution Permafrost Second International Conference* (Yakutsk, USSR, 13–28 July 1973): Washington, DC, Natl. Acad. Sci. (866 pp.)

VTYURIN, B. I., 1973, Zakonomernosti rasprostraneniya i kolichestvennaya otsenka podzemnykh l'dov na territoriy SSSR: 12–19 in *Akademiya Nauk SSSR, Sektsiya Nauk o Zemle, Sibirskoye Otdeleniye, II Mezhdunarodnaya Konferentsiya po Merzlotovedeniyu, Doklady i soobshcheniya* **3** (Genezis, sostav i stroyeniye merzlykh tolshch i podzemnyye l'dy): Yakutsk, Yakutskoye Knizhnoye Izdatel'stvo. (102 pp.)

1975, *Podzemnyye l'dy SSSR*: Akad. Nauk SSSR, Oal'nevostochnyy Nauchnyy Tsentr, Tikhookeanskiy Inst. Geografii: Moskva, Izdatel'stvo 'Nauka'. (214 pp.)

1978, Patterns of distribution and a quantitative estimate of the ground ice in the USSR [Zakonomernosti rasprostraneniya i kolichestvennaya otsenka podzemnykh l'dov na territoriy SSSR]: 159–64 in Sanger, F. J., ed., *USSR Contribution Permafrost Second International Conference* (Yakutsk, USSR, 13–28 July 1973): Washington, DC, Natl. Acad. Sci. (866 pp.)

VTYURINA, E. A., 1974, *Kriogennoye stroyeniye porod sezonno protaivauschego sloya*: Proizvodstvennyy i Nauchno-issledovatel'skiy Institut po inzhenernym izyskaniyam v stroitel'stve Gosstroya SSSR: Moskva, Izdatel'stvo 'Nauka'. (127 pp.)

ed., 1976, *Sezonno- i mnogoletnemerzlye gornye porody*: Vladivostok. (190 pp.)

VYALOV [VIALOV], S. S., *et al.*, 1962, *Prochnost' i polzuchest' merzlykh gruntov i raschety ledogruntovykh ograzhdeniy*: Moskva, Akad. Nauk SSSR, Inst. Merzlotovedeniya im. V. A. Obrucheva, Izdatel'stvo Akad. Nauk SSSR. (255 pp.)

1965, The strength and creep of frozen soils and calculations for ice-soil retaining structures [Prochnost' i polzuchest' merzlykh gruntov i raschety ledogruntovykh ograzhdeniy]: *US Army Cold Regions Research and Engineering Laboratory Translation* **76**. (301 pp.)

1973a, Printsipy upravleniya kriogennymi protsessami pri osvoyenii territorii s mnogoletnemerzlymi porodami: *Akademiya Nauk SSSR, Sektsiya Nauk o Zemle, Sibirskoye Otdeleniye, II Mezhdunarodnaya Konferentsiya po Merzlotovedeniyu, Doklady i soobshcheniya* **7**: Yakutsk, Yakutskoye Knizhnoye Izdatel'stvo. (272 pp.)

1973b, Principles of the control of cryogenic processes during the development of permafrost regions [Printsipy upravleniya kriogennymi protsessami pri osvoyenii territorii s mnogoletnemerzlymi porodami]: *US Army Corps of Engineers, Cold Regions Research and Engineering Laboratory Draft Translation* **438**. (360 pp.)

1978, Principles of the control of geocryological conditions for construction in permafrost regions [Printsipy upravleniya kriogennymi protsessami pri osvoyenii territorii s mnogoletnemerzlymi porodami]: 513–682 in Sanger, F. J., ed., *USSR Contribution Permafrost Second International Conference* (13–28 July 1973): Washington, DC, Natl. Acad. Sci. (866 pp.)

VYALOV, S. S., DOKUCHAYEV, V. V., and SHEYNKMAN, D. R., 1976, *Podzemnyye l'dy i sil'no l'distyye grunty kak osnovaniya sooruzheniy*: Leningrad, Stroyizdat. Leningradskoye Otdeleniye. (167 pp.)

WAHRHAFTIG, CLYDE, 1949, The frost-moved rubbles of Jumbo Dome and their significance in the Pleistocene chronology of Alaska: *J. Geol.* **57**, 216–31.

1965, Physiographic divisions of Alaska: *US Geol. Survey Prof. Paper* **482**. (52 pp.)

WAHRHAFTIG, CLYDE, and COX, ALLAN, 1959, Rock glaciers in the Alaska Range: *Geol. Soc. America Bull.* **70**, 383–436.

WALKER, H. J., 1973, Morphology of the North Slope: 49–92 in Britton, M. E., ed., Alaskan arctic tundra [Naval Arctic Research Laboratory, 25th Anniversary Celebration Proc.]: *Arctic Inst. North America Tech. Paper* **25**. (224 pp.)

WALKER, H. J., and ARNBORG, L., 1966, Permafrost and ice-wedge effect on riverbank erosion: 164–71 in *Permafrost International Conference* (Lafayette, Ind., 11–15 Nov. 1963) *Proc.*: Natl. Acad. Sci.–Natl. Research Council Pub. **1287**. (563 pp.)

WALLACE, R. F., 1948, Cave-in lakes in the Nabesna, Chisana, and Tanana River valleys, eastern Alaska: *J. Geol.* **56**, 171–81.

WALTERS, J. C., 1978, Polygonal patterned ground in central New Jersey: *Quaternary Research* **10**, 42–54.

WALTON, G. F., 1972, *The High Arctic environment and*

Polar Desert soils: Rutgers Univ., Ph.D. Thesis. (479 pp.)

WARD, W. H., 1959, Ice action on shores [letter]: *J. Glaciology* **3**(25), 437.

WARD, W. H., and ORVIG, S., 1953, The glaciological studies of the Baffin Island Expedition, 1950 – Part IV, The heat exchange at the surface of the Barnes Ice Cap during the ablation period: *J. Glaciology* **2**, 158–72.

WASHBURN, A. L., 1947, Reconnaisance geology of portions of Victoria Island and adjacent regions, Arctic Canada: *Geol. Soc. America Mem.* **22**. (142 pp.)

—— 1950, Patterned ground: *Rev. Canadienne Géographie* **4**(3–4), 5–59.

—— 1951, Geography and arctic lands: 267–87 in Taylor, Griffith, ed., *Geography in the twentieth century*: New York, Philosophical Library; London, Methuen. (630 pp.)

—— 1956a, Unusual patterned ground in Greenland: *Geol. Soc. America Bull.* **67**, 807–10.

—— 1956b, Classification of patterned ground and review of suggested origins: *Geol. Soc. America Bull.* **67**, 823–65.

—— 1965, Geomorphic and vegetational studies in the Mesters Vig district, Northeast Greenland – General introduction: *Medd. om Grønland* **166**(1). (60 pp.)

—— 1967, Instrumental observations of mass-wasting in the Mesters Vig district, Northeast Greenland: *Medd. om Grønland* **166**(4). (318 pp.)

—— 1969a, Weathering, frost action, and patterned ground in the Mesters Vig district, Northeast Greenland: *Medd. om Grønland* **176**(4). (303 pp.)

—— 1969b, Patterned ground in the Mesters Vig district, Northeast Greenland: *Biuletyn Peryglacjalny* **18**, 259–330.

—— 1970, An approach to a genetic classification of patterned ground: *Acta Geographica Lodziensia* **24**, 437–46.

—— 1973, *Periglacial processes and environments*: London, Edward Arnold; New York, St. Martins Press. (320 pp.)

WASHBURN, A. L., BURROUS, CHESTER, and REIN, ROBERT, JR., 1978, Soil deformation resulting from some laboratory freeze-thaw experiments: 756–62 in *Third International Conference on Permafrost* (Edmonton, Alta., 10–13 July 1978), Proc. **1**: Ottawa, Canada Natl. Research Council. (947 pp.)

WASHBURN, A. L., and GOLDTHWAIT, R. P., 1958, Slushflows (abs.): *Geol. Soc. America Bull.* **69**, 1657–8.

WASHBURN, A. L., SMITH, D. D., and GODDARD, R. H., 1963, Frost cracking in a middle-latitude climate: *Biuletyn Peryglacjalny* **12**, 175–89.

WATERS, A. C., and FLAGLER, C. W., 1929, Origin of the small mounds on the Columbia River Plateau: *Am. J. Sci., 5th Ser.*, **18**(105), 209–24.

WATERS, R. S., 1962, Altiplanation terraces and slope development in West-Spitsbergen and South-West England: *Biuletyn Peryglacjalny* **11**, 89–101.

WATSON, EDWARD, 1965, Grèzes litées ou éboulis ordonnés tardiglaciaires dans la region d'Aberystwyth, au centre du Pays de Galles: *Assoc. Géographes Français Bull.* **338–9**, 16–25.

—— 1966, Two nivation cirques near Aberystwyth, Wales: *Biuletyn Peryglacjalny* **15**, 79–101.

—— 1969, The slope deposits in the Nant Iago valley near Cader Idris, Wales: *Biuletyn Peryglacjalny* **18**, 95–113.

—— 1970, The Cardigan Bay area: 125–45 in Lewis, C. A., ed., *The glaciations of Wales and adjoining regions*: London, Longman. (378 pp.)

—— 1971, Remains of pingos in Wales and the Isle of Man: *Geol. J.* **7**(2), 381–92.

—— 1972, Pingos of Cardiganshire and the latest ice limit: *Nature* **236**, 343–4.

—— 1976, Field excursions in the Aberystwyth region, 1–10 July 1975: *Biuletyn Peryglacjalny* **26**, 79–112.

—— 1977, The periglacial environment of Great Britain during the Devensian: *Royal Soc. London Phil. Trans.* B.**280**, 183–98.

WATSON, EDWARD, and WATSON, SYBIL, 1971, Vertical stones and analogous structures: *Geografiska Annaler* **53**A, 107–14.

—— 1972, Investigations of some pingo basins near Aberystwyth, Wales: 212–23 in *Quaternary geology: Internat. Geol. Cong., 24th, Montreal, Proc. sec.* **12**. (226 pp.)

—— 1974, Remains of pingos in the Cletwr basin, south-west Wales: *Geografiska Annaler* **56**A, 213–25.

—— 1977, The mid-Wales uplands: 21–7 in *International Union for Quaternary Research (INQUA) Cong., 10th (Birmingham, England, 16–24 Aug. 1977). Guidebook, Excursion C9.* (48 pp.)

WATT, A. S., PERRIN, R. M. S., and WEST, R. G., 1966, Patterned ground in Breckland: Structure and composition: *J. Ecology* **54**, 239–58.

WAYNE, W. J., 1965, Western and central Indiana – Day 4, September 9: 29–36 in *International Assoc. for Quaternary Research (INQUA) Cong., 7th (Boulder, Col., 1965): Guidebook for field conference, G, Great Lakes – Ohio River Valley.* (110 pp.)

—— 1967, Periglacial features and climatic gradient in Illinois, Indiana, and western Ohio, east-central

United States: 393–414 in Cushing, E. J., and Wright, H. E., Jr., eds., *Quaternary palaeoecology*: New Haven, Yale Univ. Press. (433 pp.)

WEBB, P. N., and MCKELVEY, B. D., 1959, Geological investigations in South Victoria Land, Antarctica —Part 1, Geology of Victoria Dry Valley: *New Zealand J. Geol. and Geophys.* **2**, 120–36.

WEEKS, A. G., 1969, The stability of natural slopes in south-east England as affected by periglacial activity: *Quart. J. Eng. Geol.* **2**, 49–61.

WEERTMAN, JOHANNES, 1964, Rate of growth or shrinkage of nonequilibrium ice sheets: *J. Glaciology* **5**(38), 145–58.

WEIDICK, ANKER, 1968, Observations on some Holocene glacier fluctuations in West Greenland: *Medd. om Grønland* **165**(6). (202 pp.)

1975, A review of Quaternary investigations in Greenland: *Ohio State Univ. Inst. Polar Studies Rept.* **55**. (161 pp.)

WELLMAN, A. W., and WILSON, A. T., 1965, Salt weathering, a neglected geological erosive agent in coastal and arid environments: *Nature* **205**, 1097–8.

WENDLER, GERD, 1970, Some measurements of the extinction coefficients of river ice: *Polarforschung* **6**(1), 1969, 253–6.

WERENSKIOLD, W., 1922, Frozen earth in Spitsbergen: *Geofysiske Publikationer* **2**(10), 1–10.

1953, The extent of frozen ground under the sea bottom and glacier beds: *J. Glaciology* **2**(13), 197–200.

WEST, R. G., 1968, *Pleistocene geology and biology*: New York, John Wiley & Sons, Inc. (377 pp.)

WESTGATE, J. A., and BAYROCK, L. A., 1964, Periglacial structures in the Saskatchewan gravels and sands of central Alberta, Canada: *J. Geol.* **72**, 641–8.

WHALLEY, W. B., 1974, Rock glaciers and their formation as part of a glacier debris-transport system: *Univ. Reading, Dept. Geog. Geographical Papers* **24**. (60 pp.)

WHITE, E. M., 1972, Soil-desiccation features in South Dakota depressions: *Geol. J.* **80**, 106–11.

WHITE, E. M., and AGNEW, A. F., 1968, Contemporary formation of patterned ground by soils in South Dakota: *Geol. Soc. America Bull.* **79**, 941–4.

WHITE, E. M., and BONESTEEL, R. G., 1960, Some gilgaied soils in South Dakota: *Soil Sci. Soc. Am. Proc.* **24**, 305–9.

WHITE, P. G., 1976, Some observations on the origins of rock glaciers (abs.): *Geol. Soc. America Abs. with Programs* **8**(2), 299.

WHITE, S. E., 1971, Rock glacier studies in the Colorado Front Range, 1967 to 1968: *Arctic and Alpine Research* **3**, 43–64.

1972, Alpine subnival boulder pavements in Colorado Front Range: *Geol. Soc. America Bull.* **83**, 195–200.

1976a, Is frost action really only hydration shattering? A review: *Arctic and Alpine Research* **8**, 1–6.

1976b, Rock glaciers and block fields, Review and new data: *Quaternary Research* **6**, 77–97.

WHITE, W. A., 1961, Colloid phenomena in sedimentation of argillaceous rocks: *J. Sed. Petrology* **31**, 560–70.

1964, Origin of fissure fillings in a Pennsylvania shale in Vermillion County, Illinois: *Illinois Acad. Sci. Trans.* **57**, 208–15.

WICHE, KONRAD, 1953, Pleistozäne Klimazeugen in den Alpen und im Hohen Atlas: *Geog. Gesell. Wien Mitt.* **95**, 143–66.

WIEGAND, GOTTFRIED, 1965, Fossile Pingos in Mitteleuropa: *Würzburger Geog. Arbeiten* **16**. (152 pp.)

WIJMSTRA, T. A., 1969, Palynology of the first 30 metres of a 120 m deep section in northern Greece: *Acta Bot. Neerlandica* **18**, 511–27.

WILHELMY, HERBERT, 1958, *Klimamorphologie des Massengesteine*: Braunschweig, Georg Westermann Verlag. (238 pp.)

1977, Verwitterungskleinformen als Anzeichen stabiler Grossformung: 177–98 in Büdel, J., ed., *Beiträge Reliefgenese in verschiedenen Klimazonen* (*Erweiterte Vorträge der Gordon-Konferenz der Bäyerischen Akademie der Wissenschaften vom 20,–23.2.1975*): *Würzburger Geog. Arbeiten* **45**. (198 pp.)

WILKINSON, T. J., and BUNTING, B. T., 1975, Overland transport of sediment by rill water in a periglacial environment in the Canadian High Arctic: *Geografiska Annaler* **57**A, 105–16.

WILLDEN, RONALD, and MABEY, D. R., 1961, Giant desiccation fissures on the Black Rock and Smoke Creek Deserts, Nevada: *Science* **133**, 1359–60.

WILLIAMS, G. P., and GOLD, L. W., 1976, Ground temperatures: *Canada Natl. Research Council Canadian Building Digest* **180** (July 1976). (4 pp.)

WILLIAMS, J. R., 1965, Ground water in permafrost regions – an annotated bibliography: *US Geol. Survey Water-Supply Paper* **1792**. (294 pp.)

1970, Ground water in the permafrost regions of Alaska: *US Geol. Survey Prof. Paper* **696**. (83 pp.)

WILLIAMS, J. R., and VAN EVERDINGEN, R. O., 1973, Groundwater investigations in permafrost regions of North America: 435–46 in *North American Contribution, Permafrost Second International Conference* (Yakutsk, USSR, 13–28 July 1973): Washington, DC, Natl. Acad. Sci. (783 pp.)

WILLIAMS, LLEWELYN, 1964, Regionalization of freeze-thaw activity: *Assoc. Am. Geog. Annals* **54**, 597–611.

WILLIAMS, M. Y., 1936, Frost circles: *Royal Soc. Canada Trans.* **30**(4), 129–32.

WILLIAMS, P. J., 1957, Some investigations into solifluction features in Norway: *Geog. J.* **123**(1), 42–58.

1959*a*, The development and significance of stony earth circles: *Norske Vidensk.–Akad. Oslo, I. Mat.–naturv. Kl. 1959* **3**. (14 pp.)

1959*b*, Arctic 'vegetation arcs': *Geog. J.* **125**, 144–5.

1961, Climatic factors controlling the distribution of certain frozen ground phenomena: *Geografiska Annaler* **43**, 339–47.

1962, Quantitative investigations of soil movement in frozen ground phenomena: *Biuletyn Peryglacjalny* **11**, 353–60.

1963, Specific heats and unfrozen water content of frozen soils: 109–26 in *First Canadian Conference on Permafrost, Proc.: Canada Natl. Research Council Tech. Memo.* **76**. (231 pp.)

1966, Downslope soil movement at a sub-arctic location with regard to variations with depth: *Canadian Geotech. J.* **3**, 191–203.

1967, Properties and behaviour of freezing soils: *Norwegian Geotechnical Inst. Pub.* **72**. (119 pp.)

1968, Ice distribution in permafrost profiles: *Canadian J. Earth Sci.* **5**, 1381–6.

1972, Use of ice-water surface tension concept in engineering practice: 19–29 in Highway Research Board, *Frost action in soils: Natl. Acad. Sci.–Natl. Acad. Eng. Highway Research Record* **393**. (88 pp.)

1976, Volume changes in frozen soils: 233–46 in *Laurits Bjerrum memorial volume – Contributions to soil mechanics*: Oslo, Norway, Norwegian Geotech. Inst. (262 pp.)

1977, Thermodynamic conditions for ice accumulation in freezing soils: 42–53 in *Frost action in soils*: Univ. Luleå, Internat. Symposium (Luleå, Sweden, 16–18 Feb. 1977) Proc. 1. (215 pp.)

WILLIAMS, R. B. G., 1964, Fossil patterned ground in eastern England: *Biuletyn Peryglacjalny* **14**, 337–49.

1965, Permafrost in England during the last glacial period: *Nature* **205**, 1304–5.

1968, Some estimates of periglacial erosion in southern and eastern England: *Biuletyn Peryglacjalny* **17**, 311–35.

1969, Permafrost and temperature conditions in England during the last glacial period: 399–410 in Péwé, T. L., ed., *The periglacial environment*: Montreal, McGill–Queen's Univ. Press. (487 pp.)

1975, The British climate during the last glaciation; an interpretation based on periglacial phenomena: 95–120 in Wright, A. E., and Moseley, F., eds., *Ice ages: Ancient and Modern*: Liverpool, Seel House Press. (320 pp.)

WILLMAN, H. B., 1944, Resistance of Chicago area dolomites to freezing and thawing: *Illinois Geol. Survey Bull.* **68**, 249–62.

WILSON, J. W., 1952, Vegetation patterns associated with soil movement on Jan Mayen Island: *J. Ecology* **40**, 249–64.

WILSON, LEE, 1968*a*, Morphogenetic classification: 717–29 in Fairbridge, R. W., ed., *The encyclopedia of geomorphology*: New York, Reinhold Book Corp. (1295 pp.)

1968*b*, Frost action: 369–81 in Fairbridge, R. W., ed., *The encyclopedia of geomorphology*: New York, Reinhold Book Corp. (1295 pp.)

1969, Les relations entre les processus géomorphologiques et le climat moderne comme méthode de paleoclimatologie: *Revue de Géographie physique et de Géologie dynamique* **11**(3), 303–14.

WILSON, M. D., 1977, Origin of patterned ground near Boise, Idaho (abs.): *Geol. Soc. America Abs. with Programs* **9**(6), 775–6.

1978, Patterned ground of erosional origin, southwestern Idaho: *Third International Conference on Permafrost* (Edmonton, Alta., 10–13 July 1978), *Programme*, 60.

WIMAN, STEN, 1963, A preliminary study of experimental frost weathering: *Geografiska Annaler* **45**, 113–21.

WISSA, A. E. Z., and MARTIN, R. T., 1968, Behavior of soils under flexible pavements. Development of rapid frost susceptibility tests: *Mass. Inst. Technology School of Eng., Dept. Civil Eng. Res. Rept.* **R68–77** (*Soils pub. 224*). (113 pp.)

WOILLARD, GENEVIÈVE, 1975, Recherches palynologiques sur le Pléistocène dans l'Est de la Belgique et dans les Vosges Lorraines: *Acta Geographica Lovaniensia* **14**. (168 pp.)

WOLFE, P. E., 1953, Periglacial frost-thaw basins in New Jersey: *J. Geol.* **61**, 133–41.

1956, Pleistocene-periglacial frost-thaw phenomena in New Jersey: *New York Acad. Sci. Trans.* **18**, 507–15.

WOLLNY, E., 1897, Untersuchungen über die Volumveränderung der Bodenarten: *Forschungen auf dem Gebiete der Agrikultur-Physik* **20** (1897–8), 1–52.

WOODCOCK, A. H., 1974, Permafrost and climatology of a Hawaii volcano crater: *Arctic and Alpine Research* **6**, 49–62.

WORLD DATA CENTER A FOR GLACIOLOGY, 1977, Avalanches: *Inst. Arctic and Alpine Research, World*

Data Center A for Glaciology, Glaciological data Rept. **GD-1**. (134 pp.)

WORSLEY, PETER, 1966, Some Weichselian fossil frost wedges from East Cheshire: *Mercian Geologist* **1**, 357–65.

1977, Periglaciation: 205–19 in Shotton, F. W., ed., *British Quaternary studies – Recent advances*: Oxford, Clarendon Press. (298 pp.)

WORSLEY, PETER, and HARRIS, CHARLES, 1974, Evidence for Neoglacial solifluction at Okstindan, north Norway: *Arctic* **27**(2), 128–44.

WRAMNER, PER, 1972, Paliska bildningar i mineraljord – några iakttagelser från Taavauoma, Lappland (Palsa-like formations in mineral soil – some observations from Taavauoma, Swedish Lappland): *Göteborgs Universitet, Naturgeografiska Institutionen GUNI Rept.* **1**. (60 pp.)

1973, Palsmyrar i Taavauoma, Lappland (Palsa bogs in Taavauoma, Swedish Lappland): *Göteborgs Universitet, Naturgeografiska Institutionen GUNI Rept.* **3**. (140 pp.)

WRIGHT, H. E., JR., 1961, Late Pleistocene climate of Europe: A review: *Geol. Soc. America Bull.* **72**, 933–84.

YAALON, O. H., 1969, Origin of desert loess (abs.): 755 in Ters, Mireille, ed., *Études sur le Quaternaire dans le monde* **2**: International Union for Quaternary Research (INQUA) Cong., 8th (Paris, 1969).

YACHEVSKII, L. A., 1889, O vechno merzloy pochve v Sibiri: *Imperatorskoye Russkoye Geograficheskoye Obshchestvo, Izvestiya* **25**(5), 341–55.

YARDLEY, D. H., 1951, Frost-thrusting in the Northwest Territories: *J. Geol.* **59**, 65–9.

YEEND, W. E., 1972, Winter protalus mounds: Brooks Range, Alaska: *Arctic and Alpine Research* **4**, 85–7.

YEFIMOV, A. I., and DUKHIN, I. E., 1966, Kratkiye i predvaritel'nyye soobshcheniya: *Akad. Nauk SSSR, Sibirskoye Otdeleniye, Geologiya i Geofizika* **7**, 92–7.

1968, [Some permafrost thicknesses in the Arctic]: *Polar Record* **14**(88), 68.

YEGOROVA, G. N., 1962, Osobennosti rel'efa zapadnogo Verkhovyanya v svyazi s chetvertichnym oledeniyem (basseyn reki Sobopola): *Voprosy geografii Yakutii* **2**, 83–91.

YEHLE, L. A., 1954, Soil tongues and their confusion with certain indicators of periglacial climate: *Am. J. Sci.* **252**, 532–46.

YERMOLOV, V. V., 1953, O formirovanii osnovnykh elementov rel'efa okrainy Sredne-Sibirskogo Ploskogorya mezhdu rekami Kotuy a Popigay: *Nauchno-issledovatel'skogo instituta geologii Arktiki, Trudy* **72**, 50–76.

YODER, E. J., 1955, Freezing-and-thawing tests on mixtures of soil and calcium chloride: 1–11 in Soil freezing: *Natl. Acad. Sci.–Natl. Research Council, Highway Research Board Bull.* **100**. (35 pp.)

YONG, R. N., 1966, Soil freezing considerations in frozen soil strength: 315–19 in *Permafrost International Conference* (Lafayette, Ind., 11–15 Nov. 1963) *Proc.*: Natl. Acad. Sci.–Natl. Research Council Pub. **1287**. (563 pp.)

YONG, R. N., and OSLER, J. C., 1971, Heave and heaving pressures in frozen soils: *Canadian Geotech. J.* **8**, 272–82.

YONG, R. N., and WARKENTIN, B. P., 1975, *Soil properties and soil behaviour*: Amsterdam, Elsevier Scientific Publishing Co. (449 pp.)

YORATH, C. J., SHEARER, J., and HAVARD, C. J., 1971, Seismic and sediment studies in the Beaufort Sea: *Canada Geol. Survey Paper* **71-1**, Part A, 243–4.

YOSHIDA, YOSHIO, and MORIWAKI, KIICHI, 1977, Patterned ground around Syowa Station, East Antarctica (abs.): *International Union for Quaternary Research (INQUA) Cong., 10th (Birmingham, England, 16–24 Aug. 1977), Abstracts*, 508.

YOUNG, ANTHONY, 1972, *Slopes*: Edinburgh, Oliver & Boyd. (288 pp.)

YOUNG, G. M., and LONG, D. G. F., 1976, Ice-wedge casts from the Huronian Ramsay Lake Formation (> 2,300 m.y. old) near Espanola, Ontario, Canada: *Palaeogeog., Palaeoclimatol., Palaeoecol.* **19**, 191–200.

ZAGWIJN, WALDO, and PAEPE, ROLAND, 1968, Die Stratigraphie der weichselzeitlichen Ablagerungen der Niederlande und Belgiens: *Eiszeitalter und Gegenwart* **19**, 129–46.

ZAMOREUV, V. V., 1967, Nivalnyye formy relefa Kumylskogo goltsa (Yuzhnoye Zabaykalye): *Geokriologicheskiye usloviya Zabakalya i Pribaykalya*, 218–21.

ZAVARITSKIY, A. N., 1932, *Peridotitovyy massiv Rayiz v Polyarnom Urale*: Leningrad, Vsesoyuznoye geologo-razvedochnoye ob''edineniye.

ZAWADZKI, SATURNIN, 1957, Badania genezy i ewolucji gleb błotnych węglanowych Lubelszczyzny (Investigations on the origin and evolution of bog soils rich in calcium carbonate in the Lublin district): *Universitatis Mariae Curie-Skłodowska Annales, Sec. E*, **12**(1), 1–86.

ZEESE, R., 1971, Eiskeile im Keuperbergland: *Tübinger Geog. Studien* **46**, 69–91.

ZHESTKOVA [KHESTHOVA], T. N., *et al.*, 1961, Issledovaniye sloya sezonnogo Promerzaniya i protaivaniya pochv (Gornykh porod): 44–69 in *Polevye geokriologicheskiye (merzlotnyye) issledovaniya.*

Chast' I. Geokriologicheskaya s"emka: Moskva, Akad. Nauk SSSR, Inst. Merzlotovedeniya im. V. A. Obrucheva. (423 pp.)

1969, The layer of seasonal freezing and thawing of soil (rock) [Issledovaniye sloya sezonnogo Promerzaniya i protaivaniya pochv (Gornykh porod)]: *Canada Natl. Research Council Tech. Translation* **1358**. (227 pp.)

ZHIGAREV, L. A., 1960, Eksperimental'nyye issledovaniya skorostey dvizheniya gruntovykh mass na soliflyuktsionnykh skolnakh: *Inst. Merzlotovedeniya im. V. A. Obrucheva, Trudy* **16**, 183–90.

1967, *Prichiny i mekhanizm razvitiya soliflyuktsii*: Moskva, Akad. Nauk, Izdatel'stvo 'Nauka'. (158 pp.)

ZHIGAREV, L. A., and KAPLINA, T. N., 1960, Solifluktsionnye formy rel'efa na Severo-Vostoke SSSR: *Inst. Merzlotovedeniya im. V. A. Obrucheva. Trudy* **16**.

ZHIGAREV, L. A., and PLAKHT, I. R., 1974, Osobennosti stroyeniya rasprostraneniya i formirovaniya subakval'noy kriogennoy tolshchi: 115–24 in Popov, A. I., ed., 1974, *Problemy kriolitologii* **4**: Moskva, Izdatel'stvo Moskovskogo Universiteta. (253 pp.)

1976, Structure, distribution and formation of submarine permafrost [Osobennosti stroyeniya rasprostraneniya i formirovaniya subakval'noy kriogennoy tolshchi]: *US Army Corps of Engineers, Cold Regions Research and Engineering Laboratory Draft Translation* **520**. (11 pp.)

ZOLTAI, S. C., 1975, Tree-ring record of soil movements on permafrost: *Arctic and Alpine Research* **7**, 331–40.

ZOLTAI, S. C., and PETTAPIECE, W. W., 1974, Tree distribution on perennially frozen earth hummocks: *Arctic and Alpine Research* **6**, 403–11.

ZOLTAI, S. C., and TARNOCAI, C., 1974, Soils and vegetation of hummocky terrain: *Environmental-Social Comm. Northern Pipelines* [Canada], *Task Force on Northern Development Rept.* **74–5**. (86 pp.)

1975, Perennially frozen peatlands in the Western Arctic and Subarctic of Canada: *Canadian J. Earth Sci.* **12**, 28–43.

ZOLTAI, S. C., TARNOCAI, C., and PETTAPIECE, W. W., 1978, Age of cryoturbated organic materials in earth hummocks from the Canadian Arctic: 325–31 in *Third International Conference on Permafrost* (Edmonton, Alta., 10–13 July 1978), Proc. **1**: Ottawa, Canada Natl. Research Council. (947 pp.)

ZOTIKOV, I. A., 1963, Bottom melting in the central zone of the ice shield on the Antarctic continent and its influence upon the present balance of the ice mass: *Internat. Assoc. Sci. Hydrology Bull.* **8**(1), 36–44.

ZUMBERGE, J. H., and WILSON, J. T., 1953, Quantitative studies on thermal expansion and contraction of lake ice: *J. Geol.* **61**, 374–83.

ADDENDUM

BROWN, JERRY, and GRAVE, N. A., 1979*b*, Physical and Thermal disturbance and protection of permafrost: *US Army Corps of Engineers, Cold Regions Research and Engineering Research Laboratory Spec. Rept.* **79–5**. (42 pp.)

CARTER, L. D., and ROBINSON, S. W., 1978, Radiocarbon-dated episodes of activity and stabilization of large dunes, Arctic Coastal Plain, Alaska: *Am. Quat. Assoc. Abstracts, Fifth Biennial Meeting* (Edmonton, Alta., 2–4 Sept. 1978), p. 192.

DAHL, EILIF, 1946, On the origin of the strand flat: *Norsk Geografisk Tidsskrift* **11**(4), 159–72.

DEMEK, JAROMÍR, 1978, Periglacial geomorphology: Present problems and future prospects: 139–53 in Embleton, C., Brunsden, D., and Jones, D. K. C., eds., *Geomorphology: Present problems and future prospects*: Oxford, Oxford University Press. (281 pp.)

HOLTEDAHL, H., 1960, The strandflat of the Möre-Romsdal coast: 35–43 in Sømme, Axel, ed.,

Vestlandet. Geographical studies: Norges Handelshøyskole Geografiske Avh. **7**.

HOLTEDAHL, OLAF, 1929, *On the geology and physiography of some Antarctic and sub-Antarctic islands*: Scientific results Norwegian Antarctic Expeditions 1927–1928 and 1928–1929 **3**: Det Norske Videnskaps Akademi i Oslo. (172 pp.)

MABBUTT, J. A., 1977, *Desert landforms*: Cambridge, Mass., The MIT Press. (340 pp.)

SELZER, GEORG, 1936, Diluviale Lösskeile und Lösskeilnetze aus der Umgebung Göttingens: *Geol. Rundschau* **27**(3), 275–93.

SHERIF, M. A., ISHIBASHI, ISAO, and DING, W.-W., 1977, Heave of silty sands: *Am. Soc. Civil Eng. Proc., Geotech. Eng. Div. J.* **103**(GT3) Mar., 185–95.

WEERTMAN, JOHANNES, 1973, Creep of ice: 320–37 in Whalley, E., Jones, S. J., and Gold, L. W., eds., *Physics and chemistry of ice*: Ottawa, Royal Society of Canada. (403 pp.)

Index

Index

L
96

ED,